Exploiting Chemical Diversity for Drug Discovery

Exploiting Chemical Diversity for Drug Discovery

Edited by

Paul A. Bartlett
Department of Chemistry, University of California, Berkeley

Michael Entzeroth
*S*Bio Pte Ltd, Singapore*

RSCPublishing

ISBN-10: 0-85404-842-1
ISBN-13: 978-0-85404-842-7

A catalogue record for this book is available from the British Library

Published by The Royal Society of Chemistry,
Thomas Graham House, Science Park, Milton Road,
Cambridge CB4 0WF, UK

Registered Charity Number 207890

For further information see our web site at www.rsc.org

Typeset by Macmillan India Ltd, Bangalore, India
Printed by Henry Ling Ltd, Dorchester, Dorset, UK

Preface

The methods of drug discovery in the pharmaceutical industry have changed dramatically in the last two decades. By the late 1980s, a strong belief had emerged that drug development is purely a numbers game, with anticipated drop-out rates at each stage of the process such that only one compound out of 10,000 synthesized would survive to make it to the market. Technologies to accelerate both synthesis and screening were developed and adopted by virtually every pharmaceutical and biotech research division. The race for-ever higher numbers had started and the implementation of high-throughput technologies resulted in quantum changes in both chemistry and *in vitro* biology that transformed the search for new drugs.

However, a counterwind began blowing in the late 1990s, when the number of New Drug Applications (NDAs) at the regulatory authorities dropped to its all time low, and it became evident that more compounds screened were not going to translate directly into more drugs discovered. High-throughput synthesis and screening has to be informed by knowledge of what has worked and not worked in the past, in short, medicinal chemical intuition. Paradigms such as Christopher Lipinski's "Rule-of-Five,"[1] resulting from an empirical analysis of the physical properties of successful drugs, represents a milestone in the codification of this intuition. Data mining strategies, assessments of diversity and "drugability," and the design of high-content biological screens have thus become indispensable in complementing high throughput chemistry and screening technologies. The melding of a number of disciplines into the field now known as chemical biology has created the foundation and driving principle for modern drug discovery. This volume encompasses the changes that have occurred in both chemistry and screening as applied to drug discovery. In addition, we have distinguished the conceptual from the operational advances in each area, although we recognize that, like a good tango duo, neither would progress without the other.

Perhaps nowhere is the interplay of conceptual and operational advances more apparent than in the organic chemistry of drug discovery, where high-throughput concepts and automation have had a truly transforming effect. Stimulated by the automation of peptide synthesis on solid support, the concepts now embodied in combinatorial chemistry have completely changed the way in which compound collections are assembled and structure–activity relationships (SARs) are explored. Reactions are accelerated by microwave and other non-traditional techniques, reaction workups are streamlined with solid-phase reagents, and parallel or mix-and-split formats enable chemists to make hundreds of analogues in the time they used to synthesize one or two. These changes have not only altered the operational aspects of organic synthesis, but also given rise to entirely new strategies in synthetic planning.[2] The design of a chemical library will depend on whether it is destined for a discovery screen against a novel target, in which case diversity is key, or whether it will explore the SAR around a particular chemotype for a specific target. Will the library be based on a novel scaffold, or derived from a peptide or natural product? Should it be prepared by automated, parallel techniques, or can a biosynthetic pathway be pressed into service? Often there is no right answer to these questions, and debate is likely to continue for a long time on the key issue of "diversity:" how is it measured, how is it designed, and, ultimately, what does it mean in the context of a chemical collection?

High-throughput screening (HTS) has always played an important role in the lead-identification phase of drug discovery, so the impact from recent technological advances is more operational than conceptual. Nevertheless, HTS, and now ultra-HTS, has dramatically accelerated the discovery of initial hits from compound collections by increasing both the quantity and quality of information obtained. Increases in capacity have been driven by increases in the size of compound libraries, and modern screening technologies have greatly improved the analysis of the interaction of small molecules with novel pharmacological targets.

Assay methods are increasingly moving away from the use of radioisotopic labels. Modern fluorescent technologies now play an important role and are widely employed for the evaluation of more complex target systems, such as peptide–protein or protein–protein interactions. Reporter-gene assays have been developed to analyze functional responses to the activation of cellular signaling systems, such as membrane receptors, receptor tyrosine kinases, and ion channels. Using these techniques, inhibitors, allosteric, and transcriptional modulators, in addition to direct agonists, can be identified from chemical libraries. Moreover, both fluorescence- and reporter gene-based screening systems provide sufficiently high sensitivity that interactions of high-affinity ligands with targets present in low abundance can be probed.

At the other extreme, there is also interest in identifying low-affinity ligands for pharmacological targets. In fragment-based screening, the goal is to identify pharmacologically active partial structures. If different fragments for adjacent binding sites on the target are found, they can be linked to form ligands with higher affinity. As the small probe molecules normally bind with low affinity, this approach requires specific tools to detect and characterize their interaction with the biological target. NMR-based techniques have emerged as the most powerful of the methods used to detect low-affinity ligands in biological systems.

In addition to dynamic range and throughput, compound consumption remains one of the major issues in HTS. Miniaturization as well as high-density array formats have not only greatly improved the throughput of HTS campaigns, but also significantly reduced the amount of chemical material needed to determine ligand affinity. The volume of an individual assay well nowadays is in the low microliter range, down considerably from the milliliter volumes still used in the early 1980s, yet the quality of biological data obtained remains very high. Increases in daily screening throughput from automation and miniaturization have in turn required expanded capabilities in compound library storage and material handling. Technology continues to advance, with developments in nanotechnology and microfluidics opening new perspectives in reducing both time and sample amount. Chemical microarrays and bead-based applications may further enhance performance, although these formats have not yet been broadly introduced in pharma-screening laboratories because new detection capabilities are required. However, as we look to the future, it is clear that these formats will find their place in the initial hit identification process as the techniques advance.

The newer screening techniques are able to analyze many different types of ligand–target interactions and have provided research teams with a huge number of hits from HTS campaigns and, in turn, leads for the optimization phase of drug discovery. However, the output of preclinical and clinical candidates has failed to keep pace. Inadequate pharmacokinetic properties or adverse toxicological effects have been identified as the predominant reasons why drug candidates fail during development.[3,4] The pharmaceutical industry has responded by paying increased attention in the early phases of library design and lead identification to important "drug-like" attributes, such as metabolic stability, physicochemical properties, and membrane permeability. In fact, early awareness and corrective action to avoid potential liabilities has resulted in a modest decrease in the number of compounds that fail because of adverse pharmacokinetic properties.[5] With the expansions of the knowledge base, *in silico* predictive methods have become available to assess very large compound datasets, helping scientists to classify and rank compound collections or library designs according to their predicted properties or compliance with established rules. The potential for compounds to interact with cytochrome P450 enzymes[6] or to exhibit favorable absorption characteristics[7] are only two of the many examples of these predictive methods. Such machine-learning techniques are likely to play an increasingly important role in pharmaceutical screening and optimization processes in the future.

The concept of understanding more about a compound's biological behavior earlier in the process underlies the new concept of "high-content screening" (HCS). Indeed, it is in this area that the greatest conceptual advances in screening can be anticipated. The HCS approach has led to the development of a new generation of dedicated instruments that enable intracellular signaling pathways and cascades to be elucidated and analyzed. In large-scale, multiplexed, cell-based assays, the large amount of information that is generated can be captured and processed for this purpose. HCS combines specific components of assay and reagent design with robust instrumentation for automated fixed-end-point and live-cell kinetic analysis to generate information-rich data from multiple cellular targets.[8] By using multiple fluorescent reporter systems, combined with high-resolution imaging and high-throughput image processing, scientists can observe multiple intracellular events on a cellular level. HCS enables a functional analysis of how potential drug candidates modulate a particular pathway or target in living cells and has evolved into an integrated solution for accelerated drug discovery.

Over the past decade, drug discovery has undergone a dramatic evolution. It is clear that no discipline stands alone, that no breakthrough technology is "the solution," indeed, that integration of techniques and knowledge at the earliest stage is essential. The title of this volume emphasizes diversity, and we think the content of the following chapters reflects this concept in many ways. The closer integration of chemistry, biology, and technology that is transforming drug discovery has stimulated advances in numerous areas at the interfaces of these disciplines. While the individual chapters take on separate aspects of this on-going transformation, each chapter itself reflects an integration of different disciplines, techniques, or viewpoints. While the specific descriptions in the chapters that follow can only be snapshots of this evolving field, the concepts of cross-disciplinary integration, incorporation of medicinal chemical knowledge from the outset, and high-capacity information acquisition and analysis will always be relevant. It is our fervent hope that this volume will not only give readers a sense of the state of the art of drug discovery, but also stimulate the next generation of scientists to think how they could advance the field themselves.

References

1. C.A. Lipinski, *J. Pharmacol. Toxicol.*, 2000, **44**, 235.
2. M.R. Spaller, M.T. Burger, M. Fardis and P.A. Bartlett, *Curr. Opin. Chem. Biol.*, 1997, **1**, 47–53.
3. R.A. Prentis, *Br. J. Clin. Pharmacol.*, 1988, **25**, 387.
4. T. Kennedy, *Drug Discov. Today*, 1997, **2**, 436.
5. I. Kola, *Nat. Rev. Drug Discov.*, 2004, **3**, 711.
6. R. Arimoto, *J. Biomol. Screen.*, 2005, **10**, 197.
7. W.J. Egan, K.M. Merz Jr. and J.J. Baldwin, *J. Med. Chem.*, 2000, **43**, 3867–3877.
8. K.A. Giuliano, *Assay Drug Dev. Technol.*, 2003, **1**, 565.

Acknowledgements

The number of different molecules we could make – conceivably – exceeds the number of particles in the universe. How do we exploit this diversity – practically – for drug discovery? The cover image, combining a hypothetical diversity plot with representative molecules from screening libraries and a 384-well HTS plate, portrays the breadth of topics addressed in this volume. We thank Alan C. Gibbs for creating this montage and Robert Gendler (www.robgendlerastropics.com) for giving permission to use his image of the Great Galaxy in Andromeda as the background.

Contents

**Section 2   Conceptual Advances in Synthesis:
"Prospecting" – Design of Discovery
Libraries and the Search for Hits**

**Section 5   Conceptual Advances in Lead Evaluation:
                Screen Early and Often**

Contributors

D.K. Agrafiotis, *Johnson & Johnson Pharmaceutical Research and Development, LLC, 665 Stockton Drive Exton, PA 19431, USA.*

M.N. Banks, *Applied Biotechnology, Bristol Myers Squibb Co., Pharmaceutical Research Institute, Wallingford, CT 06492-7660, USA.*

P.A. Bartlett, *Department of Chemistry, University of California, Berkeley, CA 94720-1460, USA.*

M.B. Bolger, *Simulations Plus, Inc., 1220 W Avenue J, Lancaster, CA 93534, USA.*

F. Casano, *sanofi-aventis, Mail Stop: JR1-303D, P.O. Box 6800, Bridgewater, NJ 08807.*

M. Coles, *Department of Protein Evolution, Max-Planck-Institute for Developmental Biology, Spemannstrasse 35, 72076 Tübingen, Germany.*

I. Cornella, *Lead Synthesis & Chemogenetics – US, Novartis Institutes for Biomedical Research, Inc., 250 Massachusetts Avenue, Cambridge, MA 02139, USA.*

G. Dernick, *F. Hoffmann-La Roche Ltd, Grenzacherstrasse 124, Bldg 65-503, CH 4070 Basel, Switzerland.*

M. Entzeroth, *S*BIO Pte Ltd, 1 Science Park Road, #05-09 The Capricorn, Singapore Science Park 2, Singapore 117528.*

C. Fattinger, *F. Hoffmann-La Roche Ltd, Grenzacherstrasse 124, Bldg 65-503, CH 4070 Basel, Switzerland.*

R. Fraczkiewicz, *Simulations Plus, Inc., 1220 W Avenue J, Lancaster, CA 93534, USA.*

A. Ganesan, *School of Chemistry, University of Southampton, Southampton SO17 1BJ, UK.*

T. Garyantes, *sanofi-aventis, Mail Stop: JR1-303D, P.O. Box 6800, Route 202/206, Bridgewater, NJ 08807, USA.*

A.C. Gibbs, *Johnson & Johnson Pharmaceutical Research and Development, LLC, 665 Stockton Drive Exton, PA 19431, USA.*

J.G. Houston, *Applied Biotechnology, Bristol Myers Squibb Co., Pharmaceutical Research Institute, Wallingford, CT 06492-7660, USA.*

G. Hummel, *JERINI AG, Invalidenstraße 130, 10115 Berlin, Germany.*

H. Kessler, *Department Chemie, TU München, Lichtenbergstr. 4, 85747 Garching, Germany.*

C. Khosla, *Departments of Chemistry, Chemical Engineering and Biochemistry, Stanford University, Stanford, CA 94305, USA.*

J. Klages, *Department Chemie, TU München, Lichtenbergstr. 4, 85747 Garching, Germany.*

P.R. Kumaresan, *Division of Hematology & Oncology, Department of Internal Medicine, UC Davis Cancer Center, University of California Davis, 4501 X Street, Sacramento, CA 95817, USA.*

M. Ladlow, *GlaxoSmithKline, Cambridge Technology Centre, University of Cambridge, Cambridge CB2 1EW, UK.*

K.S. Lam, *Division of Hematology & Oncology, UC Davis Cancer Center, 4501 X Street, Sacramento, CA 95817, USA.*

M. Larhed, *Organic Pharmaceutical Chemistry, Department of Medicinal Chemistry, BMC, Uppsala University, Box 574, SE-751 23, Uppsala, Sweden.*

S.V. Ley, *Department of Chemistry, University of Cambridge, Lensfield Road, Cambridge CB2 1EW, UK.*

Z. Li, *sanofi-aventis, Mail Stop: JR1-303D, P.O. Box 6800, Bridgewater, NJ 08807.*

G. Mathis, *CIS bio international, BP 84175, 30204 Bagnols sur Cèze cedex, France.*

P. Nilsson, *Biolipox AB, Medicinal Chemistry, Box 6280, SE-102 34, Stockholm, Sweden.*

K. Olofsson, *Biolipox AB, Medicinal Chemistry, Box 6280, SE-102 34, Stockholm, Sweden.*

U. Reimer, *JERINI AG, Invalidenstraße 130, 10115 Berlin, Germany.*

U. Reineke, *JERINI AG, Invalidenstraße 130, 10115 Berlin, Germany.*

R. Richardson, *ChemBridge Europe, 4 Clarks Hill Rise, Evesham, Worcs WR11 2FW, UK.*

R. Schnitzer, *Drug Discovery, Boehringer Ingelheim Austria, Dr. Boehringergassse 5-11, A-1121 Vienna, Austria.*

W. Sommergruber, *Drug Discovery, Boehringer Ingelheim Austria, Dr. Boehringergassse 5-11, A-1121 Vienna, Austria.*

B. Steere, *Simulations Plus, Inc., 1220 W Avenue J, Lancaster, CA 93534, USA.*

Y. Tang, *Department of Chemical and Biomolecular Engineering, University of California, Los Angeles, CA 90095, USA.*

R. Tommasi, *Lead Synthesis & Chemogenetics – US, Novartis Institutes for Biomedical Research, Inc., 250 Massachusetts Avenue, Cambridge, MA 02139, USA.*

E. Trinquet, *CIS bio international, BP 84175, 30204 Bagnols sur Cèze cedex, France.*

E. Vickerstaffe, *GlaxoSmithKline, Cambridge Technology Centre, University of Cambridge, Cambridge CB2 1EW, UK.*

L. Zhang, *Lead Evaluation, Applied Biotechnology, Bristol Myers Squibb Co., Pharmaceutical Research Institute, Pennington, NJ 08534, USA.*

Section 1

Operational Developments in Chemistry

The Use of Polymer-Assisted Solution-Phase Synthesis and Automation for the High-Throughput Preparation of Biologically Active Compounds

STEVEN V. LEY[a], MARK LADLOW[b] AND
EMMA VICKERSTAFFE[b]

[a]Department of Chemistry, University of Cambridge, Lensfield Road, Cambridge
CB2 1EW, UK
[b]GlaxoSmithKline, Cambridge Technology Centre, University of Cambridge,
Cambridge CB2 1EW, UK

1 Introduction

In recent years, the drug discovery process has been revolutionised by progress in the areas of proteomics and genomics, together with advances in high-throughput screening (HTS) of compounds for activity in various biological assays.[1] This in turn has created a much higher demand for the rapid production of novel and functionally diverse compounds, thereby driving chemists to look for new ways to simplify, expedite and automate the process of organic synthesis. The importance of synthesising high-quality arrays of discrete compounds, from which meaningful structure activity relationships can be derived and also act as assets for future screening protocols is now well recognised.[2]

The origins of high-throughput organic chemistry can be traced back to the work of Merrifield[3] who pioneered solid-phase peptide synthesis. The development of this approach enabled the subsequent automation of peptide synthesis.[4] Early attempts at high-throughput chemistry utilised this strategy,[5–7] exploiting the advantages of solid-phase chemistry, which facilitates reaction work-up and rapid sample processing.

Today, although solid-phase organic synthesis (SPOS) remains a powerful technique for some aspects of high-throughput chemistry, there are a number of limitations that restrict its application. For example, while in some instances, by-product formation on the resin can be monitored without cleavage from the resin, separation of these materials from the desired product is not feasible until the end of the reaction sequence. The consequences are often cumulative, resulting in a complex final purification step. Additionally, even though a number of well-established methods for monitoring resin-bound intermediates have been developed, accurate quantitative measurement of immobilised material still represents a major challenge. Techniques such as magic angle spinning (MAS)-NMR[8–10] and IR[11,12] enable the analysis of molecules covalently bound to the resin, whereas MS techniques[8,13] or the use of analytical constructs,[14,15] require that the analyte be cleaved from the resin. The development of new chemistry on the solid support is therefore difficult and often protracted, requiring extended reaction rehearsal and optimisation.

Although SPOS is well suited to the synthesis of large, but relatively simple compound libraries, typically using combinatorial methods, the time required for route development and limitations as to the synthetic transformations that can be reliably performed is often restrictive. Increasingly, therefore, higher quality, smaller arrays are being synthesised by employing a much wider range of precedented solution-phase chemistries.[16] Historically, attempts to increase the throughput of solution-phase chemistries have been confounded by the need for extensive work-up and purification procedures. An increasingly popular approach to circumvent many of these drawbacks involves the use of supported reagents.[17–19] In this way, the advantages of performing chemistry in solution, namely, straightforward reaction monitoring and optimisation, are maintained. Excess reagents and their associated by-products may be readily separated from the desired solution-phase reaction product by simple filtration. A wide variety of reagents and scavenger resins have been developed in recent years to extend the scope of this new paradigm.

The concept of a reagent being immobilised on a solid-support was first exploited in catalytic applications as early as 1946.[20] However, it was not until the end of the twentieth century that solid-supported nucleophiles and electrophiles immobilised on a polystyrene (PS) support were applied to the rapid purification of compounds prepared using standard solution-phase procedures.[21] This application and its subsequent extension to the immobilisation of reagents has stimulated an explosion in the number of publications describing the development of novel polymer-bound reagents, catalysts and scavengers.[17,22,23] As an ever-increasing number of these reagents becomes commercially available, interest in applying polymer-assisted solution-phase (PASP) techniques within industrial settings escalates.

PASP synthesis can be divided into two main approaches; these being (a) the use of supported reagents and scavengers, and (b) the adoption of a catch-and-release strategy (Figure 1). Both of these techniques allow the production of clean products, without the need to resort to traditional purification techniques, such as column chromatography. In an ideal case, the incubation of a substrate with a supported reagent causes its complete transformation into a new chemical entity, with any

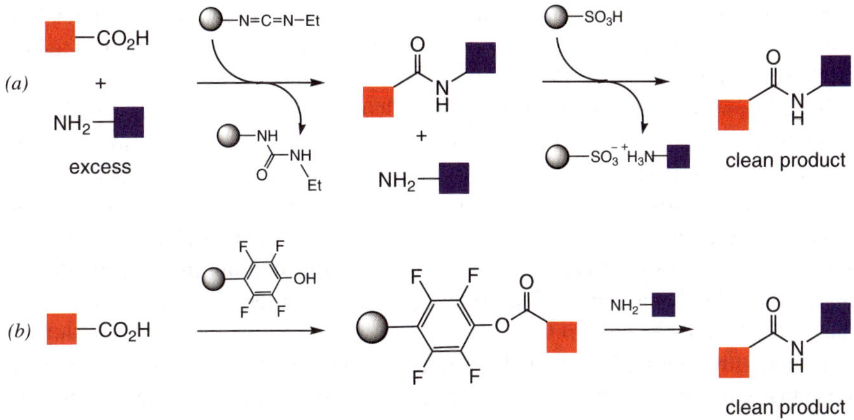

Figure 1 *Schematic representation of PASP strategies for the synthesis of carboxamides. (a) Use of supported reagents and scavengers. (b) Catch-and-release strategy*

excess or spent reagent being removed by a simple filtration. This process circumvents the need for further purification and allows reactions to be driven to completion through the use of reagent excesses. Furthermore, reaction progress can conveniently be monitored by standard solution-phase techniques (TLC, LC/MS *etc.*), thereby minimising the time required to optimise a transformation. However in reality, many reactions are not totally selective and by-products are often formed. In many cases, these can be removed by incubation with an appropriate scavenger resin, which targets a specific functional group, unique to these by-products. For example, in Figure 1a, following condensation of the amine and carboxylic acid in the presence of a polymer-supported carbodiimide, remaining excess amine is removed by attachment to the strongly acidic polymer-supported sulfonic acid resin, which in turn affords the desired carboxamide product in high purity.

Conversely, purification can sometimes be conveniently performed by selectively capturing the desired reaction product using a suitable resin, washing away the by-products, and then releasing the immobilised product back into solution. This is typically referred to as solid-phase extraction (SPE) and is particularly useful when the target molecule contains acidic or basic functionality.[24]

An alternative approach utilises the concept of catch-and-release whereby one of the reactants is first attached to the solid phase by an activated bond that is subsequently cleaved by exposure to a conjugate reactant. For example, in Figure 1b, the carboxylic acid is initially immobilised as a tetrafluorophenyl ester. Although intrinsically stable, the activated ester is highly susceptible to nucleophilic cleavage, and in the presence of a substoichiometric amount of amine, the desired carboxamide product is cleanly released into solution.

Immobilisation also allows the combination of two or more otherwise mutually incompatible reagents to be used in the same pot. This was elegantly demonstrated by Parlow,[25] who, in an important early contribution, utilised a combination of three different polymeric reagents simultaneously in a single reaction vessel to perform a three-step synthesis of pyrazoles. Notably, the yield of the pyrazole product **1** was greater when the supported reagents were combined in one flask as opposed to being used sequentially (Scheme 1).

In addition, supported reagents have been demonstrated to be effective under reaction conditions when either thermal[17] or microwave heating[26,27] is employed. They have also been utilised in traditional batch synthesis, stop-flow methods and continuous flow processes.[28–30] However, one caveat is that the immobilisation of reagents can change their reactivity. For example, polymer-supported borohydride selectively reduces α,β-unsaturated carbonyl compounds to the α,β-unsaturated alcohol[31] in contrast to the behaviour of the solution-phase counterpart, which additionally causes double bond reduction.

To date, immobilised reagents have mainly been PS based. PS supports have the advantages of being inexpensive, relatively easy to handle, have reasonably high loadings and, additionally, there is a vast amount of literature precedent[17] available for their use. Reagents can either be covalently attached to the polymer, as in the polymer-supported Schwesinger base, 2-tert-butylimino-2-diethylamino-1,3-dimethylperhydro-1,3,2 diazaphosphorin (PS-BEMP)[32] **2**, or electrostatically bound, for example, in the macroporous polymer-supported tetraalkylammonium triacetoxyborohydride **3** (Figure 2).[33] Macroporous resins contain a higher degree of cross-linking than PS supports. In practice, this means that in contrast to microporous PS supports, the resin-bound functional groups of a macroporous resin come into contact with reagents by diffusion through the network of pores and therefore do not require the use of a solvent that will normally swell the resin. More recently, catalysts have been microencapsulated within a polymeric matrix.[34,35] However, issues with the degradation of resins under certain reaction conditions have led to the

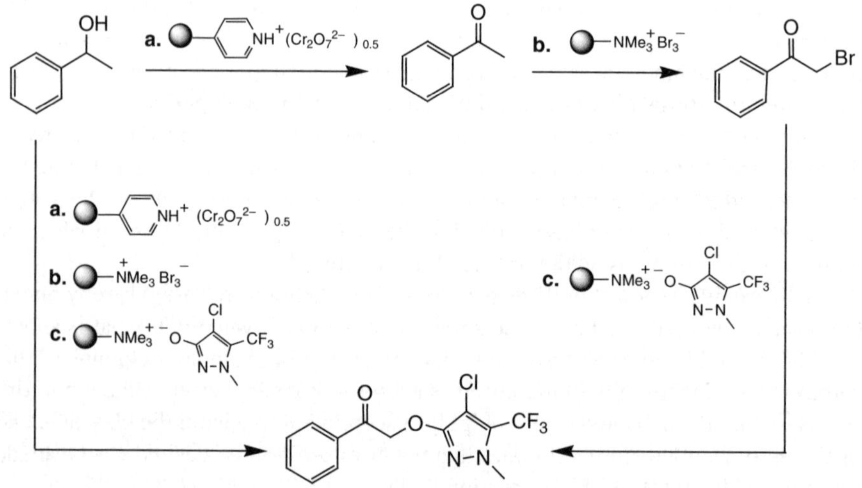

Scheme 1 *Reagents and conditions: (a) cyclohexane, 65 °C, 12 h, 55%; (b) cyclohexane, 65 °C, 12 h, 78%; (c) cyclohexane, 65 °C, 12 h, 98%; (a,b,c) cyclohexane, 65 °C, 16 h, 48% overall*

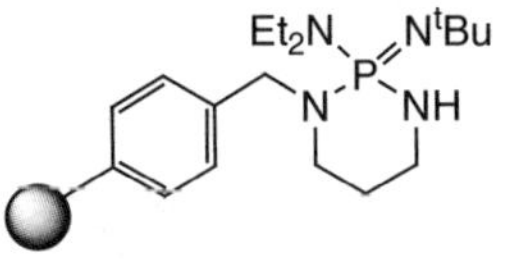

Et₂N   NᵗBu

PS-BEMP **2** 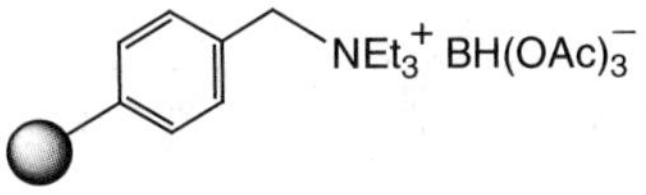   MP-Triacetoxyborohydride **3**

Figure 2 *Some examples of commercial polymer-supported reagents*

development of alternative supports such as controlled pore glass,[36] monoliths,[37–41] cellulose,[42] zeolites[43] and silicas.[44]

2 PASP Synthesis Approaches to Biologically Active Compounds

2.1 Applications to the Synthesis of Commercial Drug Molecules

A number of syntheses targeting commercially available drugs have been reported, which demonstrate the utility and effectiveness of supported reagents for the rapid and efficient preparation of drug-like scaffolds. The introduction of Sildenafil[45] **4** for the treatment of male erectile dysfunction has been incredibly successful, resulting in it becoming one of the largest selling globally marketed prescription drugs.[46] Sildenafil acts by inhibiting the phosphodiesterase enzyme PDE5, which is the main phosphodiesterase present in the smooth muscle of the corpus cavernosum. Upon sexual stimulus, nitric oxide is released from nerve terminals in the corpus cavernosum. The nitric oxide activates guanylate cyclase to produce cyclic guanosine monophosphate (cGMP), causing the intracellular levels of cGMP within the smooth muscle cells of the penis to increase. In healthy tissue, the elevated cGMP is returned to basal levels by the action of the PDE5 enzyme. Inhibition of the PDE5 enzyme prevents the breakdown of cGMP and thus potentiates the smooth muscle relaxation. This increases the blood flow in the cavernosum causing an erection.[47]

The polymer-assisted synthesis[48] to Sildenafil (Scheme 2) follows a precedented route,[45,49] which concludes with the convergent coupling of the two key fragments **5** and **6**. These fragments were prepared using PASP techniques, without the need for column chromatography, the former in a two-step sequence and the latter utilising seven different transformations. Fragment **5** was found to be contaminated with approximately 10% of the *bis*-esterified material **7**, which could be removed in the subsequent catch-and-release step. Notably by using a catch-and-release strategy in the penultimate amide-coupling step, this transformation acts as an in-line purification step while concomitantly activating the acid group to nucleophilic attack. Introduction of **6**, followed by scavenging with isocyanate resin to remove any unreacted amine, cleanly yielded the amide. Assembly of the pyrimidine ring system is performed using microwave heating to effect the rapid dehydration of **8**. A simple removal of the water formed during the cyclisation step was achieved with $MgSO_4$

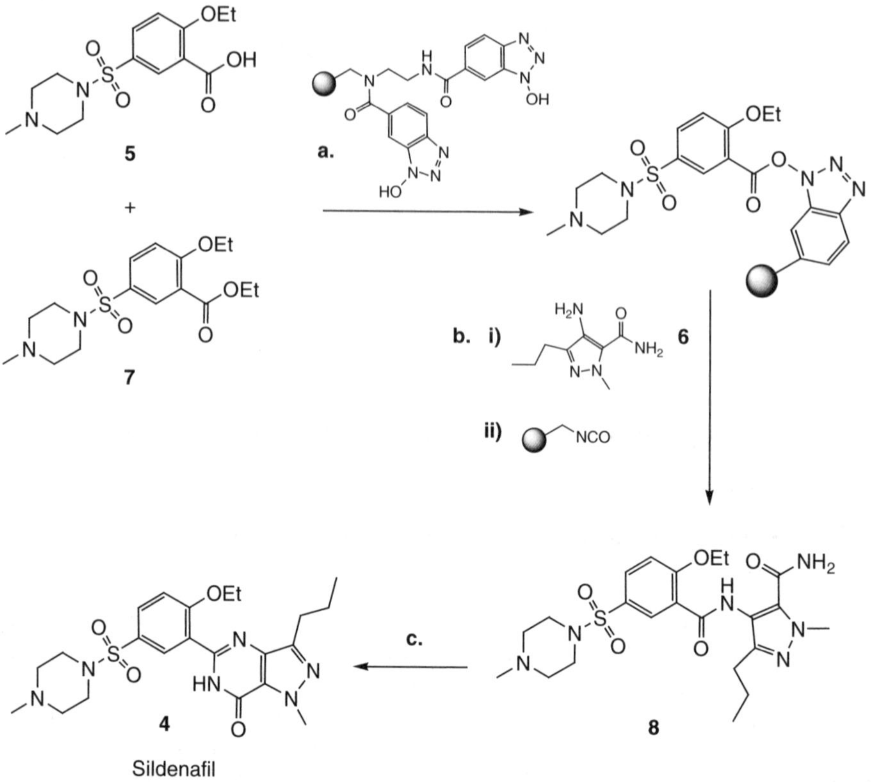

Scheme 2 *Reagents and conditions: (a) PyBrOP, DMF; (b) (i) 6, THF, (ii) THF; (c) EtOH/NaOEt, MW 10 min, 120 °C*

and the subsequent evaporation of the solvent gave a quantitative yield of Sildenafil **4**. This convergent synthetic pathway is clearly amenable to library production.

Rosiglitazone **9**, an agonist of peroxisome proliferator activated receptor-γ (PPARγ), is a recently introduced antihyperglycemic thiazolidinedione effective in the treatment of noninsulin dependant diabetes mellitus (type II diabetes).[50] A seven-step synthesis was developed (Scheme 3), which utilised supported reagents in combination with in-line SPE purifications.[51] The introduction of the pyridine moiety provided a convenient molecular handle by which to purify the molecule throughout the synthesis. Notably, the overall yield (46%) for this synthesis was higher than the yield reported in the initial shorter solution-phase synthesis (31%).[52]

The synthesis of chiral drug molecules has only recently been reported using a supported reagent approach. Although many supported reagents systems have been reported for use in enantioselective reactions, the paucity of the enantioselectivity achieved has prevented their extensive application in synthesis. A single isomer of the drug Salmeterol[53,54] **10**, a potent and long-acting β₂ adrenoceptor agonist,[55] has been prepared via a chiral auxiliary approach (Scheme 4). The key chiral reduction of

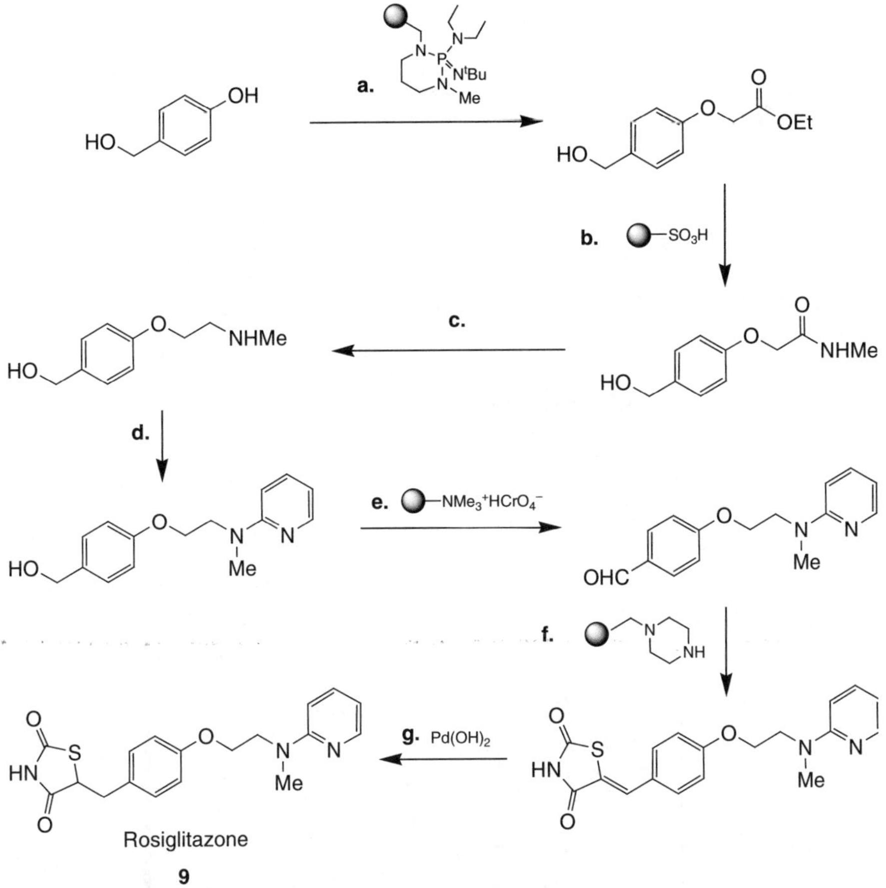

Scheme 3 *Reagents and conditions: (a) ethyl iodoacetate, DMF, rt, 93% (or ethyl iodoacetate, K$_2$CO$_3$, DMF, 90 °C, 96%); (b) MeNH$_2$, THF, 100%; (c) (i) BH$_3$:THF, 65 °C, (ii) Et$_2$NH, (iii) SCX-2, 84% overall; (d) 2-fluoropyridine, 120 °C, 82%; (e) THF, D, 86%; (f) (i) 2,4 thiazolidinedione, PhMe, 88 °C, (ii) SCX-2, 99%; (g) H$_2$, 1,4-dioxane, 80 °C, 82%*

ketone **11** to alcohol **12** is dependent upon the introduction of the (*S*)-phenylglycinol functionality. Treatment of **11** with calcium chloride at 0 °C followed by the addition of PS-borohydride resin delivered the desired amino alcohol **12** as a 10:1 mixture of diastereoisomers favouring the desired (*R*)-alcohol. A single recrystallisation afforded diastereomerically pure material. It is hypothesised on the basis of ^{1}H NMR shift studies that the reduction proceeds via a chelated intermediate **13** where the phenyl substituent points away from the crowded centre. The approach of the reducing agent then occurs preferentially from the convex face of the complex. From here, introduction of the long liphophilic chain and selective removal of the chiral auxiliary and acetonide-protecting group affords (*R*)-Salmeterol **10** in >97% ee.

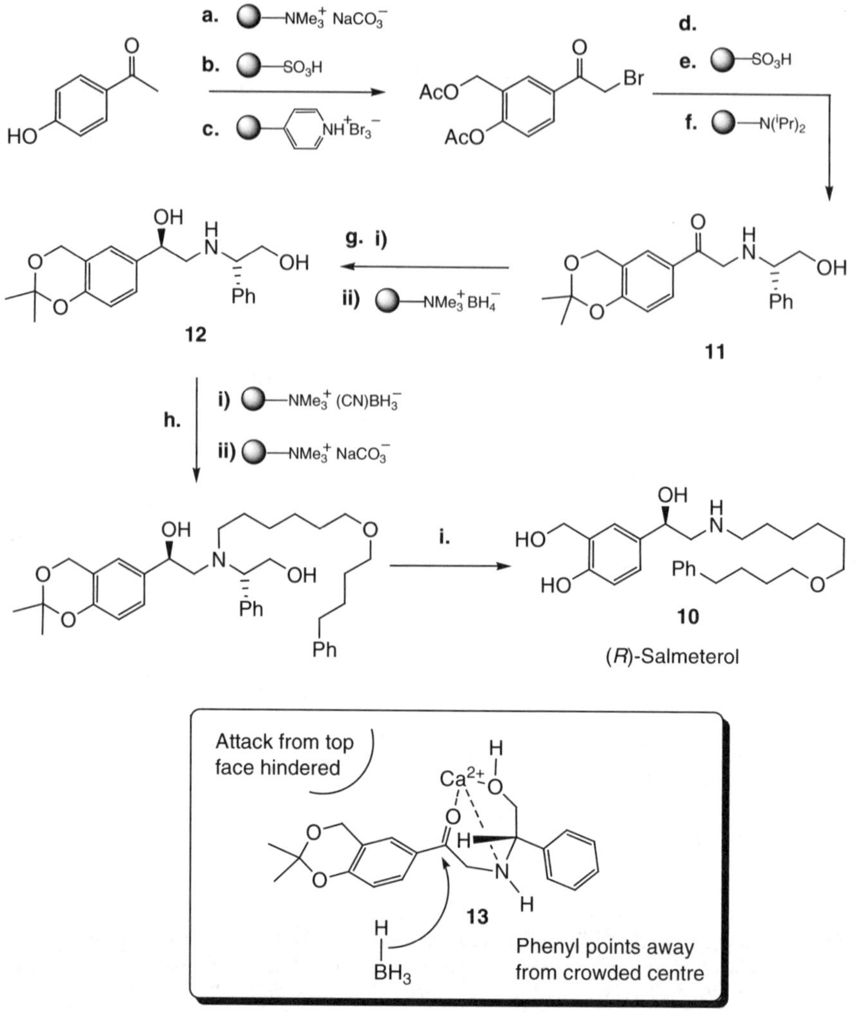

Scheme 4 *Reagents and conditions: (a) Me$_2$NCH$_2$I, CH$_2$Cl$_2$; (b) Ac$_2$O, toluene, 80 °C; (c) CH$_2$Cl$_2$; (d) 1N HBr:THF 1:1, 75 °C; (e) CH$_2$Cl$_2$, CH$_3$CH(CH$_2$)OCH$_3$; (f) THF, NH$_2$CH(Ph)CH$_2$OH; (g) (i) CaCl$_2$, MeOH; (ii) MeOH; (h) (i) Ph(CH$_2$)$_4$O(CH$_2$)$_5$ CHO, 5% AcOH, CH$_2$Cl$_2$; (ii) CH$_2$Cl$_2$; (i) (i) Pd(OH)$_2$, H$_2$, EtOAc; (ii) SCX-2*

2.2 Applications of PASP to the Synthesis of Biologically Active Natural Products

Some of the most extensive applications of PASP strategies have been directed towards the synthesis of biologically active natural products. The total synthesis of the cytotoxic antitumour natural product Epothilone C[56] **14** is a *tour de force*, the synthetic strategy demanding the exploitation and development of new immobilised reagent methods, to meet the goal of a chromatography free synthesis. The route to the 16-membered macrocycle Epothilone C, followed similar published strategies[57,58] involving coupling of three core fragments via a stereoselective C6–C7 aldol reaction, prior to either C1–C15 macrolactonisation or C12–C13 ring closing metathesis (Figure 3). The target molecule was prepared in high overall yield in 29 steps, with the longest sequence of linear steps being just 17. Impressively, considering the size, stereogenic centres and complexity of the molecule, only a single short column chromatography purification step was necessary at the end of the sequence to remove residual impurities and small quantities of the minor diastereoisomers that had been carried through the synthesis. Moreover, routes to fragments 1 and 3 are reported where the stereogenic centres are installed during the synthesis, via Oppolzer's sultam chiral auxiliary approach or an asymmetric Mukaiyama aldol reaction for fragment 1 and an asymmetric Brown allylation for fragment 3. However, the optimum synthesis to fragment 2 and ultimately fragment 3 relied on the use of chiral-starting materials to generate the desired stereochemistry.

The use of supported reagents has also been applied to the total synthesis of natural products, which have not previously been synthesised by traditional methods. An example of this is the first total synthesis of (+)-plicamine[59,60] **15** and its unnatural enantiomer (Figure 4). (+)-Plicamine is a member of the amaryllidaceae alkaloids, which exhibit antitumour, immunosuppresive and analgesic activity and have recently found application in the therapeutic treatment of Alzheimer's disease.[61] Other examples of biologically active natural products synthesised using supported reagents include the alkaloids (±)-oxomaritidine[62] **16** and (±)-epimaritidine[62] **17**, and the potent analgesic (±)-epibatidine[63] **18** isolated from the Ecuadorian poison frog *Epipedobates tricolor*.

The neolignan polysphorin **19**, isolated from *Piper polysphorum C* in China and the leaves and stems of *Rhaphidopora decursiva* in Vietnam, has been shown to possess *in vitro* antimalarial activity. In order to synthesise the neolignans, the general synthetic strategy has utilised an oxidative coupling, which afforded racemic mixtures of *syn* and *anti* products. An asymmetric route to both enantiomers of polysphorin, together with a small array of unnatural analogues has been reported, which utilises polymer-supported reagents and scavengers.[64] Noteworthy is the installation of the two stereogenic centres, through the use of a Sharpless asymmetric dihydroxylation

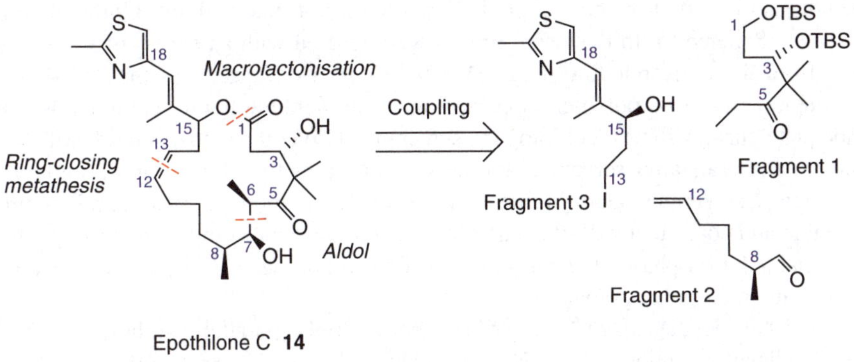

Figure 3 *Retrosynthetic analysis of Epothilone C 14*

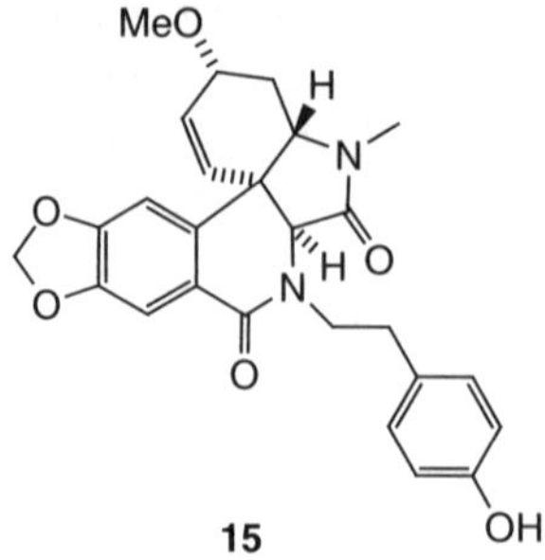

15

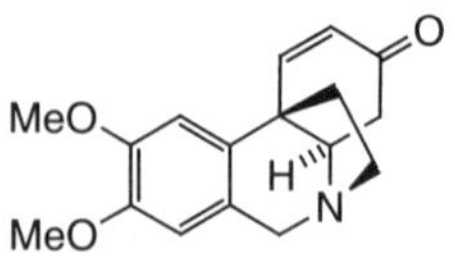

16

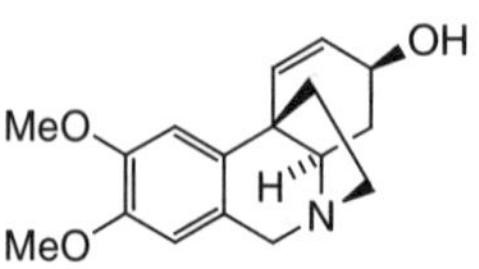

17

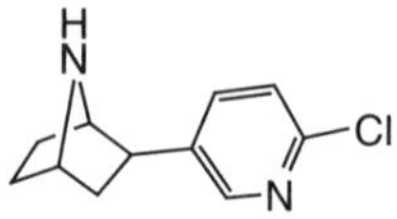

18

Figure 4 *Examples of biologically active natural products synthesised using supported reagents*

(AD) and the application of a catch-and-release procedure using a supported boronic acid to effect in-line purification of the diol **20** (Scheme 5). The enantiomer of **20** could readily be obtained through the use of AD-mix-β in the Sharpless asymmetric dihydroxylation reaction. The two alcohol groups of the diol could subsequently be differentiated using an enzymatic selective protection.

2.3 PASP Synthesis in the Library Production of Biologically Active Small Molecules

Many examples of early library production involved the use of a scavenger resin to facilitate reaction purification. An initial example was reported by Kaldor[65] for the synthesis of a 4000-membered pooled urea library of potential antirhinoviral compounds (Scheme 6). In this case, amines were reacted with an excess of isocyanate to afford the corresponding ureas **21**. The excess isocyanate was then scavenged through the use of a polymer-supported amine to yield the library compounds with adequate purities for direct biological screening. Two compounds were identified as potent antivirals after screening and deconvolution of the library. Gratifyingly, the biological screening data obtained on the library compounds was highly reproducible and consistent with the data obtained for the same compounds made by traditional solution-phase chemistry and purified to homogeneity by recrystallization and column chromatography.

In 1999, Ryder *et al.*[66] described the multiple parallel synthesis of *N,N*-dialkyldipeptidylamines **22** as *N*-type calcium channel blockers. Selective *N*-type voltage sensitive calcium channel blockers have shown utility in several models of

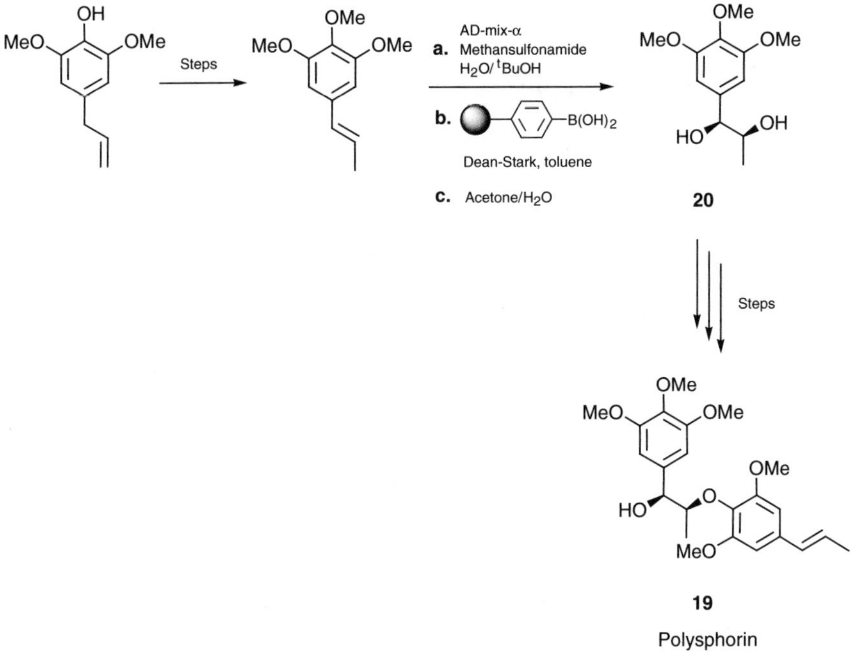

Scheme 5 *PASP synthesis to a library of natural products related to polysphorin 19, illustrating key Sharpless asymmetric oxidation*

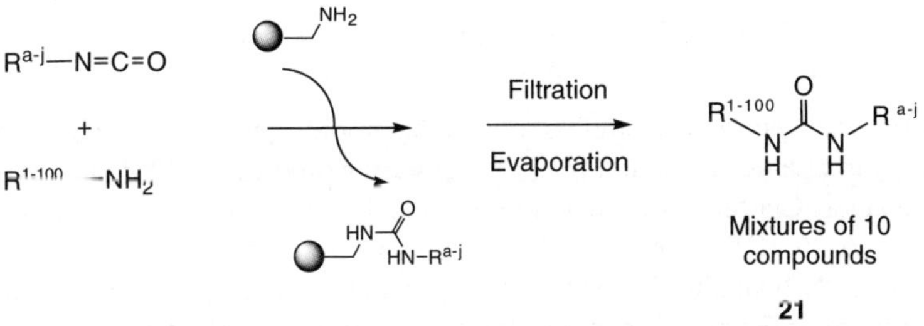

Scheme 6 *Preparation of a single-step urea library 21 utilising a PASP scavenging step*

stroke and pain. The preparation of 30 compounds surrounding a lead is described using a one-pot procedure for the coupling of an *N,N*-disubstituted leucine acid **23** with a tyrosine amine **24** (Scheme 7). All array members were isolated in >80% purity (HPLC) and good yields (60–95%). The compounds were screened in an *in vitro* assay, with the range of potencies observed between the most and least potent analogues being 325-fold. To alleviate any concerns with the reproducibility of screening data between compounds synthesised by parallel synthesis using supported reagents and traditional medicinal chemistry techniques, a number of compounds were remade. Reassuringly, when seven of these were prepared by traditional synthesis and purified to microanalytic purity, the biological screening data was identical.

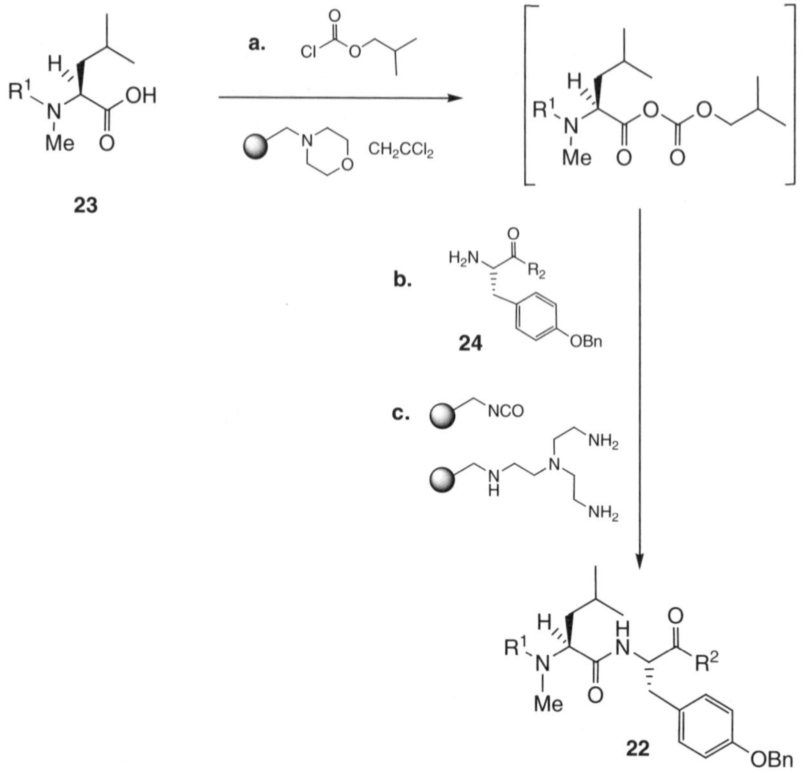

Scheme 7 *Use of PASP synthesis for the preparation of a peptidic library*

Ganesan has reported a number of combinatorial syntheses in which an ion-exchange resin serves as both a reagent and purification agent.[67,68] For example, an intramolecular Claisen-type condensation catalysed by the ion-exchange resin Amberlyst A-26 (OH⁻ form) leads to 4-hydroxyquinolin-2(1*H*)-ones **23**, which remained tightly bound to the resin, enabling impurities to be removed by simple filtration (Figure 5). Subsequent acidification then releases the desired products in high yield and purity. 4-Hydroxyquinolin-2(1*H*)-ones are known antagonists of thyroid hormone and serotonin, and also have potential in various therapeutic areas as antimicrobials and anticancer agents, local anaesthetics and antiinflammatory species.

The synthesis of an 88-membered combinatorial library **24** based upon the natural product **25** was reported by Nesterenko *et al.*[69] in their search for small molecules that selectively induce apoptosis in cancer cells. The key amide bond-forming reaction was performed in parallel using a polymer-supported carbodiimide as the dehydrating agent.

A rapid and efficient Suzuki–Miyaura coupling protocol utilising fibrous PS-based palladium catalysts (FibreCat™) and microwave irradiation has been reported by Wang and Sauer[70] (Scheme 8). The biaryl motif **26**, formed in this reaction, is an important pharmacophore in many biologically active compounds.[71] Each reaction

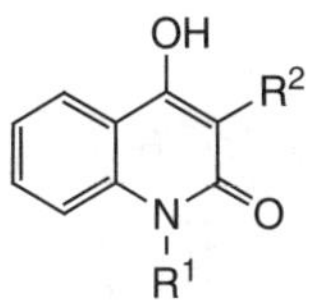

23

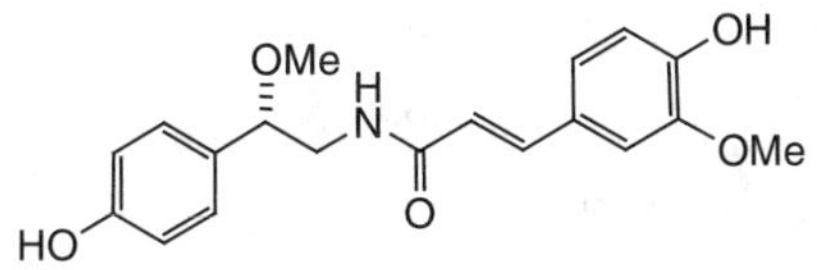

25

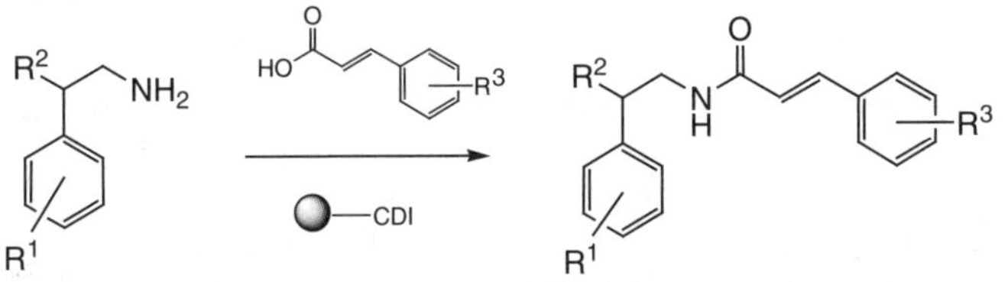

24

Figure 5 *4-Hydroxyquinolin-2(1H)-ones and the use of PS-CDI in the preparation of an 88-membered combinatorial library based upon the natural product 25*

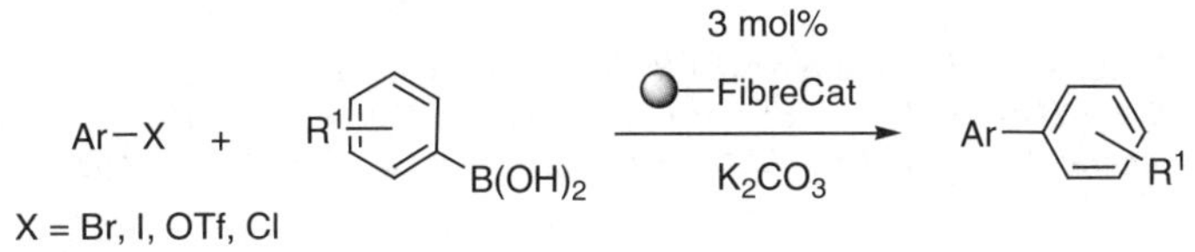

26

Scheme 8 *Reagents and conditions: EtOH, MW, 10–25 min*

was microwaved for 10–15 min at 110 °C in ethanol under an ambient atmosphere in sealed microwave tubes. This procedure is highly amenable for parallel synthesis and has been used within Abbott for routine library generation.

In addition to the synthesis of discrete compounds, there have also been multi-step parallel syntheses to drug-like compounds using polymer-supported reagents. Indeed, our group reported the early use of a PASP approach to generate an array of substituted hydroxamic acid derivatives **27**, which are known to have activity against matrix metalloproteinases.[72] These enzymes are mediators for the breakdown of structural proteins of the extracellular matrix. Their proposed pathogenic role includes tissue breakdown, metastasis and tumour angiogenesis. Five synthetic transformations (Scheme 9) were required to afford an array of final compounds **27**, all in >90% purity (LC) without the use of any chromatography.

Other examples of arrays of biologically active compounds accessed through a multi-step PASP approach, include the synthesis of cysteine protease inhibitors by Ellman,[73] and Armstrong's synthesis of a 25-membered array of analogues of the breast cancer drug Tamoxifen.[74,75]

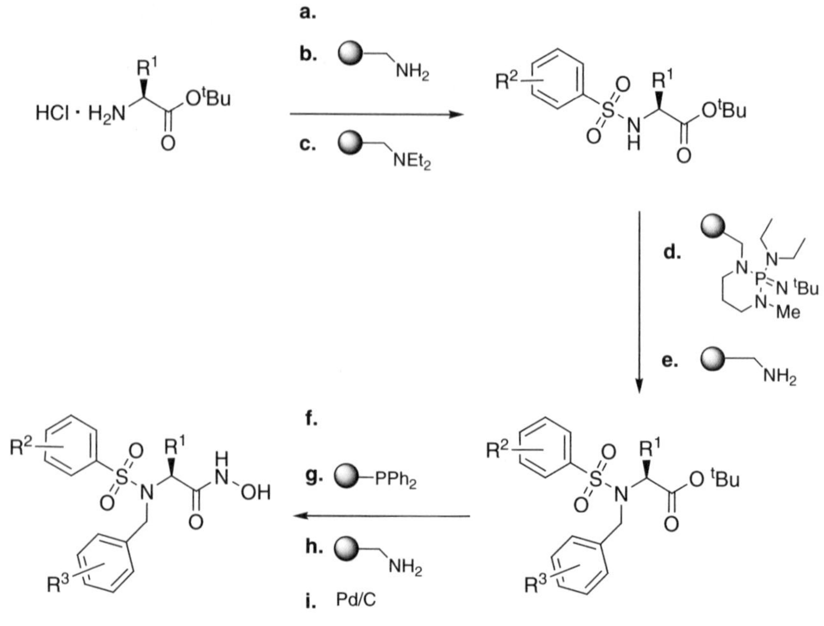

Scheme 9 *Reagents and conditions: (a) $R^2Ar\text{-}SO_2Cl$, pyridine; (b) CH_2Cl_2; (c) CH_2Cl_2; (d) R^3ArCH_2Br, CH_2Cl_2; (e) CH_2Cl_2; (f) TFA, CH_2Cl_2; (g) CBr_4, O-benzylhydroxylamine, NEt_3, CH_2Cl_2; (h) CH_2Cl_2; (i) H_2, EtOAc, ^iPrOH or MeOH*

The use of PASP synthesis in medicinal chemistry programmes to date has been extensively exploited by Parlow *et al.*,[76,77] previously at Pharmacia, in their search for tissue factor VIIa (TF/VIIa) inhibitors. Cardiovascular disease is one of the most common causes of mortality in the Western world, with Warfarin currently the only approved oral antithrombotic therapy. However, this drug has a narrow therapeutic window and its use is complicated by the need for frequent monitoring due to bleeding side effects, drug–drug interactions and food effects resulting from the narrow therapeutic window of the drug. Therefore, there remains an unmet need to develop potent and selective TF/VIIa inhibitors as safe and effective oral anticoagulants.

To this end, Parlow *et al.*[76,77] investigated two different compound series for their potential to inhibit TF/VIIa. Their initial approach was based upon the utilisation of a tripeptide, in which the scissile amide bond had been replaced by an electron-deficient carbonyl group. The design was of the general form D-Phe-L-AA-Arg-α-ketothiazole **28**, which is closely related to the known inhibitor D-Phe-L-Phe-Arg-chloromethylketone. The crystal structure of D-Phe-L-Phe-Arg-chloromethylketone had recently been determined bound to active site of TF/VIIa, allowing comparisons to be made with the closely related proteases such as thrombin and Factor Xa. In order to rapidly prepare libraries of α-ketothiazole peptidyl protease inhibitors in a parallel format, a five-step PASP synthesis was developed (Scheme 10). Initially, each of the steps in the synthesis was independently validated to identify and optimise the conditions, such that each

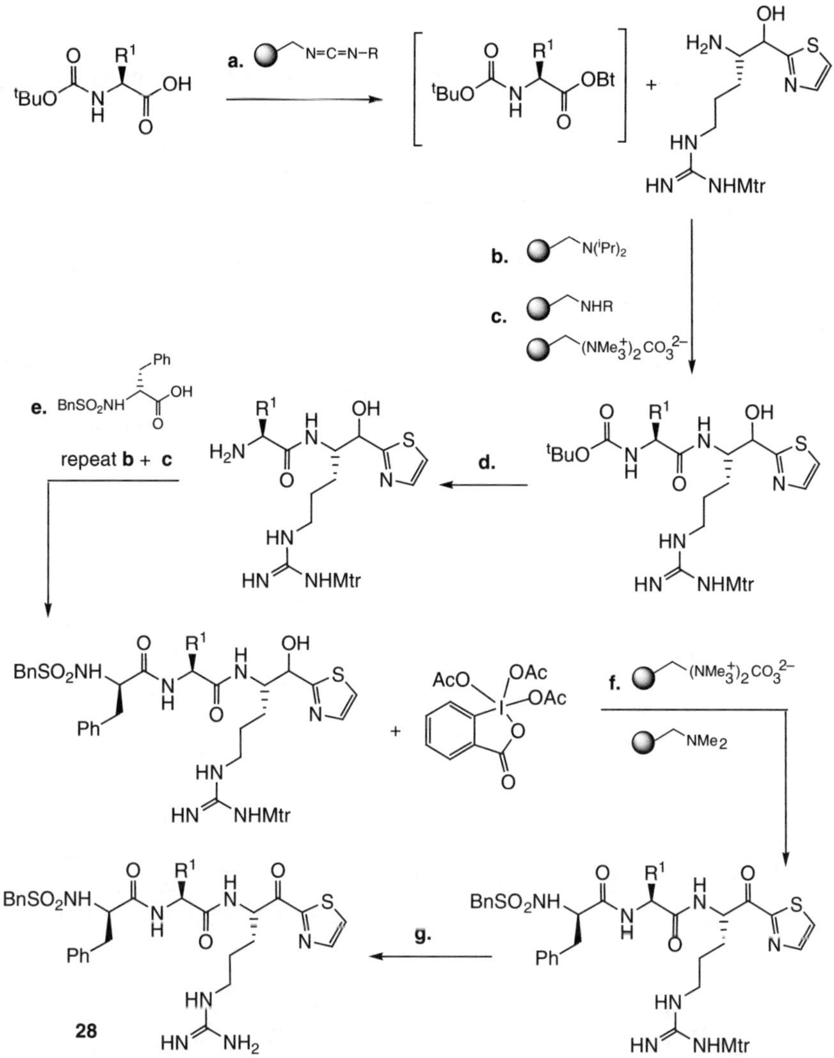

Scheme 10 *Reagents and conditions: (a) HOBt; (b) as above; (c) as above; (d) HCl; (e) as above; (f) as above; (g) TFA, thioanisole*

transformation could be performed in a high-yielding parallel format. In total, a 38-membered array was prepared using this methodology, with purity levels ranging from 70 to 99%.

The 38 compounds were screened for potency against TF/VIIa and for other enzymes affecting coagulation such as Factor Xa and Thrombin (IIa), to determine the specificity. The most potent compound from the array was **29**, which incorporated L-phenylalanine at the point of variation (Figure 6). Gratifyingly, a number of compounds demonstrated selectivity for TF/VIIa, with compound **30** having a selectivity ratio (VIIa/IIa) of over 500-fold. These initial results confirmed that a potent TF/VIIa inhibitor with selectivity versus thrombin could be obtained.

A crystal structure of **30** was obtained to gain a better understanding of the structural differences responsible for selectivity. This information, when combined with the known crystal structures facilitated the development of a series of highly selective, small molecule inhibitors of TF/VIIa that were nonpeptidic and did not interact with the catalytic apparatus of the enzyme. Based on the key features of the TF/VIIa crystal structure, a series of six-membered heterocyclic cores were docked, which led to the pyrazinone scaffold **31** as a prototype inhibitor from which to start future small molecule structure-based drug design (Figure 7).[78,79] Several hundred compounds were prepared with PASP techniques without the need for chromatography or isolation of intermediates. To maximise the efficiency of the multi-step process and to minimise manual manipulation of the reaction vessels, multiple reaction steps were carried out in the same vessel. The three points of diversity on the template

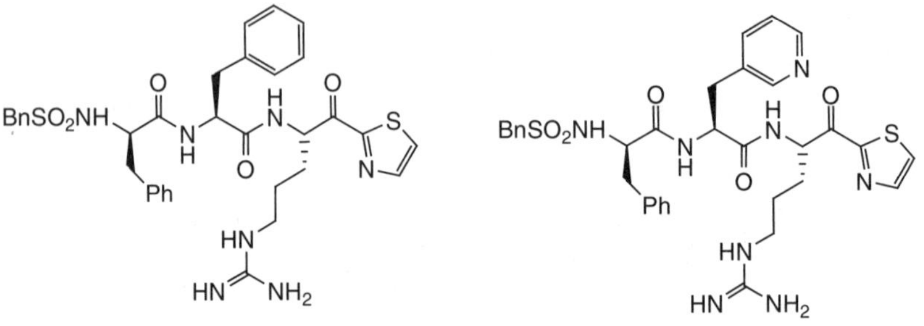

29

30

Figure 6 *TF/VIIa inhibitors identified from the PASP library production*

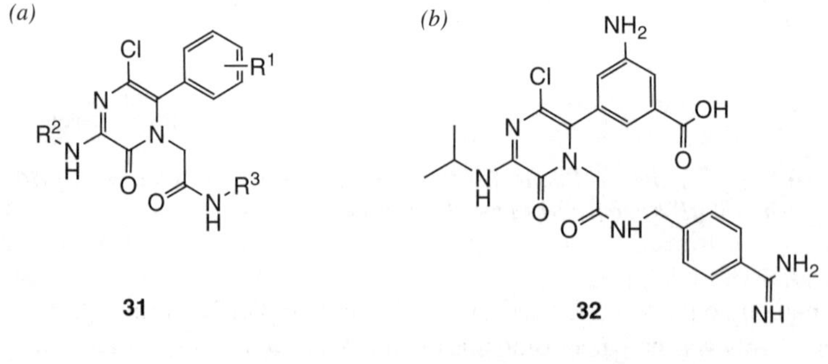

(a)

31

(b)

32

TF/VIIa IC$_{50}$ (µM): 0.016

Thrombin IC$_{50}$ (µM): >100

Xa IC$_{50}$ (µM): >100

Figure 7 *(a) Pyrazinone scaffold identified incorporating three points of variation. (b) Lead compound arising from PASP array synthesis*

were explored simultaneously and each compound screened against TF/VIIa for potency and Factor Xa and thrombin for selectivity. Compound **32** was identified as a result of these studies and displayed good potency against the TF/VIIa proteases and respective selectivity over Factor Xa and Thrombin. Compound **32** was deemed selective enough to proceed with the preclinical intravenous proof of concept studies to demonstrate the separation between antithrombotic efficacy and bleeding side effects in a nonhuman primate model of electrolytic-induced arterial thrombosis.[80]

3 Automated PASP Synthesis of Biologically Active Molecules

3.1 Stepwise Automation of PASP Synthesis in Batch Mode

The adoption of PASP synthesis presents the opportunity for in-line purification strategies, through the use of supported scavengers or reagents and catch-and-release strategies. In this way, conventional purification techniques such as aqueous work-ups and column chromatography can be eliminated. In addition, the use of PASP synthesis reduces the vast number of traditional synthetic manipulations to a series of repetitive incubations and subsequent filtrations, which are in principle well suited to automated processing.

However, early attempts at synthesising compound arrays using polymer-supported reagents and automation[81] proved to be problematic. Our group[82] used an ACT 496[83] automated synthesiser to produce a 96-membered array of compounds in a single automated step. In this array, 12 aromatic aldehydes were reacted with eight aliphatic amines in the presence of polymer-supported cyanoborohydride to produce the corresponding secondary amines **33** (Scheme 11).

In this single automated step, 88 reactions were found to have worked. The failures can be largely attributed to the variation in reactivities typically encountered across a given monomer set, and illustrates a recurring problem associated with combinatorial chemistry.

The preparation of the 3-thioalkyl-1,2,4-triazole chemotype using a 'catch, cyclise and release' approach mediated by the strong polymer-supported base PS-BEMP has been reported by Graybill.[84] Condensation of excess acyl hydrazide and isothiocyanate affords the diacylhydrazide **34**, which is immobilised by the PS-BEMP as the ion pair. After thorough resin washing to remove any impurities, cyclisation to the 3-thio-1,2,4-triazole **35** is accomplished by heating at 85 °C. Finally, treatment of **35** with a substoichiometric amount of alkylating agent releases the *S*-alkylated

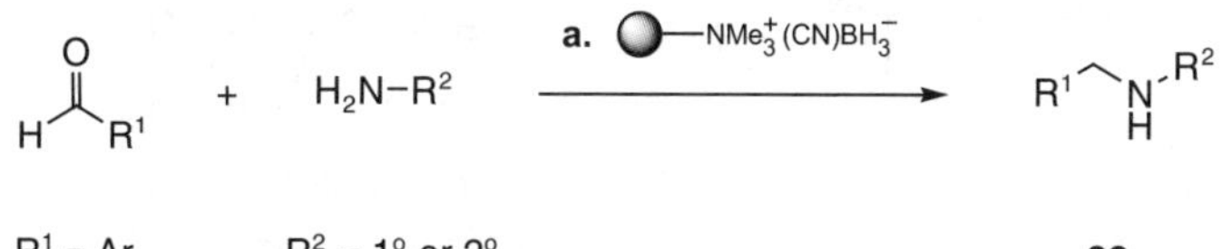

Scheme 11 *Reagents and conditions: (a) MeOH, rt, 72 h*

triazole products **36** into solution (Scheme 12). The results of 64 compounds from a larger hit identification array are exemplified. The authors state that this procedure was used successfully on a number of automated platforms, such as the Myriad Core System[85] and the Argonaut Quest 210,[33] but no details are given.

The stepwise automated PASP synthesis of a 72-membered library of 2-alkylthiobenzimidazoles **37** and benzimidazolin-2-ones **38** has been reported using a combination of manual and automated equipment.[86] By incorporating automated aqueous work-ups together with in-line scavenging and catch-and-release protocols, the desired compounds, containing three points of diversity were obtained directly in good yields and excellent purities without the need for column chromatography. The benzimidazole moiety has been exploited in many medicinal chemistry programmes and displays a range of biological activities, such as antiviral[87] and antihistamine activity[88] or the ability to modulate ion channels (Figure 8).[89]

A more highly automated synthetic route to these two classes of compounds has recently been reported utilising a Zinsser Sophas robotic synthesiser.[90,91] A single robot was used to rapidly synthesise 96 thiobenzimidazoles derivatives **39** and a 72-membered library of benzimidazolin-2-ones **40** (Scheme 13). The transformations

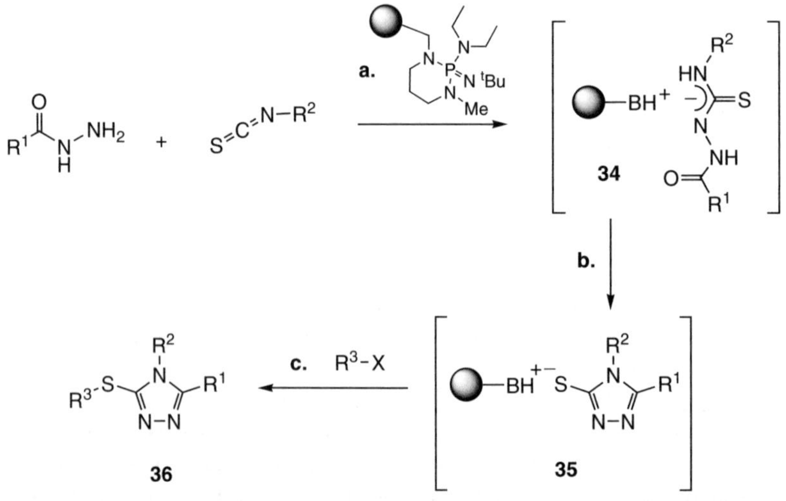

Scheme 12 *Reagents and conditions: (a) DMF, 1 h at rt then 1 h at 45 °C; (b) 1:1 dioxane/water, 16 h, 85 °C; (c) MeCN, 2 h, rt–50 °C*

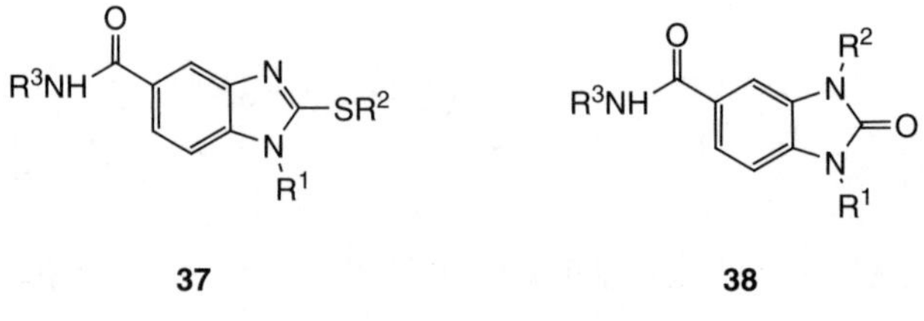

Figure 8 *Drug-like heterocyclic benzimidazole chemotypes*

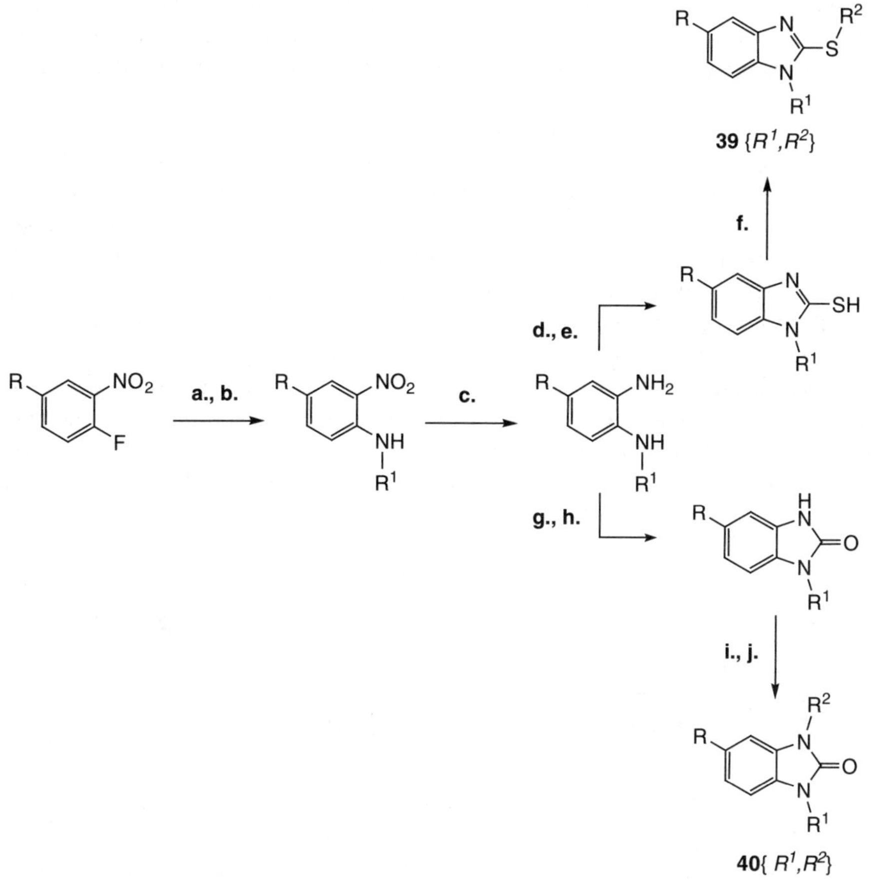

Scheme 13 *Reagents and conditions: (a) R¹NH₂, DMF, rt; (b) PS-trisamine, DMF, 50 °C; (c) H₂(g), Pd(OH)₂, DMF; (d) Im₂CS, DMF, rt; (e) Amberlyst H-15, DMF, rt; (f) R²Br, PS-BEMP, DMF, rt; (g) triphosgene, CHCl₃, rt; (h) PS-trisamine, CHCl₃, 30 °C; (i) R²Br, PS-BEMP, DMF, CHCl₃; (j) PS-methylthiourea, DMF, CHCl₃, 50 °C*

were performed sequentially by the robot, with manual intervention only necessary at the hydrogenation step, which for safety reasons was incompatible with the synthesis robot used. In both arrays, >80% of the final compounds were isolated in >80% purity.

3.2 Fully Automated PASP Synthesis of Drug-Like Molecules in Batch Mode

Recently, a number of publications have dealt with the synthesis of drug-like molecules via fully automated PASP (auto-PASP) synthesis. This scenario entails the use of a series of immobilised reagents and scavenger reagents to perform a variety of sequential synthetic transformations in a common solvent. Each of the transformations required can be automated using a commercially available robotic synthesiser. The first auto-PASP synthesis described was the preparation of a 36-membered array of Histone Deacetylase (HDAc) inhibitors[92] **41** by a 4-5 step sequence, which included a catch-and-release in-line purification step (Scheme 14).

The robot was preloaded with all the resins and reagents necessary and the 180 reactions required were allowed to run unattended over a 4-day period. A total of 34 out of 36 compounds targeted were obtained successfully, with compounds purities between 55 and 80%. Following a single auto-preparative purification step, the compounds were subjected to biological testing.[93] This publication demonstrates not only the ability to automate multi-step solution-phase chemistry successfully, but in addition that a considerable time saving can be made through the adoption of automation.

Another example of fully automated PASP synthesis is the preparation of a 192-member 2D array of 1,5-biaryl pyrazoles.[94] The 1,5-biaryl pyrazole moiety is found in a number of important pharmaceuticals, such as the selective COX-2 inhibitor Celecoxib[95] and the nonsteroidal antiinflammatory agent Tepoxaline.[96,97] The synthetic route to the 1,5-biaryl pyrazoles, exemplified by the library member **42**

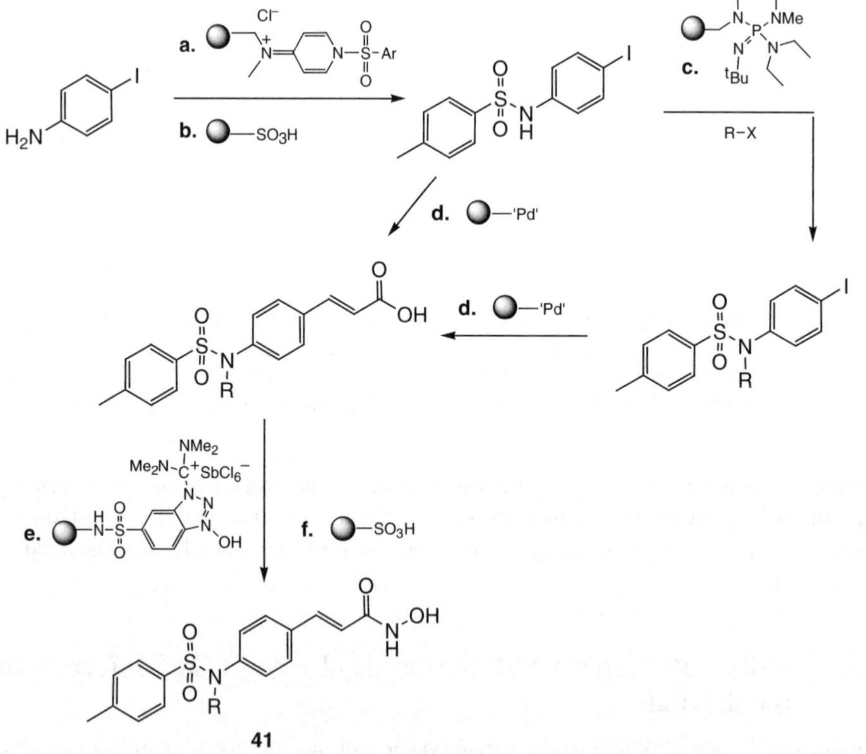

41

Scheme 14 *Reagents and conditions: (a) DMF, 40 °C, 1 h, twice; (b) DMF, rt, 2 h; (c) DMF, RX, rt, 18 h; (d) acrylic acid, ^{n}Bu$_3$N, DMF, 90 °C, 18 h; (e) DIPEA, NH$_2$OTHP, DMF, 50 °C, 18 h; (f) MeOH, DMF, rt, 3 h*

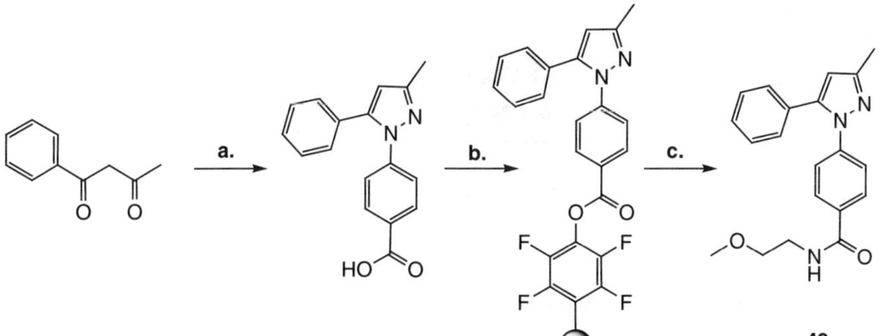

Scheme 15 *Reagents and conditions: (a) phenylhydrazine 4-benzoic acid hydrochloride, Et$_3$N, DMF, rt, 2.5 h; (b) DIC, 4-DMAP, PS-TFP, DMF-CH$_2$Cl$_2$, rt, 2 h; (c) NH$_2$CH$_2$CH$_2$OCH$_3$, DMF, rt, 2 h*

(Scheme 15), employed a solution-phase condensation reaction of a diketone and hydrazine to form the pyrazole ring. Adoption of a catch-and-release strategy for the amide bond formation enabled an in-line purification and concomitant activation of the acid functionality. The 192-membered library was synthesised in a single run using a top-filtration robotic synthesiser; after loading the synthesiser with the required starting materials and reagents, no further manual intervention was necessary. The total library synthesis took 44 h for the robot to complete and, impressively, >80% of compounds were isolated in >80% purity. This library production demonstrates that the use of a multi-step automated PASP strategy can deliver compounds rapidly in purities suitable for direct biological evaluation.

3.3 Flow Chemistry and Automation in the Synthesis of Drug-Like Molecules

As an alternative to conventional batch synthesis procedures, the use of polymer-supported reagents has been described for the implementation of flow-through processes using cartridge or column-based reactors.[98] Flow chemistry is perceived to have a number of benefits over batch reactions, including facile automation, reproducibility, safety and process reliability.[30] Indeed, flow chemistry could allow the rapid transfer of reactions from the research level to process development without time-consuming adaptation and optimisation of methods from laboratory scale to production plant scale. Scale up can be achieved by extending run time or by the use of parallel reactors (scale-out). In addition, by assembling different reactor combinations, linear, divergent or convergent syntheses are all possible (Figure 9).

An early example of the use of a polymer-supported reagent in a flow system is the oxidation of Penicillin G **43** to the corresponding sulfoxide **44** in good yield, using a polymer-supported peroxy acid (Figure 10).[99] Notably, the column could be regenerated *in situ* by treatment for 12 h with hydrogen peroxide and methanesulfonic acid at 20 °C. The ability to regenerate expensive supported reagents is a highly desirable attribute that can be facilitated using flow processes.

(a) Scale out:

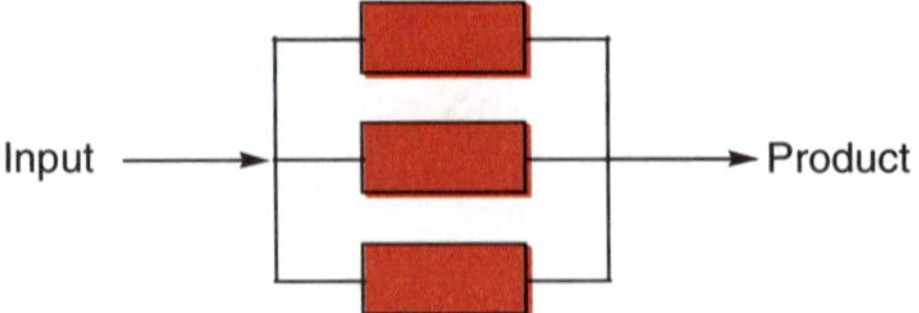

(b) Linear set-up of flow-through reactors:

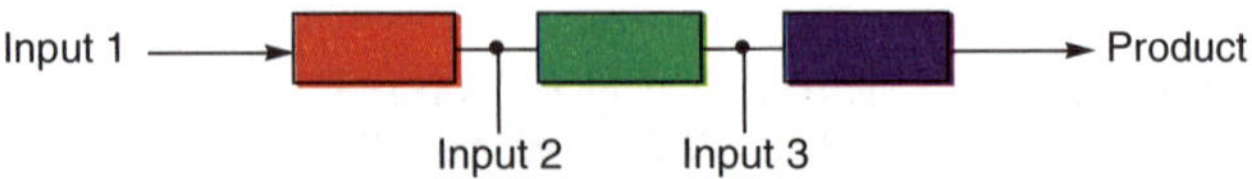

(c) Divergent set-up of flow-through reactors:

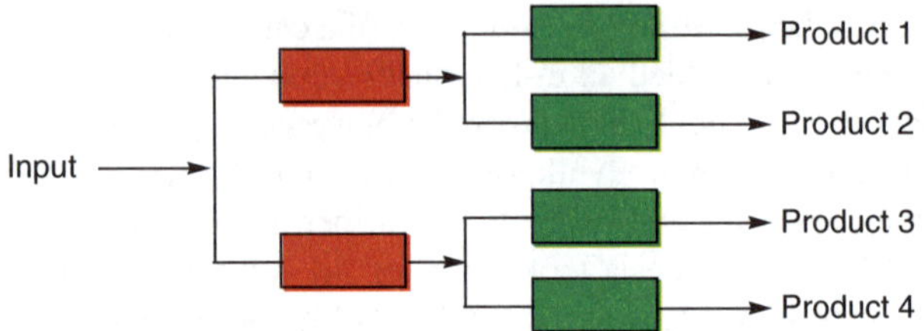

(d) Convergent set-up of flow-through reactors:

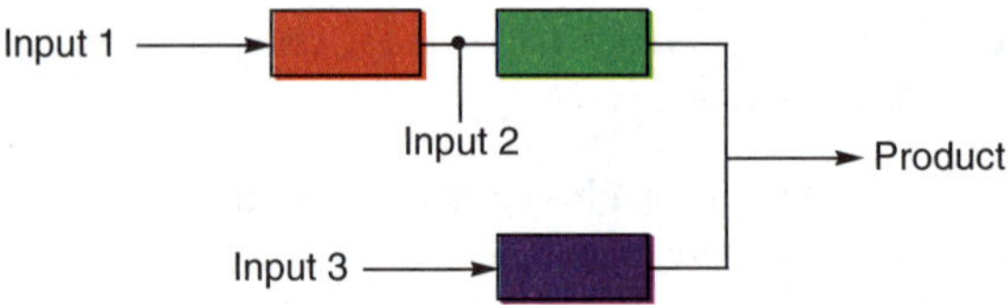

Figure 9 *Concepts for assembling flow-through reactors to enable different modes of synthesis*

Figure 10 *Polymer-supported oxidation of Penicillin G 43 in a flow system*

The derivatisation of the *O*-silylated steroid **45** into the amino derivative **46** using three different supported reagents columns (PASSflow reactors) under flow conditions is illustrated in Scheme 16.[100] The oxidation of alcohol **45** catalysed by 2,2,6,6-tetramethylpiperidine-1-oxyl (TEMPO) via the bound bromate (I) anion in column gave ketone **47**, which was subsequently desilylated after being flowed through a fluoride-loaded reactor. Finally, the reductive amination step was performed by the *in situ* generation of the imine using benzylamine and passing this solution through a reactor loaded with borohydride to furnish the desired amine **46**, albeit contaminated with small amounts of the diol.[101] Having only anions immobilised, ensured the simple regeneration of the reactor columns after each reaction.

A catalytic asymmetric reaction process that involves a series of reaction columns, each of which is packed with supported reagents or catalysts, has been reported by Hafez *et al.*[102] (Figure 11). This process was applied to the catalytic asymmetric synthesis of β-lactams **48**, to yield pure products with excellent enantio- and diastereoselectivity. Acyl substituted β-lactams are known inhibitors of many biological targets, including prostate-specific antigen (PSA1)[103] and cytomegalovirus protease.[104] There are four discrete steps in the β-lactam formation; (1) formation of reactive ketenes in column A; (2) formation of imines *in situ* in column B; (3) catalysis of the condensation of the ketene and imine to form a β-lactam product in column C; and finally (4) removal of unwanted by-products from the reaction stream with a scavenger resin in column D. Reaction solutions were allowed to percolate through the columns in a sequential fashion. After passing

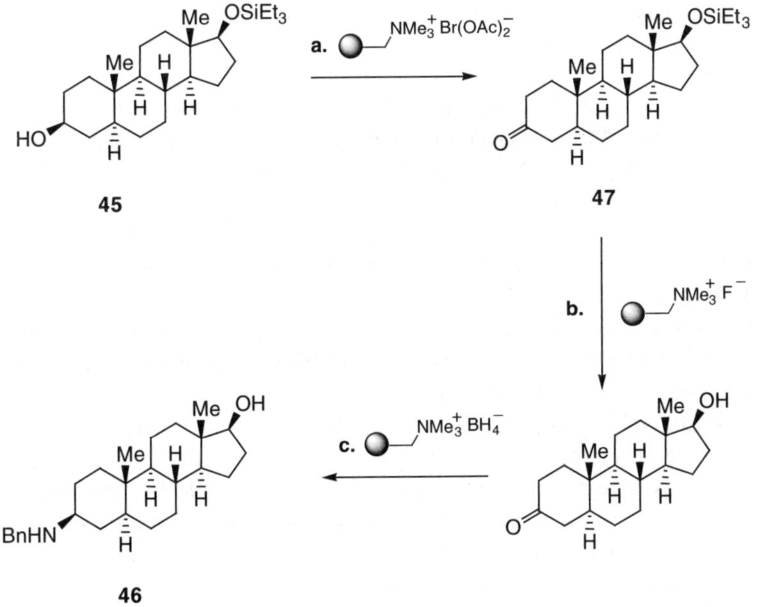

Scheme 16 *Reagents and conditions: (a) cat. TEMPO, CH$_2$Cl$_2$, rt, 6 h, >99%; (b) MeOH, rt, 24 h, 90%; (c) BnNH$_2$, MeOH, rt, 12 h, 85%*

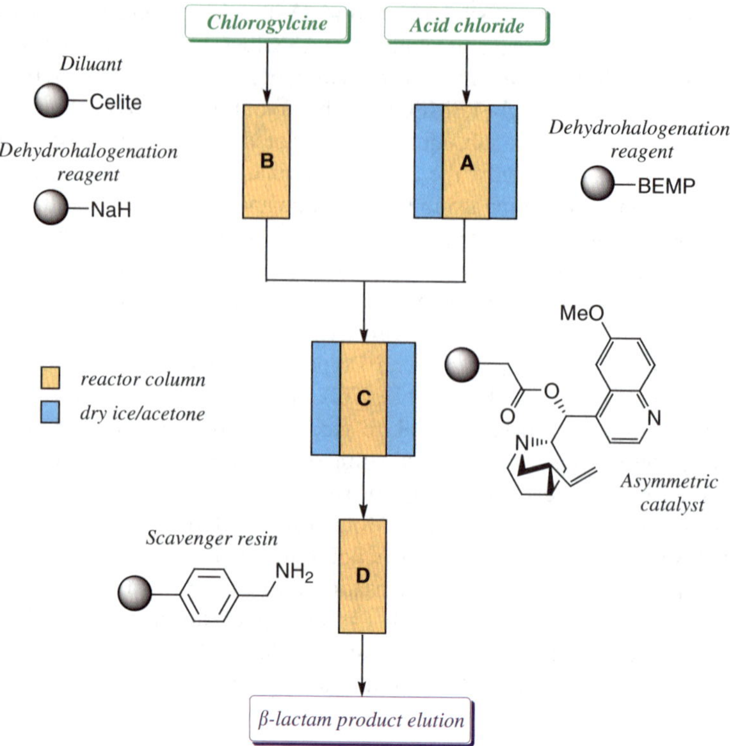

Figure 11 *Asymmetric catalysis using flow chemistry on sequentially linked columns*

through the scavenger resin column, the eluted reaction mixture was concentrated to afford β-lactams **48**. Although only two compounds are exemplified, it is stated that this flow procedure would be applicable to the production of a variety of β-lactam products.

The fully automated, sequential flow-through synthesis of a 44-member array of thioethers via a resin capture-and-release reactor column was performed in our laboratory.[105] Each of the acidic heterocycles (**49–52**) containing thiourea moieties were deprotonated by the use of a strong polymer-supported base, such as 1,5,7-triazabicyclo[4.4.0]dec-5-ene polystyrene (PS-TBD) to generate an immobilised ionic complex on the column (Figure 12).

Introduction of a substoichiometric amount of alkylating agent promoted an alkylative release of the corresponding *S*-thioether from the column (Figure 13). This process was repeated until the column was depleted. The PS-TDB column could then be regenerated by eluting with a solution of a stronger base, such as BEMP. The compounds were obtained in high yield (>75%) and excellent purities (>95% according to LC/MS and ¹H NMR).

In a further extension of this work, a 576-member combinatorial array was prepared in a sequential manner from a set of 24 heterocyclic thiourea and 24 alkylating agent building blocks, reusing the same reactor column over 40 times. Remarkably,

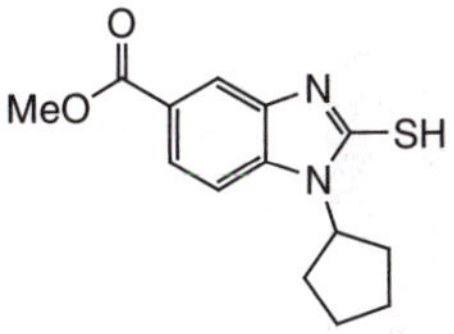

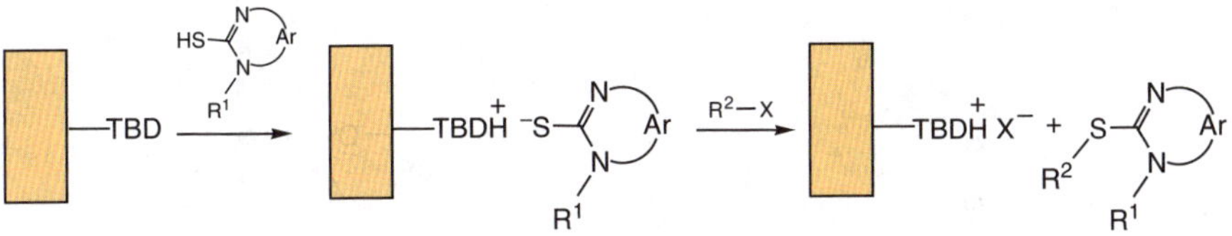

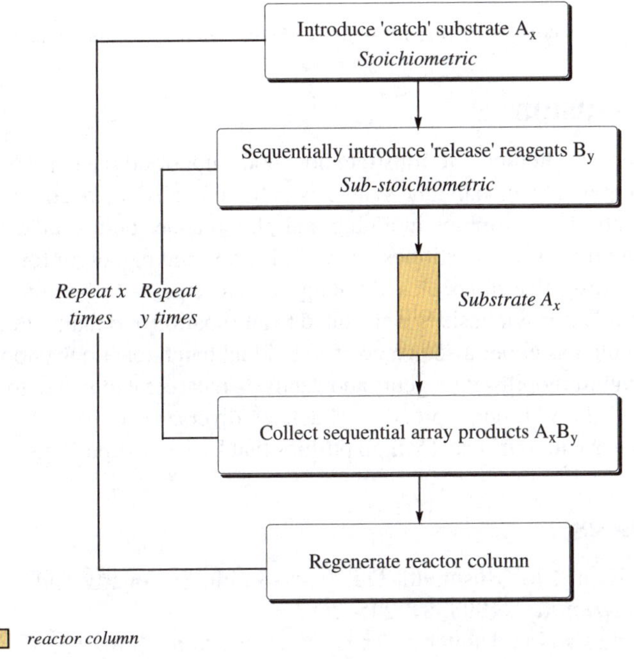

Figure 12 *Catch-and-release PASP synthesis approach to heterocyclic thioethers*

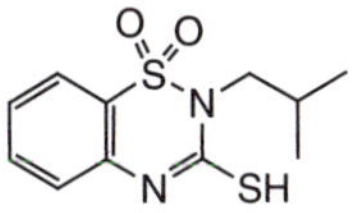

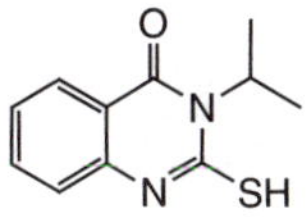

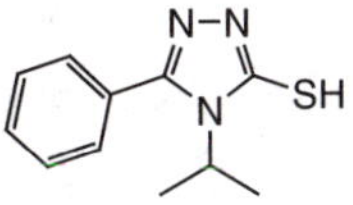

Figure 13 *Schematic representation of an automated iterative catch-and-release flow-through process for the synthesis of a heterocyclic thioether library*

less than 2% of the unpurified library members fell below the 80% purity threshold, even when the yield was less impressive (Figure 14).[106] This provides a clear demonstration of the power of flow processes to deliver high purity compound libraries, particularly for chemistries that may be implemented as a catch-and-release strategy.

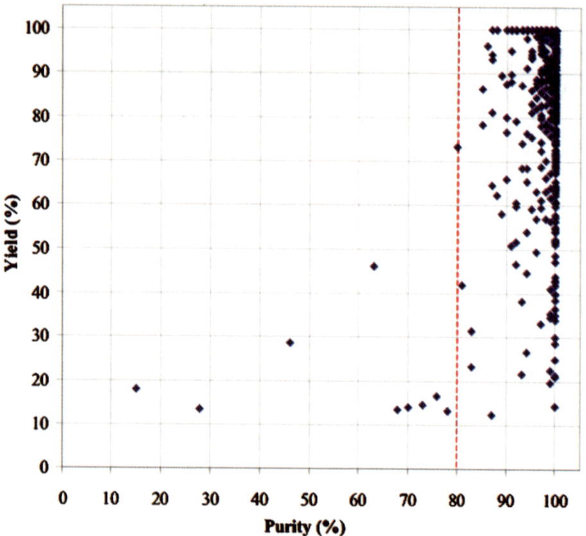

Figure 14 *Yield versus purity plot for 576-membered heterocyclic thioether library*

4 Conclusion

This chapter has attempted to illustrate how both supported reagents and scavengers may be integrated to expedite the synthesis of biologically active compounds. In particular, the ability to combine in a plug-and-play manner both synthesis and in-line purification procedures constitutes a new and simplified paradigm for the multi-step synthesis of drug-like molecules. Although much remains to be done, it is already apparent that PASP synthesis is well suited to automation and that, attractively, it can be implemented as either a sequence of individual batch reactions (stop-flow) or, by incorporating immobilised reagents and catalysts incarcerated in reactor columns, as a continuous flow-through process to deliver diverse drug-like chemotypes in a timely manner with intrinsically high purities that bypass the purification bottleneck.

References

1. B.A. Kenny, M. Bushfield, D.J. Parry-Smith, S. Fogarty and J.M. Treherne, *Prog. Drug Res.*, 1998, **51**, 245–269.
2. I. Hughes and D. Hunter, *Curr. Opin. Chem. Biol.*, 2001, **5**, 243–247.
3. R.B. Merrifield, *J. Am. Chem. Soc.*, 1963, **85**, 2149–2154.
4. R.B. Merrifield and J.M. Stewart, *Nature*, 1965, **207**, 522–523.
5. C.C. Leznoff and W. Sywanyk, *J. Org. Chem.*, 1977, **42**, 3203–3205.
6. C.C. Leznoff and J.Y. Wong, *Can. J. Chem.*, 1972, **50**, 2892–2893.
7. C.C. Leznoff and J.Y. Wong, *Can. J. Chem.*, 1973, **51**, 3756–3764.
8. W.L. Fitch, *Mol. Div.*, 1999, **4**, 39–45.
9. M.J. Shapiro and J.S. Gounarides, *Biotech. Bioeng.*, 2001, **71**, 130–148.
10. P.A. Keifer, *Prog. Drug Res.*, 2000, **55**, 137–211.

11. B. Yan, G. Kumaravel, H. Anjaria, A. Wu, R.C. Petter, C.F. Jewell Jr. and J.R. Wareing, *J. Org. Chem.*, 1995, **60**, 5736–5738.

12. B. Yan, J.B. Fell and G. Kumaravel, *J. Org. Chem.*, 1996, **61**, 7467–7472.

13. V. Swali, G.J. Langley and M. Bradley, *Curr. Opin. Chem. Biol.*, 1999, **3**, 337–341.

14. M.S. Congreve, S.V. Ley and J.J. Scicinski, *Chem. Eur. J.*, 2002, **8**, 1768–1776.

15. M.S. Congreve, M. Ladlow, P. Marshall, N. Parr, J.J. Scicinski, T. Sheppard, E. Vickerstaffe and R.A. Carr, *Org. Lett.*, 2001, **3**, 507–510.

16. C.M. Baldino, *J. Comb. Chem.*, 2000, **2**, 89–103.

17. S.V. Ley, I.R. Baxendale, R.N. Bream, P.S. Jackson, A.G. Leach, D.A. Longbottom, M. Nesi, J.S. Scott, R.I. Storer and S.J. Taylor, *J. Chem. Soc., Perkin Trans. 1*, 2000, 3815–4195.

18. R.J. Booth and J.C. Hodges, *Acc. Chem. Res.*, 1999, **32**, 18–26.

19. A. Kirschning, H. Monenschein and R. Wittenberg, *Angew. Chem. Int. Edn.*, 2001, **40**, 650–679.

20. S. Sussman, *Ind. Eng. Chem. Res.*, 1946, **38**, 1228.

21. S.W. Kaldor, M.G. Siegel, J.E. Fritz, B.A. Dressman and P.J. Hahn, *Tet. Lett.*, 1996, **37**, 7193–7196.

22. A.T. Merritt, *Comb. Chem. HTS*, 1998, **1**, 57–72.

23. J.J. Parlow, R.V. Devraj and M.S. South, *Curr. Opin. Chem. Biol.*, 1999, **3**, 320–336.

24. M.G. Siegel, P.J. Hahn, B.A. Dressman, J.E. Fritz, J.R. Grunwell and S.W. Kaldor, *Tet. Lett.*, 1997, **38**, 3357–3360.

25. J.J. Parlow, *Tet. Lett.*, 1995, **36**, 1395–1396.

26. C.T. Brain, J.M. Paul, Y. Loong and P.J. Oakley, *Tet. Lett.*, 1999, **40**, 3275–3278.

27. S.V. Ley, A.G. Leach and R.I. Storer, *J. Chem. Soc., Perkin Trans. 1*, 2001, 358–361.

28. E. Angeletti, C. Canepa, G. Martinetti and P. Venturello, *J. Chem. Soc., Perkin Trans. 1*, 1989, 105–107.

29. E. Angeletti, C. Canepa, G. Martinetti and P. Venturello, *Tet. Lett.*, 1988, **29**, 2261–2264.

30. G. Jas and A. Kirschning, *Chem. Eur. J.*, 2003, **9**, 5708–5723.

31. Ex-Novabiochem Catalogue, 2002, 359.

32. R. Schwesinger, *Chimia*, 1985, **39**, 269–272.

33. Commercially Available from Argonaut Technologies, Inc., See www. argotech.com.

34. S.V. Ley, C. Ramarao, R.S. Gordon, A.B. Holmes, A.J. Morrison, I.F. McConvey, I.M. Shirley, S.C. Smith and M.D. Smith, *Chem. Commun.*, 2002, 1134–1135.

35. Commercially available from Biotage AB, See www.biotage.com.

36. Commercially Available from Fluka, See www.sigmaaldrich.com.

37. N. Hird, I. Hughes, D. Hunter, M.G.J.T. Morrison, D.C. Sherrington and L. Stevenson, *Tetrahedron*, 1999, **55**, 9575–9584.

38. F. Svec and J.M.J. Frechet, *Anal. Chem.*, 1992, **64**, 820–822.

39. C. Viklund, F. Svec, J.M.J. Frechet and K. Irgum, *Chem. Mat.*, 1996, **8**, 744–750.

40. J.A. Tripp, J.A. Stein, F. Svec and J.M.J. Frechet, *Org. Lett.*, 2000, **2**, 195–198.

41. J.A. Tripp, F. Svec and J.M.J. Frechet, *J. Comb. Chem.*, 2001, **3**, 604–611.

42. A. Akelah and D.C. Sherrington, *Eur. Polym. J.*, 1982, **18**, 301–305.

43. M.E. Davis, *Micropor. Mesop. Mat.*, 1998, **21**, 173–182.

44. C. Bolm and T. Fey, *Chem. Commun.*, 1999, 1795–1796.

45. N.K. Terrett, A.S. Bell, D. Brown and P. Ellis, *Biorg. Med. Chem. Lett.*, 1996, **6**, 1819–1824.

46. R.G. Manecke and J.P. Mulhall, *Annal. Med.*, 1999, **31**, 388–398.

47. C. Page, C. Schudt, G. Dent and K. Rabe (eds), *Phosphodiesterase Inhibitors*, Academic Press Ltd., London, 2006.

48. I.R. Baxendale and S.V. Ley, *Biorg. Med. Chem. Lett.*, 2000, **10**, 1983–1986.

49. D.J. Dale, P.J. Dunn, C. Golightly, M.L. Hughes, P.C. Levett, A.K. Pearce, P.M. Searle, G. Ward and A.S. Wood, *Org. Proc. Res. Dev.*, 2000, **4**, 17–22.

50. B.C.C. Cantello, M.A. Cawthorne, D. Haigh, R.M. Hindley, S.A. Smith and P. Thurlby, *Biorg. Med. Chem. Lett.*, 1994, **4**, 1181–1184.

51. X. Li, C. Abell, B.H. Warrington and M. Ladlow, *Org. Biomol. Chem.*, 2003, **1**, 4392–4395.

52. B.C.C. Cantello, M.A. Cawthorne, G.P. Cottam, P.T. Duff, D. Haigh, R.M. Hindley, C.A. Lister, S.A. Smith and P.L. Thurlby, *J. Med. Chem.*, 1994, **37**, 3977–3985.

53. R.N. Bream, S.V. Ley and P.A. Procopiou, *Org. Lett.*, 2002, **4**, 3793–3796.

54. R.N. Bream, S.V. Ley, B. McDermott and P.A. Procopiou, *J. Chem. Soc., Perkin Trans. 1*, 2002, 2237–2242.

55. A.T. Nials, D.I. Ball, P.R. Butchers, R.A. Coleman, A.A. Humbles, M. Johnson and C.J. Vardey, *Eur. J. Pharm.*, 1994, **251**, 127–135.

56. R.I. Storer, T. Takemoto, P.S. Jackson, D.S. Brown, I.R. Baxendale and S.V. Ley, *Chem. Eur. J.*, 2004, **10**, 2529–2547.

57. K.C. Nicolaou, N. Winssinger, J. Pastor, S. Ninkovic, F. Sarabia, Y. He, D. Vourloumis, Z. Yang, T. Li, P. Giannakakou and E. Hamel, *Nature*, 1997, **387**, 268–272.

58. K.C. Nicolaou, F. Sarabia, S. Ninkovic and Z. Yang, *Angew. Chem. Int. Edn.*, 1997, **36**, 525–527.

59. I.R. Baxendale, S.V. Ley, M. Nessi and C. Piutti, *Tetrahedron*, 2002, **58**, 6285–6304.

60. I.R. Baxendale, S.V. Ley and C. Piutti, *Angew. Chem. Int. Edn.*, 2002, **41**, 2194–2197.

61. C.W. Fennell and J. van Staden, *J. Ethnopharm*, 2001, **78**, 15–26.

62. S.V. Ley, O. Schucht, A.W. Thomas and P.J. Murray, *J. Chem. Soc., Perkin Trans. 1*, 1999, 1251–1252.

63. J. Habermann, S.V. Ley and J.S. Scott, *J. Chem. Soc., Perkin Trans. 1*, 1999, 1253–1256.

64. A.L. Lee and S.V. Ley, *Org. Biomol. Chem.*, 2003, **1**, 3957–3966.

65. S.W. Kaldor, J.E. Fritz, J. Tang and E.R. McKinney, *Biorg. Med. Chem. Lett.*, 1996, **6**, 3041–3044.

66. T.R. Ryder, L.Y. Hu, M.F. Rafferty, E. Millerman, B.G. Szoke and K. Tarczy-Hornoch, *Biorg. Med. Chem. Lett.*, 1999, **9**, 1813–1818.

67. B.A. Kulkarni and A. Ganesan, *Angew. Chem. Int. Edn.*, 1997, **36**, 2454–2455.
68. B. Kulkarni and A. Ganesan, *Chem. Commun.*, 1998, 785–786.
69. V. Nesterenko, K.S. Putt and P.J. Hergenrother, *J. Am. Chem. Soc.*, 2003, **125**, 14672–14673.
70. Y. Wang and D.R. Sauer, *Org. Lett.*, 2004, **6**, 2793–2796.
71. P.J. Hajduk, M. Bures, J. Praestgaard and S.W. Fesik, *J. Med. Chem.*, 2000, **43**, 3443–3447.
72. M. Caldarelli, J. Habermann and S.V. Ley, *Biorg. Med. Chem. Lett.*, 1999, **9**, 2049–2052.
73. A. Lee and J.A. Ellman, *Org. Lett.*, 2001, **3**, 3707–3709.
74. J. Higginson and E.P. Gelmann (eds), Scientific Reviews of Tamoxifen, *Sem. Oncol.*, 1997, **24**, 156.
75. S.D. Brown and R.W. Armstrong, *J. Org. Chem.*, 1997, **62**, 7076–7077.
76. J.J. Parlow, T.A. Dice, R.A. Lachance, T.J. Girard, A.M. Stevens, R.A. Stegeman, W.C. Stallings, R.G. Kurumbail and M.S. South, *J. Med. Chem.*, 2003, **46**, 4043–4049.
77. M.S. South, T.A. Dice, T.J. Girard, R.M. Lachance, A.M. Stevens, R.A. Stegeman, W.C. Stallings, R.G. Kurumbail and J.J. Parlow, *Biorg. Med. Chem. Lett.*, 2003, **13**, 2363–2367.
78. J.J. Parlow, B.L. Case, T.A. Dice, R.L Fenton, M.J. Hayes, D.E. Jones, W.L. Neumann, R.S. Wood, R.M. Lachance, T.J. Girard, N.S. Nicholson, M. Clare, R.A. Stegeman, A.M. Stevens, W.C. Stallings, R.G. Kurumbail and M.S. South, *J. Med. Chem.*, 2003, **46**, 4050–4062.
79. M.S. South, B.L. Case, R.S. Wood, D.E. Jones, M.J. Hayes, T.J. Girard, R.M. Lachance, N.S. Nicholson, M. Clare, A.M. Stevens, R.A. Stegeman, W.C. Stallings, R.G. Kurumbail and J.J. Parlow, *Biorg. Med. Chem. Lett.*, 2003, **13**, 2319–2325.
80. O.D. Suleymanov, J.A. Szalony, A.K. Salyers, R.M. Lachance, J.J. Parlow, M.S. South, R.S. Wood and N.S. Nicholson, *J. Pharm. Exp. Ther.*, 2003, **306**, 1115–1121.
81. J.C. Reader, *Curr. Top. Med. Chem.*, 2004, **4**, 671–686.
82. S.V. Ley, M.H. Bolli, B. Hinzen, A.-G. Gervois and B.J. Hall, *J. Chem. Soc., Perkin Trans. 1*, 1998, 2239–2242.
83. Commercially Available from Advanced ChemTech., See www.advanced-chemtech.com.
84. T.L. Graybill, S. Thomas and M.A. Wang, *Tet. Lett.*, 2002, **43**, 5305–5309.
85. Commercially Available from Mettler-Toledo Ltd., See www.bohdan.com.
86. S.P. Andrews, D. Jönsson, B.H. Warrington and M. Ladlow, *Comb. Chem. HTS*, 2004, **7**, 163–178.
87. E. De Clercq, *J. Clin. Vir.*, 2001, **22**, 73.
88. P. Lindberg, D. Keeling, J. Fryklund, T. Andersson, P. Lundborg and E. Carlsson, *Aliment. Pharm. Ther.*, 2003, **17**, 481–488.
89. S. Singh, C.A. Syme, A.K. Singh, D.C. Devor and R.J. Bridges, *J. Pharm. Exp. Ther.*, 2001, **296**, 600–611.

90. Commercially Available from www.zinsser-analytic.com.

91. E. Vickerstaffe, B.H. Warrington, M. Ladlow and S.V. Ley, *J. Comb. Chem.*, 2005, **7**, 385–397.

92. E. Vickerstaffe, B.H. Warrington, M. Ladlow and S.V. Ley, *Org. Biomol. Chem.*, 2003, **1**, 2419–2422.

93. A. Bapna, E. Vickerstaffe, B.H. Warrington, M. Ladlow, T.-P.D. Fan and S.V. Ley, *Org. Biomol. Chem.*, 2004, **2**, 611–620.

94. E. Vickerstaffe, B.H. Warrington, M. Ladlow and S.V. Ley, *J. Comb. Chem.*, 2004, **6**, 332–339.

95. T.D. Penning, J.J. Talley, S.R. Bertenshaw, J.S. Carter, P.W. Collins, S. Docter, M.J. Graneto, L.F. Lee, J.W. Malecha, J.M. Miyashiro, R.S. Rogers, D.J. Rogier, S.S. Yu, G.D. Anderson, E.G. Burton, J.N. Cogburn, S.A. Gregory, C.M. Koboldt, W.E. Perkins, K. Seibert, A.W. Veenhuizen, Y.Y. Zhang and P.C. Isakson, *J. Med. Chem.*, 1997, **40**, 1347–1365.

96. D.C. Argentieri, D.M. Ritchie, M.P. Ferro, T. Kirchner, M.P. Wachter, D.W. Anderson, M.E. Rosenthale and R.J. Capetola, *J. Pharm. Exp. Ther.*, 1994, **271**, 1399–1408.

97. T. Kirchner, B. Aparicio, D.C. Argentieri, C.Y. Lau and D.M. Ritchie, *Prostaglandins, Leukotrienes, and Essential Fatty Acids*, 1997, **56**, 417–423.

98. P. Hodge, *Curr. Opin. Chem. Biol.*, 2003, **7**, 362–373.

99. C.R. Harrison and P. Hodge, *J. Chem. Soc., Perkin Trans. 1*, 1976, 2252–2254.

100. A. Kirschning, C. Altwicker, G. Drager, J. Harders, N. Hoffmann, U. Hoffmann, H. Schonfeld, W. Solodenko and U. Kunz, *Angew. Chem. Int. Edn.*, 2001, **40**, 3995–3998.

101. S. Saaby, K. Rahbek Knudsen, M. Ladlow and S.V. Ley, *Chem. Commun.*, 2005, 2909–2911.

102. A.M. Hafez, A.E. Taggi, T. Dudding and T. Lectka, *J. Am. Chem. Soc.*, 2001, **123**, 10853–10859.

103. R.M. Adlington, J.E. Baldwin, G.W. Becker, B. Chen, L. Cheng, S.L. Cooper, R.B. Hermann, T.J. Howe, W. McCoull, A.M. McNulty, B.L. Neubauer and G.J. Pritchard, *J. Med. Chem.*, 2001, **44**, 1491–1508.

104. P.R. Bonneau, F. Hasani, C. Plouffe, E. Malenfant, S.R. LaPlante, I. Guse, W.W. Ogilvie, R. Plante, W.C. Davidson, J.L. Hopkins, M.M. Morelock, M.G. Cordingley and R. Deziel, *J. Am. Chem. Soc.*, 1999, **121**, 2965–2973.

105. D. Jönsson, B.H. Warrington and M. Ladlow, *J. Comb. Chem.*, 2004, **6**, 584–595.

106. D. Jönsson and M. Ladlow, Unpublished results.

Accelerated Chemistry: Microwave, Sonochemical, and Fluorous Phase Techniques

KRISTOFER OLOFSSON[a], PETER NILSSON[a] AND
MATS LARHED[b]

[a]Biolipox AB, Medicinal Chemistry, Box 6280, SE-102 34, Stockholm, Sweden
[b]Organic Pharmaceutical Chemistry, Department of Medicinal Chemistry, BMC,
Uppsala University, Box 574, SE-751 23, Uppsala, Sweden

1 Introduction

The difference in concept between drug discovery today and before has been described many times and most popularly by underlining the difference in the number of compounds produced.[1] Typically, and perhaps a little simplified, if the focus used to be on pushing a small number of rationally designed drugs through the not always easily understood maze of preclinical and clinical studies, there is today a clear trend and willingness to improve the chances of presenting a successful drug by one of the simplest methods available: increasing the number of synthesized compounds. There have been a number of practical solutions on how to address the problem of enhancing the output of unique chemical identities (*e.g.* combinatorial synthesis, parallel synthesis, and library production) and their subsequent testing (*e.g.* high-throughput screening (HTS)), many of which are reviewed elsewhere in this book.

Even though many of these techniques in themselves are modern, there has been an interest to develop these methods, and invent new techniques, to further increase the speed of hit identification and lead optimization. In large, and in direct application to this chapter, these efforts can be divided into techniques that increase either the speed of *synthesis*, such as microwave and sonochemical approaches, or the speed of separation,[2] such as fluorous techniques.

Microwave heating is today an important and accepted method of heating in preparative chemistry.[3–5] The attractions of microwave-heated technologies are many. For example, reaction times can in numerous cases be shortened from hours or days to minutes or seconds. There has also been a development of the microwave reactors. Reactions can nowadays be executed safely, with pressure- and temperature control, in an interlaboratory reproducible manner. The number of publications in this field is large and the focus of this section will be limited to a few selected examples, describing high-speed synthesis with biological evaluation of the produced compounds. Reactions made under sonochemical methods are also briefly covered. This unconventional technology can in many times offer cheap and reliable alternatives to classical methods of reaction heating, but the number of research articles directed towards drug discovery and diversity-oriented synthesis is so far not large. Thus, the presented material will be of a more general character. Fluorous chemistry and separation, which today is an umbrella term for several interrelated subcategories of chemistry using the unique properties of compounds with a high fluorine content, is an area where there is an invigorating development, albeit with the drawback of a fewer number of contributors than in microwave chemistry.[6–9] One sign of the promise of fluorous chemistry is that many of the most important papers have been published within the last 3 years. Even though the most immediate gain of fluorous chemistry is in the separation steps, it could be argued that fluorous chemistry offers a strategic alternative in synthesis that covers all steps from the beginning of a reaction to the final purification of the end product.

2 Microwave Enhanced Chemistry

2.1 General

The microwave story in the drug-discovery area began in the late 1990s, although a number of pioneers had started using microwave irradiation for organic synthesis much earlier.[10–15] In terms of personal dynamics in the integration of microwave technology with medicinal chemistry, important contributions were made by Anders Hallberg,[16] who early emphasized the potential of combining microwave heating and transition metal catalysis in the development of sophisticated lead molecules. Exploiting high-density microwave heating, Hallberg's[17] research group demonstrated that it was possible to drive palladium-catalysed transformations to completion in minutes instead of hours, and that novel HIV-inhibitors could be equally rapidly generated.[18,19] When the benefits of combining microwave heating with solid-phase chemistry and polymer-supported-reagents were recognized, these concepts were investigated in detail by especially Oliver Kappe[20] and Steven Ley.[21] Convincing evidence that dedicated microwave equipment indeed accelerated both reaction development and compound-production accumulation, and today, microwave reactors are an integrated part of the standard instrumentation in most modern high-throughput and medicinal chemistry laboratories.

Since a large number of recent reviews on microwave-assisted organic and combinatorial chemistry are available,[3,22] no introduction to the theory behind microwave heating will be given.[23] We will only conclude that high-density microwave irradiation produces efficient internal heating of polar liquid systems, resulting in an even and rapid heating throughout the sample. Importantly, almost all organic reaction systems can be accelerated utilizing this energy source. The different types of commercial reactors available will not be discussed, although all reactions reported in this section were conducted in 0.5–5.0 mL volume, septum-sealed reaction vessels employing temperature-controlled single-mode synthesizers.[24]

Until today, most applications of microwave-promoted discovery chemistry reporting biological activities involve small-scale lead-optimizations of peptidomimetic or transition state-mimicking core structures using palladium-catalysed coupling reactions.[18,25–27] The success of this strategy is based on the high chemoselectivity associated with transition-metal catalysis together with the simplicity, high likelihood of success, and short reaction times achieved with modern single-mode microwave reactors. Furthermore, to allow fine-tuning of the molecular composition of a lead scaffold with the ultimate goal to obtain improved biological activity, solubility, metabolic stability, and pharmacokinetic properties, very large sets of modified analogues are required.

2.2 Applications in Medicinal Chemistry

As a result of the emergence of drug resistant HIV-1 strains and severe side effects of the anti-HIV drugs used, continued interest in HIV-1 protease inhibitor research is still highly motivated. In this field, the activity of tetra-substituted cyclic sulfamide protease inhibitors incorporating the central dihydroxyethylene as the transition isostere, has been well documented.[25,28,29] Thus, in an attempt to develop a new class of HIV-1 protease inhibitors, the benzylic P2/P2′ groups were substituted in the *ortho*-position to reach also the S1/S1′ pockets.[30] The dibromo starting material was first smoothly prepared from tartaric acid in a four-step procedure. In Reaction Scheme 1 below, a set of successful microwave-enhanced Suzuki- and Heck couplings executed by 5–20 min irradiation to elongate the P2/P2′ side chains of the cyclic HIV-1 protease inhibitor template is presented. The diversified products were after standard deprotection, screened for HIV-1 protease activity, identifying a relatively potent dibenzofurane inhibitor that possessed a K_i value of 530 nM.[30]

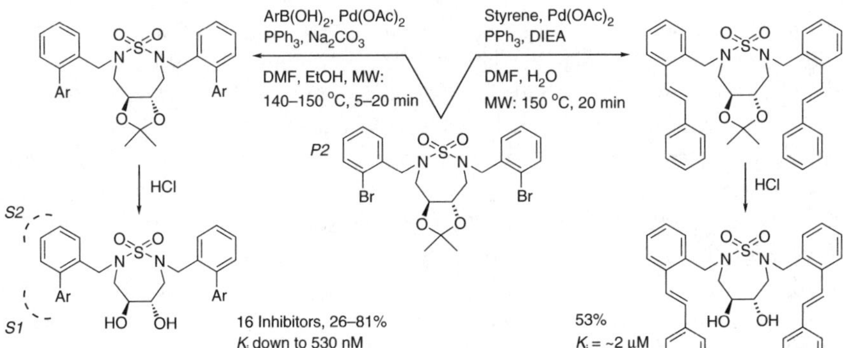

Scheme 1

Another class of HIV-1 protease inhibitors is the linear 1,2-dihydroxyethylene-based compounds.[31] In a project searching for novel binding modes to the protease, the 2-iodobenzyloxy and 3-bromobenzyloxy P1/P1′-substituted structures depicted in Reaction Scheme 2 were used as arylpalladium precursors.[32] Two series of microwave-promoted aminocarbonylations were performed in which $Mo(CO)_6$ served as a convenient solid source of carbon monoxide.[33,34] The discovery of active single-digit nanomolar inhibitors containing large phenyl amide *ortho*-substituents in the P1/P1′ positions indicates that larger groups might be tolerated in this part of the S1/S1′-pocket than previously believed.[32]

The human octapeptide angiotensin II mediates its effects by two major receptors, named AT_1 and AT_2. While the much-studied AT_1 receptor is involved in the control of blood pressure, electrolyte, and fluid balance, the function of the AT_2 receptor has been more difficult to elucidate. Recent results suggest, however, that activation of the AT_2 receptor, among other effects, stimulates alkaline secretion by the duodenal mucosa in rats.[35] Based on this information, Alterman[36] has published the discovery of selective AT_2 receptor agonists using a number of microwave-mediated transformations. A small part of this extensive project is disclosed in Reaction Scheme 3, where two compounds of modest activity were prepared by microwave heated coupling reactions.[36]

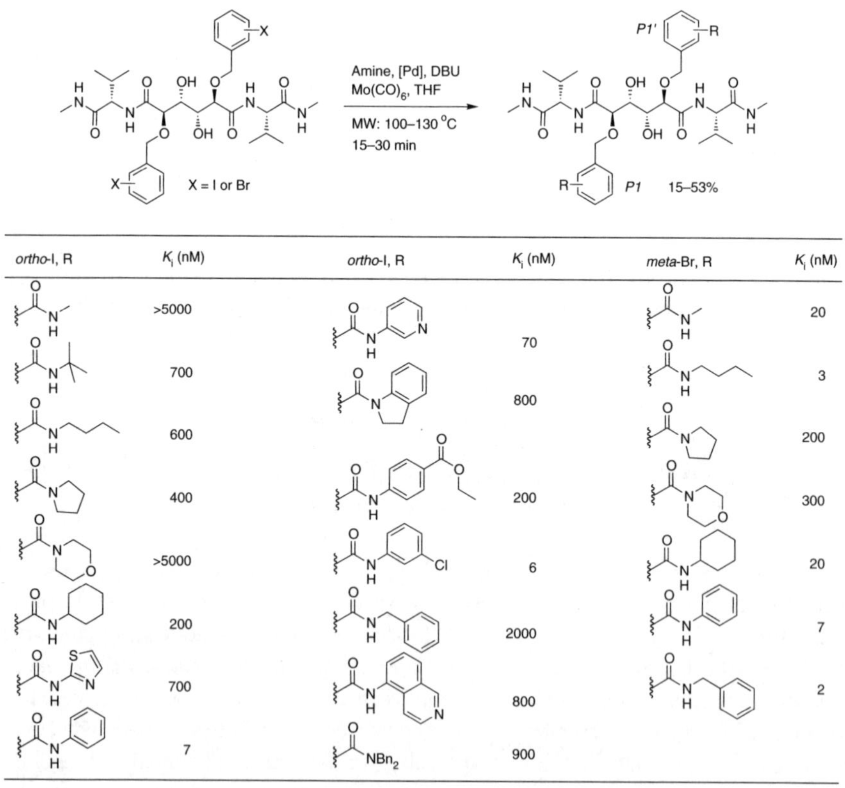

Scheme 2

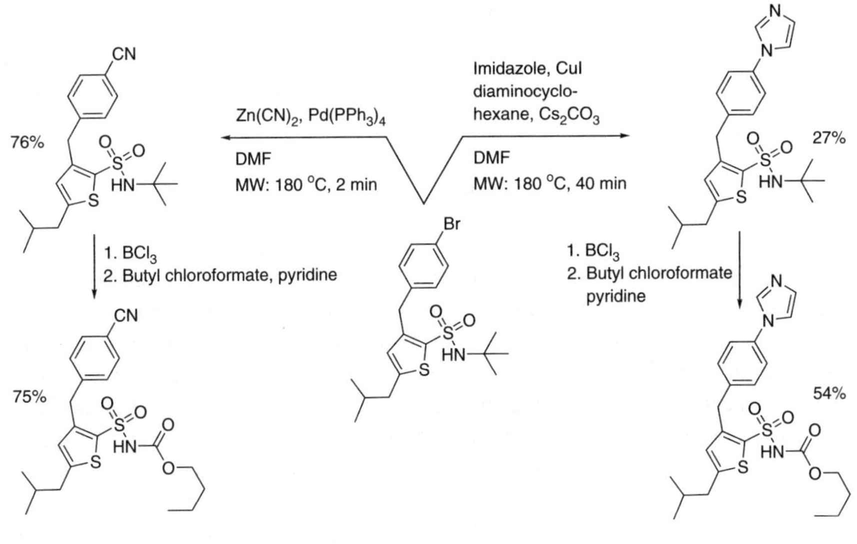

Scheme 3

In the top left case, a palladium-catalysed nitrile coupling was rapidly executed by 2 min of microwave irradiation, while the phenyl-imidazole moiety was created using a very sluggish Ullman-type of reaction.

2.3　Applications in Solid-Phase Chemistry

Microwave chemistry is beginning to play a greater role also in solid-phase chemistry and in cases where polymer-supported reagents are utilized.[5,22] The main drawback with these methodologies concerns the slow kinetics, which makes reaction scouting and optimization tedious. An impressive example of a microwave accelerated and polymer assisted process was recently described by Linclau.[37] In this report, a set of carboxylic acids were fully *O*-alkylated after only 3–5 min of microwave heating (Reaction Scheme 4).

3　Sonochemistry as a Means to Accelerate Synthesis

3.1　General

Ultrasound can be described as alternating pressure waves (frequency higher than *ca.* 16 kHz) that in contrast to electromagnetic waves need an elastic medium to propagate. Sonochemistry in a liquid is a result of an acoustic cavitation occurring when ultrasound travels through the media with a concomitant formation of gas or vapour bubbles. The formation, expansion, and implosive collapse of such bubbles create local hot spots with temperatures estimated to reach several thousand Kelvins.[38] The cavitation phenomenon is believed to be responsible for the various chemical and mechanical effects seen in sonochemical applications. The chemical outcome of ultrasonic irradiation is influenced by a number of factors including the choice of solvent, bulk

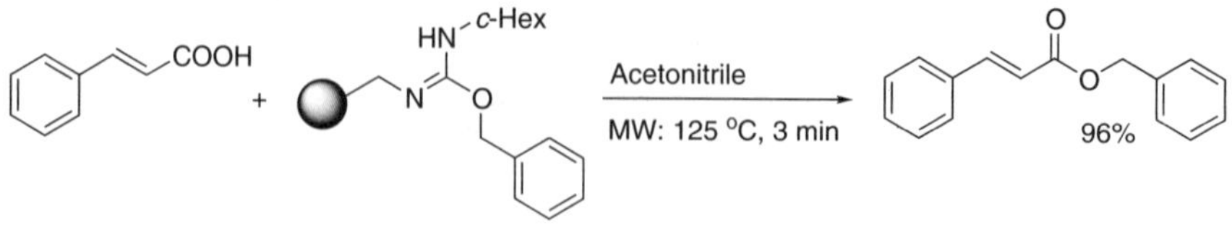

Scheme 4

solution temperature, and the applied ultrasound frequency. Even though the physical fundamentals of ultrasound-enhanced chemistry have been recognized for several years and despite the recent progress, the underlying mechanisms of heterogeneous and homogeneous sonochemistry are not fully perceived.[39] However, this emerging method of speeding up chemical reactions has become a widespread technique in modern laboratories due to the simplicity of use and the relatively inexpensive equipment. The objective here is to present some selected reactions that are of importance in library generation, medicinal chemistry, and natural product synthesis[40] and which are reported to be significantly accelerated by ultrasound. More comprehensive books and reviews, dealing with synthesis and ultrasound in general, can be found elsewhere.[41–44]

3.2 Organometallic Sonochemistry

Rate improvements have been realized mainly in organometallic reactions and radical processes, but also in reactions run under phase-transfer conditions.[45] Mechanical effects such as cavitational erosion, causing activation of the metal surface, are believed to be the reason behind the suppression of the usual induction time. A vast number of different heterogeneous organometallic reactions are reported to be improved by ultrasound irradiation, such as Grignard (Mg),[46] permanganate oxidation,[47] lithiation,[48] Reformatsky (Zn),[49] Ullmann coupling (Cu),[50] Dieckmann condensation (K),[51] Simmon–Smith cyclopropanation (Zn),[52] and Clemmensen reduction (Zn)[53] to name a few. Often, applying ultrasound in organometallic chemistry allows the synthesis to be performed under much milder conditions than the corresponding conventional procedure. Furthermore, *in situ* ultrasound preparation of the organometallic reagent allows for time saving and easier handling of sensitive intermediates as exemplified in the classical Barbier[54] and Reformatsky protocols. Asymmetric heterogeneous metal-catalysed reactions are often temperature sensitive, but selectivities can be improved with sustained reaction rate employing ultrasound at a lowered temperature.[55] Accordingly, heterogeneous organometallic sonochemistry has been advantageously used in the total synthesis of complex structures as steroids[52,53] and vitamin D analogues[56]. Recently, several successful examples of ultrasound promoted homogeneous metal-catalysed processes such as the Heck[57] and Suzuki[58,59] couplings and the Wolff rearrangement[60] have been reported.

3.3 Heterocyclic and Pericyclic Chemistry

As a result of the plethora of pharmacologically active structures that contain heterocyclic moieties, an important part of modern preparative organic chemistry involves heterocyclic synthesis. Ultrasound has accelerated the synthesis of many privileged heterocyclic scaffolds, *e.g.* pyridazine,[61] pyridoquinazolinones,[62] hydantoins,[63] and pyrazoles.[64]

The dihydropyrimidinone core is found in a large number of biologically potent compounds and the synthesis is carried out most straightforwardly via the multi-component Biginelli condensation.[65] For productivity, a one-pot procedure that forms several chemical bonds in one synthetic step is favourable. However, despite extensive research, the one-pot Biginelli protocol often suffers from low yields, long reaction times, and difficult purification. Ultrasound provides an attractive tool to circumvent some of these limitations. Examples of mild- and high-yielding procedures[66–70] have been reported with reaction times down to a few minutes (Reaction Scheme 5)[71] compared to the classical method requiring several days for completion.

Pericyclic reactions, *e.g.* Diels–Alder type reactions, constitute some of the most attractive methods for the creation of cyclic structures, allowing both carbo- and heterocycles to be formed in a single step. Somewhat surprisingly, even this concerted reaction has proved to benefit from ultrasonic treatment.[72,73] A Diels–Alder reaction with a thermally labile substrate (*o*-quinone) has been reported to run to completion within 6 h even at the low temperature of 6 °C.[74] The endo:exo ratio[75] and regioselectivity[76] of the Diels–Alder products can be ameliorated utilizing sonochemistry. In the former case, the endo:exo ratio was found to be influenced by a secondary sonochemical effect derived from cleavage of the C–X bond in the halogenated solvent used. This effect could be deduced since addition of a small amount of HCl produced the same result in the corresponding conventional protocol.

3.4   Applications in Medicinal Chemistry

In the area of medicinal chemistry, ultrasound has proven to be a useful technique to promote desired transformations. An innovative application intended to boost the early drug-discovery process has been reported by the Whitesides research group.[77] They generated a library of polymeric *N*-acetylneuraminic acid analogues by direct sonication in a microtiter plate, obtaining potent inhibitors of erythrocytes hemagglutination. This *in situ* generation and evaluation of a library represents a very interesting approach to increase throughput in both the lead generation and lead optimization processes. Fluorescent ligands for the estrogen receptor have been created utilizing a ultrasound-enhanced cyclization procedure according to Reaction Scheme 6 in the total synthesis of 2,3-bis(4-hydroxyphenyl)indole derivatives.[78]

The application of ultrasound in solid-phase synthesis has been shown to be useful in several cases such as carboxylic acid attachment to Merrifield resin,[79] condensation

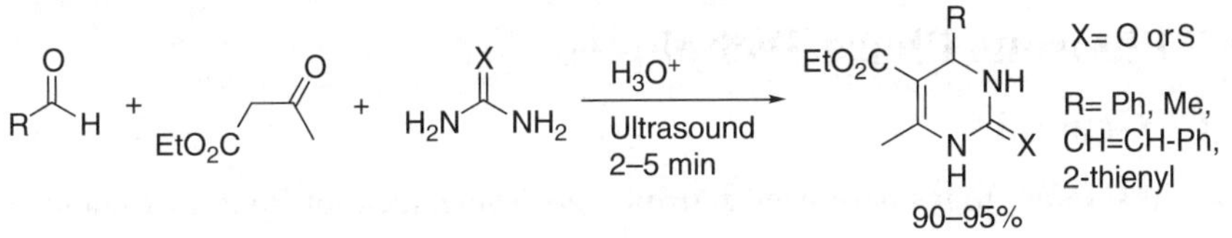

Scheme 5

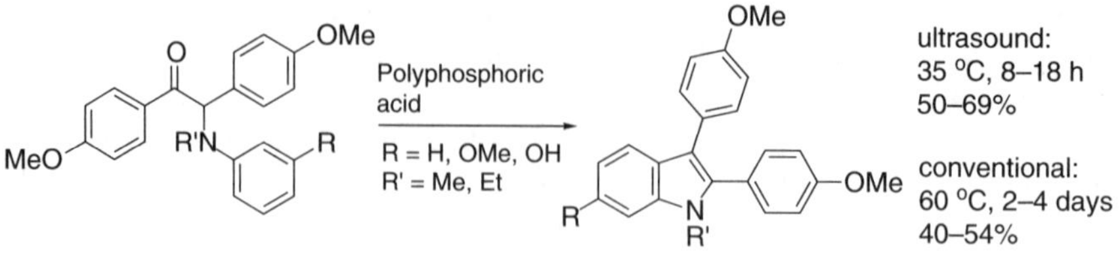

Scheme 6

of imines with H-phosphonates bound to Wang resin,[80] and linking of proline to magnetic support.[81] The latter example by Sucholeiki and co-workers reports on the generation of a library targeted against the κ-opioid receptor and, interestingly, ultrasound is employed in all the three steps; attachment of support, synthesis, and finally, detachment of the support. A small library of carbamates and sulfonamides, generated via solution phase combinatorial chemistry using azides and various electrophiles has been reported by Chandrasekhar and Narsihmulu[82] (Reaction Scheme 7). The authors make a comparison between ultrasound (25 °C) and conventional (reflux) conditions; the versatility of the ultrasound methodology is clearly demonstrated in this case involving thermally labile substrates, since both better yields and shorter reaction times are obtained.

Lonapalene, a selective 5-lipoxygenase inhibitor, has been prepared as shown in Reaction Scheme 8. The rate enhancement from ultrasound in the Diels–Alder cyclization and aromatization steps is substantial and represents a typical example of accelerated synthesis.[83]

Several improvements have been realized in the branch of carbohydrate chemistry using ultrasound, including important procedures such as acetalization, glycosylation, oxidation, and C–C bond formation. The reaction rates and yields are increased, and occasionally, also the chemo-, regio-, and stereoselectivities.[84] In the total synthesis of Caloporoside, a potent and selective inhibitor of phospholipase C, ultrasound was advantageously used in the key transformation of β-D-pyranoside to β-D-mannopyranoside furnishing 95% yield using the phase transfer reagent Bu$_4$NOAc.[85]

Successful acceleration by sonication has also been reported in the field of organic biocatalysis, employing whole cell and immobilized enzymes.[86] The improved yields and retained stereoselectivities[87] obtained under ultrasound irradiation emphasize the broad range of chemical transformations that benefits from ultrasonic treatment.

In summary, ultrasound appears to be a very promising technology that opens for numerous sophistications to improve and accelerate organic compound production and, accordingly, further accomplishments in this field are to be expected.

4   Fluorous Phase Techniques

4.1   General

Even if several groups have been active in the development of fluorous techniques, many of the early, seminal papers were presented by just a few groups, most notably Horváth, Gladysz, and Curran.[6,7] In the last few years several excellent reviews have

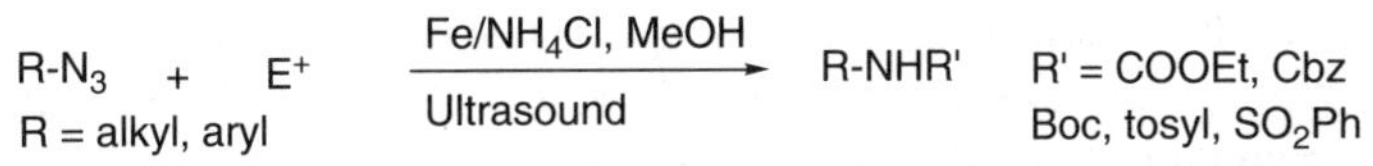

R-N$_3$ + E$^+$ $\xrightarrow[\text{Ultrasound}]{\text{Fe/NH}_4\text{Cl, MeOH}}$ R-NHR' R' = COOEt, Cbz

R = alkyl, aryl Boc, tosyl, SO$_2$Ph

Scheme 7

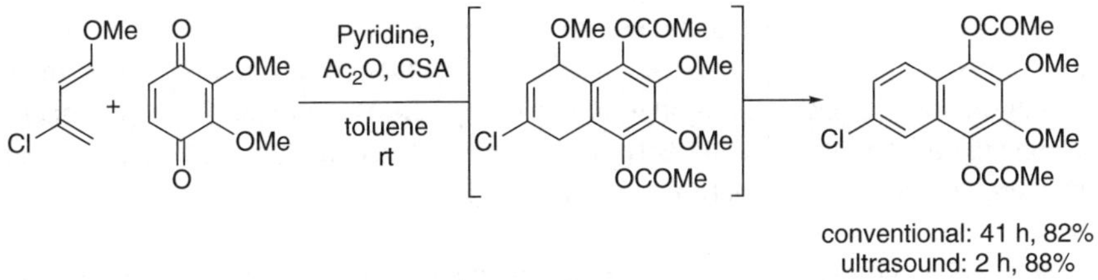

Pyridine, Ac$_2$O, CSA, toluene, rt

conventional: 41 h, 82%
ultrasound: 2 h, 88%

Scheme 8

also been published by Zhang.[8,88–90] The physical and chemical bases behind the special properties of perfluorinated or highly fluorinated compounds have been reviewed thoroughly and will not be dealt with in detail in this chapter.[9]

In a broad sense, fluorous chemistry takes advantage of the special characteristics of compounds with a high content of fluorine atoms. These compounds prefer to be solubilized in perfluorinated or highly fluorous solvents, instead of water or non-fluorous organic solvents. Perfluorinated solvents form a separate phase when mixed with these other solvents and highly fluorinated compounds can usually be extracted easily from a reaction mixture if a fluorous solvent is added. The fluorous content is usually introduced to a molecule by labelling with a fluorous tag or protecting group.[90,91] Products and/or non-reacted starting material can then be separated handily from non-tagged compounds by fluorous extraction or chromatography. The earlier papers often took advantage of two- or three-phase liquid extractions to separate reaction mixtures. In these cases, the choice of tag was vital to make certain that the tagged compound would partition to the fluorous phase. The ratio of fluorine atoms to other atoms is often discussed in terms of "fluoricity";[7] to ensure that a full partitioning takes place, a fluoricity of at least 60% w w^{-1} has been recommended.[9] The size of the fluorous tag required is directly dependent on the size of the molecule attached – a large molecule needs a larger number of fluorine atoms in the tag than a small molecule. This requirement in some cases leads to problems in solubility during the reaction, since highly fluorinated compounds have a tendency to be insoluble in both fluorous and classical organic solvents, although this insolubility could also be advantageous in the isolation step.[92]

An elegant way out of the problem of low solubility was the advent of light fluorous chemistry (usually employing a fluoricity of below 40% w w^{-1})[93,94] and fluorous silica separation techniques.[95,96] Instead of relying on a high fluoricity to guarantee liquid partitioning, separations are now carried out on fluorous reverse-phase silica simply by taking advantage of the differing fluoricities of the compounds. In other words, a compound with a C$_4$F$_9$ tag can easily be separated from a similar compound with a C$_6$F$_{13}$ tag. A wide range of separation tools are now available commercially, including standard reverse-phase fluorous silica gel and HPLC columns, solid-phase cartridges,[8] a wide range of scavengers[97] and capping reagents, and also recently reverse fluorous solid-phase extraction (F-SPE) tools.[98] β-Cyclodextrin columns have also been shown to be of interest in separations of fluorous compounds.[99,100]

A tempting comparison is that between standard solid-phase chemistry and fluorous chemistry. Both of these techniques have several attributes in common, including the use of linkers, frequent use of scavengers, and utility in many similar applications. Even though the use of standard solid-phase chemistry has many advantages, some aspects of polymer-supported synthesis strategies have drawbacks. Fluorous chemistry has in many ways marketed itself as an alternative to solid-support chemistry due to its superior performance in a number of respects.[8]

1. Fluorous chemistry is in essence homogeneous and thus results in better reaction kinetics than solid-phase chemistry.
2. Large excesses of reagents are seldom needed in fluorous chemistry, but often in standard solid-phase chemistry.
3. Fluorous reactions can be monitored easily by TLC, HPLC, and NMR.
4. Intermediates as well as end products can be purified by reverse- or straight-phase fluorous chromatography as well as by ordinary chromatographic methods.
5. Literature conditions are in many cases easy to apply or adjust to fluorous chemistry.
6. Fluorous tags are essentially inert towards unwanted chemical reactions.
7. Due to their special solubility characteristics, fluorous tags and reagents can be collected and recycled.

The applications of fluorous chemistry in high-throughput or parallel synthesis are many since fluorous reagents and separation techniques can be used in all stages of synthesis and work-up. As a result, fluorous chemistry can today offer a whole parallel synthetic strategy to traditional chemical methods.

4.2   Reagents, Linkers, and Scavengers

Several fluorous counterparts of common reagents have been reported and many of these are also commercially available. Of some interest is the development of fluorous Mitsunobu reagents.[101] A first generation fluorous DEAD reagent was found to have poor reactivity when difficult Mitsunobu reactions were encountered; it has since been replaced with two enhanced, second-generation reagents, (1) and (2), for use in parallel or sequential fluorous chromatography and rapid F-SPE isolation, respectively.

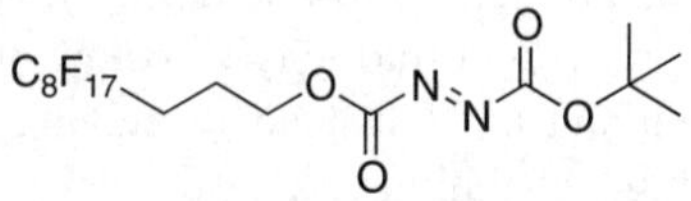

$$\text{(1)}$$

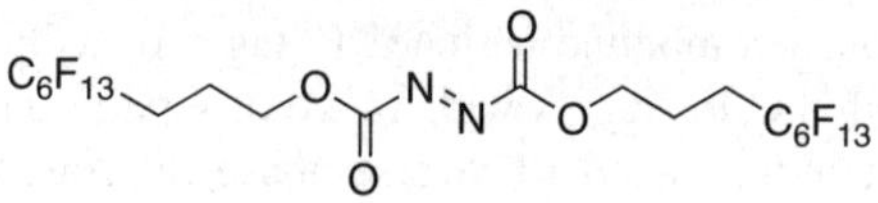

$$\text{(2)}$$

A fluorous imine reagent (3), also commercially available, was introduced by Herr as an ammonia equivalent in the Buchwald–Hartwig amination of aryl bromides, iodides, and triflates (Reaction Scheme 9).[102] In this case, the problem of introducing a primary amine was solved by the use of (3) as a synthon for ammonia.

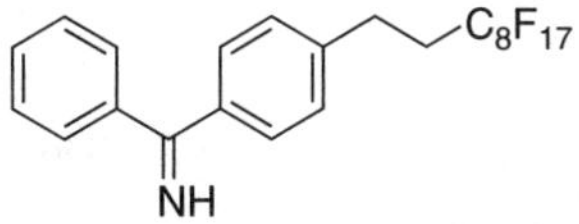

$$(3)$$

An interesting application of the unique possibilities inherent in fluorous chemistry is the phase-vanishing reactions reported by Ryu and co-workers.[103] This strategy encompasses no fluorous reagents as such but uses a fluorous phase as a physical barrier for passive transport between an organic phase (hexane) and a reagent (BBr_3) in brominations of alkenes. The completion of the reaction is easily monitored by the disappearance of the reagent.

An important reagent in fluorous chemistry is the fluorous version of the Marshall resin, dubbed FluoMar (4). This separation tag is reported to dissolve readily in dichloromethane, tetrahydrofuran, and ethyl acetate and can, as many other fluorous reagents, be monitored by traditional chromatographic and spectroscopic methods. The usefulness of (4) was demonstrated in a multistep parallel synthesis of a 3×3 array of diamides, where the final products were efficiently purified by F-SPE and cleaved from the FluoMar tag. Tentative results indicated that the homogeneous kinetics of the soluble (4) resulted in reactions that proceeded approximately three times faster than polymer-support bound reactions using standard Marshall resin.[104]

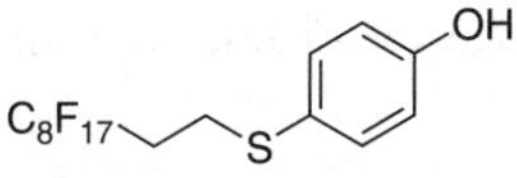

$$(4)$$

A traceless perfluoroalkylsulfonyl linker for the deoxygenation of phenols has been reported by Holmes.[105] A more lightly fluorous variant has also been presented by Zhang,[106] where microwave heating was applied to increase the speed of the reaction. The traceless tag was exemplified in syntheses of triaryl-substituted pyrimidines and hydantoins (Reaction Scheme 10).

Fluorous scavengers have been reported by many groups.[97] Lindsley[107] reported a number of fluorous reagents for scavenging of both nucleophiles and electrophiles, using the higher reactivity of homogeneous reagents and the easy separation of fluorous compounds with fluorous chromatography. Curran[108] reported, independently,

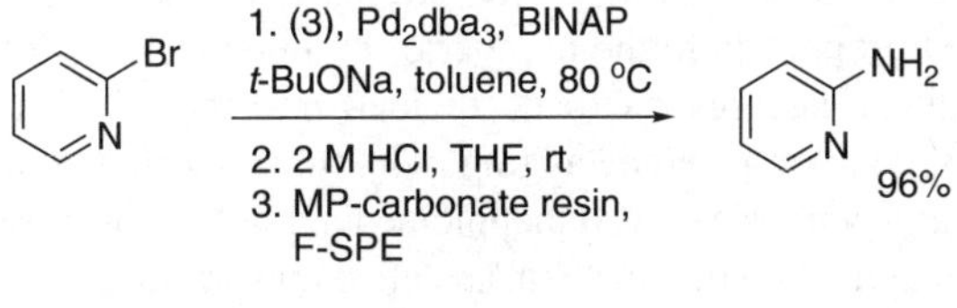

Scheme 9

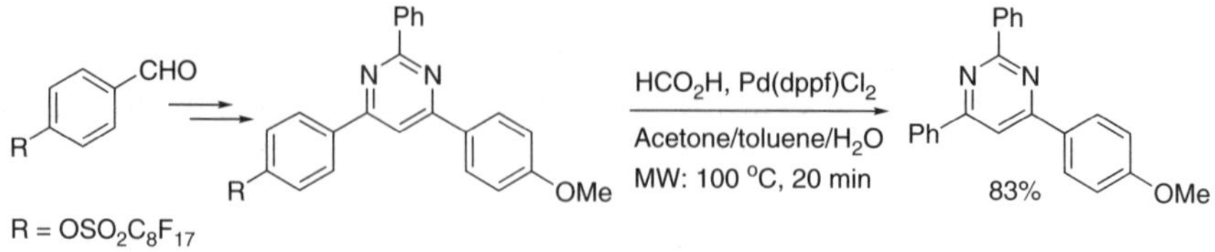

Scheme 10

a fluorous thiol scavenger also found in the previous publication. Zhang[109] later showed that fluorous quenching can be 5–10 times faster than the corresponding polymer-based methods and that less fluorous scavenger was needed. The same group also recently presented two new fluorous electrophilic scavengers based on isatoic anhydride and isocyanate moieties.

4.3 Fluorous Protecting Groups

The development of fluorous protecting groups has attracted a lot of interest. This stems partly from the ubiquitous need for protecting groups, but also because the use of a fluorous protecting group offers a convenient way to introduce the desired fluorous label, which later can be removed as a natural part of the synthetic scheme. The scope of this chapter does not allow a thorough discussion of the different protecting groups, hence the reader is advised to consult the well-written reviews on the subject.[8,89,90] A short, and by no means complete, list of fluorous protecting groups could include BOC,[110] THP,[111] Z,[112, 113] acetals,[114] Msc,[115] and Cbz.[116] A recent addition to the field is a report by Ladlow on an acid-labile protecting group for synthesis of carboxamides and sulfonamides.[117]

4.4 Fluorous Mixture Synthesis

A highly relevant development of the light fluorous approach for parallel synthesis and library production is the fluorous mixture synthesis (FMS) strategy by Curran.[118] FMS offers an alternative that combines the positive aspects of homogeneous kinetics with agreeable and practical separation protocols. While solid-phase mixture techniques (split-pool) intermingle beads or containers, they do not mix compounds. The corresponding solution-phase mixture strategies often have the disadvantage that deconvolution steps and resynthesis of the most active compounds must be undertaken.[119] The potential problem of keeping and handling large numbers of samples and reagents that are often encountered in solution-phase strategies is also not an issue in FMS. Indeed, one of the strengths of FMS is that its usefulness increases in direct proportion to the number of compounds mixed.[119]

The FMS synthetic strategy has been divided into five distinct phases: tagging, mixing, synthesis, demixing, and detagging, with the work flow as depicted in Figure 1.[118] A small library synthesis of 100 mappicine analogues was presented using FMS as well as an elegant method of what Curran called quasiracemic synthesis. Using this method, two enantiomers of starting material yielded two enantiomers of mappicine by FMS methods, without the use of separation on a chiral support.[118]

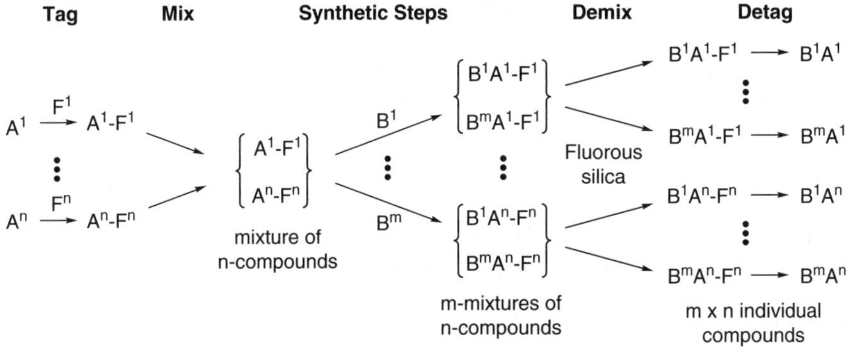

Figure 1 *Representative scheme for fluorous mixture synthesis*

Further examples of the usefulness of FMS include the application of the system in exploratory chemistry in the synthesis of natural products[120] and in the creation of a 560-compound library of mappicine analogues (5).[119] As the synthesis of this library clearly illustrates the practical use of FMS, it will be described in detail. In this report, seven pyridyl alcohols were attached to seven different fluorous tags, mixed together and reacted further to yield a mixture of tagged pyridones. This mixture was in turn divided into eight portions, which were then reacted with eight different propargyl bromides in parallel. These reaction mixtures were then split into 10 portions and reacted with 10 different isonitriles under free radical conditions to generate 80 mixtures of F-tagged mappicines. Each mixture, containing seven derivatives, was then demixed by fluorous HPLC to give a total of 560 mappicine analogues after detagging.[119] As is seen in this example, even though FMS is apparently limited by the number and availability of fluorous tags, this problem can be relatively easily dealt with by a split and mix approach.

$$(5)$$

A stereoisomer library of 16 murisolin analogues was recently reported[121] and early in 2005, an FMS strategy for the production of cyclopentenone scaffolds with Pauson–Khand-type reactions was published.[122]

4.5 Peptides and Oligosaccharides

Polymer-supported technologies have been widely used in peptide and protein chemistry. Especially in automated solid-phase synthesis, there is a need to separate products from side products deficient in one or several units. In particular, the $(n-1)$ products are notoriously difficult to separate from the end products. As fluorous chemistry is well suited for multistep reactions, it would be attractive to develop peptide syntheses with fluorous methods. Generally, the techniques that have received the most attention are either the use of fluorous tags to mark the products of interest or the use of capping reagents to mark unreacted reagents or peptide sequences for easier removal. One of the earliest examples of fluorous capping techniques in synthesis of oligosaccharides was published by Seeberger.[123] Recently, a trivalent iodonium reagent was reported by Kumar, which requires only the addition of water to precipitate short peptides (8–10 residues of length). Longer sequences need fluorous flash silica separation for efficient purification.[124]

Fluorous dialkoxy chlorotriazine (^{F}CDMT) (6) was recently reported as an activating reagent for carboxylic acids in peptide synthesis.[125]

$$\text{F}_3\text{C}(\text{F}_2\text{C})_7\text{H}_2\text{CO} \quad \overset{\displaystyle \text{Cl}}{\underset{\displaystyle \text{N}}{\bigtriangleup}} \quad \text{OCH}_2(\text{CF}_2)_7\text{CF}_3 \tag{6}$$

Inazu[126] has published details on the synthesis and use of three fluorous supports for peptide synthesis, the fluorous trialkoxybenzhydryl, Wang-type, and *tert*-butyl supports. The fluorous synthesis of Leu-enkephalin was later reported by the same group.[127]

The synthesis of oligosaccharides is known to be challenging and not easily applied under solid-phase reaction conditions or with automated synthesizers. Although perhaps somewhat outside the scope of this chapter, it is interesting to note that considerable progress from many groups has been made in applying fluorous techniques in oligosaccharide synthesis.[128–134]

4.6 Fluorous Applications in High-Throughput Chemistry

A fluorous phase Pummerer reaction was introduced by Procter. This report was based on the need to address problems associated with difficulties in optimization and monitoring of solid-phase processes. Good to excellent yields of heterocyclic scaffolds, mainly oxindoles, were reported after the development of a high-throughput fluorous phase synthesis strategy.[135]

A somewhat different approach was used by Vincent in phase-switching reactions using pyridyl-labelled substrates and products. The pyridyl-containing tag is here thought of as a masked phase tag, which allows for phase switching with the help of a heavy fluorous copper(II)-carboxylate complex. Comparison with a non-fluorous system indicated that a problem of release of the strongly coordinating pyridine linker was avoided in the fluorous approach.[136]

A fluorous catch and release method was used by Zhang in the synthesis of di-substituted pyrimidines.[137] A fluorous route to hydantoins and thiohydantoins was also reported by the same group.[138]

4.7 Microwave-Enhanced Fluorous Chemistry

The combination of microwave heating and fluorous chemistry is of considerable interest, as this approach combines fast chemistry and easy separation.

This area has been reviewed recently[139] and only some of the latest papers will be discussed here.

A fluorous palladium pincer complex for use in Heck reactions was reported by Curran (Reaction Scheme 11). The tridentate ligand ensured stability under the high temperatures (140 °C) needed for the reaction to go to completion. The complex was recovered and reused three times without any sign of lowered catalytic activity.[140]

An *in situ* carbonylation procedure was shown by Larhed to be compatible with recycling of the fluorous catalyst used in the reaction (Reaction Scheme 12). The catalyst was collected five times by a two-phase liquid fluorous extraction. The yields were shown to vary only slightly between the experiments.[141]

Zhang has explored the possibilities to use a more highly fluorous version of the triflate group in palladium catalysed reactions. The $C_8F_{17}O_2SO$- leaving group worked efficiently with organoboronic acids in Suzuki couplings,[142] and perhaps more interestingly, with aryl sulfides using a Pd/dppf-based catalytic system (Reaction Scheme 13).[143] The higher degree of fluoricity enabled purification by F-SPE. The authors also underlined the higher speed of the reaction under microwaves as compared to classic heating as well as the ease of separation connected with the use of fluorous tags.

Microwave-assisted fluorous Ugi reactions were presented where the reaction times and convenient separation techniques appeared more attractive than the corresponding room temperature methods with traditional scavenging techniques.[144]

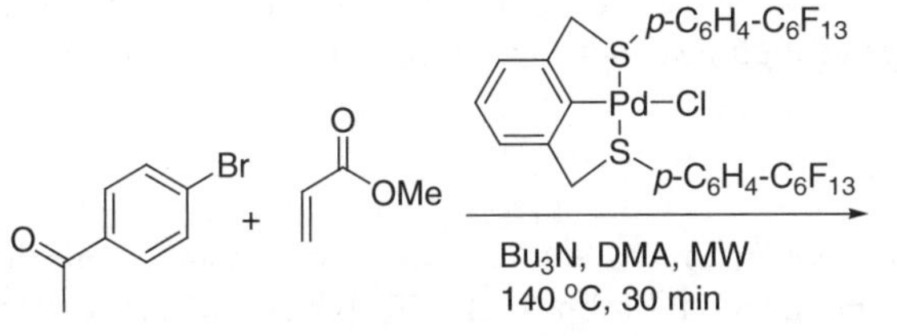

Scheme 11

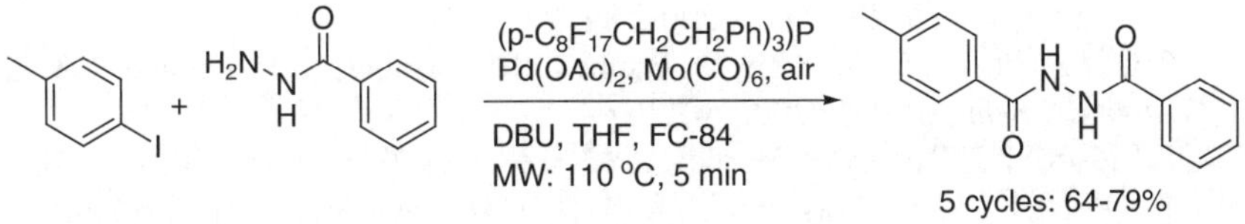

Scheme 12

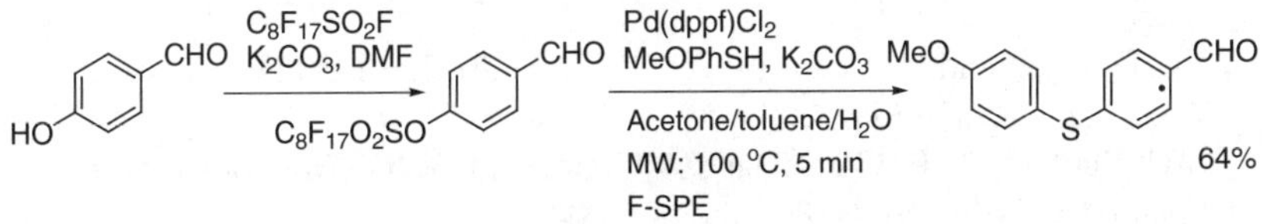

Scheme 13

A parallel synthesis method for the synthesis of a dihydropteridine library in five steps was recently published by Zhang.[145] The synthetic route was based on the use of fluorous amino acids, which were prepared from *N*-Boc-amino acids. The final cyclization was promoted by the use of microwave heating.

5 Conclusion

In this chapter, we have discussed the potential of unconventional approaches to increase the preparative throughput in drug-discovery efforts. Selected applications of microwave heating and ultrasonic irradiation for acceleration of organic transformations in diversity-oriented and medicinal chemistry applications have been discussed. In all examples of irradiation-enhanced chemistry presented, the main synthetic effort was directed towards the development of rapid, convenient, and reliable procedures. Fluorous strategies were discussed as an alternative approach to the very common use of solid polymers for fast purifications in small molecule synthesis, yet fluorous chemistry retains many of the advantages of traditional solution-phase synthesis. An impressive array of different fluorous technologies is today available for various high-throughput chemistry applications.

The combination of microwave and ultrasound heating with purification methods by phase switching will further increase the benefits due to the generally low reactivity of both polymer-supported and fluorous reagents. Clearly, more research is needed to be able to fully incorporate these novel high-speed technologies into the industrial drug-discovery toolbox, but the examples thus far are very promising. With these advances, the age of high-speed chemistry continues to evolve.

Acknowledgements

We would like to express our sincere gratitude to the Swedish Research Council, to Knut and Alice Wallenberg's Foundation, Biolipox AB, Medivir AB, and Biotage AB.

References

1. D. Hunter, *J. Cell. Biochem.*, 2001, **84**, 22.
2. C.C. Tzschucke, C. Markert, W. Bannwarth, S. Roller, A. Hebel and R. Haag, *Angew. Chem. Int. Ed.*, 2002, **41**, 3964.
3. P. Lidström, J. Tierney, B. Wathey and J. Westman, *Tetrahedron*, 2001, **57**, 9225.
4. K. Olofsson, A. Hallberg and M. Larhed, in *Microwaves in Organic Synthesis*, A. Loupy (ed), Wiley-VCH, Weinheim, 2002, 379.
5. C.O. Kappe, *Angew. Chem. Int. Ed.*, 2004, **43**, 6250.
6. A. Studer, S. Hadida, R. Ferritto, S.Y. Kim, P. Jeger, P. Wipf and D.P. Curran, *Science*, 1997, **275**, 823.
7. J.A. Gladysz and D.P. Curran, *Tetrahedron*, 2002, **58**, 3823.
8. W. Zhang, *Tetrahedron*, 2003, **59**, 4475.
9. J.A. Gladysz, D.P. Curran and I.T. Horváth (eds), *Handbook of Fluorous Chemistry*, Wiley-VCH, Weinheim, 2004.
10. D.M.P. Mingos and D.R. Baghurst, *Chem. Soc. Rev.*, 1991, **20**, 1.

11. S. Caddick, *Tetrahedron*, 1995, **51**, 10403.

12. C.R. Strauss and R.W. Trainor, *Aust. J. Chem.*, 1995, **48**, 1665.

13. N. Elander, J.R. Jones, S.-Y. Lu and S. Stone-Elander, *Chem. Soc. Rev.*, 2000, **29**, 239.

14. A. Loupy (ed), *Microwaves in Organic Synthesis*, Wiley-VCH, Weinheim, 2002.

15. A.K. Bose, M.S. Manhas, S.N. Ganguly, A.H. Sharma and B.K. Banik, *Synthesis*, 2002, 1578.

16. M. Larhed and A. Hallberg, *Drug Discov. Today*, 2001, **6**, 406.

17. M. Larhed, C. Moberg and A. Hallberg, *Acc. Chem. Res.*, 2002, **35**, 717.

18. M. Alterman, H.O. Andersson, N. Garg, G. Ahlsen, S. Lövgren, B. Classon, U.H. Danielson, I. Kvarnström, L. Vrang, T. Unge, B. Samuelsson and A. Hallberg, *J. Med. Chem.*, 1999, **42**, 3835.

19. K. Ersmark, M. Larhed and J. Wannberg, *Curr. Opin. Drug Discov. Devel.*, 2004, **7**, 417.

20. C.O. Kappe, *Curr. Opin. Chem. Biol.*, 2002, **6**, 314.

21. S.V. Ley and I.R. Baxendale, *Nature Rev. Drug Discov.*, 2002, **1**, 573.

22. A. Lew, P.O. Krutzik, M.E. Hart and A.R. Chamberlin, *J. Comb. Chem.*, 2002, **4**, 95.

23. C. Gabriel, S. Gabriel, E.H. Grant, B.S.J. Halstead and D.M.P. Mingos, *Chem. Soc. Rev.*, 1998, **27**, 213.

24. A. Stadler and C.O. Kappe, *J. Comb. Chem.*, 2001, **3**, 624.

25. W. Schaal, A. Karlsson, G. Ahlsen, J. Lindberg, H.O. Andersson, U.H. Danielson, B. Classon, T. Unge, B. Samuelsson, J. Hulten, A. Hallberg and A. Karlen, *J. Med. Chem.*, 2002, **45**, 752.

26. D. Nöteberg, E. Hamelink, J. Hultén, M. Wahlgren, L. Vrang, B. Samuelsson and A. Hallberg, *J. Med. Chem.*, 2003, **46**, 734.

27. K. Ersmark, I. Feierberg, S. Bjelic, E. Hamelink, F. Hackett, M.J. Blackman, J. Hulten, B. Samuelsson, J. Qvist and A. Hallberg, *J. Med. Chem.*, 2004, **47**, 110.

28. J. Hulten, N.M. Bonham, U. Nillroth, T. Hansson, G. Zuccarello, A. Bouzide, J. Åqvist, B. Classon, H. Danielson, A. Karlén, I. Kvarnström, B. Samuelsson and A. Hallberg, *J. Med. Chem.*, 1997, **40**, 885.

29. K. Bäckbro, S. Löwgren, K. Österlund, J. Atepo, T. Unge, J. Hulten, N.M. Bonham, W. Schaal, A. Karlen and A. Hallberg, *J. Med. Chem.*, 1997, **40**, 898.

30. A. Ax, W. Schaal, L. Vrang, B. Samuelsson, A. Hallberg and A. Karlen, *Bioorg. Med. Chem.*, 2005, **13**, 755.

31. M. Alterman, M. Björsne, A. Mühlman, B. Classon, I. Kvarnström, H. Danielson, P.-O. Markgren, U. Nillroth, T. Unge, A. Hallberg and B. Samuelsson, *J. Med. Chem.*, 1998, **41**, 3782.

32. J. Wannberg, N.F.K. Kaiser, L. Vrang, B. Samuelsson, M. Larhed and A. Hallberg, *J. Comb. Chem.*, 2005, **7**, 611.

33. N.F.K. Kaiser, A. Hallberg and M. Larhed, *J. Comb. Chem.*, 2002, **4**, 109.

34. J. Wannberg and M. Larhed, *J. Org. Chem.*, 2003, **68**, 5750.

35. Y. Wan, C. Wallinder, B. Johansson, M. Holm, A.K. Mahalingam, X. Wu, M. Botros, A. Karlen, A. Pettersson, F. Nyberg, L. Fändriks, A. Hallberg and M. Alterman, *J. Med. Chem.*, 2004, **47**, 1536.

36. Y. Wan, C. Wallinder, B. Plouffe, H. Beaudry, A.K. Mahalingam, X. Wu, B. Johansson, M. Holm, M. Botoros, A. Karlen, A. Pettersson, F. Nyberg, L. Fändriks, N. Gallo-Payet, A. Hallberg and M. Alterman, *J. Med. Chem.*, 2004, **47**, 5995.

37. S. Crosignani, P.D. White and B. Linclau, *J. Org. Chem.*, 2004, **69**, 5897.

38. Y.T. Didenko, W.B. McNamara III and K.S. Suslick, *J. Am. Chem. Soc.*, 1999, **121**, 5817.

39. L.H. Thompson and L.K. Doraiswamy, *Ind. Eng. Chem. Res.*, 1999, **38**, 1215.

40. C.M.R. Low, *Ultrason. Sonochem.*, 1995, **2**, S153.

41. J.-L. Luche, *Synthetic Organic Sonochemistry*, Plenum Press, New York, 1998.

42. Y.G. Adewuyi, *Ind. Eng. Chem. Res.*, 2001, **40**, 4681.

43. T.J. Mason, *Chem. Soc. Rev.*, 1997, **26**, 443.

44. P. Cintas and J.L. Luche, *Green Chem.*, 1999, **1**, 115.

45. C. Einhorn, J. Einhorn and J.L. Luche, *Synthesis*, 1989, 787.

46. C. Felix, A. Laurent and P. Mison, *Tetrahedron Lett.*, 1990, **31**, 4143.

47. R.S. Varma and K.P. Naicker, *Tetrahedron Lett.*, 1998, **39**, 7463.

48. M.J. Aurell, C. Einhorn, J. Einhorn and J.L. Luche, *J. Org. Chem.*, 1995, **60**, 8.

49. B.H. Han and P. Boudjouk, *J. Org. Chem.*, 1982, **47**, 5030.

50. M. Xi and B.E. Bent, *J. Am. Chem. Soc.*, 1993, **115**, 7426.

51. J.L. Luche, C. Petrier and C. Dupuy, *Tetrahedron Lett.*, 1984, **25**, 753.

52. H.J. Siemann, P. Droescher, B. Undeutsch and S. Schwarz, *Steroids*, 1995, **60**, 308.

53. M. Melo, M. Moreno, S.C.P. Dacosta, J.A.R. Salvador and A.S.C. Neves, *Ultrason. Sonochem.*, 1994, **1**, S37.

54. M.J. Aurell, Y. Danhui, C. Einhorn and J.L. Luche, *Synlett*, 1995, 459.

55. R.M. Suarez, J.P. Sestelo and L.A. Sarandeses, *Chem. Eur. J.*, 2003, **9**, 4179.

56. I. Cornella, R.M. Suarez, A. Mourino, J.P. Sestelo and L.A. Sarandeses, *J. Steroid Biochem. Mol. Biol.*, 2004, **89–90**, 19.

57. R.R. Deshmukh, R. Rajagopal and K.V. Srinivasan, *Chem. Commun.*, 2001, 1544.

58. R. Rajagopal, V.J. Dilip and K.V. Srinivasan, *Chem. Commun.*, 2002, 616.

59. V. Polackova, M. Hut'ka and T. Toma, *Ultrason. Sonochem.*, 2005, **12**, 99.

60. A. Muller, C. Vogt and N. Sewald, *Synthesis*, 1998, 837.

61. M. Jie and P. Kalluri, *J. Chem. Soc. Perkin Trans. 1*, 1997, **23**, 3485.

62. M.L.D. Palacios and R.F.P. Comdom, *Synth. Commun.*, 2003, **33**, 1777.

63. J.T. Li, L.J. Li, T.S. Li, H.Z. Li and J.K. Liu, *Ultrason. Sonochem.*, 1996, **3**, S141.

64. M. Jie and M.M.L. Lau, *Chem. Phys. Lipids*, 1999, **101**, 237.

65. C.O. Kappe, *Tetrahedron*, 1993, **49**, 6937.

66. M.S. Zhidovinova, O.V. Fedorova, G.L. Rusinov and I.G. Ovchinnikova, *Mol. Divers.*, 2003, **6**, 323.

67. J.T. Li, Z.P. Lin, J.F. Han and T.S. Li, *Synth. Commun.*, 2004, **34**, 2623.

68. A.R. Gholap, K. Venkatesan, T. Daniel, R.J. Lahoti and K.V. Srinivasan, *Green Chem.*, 2004, **6**, 147.

69. J.T. Li, J.F. Han, J.H. Yang and T.S. Li, *Ultrason. Sonochem.*, 2003, **10**, 119.

70. J.S. Yadav, B.V.S. Reddy, K.B. Reddy, K.S. Raj and A.R. Prasad, *J. Chem. Soc. Perkin Trans. 1*, 2001, 1939.

71. M.S. Zhidovinova, O.V. Fedorova, G.L. Rusinov and I.G. Ovchinnikova, *Russ. Chem. Bull.*, 2003, **52**, 2527.

72. M. Avalos, R. Babiano, N. Cabello, P. Cintas, M.B. Hursthouse, J.L. Jimenez, M.E. Light and J.C. Palacios, *J. Org. Chem.*, 2003, **68**, 7193.

73. G. Abbiati, F. Clerici, M.L. Gelmi, A. Gambini and T. Pilati, *J. Org. Chem.*, 2001, **66**, 6299.

74. J.G. Zhang, W.H. Duan and J.C. Cai, *Tetrahedron*, 2004, **60**, 1665.

75. T.P. Caulier and J. Reisse, *J. Org. Chem.*, 1996, **61**, 2547.

76. J. Lee and J.K. Snyder, *J. Am. Chem. Soc.*, 1989, **111**, 1522.

77. S.K. Choi, M. Mammen and G.M. Whitesides, *J. Am. Chem. Soc.*, 1997, **119**, 4103.

78. S.D. Koulocheri and S.A. Haroutounian, *Eur. J. Org. Chem.*, 2001, **9**, 1723.

79. M.V. Anuradha and B. Ravindranath, *Tetrahedron*, 1995, **51**, 5671.

80. C.Z. Zhang and A.M.M. Mjalli, *Tetrahedron Lett.*, 1996, **37**, 5457.

81. J.M. Perez, E.J. Wilhelm and I. Sucholeiki, *Bioorg. Med. Chem. Lett.*, 2000, **10**, 171.

82. S. Chandrasekhar and C. Narsihmulu, *Tetrahedron Lett.*, 2000, **41**, 7969.

83. T. Javed, T.J. Mason, S.S. Phull and N.R. Baker, *Ultrason. Sonochem.*, 1995, **2**, S3.

84. N. Kardos and J.L. Luche, *Carbohydr. Res.*, 2001, **332**, 115.

85. A. Furstner and I. Konetzki, *J. Org. Chem.*, 1998, **63**, 3072.

86. J.V. Sinisterra, *Ultrasonics*, 1992, **30**, 180.

87. G.L. Lin and H.C. Liu, *Tetrahedron Lett.*, 1995, **36**, 6067.

88. W. Zhang, *Chem. Rev.*, 2004, **104**, 2531.

89. W. Zhang, *Curr. Opin. Drug Discov. Devel.*, 2004, **7**, 784.

90. W. Zhang, in *Handbook of Fluorous Chemistry*, J.A. Gladysz, D.P. Curran and I.T. Horváth (eds), Wiley-VCH, Weinheim, 2004, 222.

91. J. Yoshida and K. Itami, *Chem. Rev.*, 2002, **102**, 3693.

92. K. Olofsson, S.Y. Kim, M. Larhed, D.P. Curran and A. Hallberg, *J. Org. Chem.*, 1999, **64**, 4539.

93. D.P. Curran and Z.Y. Luo, *J. Am. Chem. Soc.*, 1999, **121**, 9069.

94. D.P. Curran, in *Handbook of Fluorous Chemistry*, J.A. Gladysz, D.P. Curran and I.T. Horváth (eds), Wiley-VCH, Weinheim, 2004, 128.

95. D.P. Curran, *Synlett*, 2001, 1488.

96. D.P. Curran, in *Handbook of Fluorous Chemistry*, J.A. Gladysz, D.P. Curran and I.T. Horváth (eds), Wiley-VCH, Weinheim, 2004, 101.

97. C.W. Lindsley and W.H. Leister, in *Handbook of Fluorous Chemistry*, J.A. Gladysz, D.P. Curran and I.T. Horváth (eds), Wiley-VCH, Weinheim, 2004, 236.

98. M. Matsugi and D.P. Curran, *Org. Lett.*, 2004, **6**, 2717.

99. K. Mikami, *Synlett*, 2004, 2713.

100. Y. Nakamura, S. Takeuchi, K. Okumura, Y. Ohgo, H. Matsuzawa and K. Mikami, *Tetrahedron Lett.*, 2003, **44**, 6221.

101. S. Dandapani and D.P. Curran, *J. Org. Chem.*, 2004, **69**, 8751.

102. C.L. Cioffi, M.L. Berlin and R.J. Herr, *Synlett*, 2004, 841.

103. I. Ryu, H. Matsubara, S. Yasuda, H. Nakamura and D.P. Curran, *J. Am. Chem. Soc.*, 2002, **124**, 12946.

104. C.H.T. Chen and W. Zhang, *Org. Lett.*, 2003, **5**, 1015.

105. Y.J. Pan, B. Ruhland and C.P. Holmes, *Angew. Chem. Int. Ed.*, 2001, **40**, 4488.

106. W. Zhang, T. Nagashima, Y.M. Lu and C.H.T. Chen, *Tetrahedron Lett.*, 2004, **45**, 4611.

107. C.W. Lindsley, Z. Zhao and W.H. Leister, *Tetrahedron Lett.*, 2002, **43**, 4225.

108. W. Zhang, D.P. Curran and C.H.T. Chen, *Tetrahedron*, 2002, **58**, 3871.

109. W. Zhang, C.H.T. Chen and T. Nagashima, *Tetrahedron Lett.*, 2003, **44**, 2065.

110. Z.Y. Luo, J. Williams, R.W. Read and D.P. Curran, *J. Org. Chem.*, 2001, **66**, 4261.

111. P. Wipf and J.T. Reeves, *Tetrahedron Lett.*, 1999, **40**, 4649.

112. D.V. Filippov, D.J. van Zoelen, S.P. Oldfield, G.A. van der Marel, H.S. Overkleeft, J.W. Drijfhout and J.H. van Boom, *Tetrahedron Lett.*, 2002, **43**, 7809.

113. D. Schwinn and W. Bannwarth, *Helv. Chim. Acta*, 2002, **85**, 255.

114. R.W. Read and C.T. Zhang, *Tetrahedron Lett.*, 2003, **44**, 7045.

115. P.C. de Visser, M. van Helden, D.V. Filippov, G.A. van der Marel, J.W. Drijfhout, J.H. van Boom, D. Noort and H.S. Overkleeft, *Tetrahedron Lett.*, 2003, **44**, 9013.

116. D.P. Curran, M. Amatore, D. Guthrie, M. Campbell, E. Go and Z.Y. Luo, *J. Org. Chem.*, 2003, **68**, 4643.

117. A.L. Villard, B.H. Warrington and M. Ladlow, *J. Comb. Chem.*, 2004, **6**, 611.

118. Z.Y. Luo, Q.S. Zhang, Y. Oderaotoshi and D.P. Curran, *Science*, 2001, **291**, 1766.

119. W. Zhang, Z.Y. Luo, C.H.T. Chen and D.P. Curran, *J. Am. Chem. Soc.*, 2002, **124**, 10443.

120. D.P. Curran and T. Furukawa, *Org. Lett.*, 2002, **4**, 2233.

121. Q.S. Zhang, H.J. Lu, C. Richard and D.P. Curran, *J. Am. Chem. Soc.*, 2004, **126**, 36.

122. S. Manku and D.P. Curran, *J. Comb. Chem.*, 2005, **7**, 63.

123. E.R. Palmacci, M.C. Hewitt and P.H. Seeberger, *Angew. Chem. Int. Ed.*, 2001, **40**, 4433.

124. V. Montanari and K. Kumar, *J. Am. Chem. Soc.*, 2004, **126**, 9528.

125. M.W. Markowicz and R. Dembinski, *Synthesis*, 2004, 80.

126. M. Mizuno, K. Goto, T. Miura, D. Hosaka and T. Inazu, *Chem. Commun.*, 2003, 972.

127. M. Mizuno, K. Goto, T. Miura, T. Matsuura and T. Inazu, *Tetrahedron Lett.*, 2004, **45**, 3425.

128. T. Miura, Y. Hirose, M. Ohmae and T. Inazu, *Org. Lett.*, 2001, **3**, 3947.

129. T. Miura, K.T. Goto, D. Hosaka and T. Inazu, *Angew. Chem. Int. Ed.*, 2003, **42**, 2047.

130. T. Miura, *Trends Glycosci. Glycotechnol.*, 2003, **15**, 351.

131. T. Miura and T. Inazu, *Tetrahedron Lett.*, 2003, **44**, 1819.

132. T. Miura, K. Goto, H. Waragai, H. Matsumoto, Y. Hirose, M. Ohmae, H. Ishida, A. Satoh and T. Inazu, *J. Org. Chem.*, 2004, **69**, 5348.

133. L. Manzoni and R. Castelli, *Org. Lett.*, 2004, **6**, 4195.

134. L. Manzoni, *Chem. Commun.*, 2003, 2930.

135. L.A. McAllister, R.A. McCormick, S. Brand and D.J. Procter, *Angew. Chem. Int. Ed.*, 2005, **44**, 452.

136. M. El Bakkari and J.M. Vincent, *Org. Lett.*, 2004, **6**, 2765.

137. W. Zhang, *Org. Lett.*, 2003, **5**, 1011.

138. W. Zhang and Y.M. Lu, *Org. Lett.*, 2003, **5**, 2555.

139. K. Olofsson and M. Larhed, in *Handbook of Fluorous Chemistry*, J.A. Gladysz, D.P. Curran and I.T. Horváth (eds), Wiley-VCH, Weinheim, 2004, 359.

140. D.P. Curran, K. Fischer and G. Moura-Letts, *Synlett*, 2004, 1379.

141. M. A. Herrero, J. Wannberg and M. Larhed, *Synlett*, 2004, 2335.

142. W. Zhang, C.H.T. Chen, Y.M. Lu and T. Nagashima, *Org. Lett.*, 2004, **6**, 1473.

143. W. Zhang, Y.M. Lu and C.H.T. Chen, *Mol. Divers.*, 2003, **7**, 199.

144. W. Zhang and P. Tempest, *Tetrahedron Lett.*, 2004, **45**, 6757.

145. T. Nagashima and W. Zhang, *J. Comb. Chem.*, 2004, **6**, 942.

Section 2

Conceptual Advances in Synthesis: "Prospecting" – Design of Discovery Libraries and the Search for Hits

Biosynthesis of "Unnatural" Natural Products

YI TANG[a] AND CHAITAN KHOSLA[b]

[a]Department of Chemical and Biomolecular Engineering, University of California, Los Angeles, CA 90095, USA
[b]Departments of Chemistry, Chemical Engineering and Biochemistry, Stanford University, Stanford, CA 94305, USA

1 Introduction

Natural products biosynthesized by bacteria, plants and fungi display a wide spectrum of pharmaceutical properties and have long been a rich source of clinically important compounds.[1] Penicillin produced by *Penicillium chrysogenum* and erythromycin produced by *Saccharopolyspora erythraea* have been used as broad-spectrum antibiotics since the middle of last century. Taxol (paclitaxel) is a frontline anticancer drug used in the treatment of carcinomas, melanomas and sarcomas and was first isolated from the bark of the Pacific yew (*Taxus brevifolia* Nutt.).[2] The importance of natural products to human health has prompted significant efforts toward the total and semisynthesis of these compounds. The structural complexity and chemical lability of many natural products, however, have prevented the facile synthesis and analoguing of these compounds using synthetic chemistry. Furthermore, the scaffolds of most natural products do not allow selective chemical modifications, and have remained relatively inaccessible as semisynthetic starting materials. Herein lies the relevance of emerging methods for the engineered biosynthesis of natural products.[3] Rational and combinatorial manipulation of biosynthetic pathways can afford "unnatural" natural products that (1) contain diverse, reactive substituents that can facilitate chemo- and regioselective synthetic transformations, and (2) display enhanced pharmacological properties.

The power of combinatorial biosynthesis has been best demonstrated through the engineered biosynthesis of polyketides.[4] Polyketides consist of a structurally diverse family of natural products and are mostly biosynthesized by soil-borne actinomyces as secondary metabolites.[5] Fungi and plants have also been sources of polyketides.

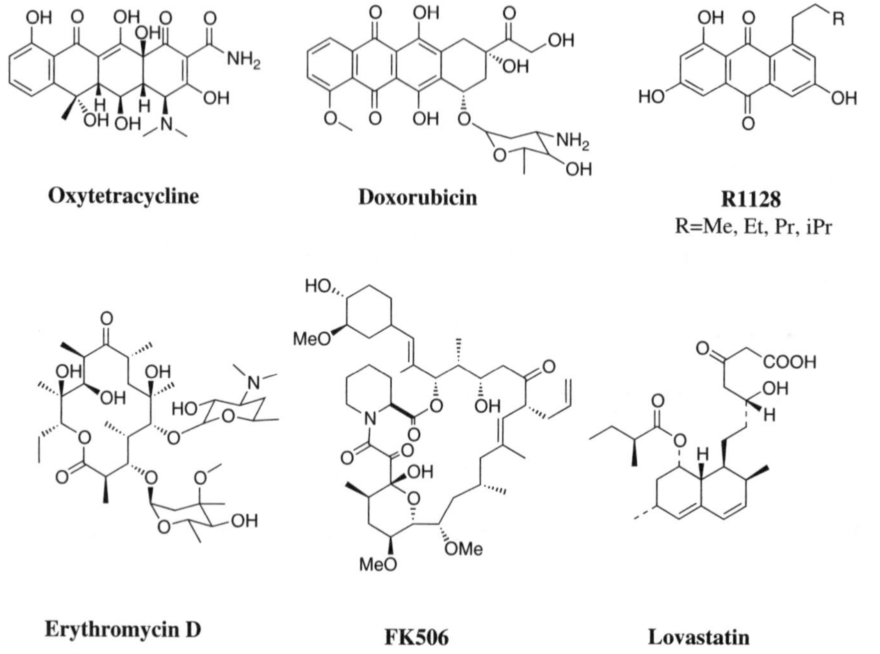

Oxytetracycline

Doxorubicin

R1128

R=Me, Et, Pr, iPr

Erythromycin D

FK506

Lovastatin

Figure 1 *Polyketide natural products*

Pharmaceutically important polyketides (Figure 1) include antibacterial (such as erythromycin and tetracycline), antitumor agents (daunorubicin and epothilone), immunosuppressants (FK506 and rapamycin), cholesterol-lowering agents (lovastatin) and antifungal compounds (amphotericin and griseofulvin). During the last decade, the biosynthetic pathways of these clinically relevant polyketides have been elucidated, largely aided by DNA sequencing of the corresponding gene clusters.[6] Three central themes regarding polyketide biosynthesis have emerged: (1) Genes encoding different enzymes in the same biosynthetic pathway are arranged in clusters on the host chromosomes. The organizations and sequences of biosynthetic genes from structurally related polyketide gene clusters are similar. (2) The assembly of a basic polyketide scaffold from simple, primary metabolites resembles that of fatty acid biosynthesis: repeated Claisen-like condensations between a growing polyketide chain and a substituted malonyl monothioester. (3) The structural diversity observed within the polyketide family is amplified via modification of the polyketide chain by a group of tailoring enzymes. Reprograming of the tailoring enzymes can therefore result in greater structural diversity than observed in nature.

1.1 Polyketide Assembly

Polyketides are biosynthesized by a group of enzymes collectively known as polyketide synthase (PKS). A polyketide chain is assembled through step-wise decarboxylative condensation between the electrophile (growing polyketide chain) and a nucleophile (the extender unit), catalyzed by the ketosynthase (KS) (see Figure 2).[7]

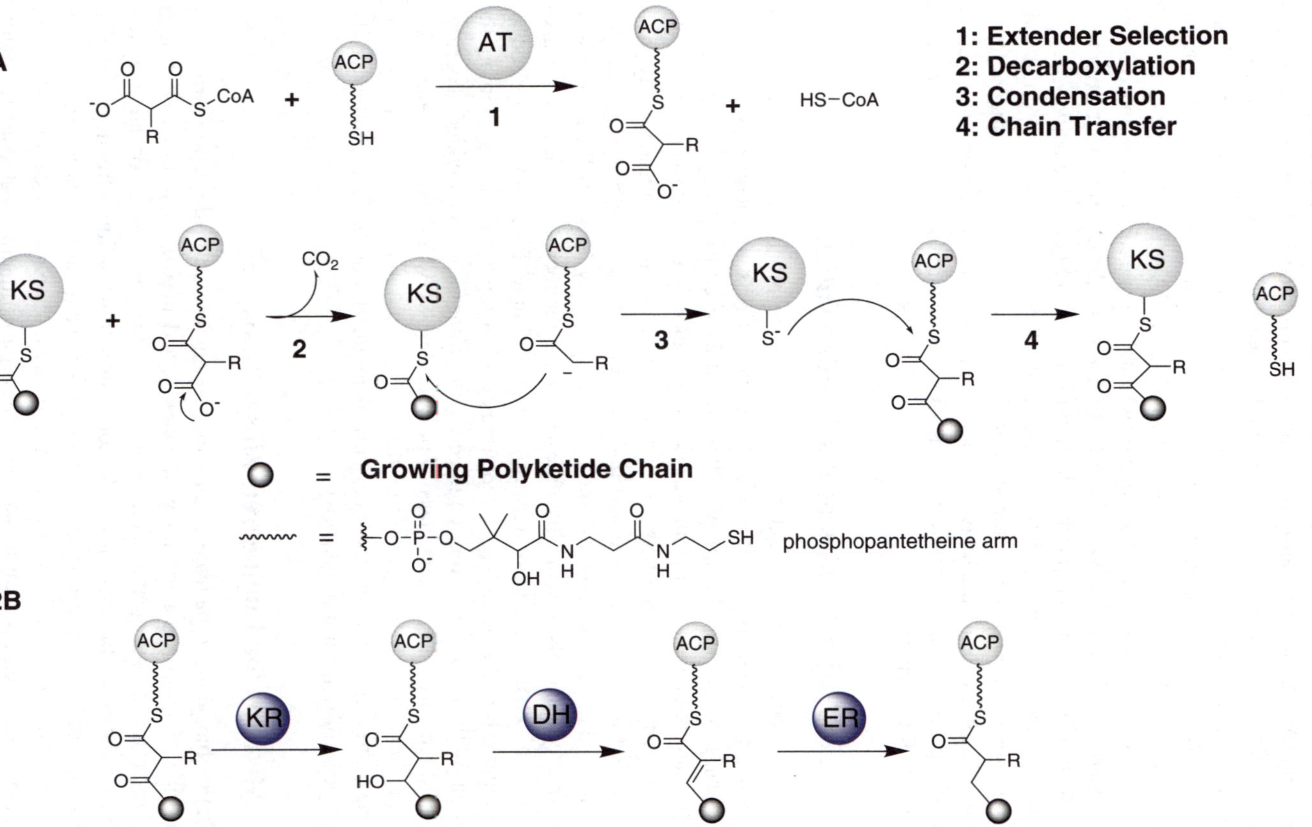

Figure 2 *(A) Decarboxylative condensation during polyketide synthesis. (B) Reductive steps in type I PKS*

The nucleophile is a substituted malonate tethered to a pantetheinyl arm via a thioester linkage. The pantetheinyl arm is part of either a coenzyme A or an acyl carrier protein (ACP), the latter being post-translationally modified at a universally conserved serine. Examples of extender units commonly utilized by PKSs are malonyl-CoA, methylmalonyl-CoA, and to a lesser extent, ethylmalonyl-CoA[8] and methoxymalonyl-CoA.[9] Specific selection of the extender unit is controlled by an acyltransferase (AT) component of a PKS (Figure 2A). The KS, AT and ACP enzymes make up a minimal PKS module and catalyze one round of chain extension in four steps: (1) the AT catalyzes thioester exchange of the extender unit from CoA to the ACP; (2) the growing polyketide chain is transferred onto the active-site cysteine of the KS; (3) decarboxylative condensation between the growing polyketide chain and the α-carboxylated extender unit yields a polyketide chain tethered to the ACP that has been elongated by two carbon atoms; and (4) transfer or release of the newly elongated polyketide chain frees the ACP, thereby allowing the next round of condensation to take place.

1.2 Three Major Classes of Polyketide Synthases

Although variations exist, polyketide synthases can be broadly classified into three categories based on their overall architecture. Type I PKSs are encoded as large multifunctional proteins in which active sites are either used iteratively (*e.g.* in fungal PKSs) or are organized as enzymatic assembly lines (*e.g.* erythromycin, epothilone and rapamycin).[10] Type II PKSs are of bacterial origin and synthesize polycyclic compounds such as tetracycline and daunorubicin.[11] Individual enzymes are typically encoded as separate proteins and are used iteratively. Type III PKSs are found among bacteria and plants and biosynthesize simple monocyclic or bicyclic metabolites such as chalcone and resveratrol.[12] These simple PKSs are comprised of homodimeric KSs that iteratively condense acyl-CoA thioesters directly without the aid of an ACP. Our discussion of engineered biosynthesis will be divided according to these three classes. For each class of polyketides, we will detail the genetic features of the biosynthetic genes and the fundamental biochemistry of the core enzymes, and follow with examples of how manipulation of the PKS can lead to the biosynthesis of unnatural polyketides.

1.3 Methods for Engineered Biosynthesis

Polyketide biosynthetic pathways can either be engineered in the native producing organism[8,13–15] or in a genetically well-characterized heterologous host.[16–18] Native hosts offer the advantages of coordinated functional expression of all enzymes involved in the biosynthetic pathway. Homologous recombination is frequently used in constructing deletion, insertion and hybrid PKS mutants at the chromosome level. However, there are limitations in using native hosts: (1) native hosts are difficult to manipulate and often resistant to standard DNA transformation protocols; (2) native hosts are often difficult and in some cases virtually impossible (*e.g.* marine organisms) to culture; and (3) marker-independent gene replacement by homologous recombination is a low-frequency and consequently time-consuming process.

Genetically well-defined heterologous hosts combined with versatile shuttle vectors can address these limitations. For example, a host/vector system was developed in a derivative of *Streptomyces coelicolor* (CH999)[18] from which the primary endogenous polyketide pathway (actinorhodin, *act*) had been deleted, hence offering a "clean" background to express and analyze heterologous PKS genes. *S. coelicolor* is the best-studied actinomycete and can be subjected to genetic modifications with well-established protocols. DNA transformation and conjugation can be performed with high efficiencies and a range of selectable markers can be used. PKS genes can be expressed in a "natural" manner under control of the engineered *act* promoters. Utilizing a shuttle vector, PKS genes can be rapidly manipulated in *Escherichia coli* and transformed into CH999 for polyketide biosynthesis. More recently, *E. coli* has been metabolically engineered to produce complex polyketides.[19] *E. coli* is extremely simple to handle with excellent growth characteristics. Moreover, *E. coli* has no endogenous PKS-production capabilities and thus provides an excellent opportunity for fundamental studies of PKSs. In addition to the PKS genes, post-translational modification genes and genes encoding specialty precursor biosynthetic pathways must also be introduced. Other examples of heterologous hosts include *Streptomyces lividans*[17] and *Myxococcus xanthus*.[20]

2   Type I Polyketide Synthases

2.1   Modular Architecture

Type I PKSs are characterized by the organization of catalytic domains along large polypeptides known as megasynthases (Figure 3).[21] A megasynthase may house a single module used iteratively or multiple modules used sequentially in an assembly-line manner. Each module is approximately 150–200 kDa in size, and contains minimally the KS, AT and ACP domains. Since multimodular PKSs have been especially fertile targets for engineering, most of this section will focus on a discussion of these systems. Polyketide biosynthesis by a multimodular PKS is initiated through a loading module, typically installed at the N-terminus of the first module.[22] The loading module primes the assembly line with a specific acyl substrate, such as acetate, propionate, cyclohexanoate, etc. A thioesterase (TE) domain is often present at the C-terminus of the final module, catalyzing the release and macrocyclization of the polyketide.[23] Completion of chain extension and macrocyclization yields a macrolide aglycon. An additional thioesterase (TEII), frequently found as a free-standing enzyme, is involved in the hydrolysis of incorrectly loaded, truncated polyketide intermediates formed during chain elongation.[24,25]

The structural diversity observed among macrolides is largely due to the reductive domains (Figure 2B) inserted between the AT and ACP domains in each module (Figure 3). Ketoreductase (KR) reduces the β-ketoacyl thioester to a β-hydroxyl thioester. Subsequently, the dehydratase (DH) domain dehydrates the β-hydroxyl thioester to give an α-β unsaturated thioester. Enoylreductase (ER) finally reduces the resulting olefin to an alkyl moiety. The combinations (or the lack) of the three reductive domains can tailor the β-carbons with four different oxidation states, resulting in the large structural variation observed at alternating carbons within

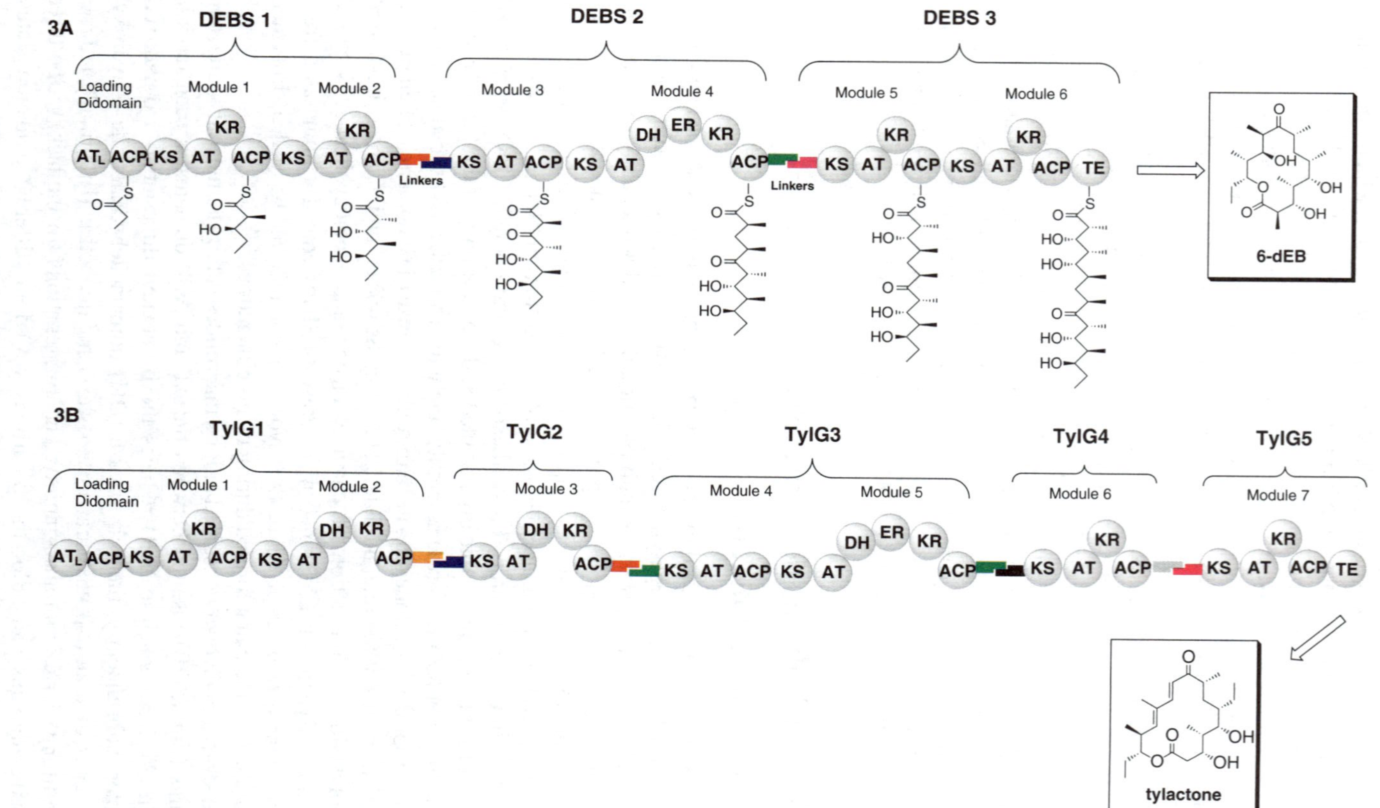

Figure 3 *Modular architecture of (A) erthromycin and (B) tyllosin PKS*

macrolides. Methyltransferase domains are also found within a PKS module.[26] Additional tailoring enzymes,[27] including oxygenases and transferases, can further decorate the aglycon to give structurally diverse compounds. Methyltransferase, aminotransferase and glycotransferase are the most common transferases associated with PKSs. The post-PKS enzymes are separate and are not part of the modular assembly line.

Two elegant biochemical features of multimodular PKSs are especially important from the viewpoint of combinatorial biosynthesis. First, the organization of catalytic domains and the polyketide structures are colinear.[22] Each module is responsible for a specific step in the assembly of the polyketide chain and is utilized only once during chain extension. As a result, the length of the polyketide is dictated by the overall number of modules present in the PKS. The oxidation states of the different β-carbons within a macrolide can be readily predicted based on the combination of reductive domains present in the corresponding module. Second, communication between large modules is facilitated by short oligopeptide linkers at both the N- and C-termini of the modules (Figure 3).[28] The linkers are especially important for the correct transfer of the polyketide intermediate from an upstream module to a cognate downstream module. Compatible linkers interact adhesively to facilitate module–module crosstalk. Modules juxtaposed in the same polypeptide communicate through intrapeptide linkers, while two interacting modules housed in different polypeptides communicate through interpeptide linkers. In the absence of the linkers, chain transfer between two modules is severely impeded. Two modules that are not programed to communicate can be engineered to function in tandem by placing compatible linkers at the C-terminus of the upstream module, and at the N-terminus of the downstream module. The discovery of linker–linker interactions has allowed the heterologous recombination of entire modules for the biosynthesis of hybrid polyketides.[29]

2.2   The Erythromycin Synthase

The best-characterized modular PKS is the erythromycin synthase (DEBS, Figure 3A) from *S. erythrea*.[30] The DEBS PKS contains three megasynthases (DEBS1, DEBS2 and DEBS3). Each megasynthase houses two modules. The loading didomain consists of AT_L and a loading ACP_L, and selects propionyl-CoA as the starter unit in preference to acetyl- and butyryl-CoA.[31] In all six modules, the AT domains are highly specific for methylmalonyl-CoA as extender units.[32] The six modules are outfitted different reductive domains.[22] M1 (module 1), M2, M5 and M6 contain only KR, while M4 contains the entire KR/DH/ER machinery. M3 contains an inactive KR domain. The linker regions that dictate correct substrate transfer have been identified for each pair of interacting modules.[28] A TE is positioned at the C-terminus of DEBS3 and catalyzes the cyclization of the 14-membered ring.[33] Together, the DEBS megasynthases contain 28 catalytic domains[34] and synthesize 6-deoxyerythronolide B (6-dEB) using one propionyl-CoA and six methylmalonyl-CoAs. The overall turnover of DEBS occurs approximately at the rate of one molecule of 6-dEB per minute.[35]

2.3 Engineered Biosynthesis of Multimodular PKS Products

The differences among the various Type I PKSs that give rise to the diverse aglycons shown in Figure 3 are: (1) the specificity of the loading module; (2) the number of modules present in the PKS; (3) the specificity of the individual AT domain in each module; (4) the combinations of reductive domains in each module; and (5) the linear combination of modules within a PKS. The modular architecture of Type I PKSs provides a powerful platform for combinatorial biosynthesis through domain shuffling, module shuffling and pathway shuffling.

2.3.1 Domain Engineering

Domain engineering involves the specific domain inactivation and interchange of individual catalytic domains between different PKS modules (Figure 4). Because the AT and reductive domains contribute extensively to polyketide diversity, these domains have been subjected to intense shuffling experiments.[10] A partial compilation of AT domains with different CoA specificity and reductive domains with different tailoring properties are listed in Table 1.

Domain swapping between different DEBS modules or different PKSs has been accomplished mostly in DEBS M2,[36–40] M5[8,22] and M6,[41] and also in other DEBS modules. Chimeric DEBS PKSs containing different AT and/or reductive domains have yielded a library of unnatural 6-DEB analogs, demonstrating the surprising tolerance of the overall DEBS PKS to individual domain changes in these particular modules.[42]

Homologous recombination in *S. erythraea* has been successfully used in the construction of hybrid modules. Leadlay and co-workers[15] replaced the methylmalonyl-specific DEBS AT1 with the malonyl-specific RAPS (rapamycin PKS) AT2. The resulting DEBS1 mutant synthesized the expected triketide lactone (2S, 3S, 5R)-2-methyl-3,5-dihydroxy-*n*-heptanoic acid δ-lactone in *S. erythrea*. Similarly, Katz and co-workers[8] replaced the methylmalonyl-specific DEBS AT2 with an ethylmalonyl-specific domain found in the niddamycin PKS cluster. The resulting host was able to produce the expected 6-desmethyl-6-ethylerythromycin A in addition to erythromycin A when precursor molecules such as diethyl ethylmalonate were included in the growth media. The authors then performed a second chromosomal modification in which the *Streptomyces collinus* crotonyl-CoA reductase (*ccr*) gene was inserted under the control of a strong *ermE** promoter separate from the DEBS PKS genes. When overexpressed *in vivo*, *ccr* was able to synthesize butyryl-CoA, which can be converted to ethylmalonyl-CoA for utilization by the hybrid PKS. The *S. erythraea* strain equipped with the *ccr* gene produced the modified erythromycin A at levels comparable to that of wild type. Similar experiments have also been performed with malonyl-CoA-specific acyltransferase domains inserted into various DEBS modules in place of the cognate methylmalonyl-CoA-specific domains.[43] More recently, the methoxymalonyl-CoA-specific domains of the FK520 PKS have been replaced with both malonyl-CoA- and methylmalonyl-CoA-specific domains in *Streptomyces hygroscopicus*.[44]

In order to make chimeric PKSs in an efficient and combinatorial fashion, one must be able to clone various AT or reduction domains into DEBS PKS as cassettes.

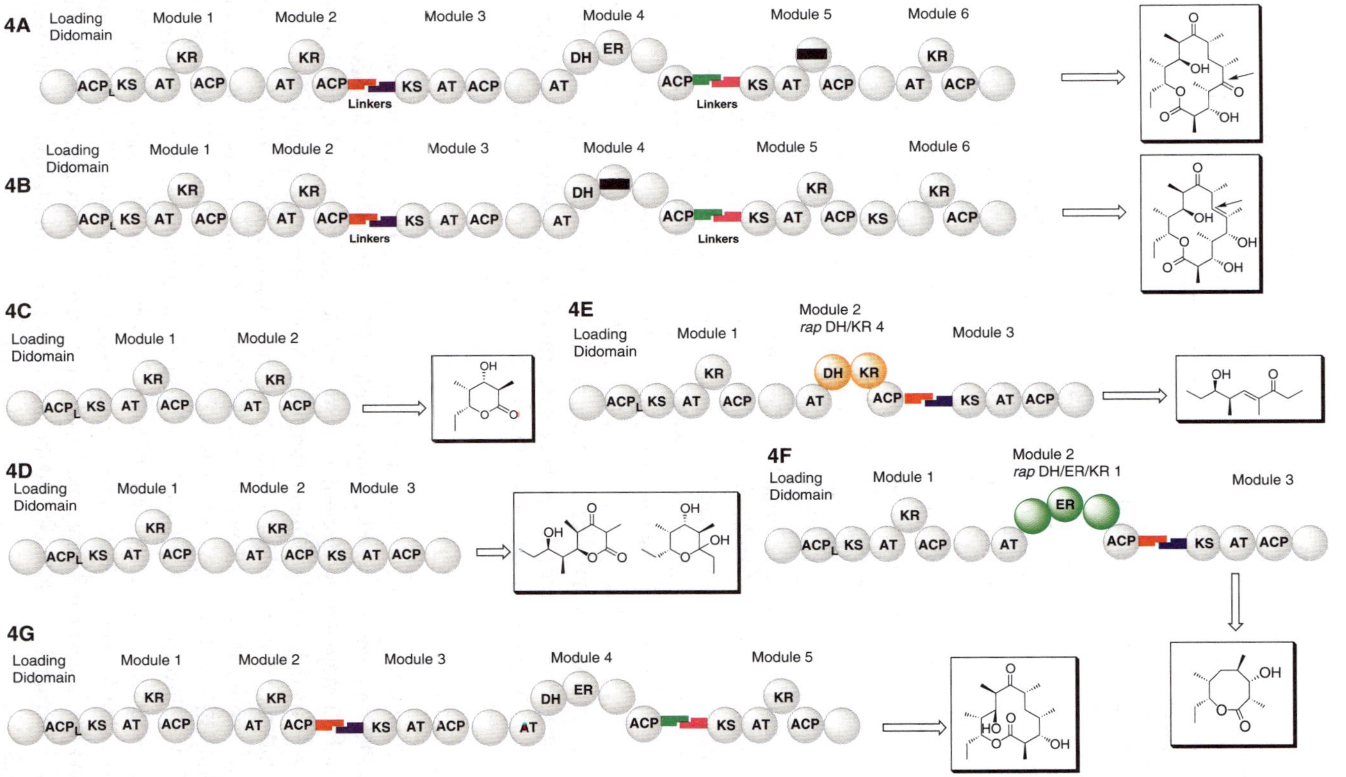

Figure 4 *Engineered biosynthesis of unnatural macrolactones*

Table 1 *AT domains with different extender unit specificity and reductive domains with different catalytic functions found in Type I PKSs*

AT Domains		Reductive Domains	
Malonyl-CoA	RAPS AT2	No reduction	DEBS KR3[b]
Methylmalonyl-CoA	DEBS ATs	β-keto reduction	DEBS KR5
			RAPS KR2[c]
Ethylmalonyl-CoA	Nid AT5[a]	β-OH dehydration	RAPS DH/KR4
Methoxymalonyl-CoA	FK520 AT8	Enoyl reduction[d]	RAPS DH/ER/KR1

[a] Nid: Niddamycin.
[b] A nonfunctional ketoreductase.
[c] RAPS KR2 reduces a β-keto moiety with opposite stereochemistry to that of DEBS KR5.
[d] Complete reduction from a β-keto moiety to an alkylacyl moiety.

Toward this end, unique restriction sites have been introduced at junctions between (1) KS and AT; (2) AT and KR/DH/ER; and (3) KR/DH/ER and ACP. The reshuffled DEBS modules can be cloned into the shuttle vector and introduced into a heterologous host. Using these flanking restriction sites, McDaniel *et al.*[45] were able to generate single, double and triple mutants of DEBS PKS containing AT and reductive domains from different sources. Most of the engineered DEBS PKSs were able to produce the predicted 6-DEB analogs at isolatable quantities. Over 50 6-DEB analogs were biosynthesized in this effort.

Notwithstanding the simplicity and power of domain shuffling, it suffers from the limitation that the heterologous AT domain perturbs domain–domain interactions in the module, thereby impairing their catalytic efficiency. An alternative approach to altering extender unit specificity of individual modules involves inactivation of the endogenous AT domain via site-specific mutagenesis, followed by co-expression of a kinetically fast Type II acyl transferase (see Section 3) *in trans.*[46] This strategy has the advantages of minimal perturbations of the tertiary and quaternary structure of modules and a clean alteration of extender unit specificity, but is restricted in scope by the limited repertoire of stand-alone acyl transferases, most of which have malonyl-CoA specificity.

2.3.2 Module Engineering

An alternative strategy for combinatorial biosynthesis is to rearrange intact modules within a PKS or combine modules from different PKSs. Experiments that have been successfully employed to generate novel polyketides include module deletion,[13,37,47–49] module rearrangement and heterologous module shuffling.[50] Leadlay and co-workers[13] constructed a bimodular system in which a 20-kb stretch of *S. erythraea* chromosomal DNA spanning from ACP2 to ACP6 was deleted. The resulting truncated DEBS1+TE was able to produce the triketide lactone (*2R, 3S, 4S, 5R*)-2,4-dimethyl-3,5-dihydroxy-*n*-heptanoic acid δ-lactone with high efficiency. A different genetic approach to generate the same triketide lactone was to insert the intact DEBS1 gene into a shuttle vector.[49] Upon transformation of the vector into CH999, the same compound was recovered. Fusing the DEBS TE to the C-terminus of DEBS1 in the latter approach significantly improved the turnover of the "mini-lactone" by the truncated PKS (Figure 4C).

Using the heterologous host/vector pair, additional DEBS deletion mutants synthesized other truncated polyketides. A trimodular PKS was constructed by fusing DEBS1 to M3 of DEBS2 and TE.[37] When expressed in CH999, the mutant produced primarily the tetraketide CK13a (Figure 4D). Furthermore, when the full set of DEBS genes except module six was expressed in CH999, the expected 12-member ring lactone (8R, 9S)-9,10-dihydro-8-methyl-9-hydroxy-10-deoxymethynolide was efficiently synthesized (Figure 4G).[48]

Domain shuffling has been combined with module shuffling to yield additional polyketide variants. For example, when KR2 of the aforementioned trimodular construct was replaced with RAPS DH-KR4, a linear tetraketide product was observed (Figure 4E).[36] Dehydration of the hydroxyl group by the new DH domain prevented the cyclization of the tetraketide (which yields CK13a). More interestingly, when the reductive domain in M2 in the trimodular construct was replaced with RAPS ER-DH-KR1, the tetraketide was able to form an eight-membered macrolactone (Figure 4F).[39] Formation of these truncated polyketides shows that (1) the TE can be fused to different modules and function properly as a thioesterase toward different substrates; (2) the downstream modules have no influence on the functions of the upstream modules; and (3) the ketosynthases have relaxed substrate specificities toward polyketide intermediates of varying length and reductive states. These features are further highlighted in the following sections.

Large-scale assembly of intact modules from heterologous PKSs is perhaps the ultimate goal of combinatorial polyketide biosynthesis. Module shuffling minimizes the disruption of domain–domain interactions observed in many domain shuffling studies. Early attempts at module-swapping experiments failed mainly because communication between modules facilitated by the natural linkers between them was ignored. Both the intramodular and the intermodular linkers are important in establishing polyketide chain transfer between two modules in DEBS PKS and other Type I PKSs. Intermodular linkers allowed fruitful interactions between M1 (outfitted with the M2 C-terminus linker) and dissociated M2, M3, M5 and M6 (each containing M3 C-terminus linker).[28] When rifamycin M5 was fitted with a DEBS M3 N-terminus linker and inserted into the DEBS PKS in place of DEBS M2, it was able to function properly in the assembly line.[28] By placing linkers between two target modules to facilitate module communication, the substrate tolerance of different modules can be examined. Assaying individual modules from DEBS, pikromycin (Pik) and rifamycin (Rif) PKSs have shown that while some modules are perfectly competent when transferred into a foreign context, other modules are incapable of processing unnatural substrates.[51]

Researchers from Kosan Biosciences performed module shuffling between three PKSs that synthesize 16-member macrolides.[50] The PKSs of tylosin (*tyl*G) (see Figure 3B), spiramycin (*srm*G) and chalcomycin (*chm*G) all contain the same module organizations (seven modules, M1–M7), but differ in loading module and AT domain specificity. Chimeric PKSs were constructed by combining modules of different substrate specificities from the three PKSs. Chain transfer between heterologous modules was optimized by adopting a pair of naturally interacting linkers. The hybrid PKSs were able to synthesize novel 16-membered macrolides exhibiting structural features from parent compounds in *Streptomyces fradiae*. The unnatural macrolides were further modified combinatorially by the downstream tailoring enzymes associated with the different PKSs to yield a library of bioactive polyketides.

2.3.3 Primer Unit Engineering and Precursor-Directed Biosynthesis

The primer units of macrolides have been modified using genetic approaches as well. Swapping loading modules with alternative substrate specificities can lead to the incorporation of new starter units. The loading module specificity is governed by the gatekeeping AT_L.[52] In a typical swapping experiment, the entire AT_L–ACP_L didomain is exchanged. For example, Kuhstoss *et al.*[53] replaced the acetate-specific loading module of the spiramycin PKS in *Streptomyces ambofaciens* with that of the tylosin PKS (propionate-specific) to yield a propionyl-primed spiramycin. Similarly, Leadlay and co-workers[54] replaced the loading didomain of DEBS1 with the broadly specific loading domain from the avermectin PKS found in *Streptomyces avermitilis*. In addition to propionate, the hybrid PKS inserted a large assortment of α-branched starter units into 6-dEB (Figure 5A). Downstream tailoring enzymes were able to transform the unnatural aglycons into the corresponding erythromycin A analogs. The rifamycin-loading module, which contains an adenylation and a thiolation domain typically found in nonribosomal peptide synthase (NRPS), was fused to the N-terminus of DEBS1 in place of its endogenous loading domain. The hybrid PKS was able to activate benzoic acid as the starter unit and synthesize the expected benzyl-substituted 6-dEB in *E. coli*.[19]

Although domain and module shuffling can afford new polyketides that are not produced naturally, the variations introduced via those techniques are still limited by the relatively small pool of natural building blocks. To combine the vast chemical space accessible through synthetic chemistry and the exquisite precision of biosynthesis, precursor-directed biosynthesis was developed for the Type I PKSs.[55] Briefly, precursor-directed biosynthesis is the process in which novel primer units mimicking an early polyketide intermediate are supplied exogenously to the PKS. The synthetic substrate is elongated by the PKS via the downstream modules and yields an unnatural polyketide. This concept has been thoroughly explored with the DEBS PKS (Figure 5B). The KS of M1 was first inactivated through site-directed mutagenesis, which stalls the entire assembly line since the first decarboxylative condensation reaction can no longer take place. Bypassing the defective M1 requires the natural substrate of KS2, which is a diketide thioester, to be made available in an alternative form. Administration of natural diketide in the form of a *N*-acetylcysteamine (NAC) thioester to KS1° mutant restored polyketide biosynthesis. An array of unnatural diketides containing novel functional groups was introduced to the KS1° mutant and the corresponding 6-dEB analogs were synthesized. Aromatic,[55] unsaturated[56] and halogenated[57] polyketides were synthesized using the precursor-directed approach, demonstrating the surprising tolerance of the PKS toward unnatural starter units. The most unexpected result was that in addition to diketides, KS2 was also able to accept a triketide intermediate presented as an NAC thioester. Incorporation of the triketide at M2 led to the biosynthesis of a 16-member macrolactone.[58]

Using precursor-directed biosynthesis, Boddy *et al.*[59] were able to effectively reconstitute epothilone biosynthesis in *E. coli*. A pentaketide NAC thioester intermediate

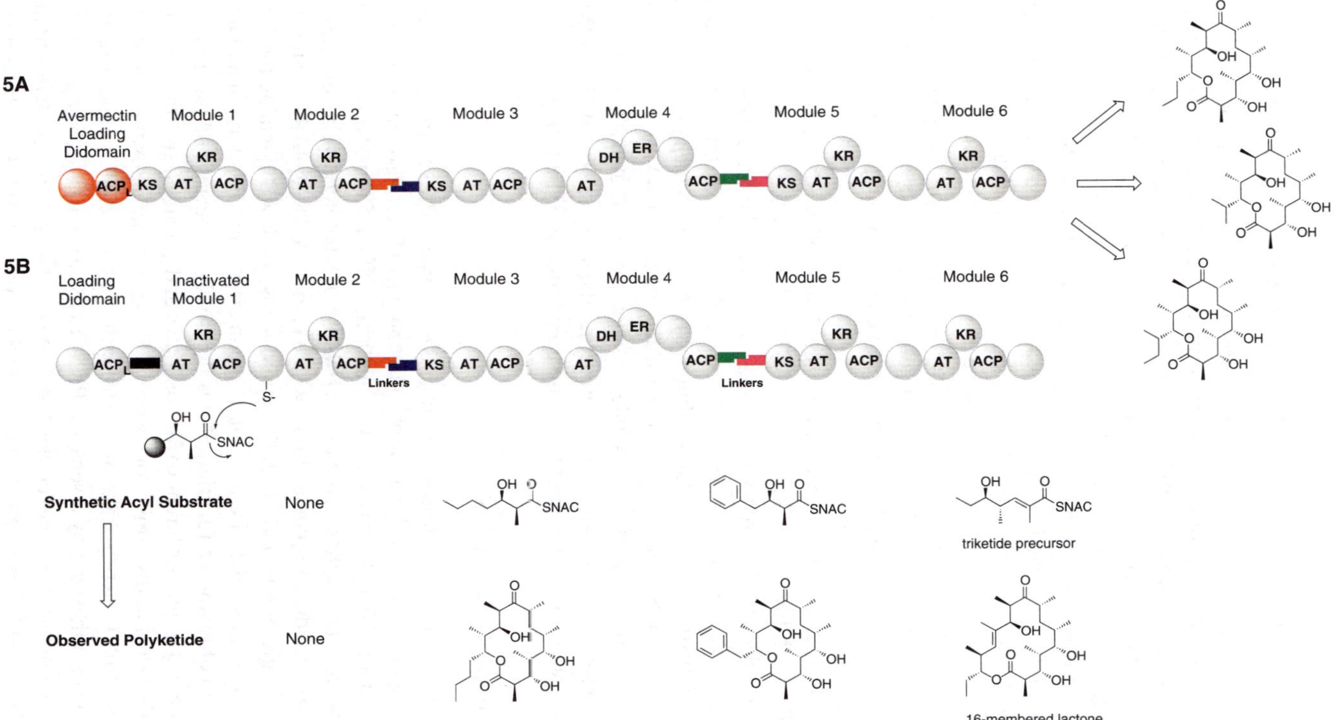

Figure 5 *Starter unit modification in Type I PKS. (A) Didomain replacement. (B) Precursor directed feeding*

was presented to the last three modules of *epo* PKS. Incorporation of the pentaketide substrate resulted in efficient biosynthesis of the important antitumor drug in *E. coli*. Supplementing unnatural pentaketides modified at the starter position led to the engineered biosynthesis of epothilone analogs.

2.4 Multimodular PKSs that Exhibit Special Features

The colinearity of the modular architecture and the minimal catalytic domains within a module are observed in most multimodular PKSs known to date. However, there are several subclasses of PKSs that exhibit intriguing structural and biochemical features not observed among the best-known multimodular PKSs.

The PKS of the angiogenesis inhibitor borrelidin from *Streptomyces parvulus* Tü4055 is an exception to the colinearity paradigm.[60,61] Borrelidin is an 18-membered macrolide assembled from a cyclopentane-1,2-dicarboxylic acid primer unit and eight extender units (Figure 6A). However, only six modules were found in the *Bor* PKS. Comparing the organization of the modules and the structure of borrelidin, it is evident that *Bor* M5 must catalyze three rounds of chain elongation to account for the size of the macrolide. Each of the other modules account for one round of condensation. The exact mechanism by which chain recycling occurs three times at the same module is not known.

The leinamycin[26] and pederin[62] PKSs are both "AT-less" multimodular PKSs (Figure 6B). Both PKSs contain only KS, ACP and the reductive domains in each of the modules. Discrete ATs are found in both the *lnm* and the *ped* gene cluster as separate enzymes. The dissociated AT iteratively catalyze malonyl transfer *in trans* onto all the ACP domains. Decarboxylation of malonyl-ACP by KS then proceeds normally as in a typical module. Surprisingly, the fatty acid malonyl-CoA:ACP acyltransferase is not able to restore leinamycin biosynthesis when the dissociated AT is deleted from the gene cluster.[26] Understanding and engineering the substrate specificity of the freestanding AT can lead to a larger extender unit repertoire in Type I PKS.

2.5 Fungal Type I PKSs

Fungal PKSs represent an important subclass of Type I PKSs. Important pharmaceutical compounds such as lovastatin[63] and compactin[64] are produced via the iterative function of a single module. The lovastatin PKS from *Aspergillus terreus* consists of a core nonaketide synthase (LNKS, LovB) that iteratively condenses nine malonates to form dihydromonacolin L, which is modified to yield lovastatin (Figure 6C).[65] The single polypeptide (>3000 amino acids) of LovB contains the KS, AT, DH, MT, KR and the ACP domains. It also contains a null ER domain and a C-terminal domain of unknown function. LovB iteratively elongates and precisely tailors a polyketide chain with different oxidation states and α-carbon substitutions at each step with remarkable discriminatory ability. The exact mechanism of substrate differentiation at each step is unknown. Furthermore, an enzyme-bound 2-methylbutyryl group is synthesized by a separate PKS (LDKS, LovF) and is transacylated to lovastatin precursors via a novel esterase LovD.[65] The protein–protein interaction between LovF and LovD represents a novel mode of substrate transacylation.

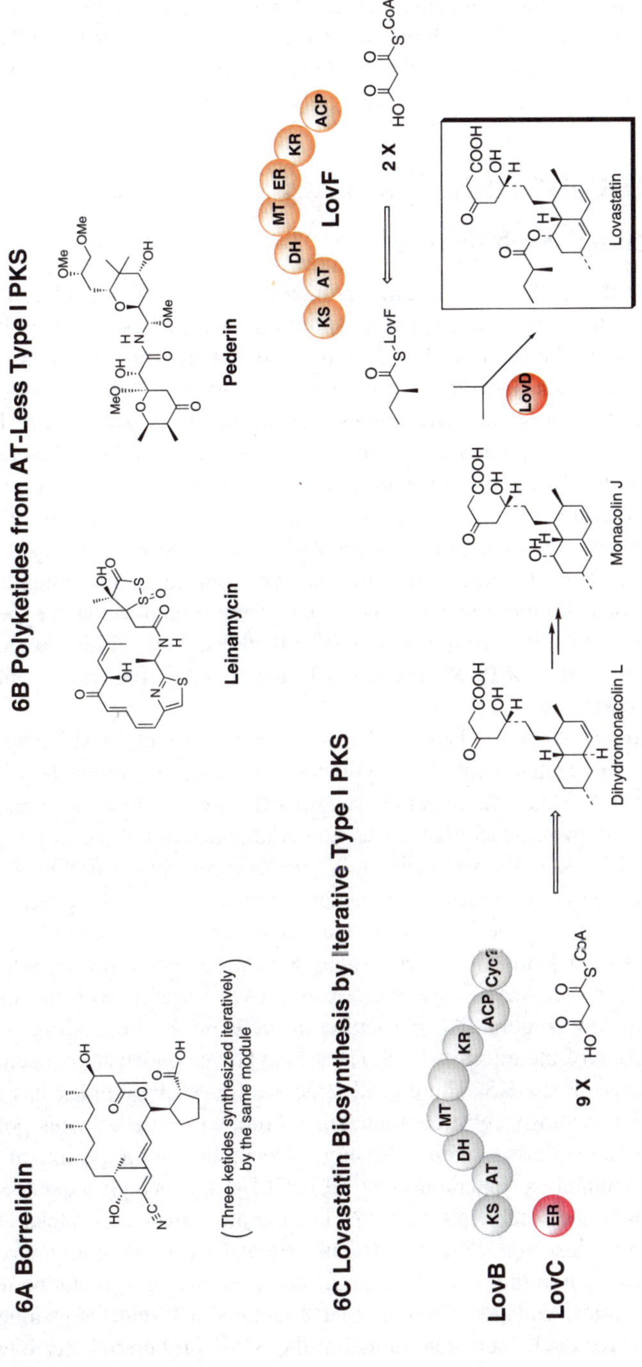

Figure 6 *Unusual and iterative Type I PKSs*

Equally intriguing is an intramolecular Diels–Alder cyclization after hexaketide formation. The domain in LovB that catalyzes the cyclization has not been discovered.

Uncovering the biochemical principles governing these mechanisms is therefore important for the rational biosynthesis of lovastatin and other fungal polyketides. It will also allow the enzymatic components of iterative PKSs to be used as tools in the combinatorial biosynthesis of entirely new polyketide scaffolds.

3 Type II Polyketide Synthases

3.1 Dissociated Architecture

Type II polyketide synthases synthesize aromatic polyketides, including the clinically important tetracycline, daunorubicin and mithramycin.[66] Aromatic polyketides are characterized by the multiply fused, richly substituted aromatic rings. The aromatic rings can be fused linearly as in the scaffold of tetracycline,[67] or angularly, as in the scaffold of jadomycin.[68] Biosynthesis of aromatic polyketides can be separated into three stages: (1) the assembly of the complete polyketide chain by the minimal PKS module;[18] (2) the reduction, aromatization and cyclization of the polyketide by intermediate tailoring enzymes *after* completed assembly of the nascent polyketide;[69] and (3) the full decoration of cyclized polyketides by additional enzymes specific for each PKS.[27] Much of what we know about aromatic polyketide biosynthesis, especially the first two stages, stems from mix-and-match experiments with heterologous enzymes from different Type II PKSs.[69] The ability to recombine components from different PKSs has also afforded a large collection of unnatural aromatic polyketides.

The enzymatic domains in a Type II PKS are dissociated and free standing, in contrast to the modular arrangement of catalytic domains observed among Type I PKSs.[6] As a result, the individual enzymes in a Type II PKS are smaller in size and can be easily manipulated and characterized. The genetic organization of several well-characterized Type II PKSs is shown in Figure 7. The arrangements of pertinent genes in these PKSs are highly similar, suggesting these clusters are evolutionarily closely related. The minimal PKS module[70] (also referred to as elongation module) in a Type II PKS consists of a ketosynthase/chain-length factor heterodimer (KS–CLF), an ACP, and a malonyl-CoA:ACP acyltransferase (MAT)[71] shared with the fatty acid biosynthetic pathway (Figure 8A). Formation of the KS–CLF heterodimer is essential for the function of the minimal PKS. Decarboxylative condensation occurs at the active-site cysteine of the KS. The crystal structure KS–CLF complex has recently been elucidated, revealing a substrate tunnel in KS into which the growing polyketide chain can be extruded during chain extension.[72] Depending upon the size of polyketide product, the tunnel can extend into the CLF. CLF displays high sequence homology to the KS with the notable difference of lacking the active-site cysteine, where a glutamine is found instead.[73] The catalytically silent CLF is integrally involved in determining the length of the polyketide during chain extension by dictating the number of catalytic cycles. Mutagenesis studies have revealed that selected residues of the CLF define the size of the substrate tunnel at the KS–CLF heterodimer interface.[74] MAT catalyzes the acyl transfer of malonyl units from malonyl-CoA to malonyl-ACP.

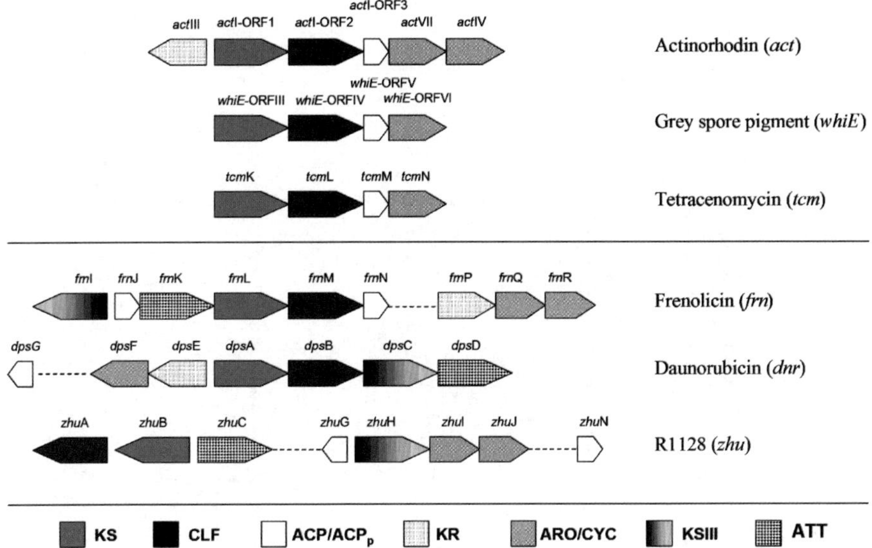

Figure 7 *Organization of Type II PKS gene cluster. Top: acetate primed PKS; bottom: nonacetete primed PKS*

MAT has stringent specificity toward malonyl-CoA, which is reflected in the exclusive utilization of malonyl extender units by Type II PKS.[75] Together, these four enzymes consist of the smallest set of enzymes required for the synthesis of a complete polyketide chain. For example, the minimal PKS from the actinorhodin (*act*) biosynthetic pathway synthesizes an octaketide (C_{16}) backbone from eight malonyl-CoA equivalents,[76] the tetracenomycin (*tcm*) minimal PKS synthesizes a decaketide (C_{20}) backbone from ten equivalents of malonyl-CoA (Figure 8A),[77] and the pradimycin (*pms*) minimal PKS synthesizes a dodecaketide (C_{24}) backbone from twelve equivalents of malonyl-CoA.[78]

Most aromatic PKSs initiate polyketidc biosynthesis through decarboxylation of malonyl-ACP, resulting in an invariant acetyl primer unit (Figure 8A). Decarboxylative priming occurs when malonyl-ACP is captured by an "unloaded" KS–CLF. Transfer of acetate to the KS active site primes the minimal PKS for the subsequent rounds of elongation.[73] Several aromatic PKSs are primed with nonacetate primer groups. Important antitumor anthracyclines such as doxorubicin and daunorubicin are each primed with propionate.[79] Tetracyclines are primed with a malonamate starter,[80] while enterocin is primed with a benzoate.[81] The estrogen receptor antagonist R1128 and the anti-parasitic agent frenolicin are primed with alkylacyl groups of varying lengths.[82] The biosynthesis and incorporation of the nonacetate starter unit is facilitated by an orthogonal set of dissociated enzymes collectively known as the initiation module.[83] The R1128 initiation module synthesizes alkylacyl starter units ranging in lengths between butyrate to hexanoate, and primes the R1128 KS–CLF with these alkylacyl groups in place of acetate.[82] Three additional enzymes (Figure 8B) form the R1128 initiation module: (1) an additional ACP (ACP_p) that is

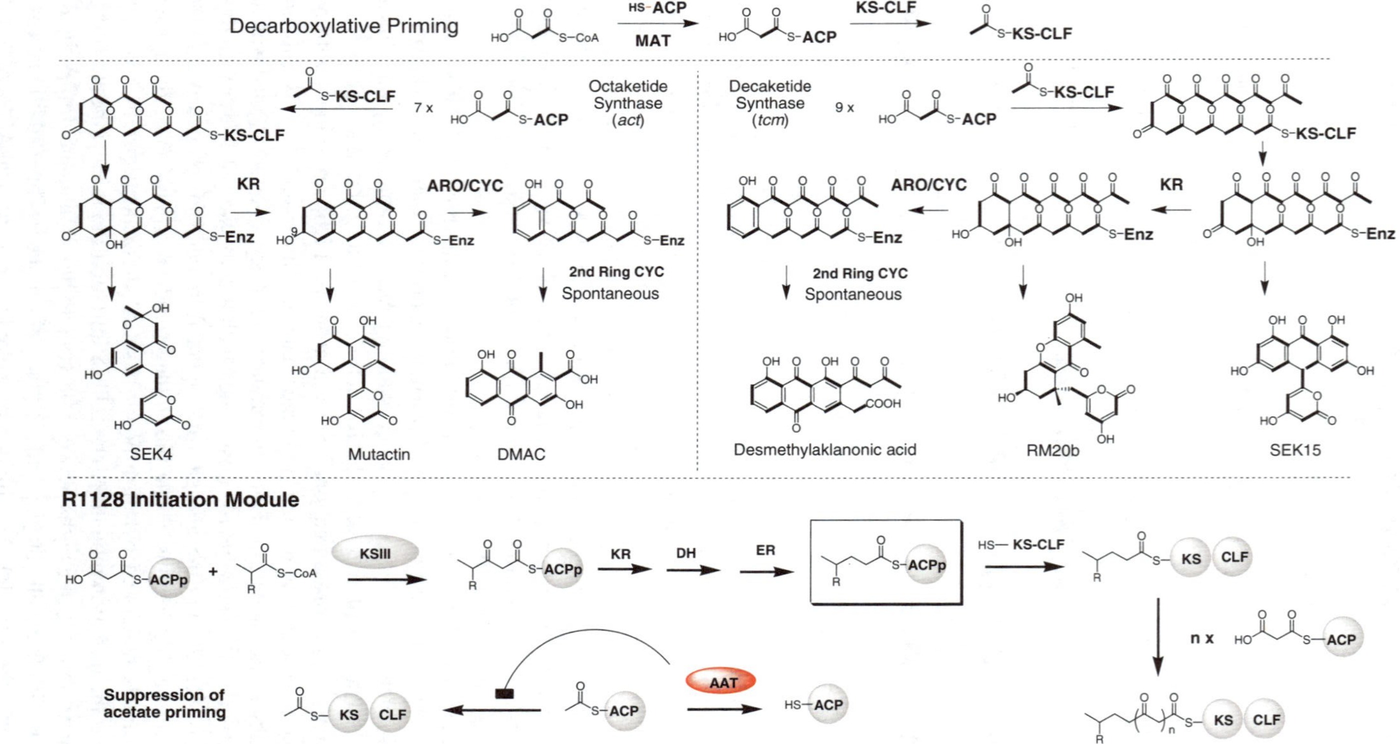

Figure 8 *Mechanism of Type II PKS and selected polyketide products*

used by the initiation module in the synthesis of the primer unit; (2) a ketosynthase III (KSIII) homolog that condenses short-chain acyl-CoAs, such as acetyl, propionyl and butyryl-CoAs with malonyl-ACP_p to yield β-ketoacyl-ACP_p. The β-ketoacyl-ACP_p is subsequently reduced and dehydrated by the KR, DH and ER associated with the endogenous fatty acid biosynthesis pathway to yield a fully reduced alkylacyl-ACP_p, which serves as the primer unit for the R1128 KS–CLF;[82] and (3) an acetyl-ACP thiolase (AAT) that selectively removes acetyl-ACP species that may compete with the alkylacyl-ACP_p for the KS active site.[84] The AAT rapidly hydrolyzes the acetyl-ACP thioester bond to yield acetate and free ACP, while leaving longer alkyl-ACP_p substrates intact. The action of the AAT effectively attenuates the decarboxylative priming pathway, resulting in efficient incorporation of the nonacetate primer.

In contrast to the Type I PKS architecture, in which individual reductive domains are dispersed throughout the megasynthases and can tailor the growing polyketide after each round of chain extension, Type II PKSs complete polyketide synthesis prior to tailoring. A fixed set of enzymes from each PKS processes the polyketide chain and transforms it into aromatic polyketides with varying oxidation states and substitutions.[69] A collection of immediate tailoring enzymes are commonly found in most Type II PKS (Figure 8A); they include: (1) KR, which is broadly specific for polyketides of varying chain lengths and regiospecifically reduces the carbonyl at C9 to a hydroxyl;[85] (2) cyclase (CYC), which catalyzes the intramolecular cyclization of the polyketide scaffold (the regioselectivity of CYCs determines the orientation of the aromatic structures);[86] and (3) aromatase (ARO), which dehydrates and aromatizes a cyclized ring to yield the conjugated ring systems observed in most aromatic polyketides.[69] CYC and ARO have stringent preferences for the number of β-carbonyl units in the polyketide chain.[69] Additional KRs specific for other carbonyls in the polyketide as well as multiple CYCs and AROs can be found in certain PKSs, depending upon the length of the polyketide and the complexity of the ring systems.[11]

The downstream tailoring enzymes include oxygenases, methyltransferases, aminotransferases and glycotransferases.[27] Oxygenases perform a wide array of modifications to aromatic polyketides, including hydroxylation, Baeyer–Villiger oxidative insertion,[87] ring cleavage, epoxidation and quinone formation. Aminotransferases are rare but are present in all tetracycline PKSs. Glycosylation is essential for the functioning of many aromatic polyketides, such as daunorubicin, pradimycin and the aureolic acids.

3.2 Combinatorial Biosynthesis of Type II Polyketides

The Type II PKS modules that can be reshuffled to introduce structural diversity among aromatic polyketides include: (1) the initiation module which specifies the starter unit; (2) the elongation module which determines chain length; (3) the immediate tailoring enzymes that dictate the oxidation state at C9 and the C–C connectivity during cyclization; and (4) the downstream tailoring enzymes. Systematic recombination of PKS genes from different clusters has provided insights into the substrate specificities of individual enzymes as well as the protein–protein interactions between different modules. The "programing rules" have allowed engineered biosynthesis of completely novel polyketide scaffolds and unnatural analogs of natural polyketides.

3.2.1   Chain-Length Variations

In the absence of any tailoring enzymes, the highly reactive nascent polyketide undergoes spontaneous cyclization. Several different structural scaffolds can be generated from the same nascent chain. It remains unclear whether the first cyclization event takes place while the polyketide chain is still positioned in the substrate tunnel of the KS–CLF complex. A common feature to all spontaneously cyclized, unreduced polyketides is a pyrone moiety that arises through the rapid cyclization of the last three ketide units (Figure 9, top). When the KS–CLFs are primed by acetate units, polyketides synthesized by octaketide-specific minimal PKS rearrange to yield SEK4 (first ring cyclization C7–C12) and SEK4b (C10–C15),[88] whereas polyketides synthesized by decaketide-specific minimal PKS cyclize to yield SEK15 (C7–C12) and SEK15b (C9–C14).[89] The frenolicin (*fren*) KS–CLF shows chain-length specificity for both octaketide and nonaketide backbones.[90] Dodecaketides produced by the *whiE*, *sch* and *pms* minimal PKSs yield TW93c–e (C7–C12), TW93f (C8–C13), TW93g (C10–C15) and TW93h (C8–C13).[91,92] The cyclization patterns of the dodecaketide TW93h is especially unusual and contains a 2,4-dioxoadamantane moiety not observed in any other polyketides. The *whiE* and *sch* minimal PKS also synthesizes small amounts of truncated polyketides, suggesting relaxed chain-length restriction by these KS–CLFs (the *pms* minimal PKS does not produce truncated polyketides).

3.2.2   Mix and Match of Tailoring Enzymes

Nascent polyketides produced by Type II PKSs can be modified by ketoreductases, cyclases and aromatases to yield a wide variety of aromatic compounds. Tailored octaketides undergo C7–C12 first-ring cyclization and can be transformed into benzoisochromanequinones and anthracyclines. Different combinations of tailoring enzymes can cyclize the first ring of decaketides with either C7–C12 or C9–C14 regiospecificity. The C7–C12 cyclized decaketides can be subsequently transformed into tetracyclines, angucyclines[93] and aureolic acids,[94] while the C9–C14 cyclized decaketides can be converted into tetracenomycins.[79]

The *act* KR has been found to be specific for polyketide chains of varying sizes. Hexaketides, octaketides, nonaketides, decaketides and dodecaketides can all be reduced regioselectively at the C9 carbonyl in the presence of the *act* KR.[95] Reduced octaketides spontaneously form mutactin,[85] while reduced decaketides form RM20,[18] RM20b[96] (major product) and RM20c. The frenolicin-minimal PKS outfitted with KR produces both reduced octaketide mutactin and reduced nonaketides RM18. Surprisingly, the *act* KR is also able to reduce the octaketide synthesized by *fren* KS–CLF at C7, which rearranges to become RM18b (Figure 9).[90] The apparent relaxed substrate specificity of *act* KR in the presence of *fren* KS–CLF has been attributed to a "slippage" mechanism, in which the octaketide chain is fixed in an alternative position inside the *fren* KS substrate tunnel. The *act* KR is also able to reduce dodecaketides at C9 to yield TW94d[91] and YT231.[78]

All C9-reduced polyketides first undergo cyclization between C7 and C12. The reduced ring is unable to aromatize in the absence of additional enzymes, as seen in the structures of mutactin, RM20b and RM20c. Introduction of chain-length-specific

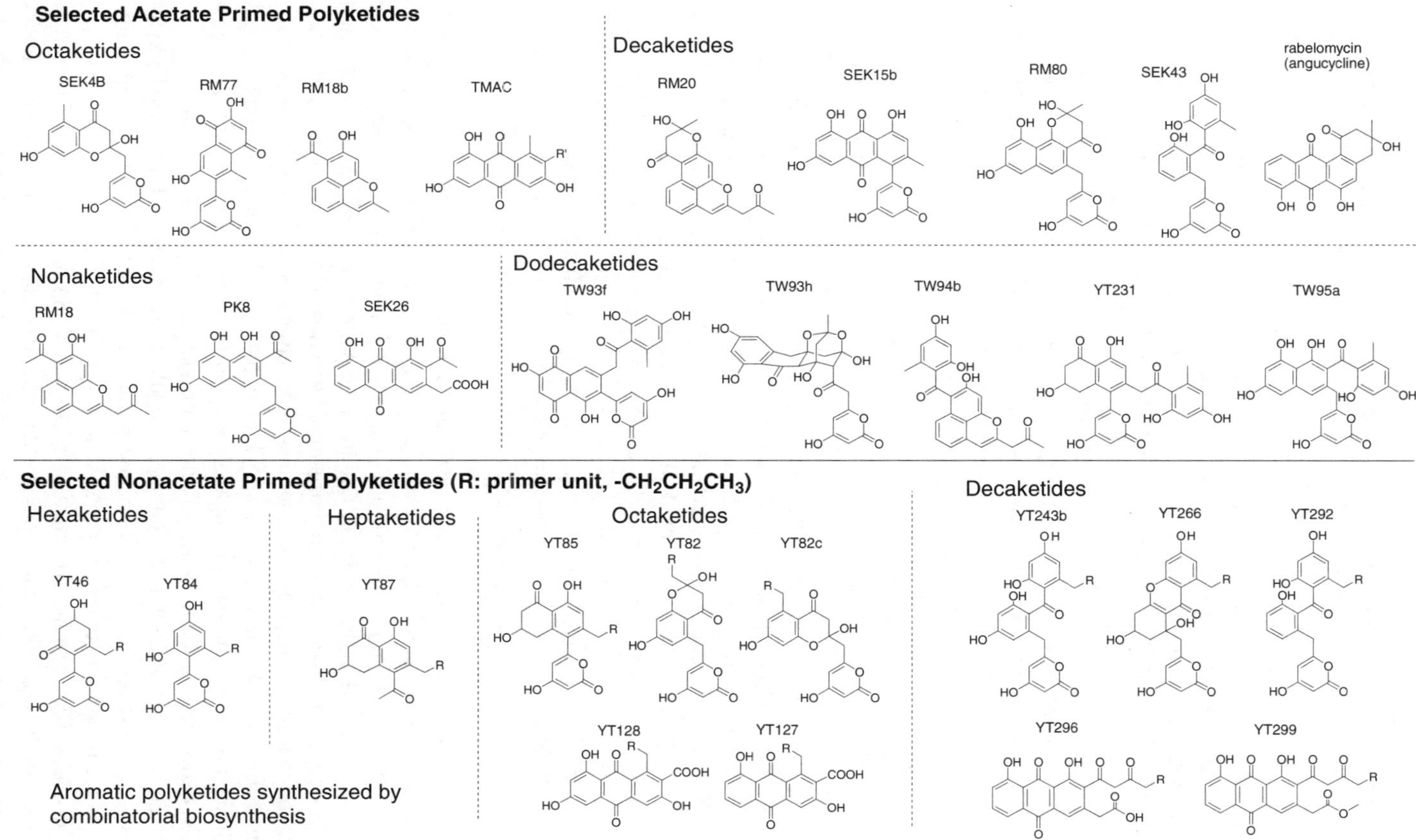

Figure 9 *Aromatic polyketides synthesized by combinatorial biosynthesis*

ARO leads to dehydration and aromatization of the first ring. The *act* ARO aromatizes the first ring of reduced octaketides to form SEK34 and SEK34b,[86] while the griseosin (*gris*) or daunorubicin (*dnr*) ARO aromatize the first ring of reduced decaketides to yield SEK43 (Figure 9).[97] The *act* ARO can only process reduced octaketides, while the *gris* ARO is able to aromatize both nonaketide and decaketides.

Cyclization of the second ring (C5–C14) in C9-reduced aromatic polyketides is catalyzed by specific second-ring CYCs. These CYCs are unable to process a substrate without an aromatized first ring. After cyclization and aromatization of the first two rings, the third ring cyclizes spontaneously (C2–C15 in octaketides and C3–C16 in decaketides). This pattern is evident in the biosynthesis of both 3,8-dihydroxy-1-methylanthraquinone-2-carboxylic acid (DMAC)[18] and aklanonic acid (Figure 9):[98,99] the combination of *act* KR, *act* ARO and *act* CYC with a octaketide-specific minimal PKS yields the anthraquinone DMAC and the decarboxylated variant aloesaponorin; the combination of *act* KR, *gris* ARO and *Dps* CYC with the a decaketide-specific minimal PKS yields desmethylaklanonic acid. Oxidation of the second ring in these compounds (also seen in SEK15b) presumably arises through the action of endogenous oxygenases.

The AROs and CYCs associated with reduced polyketides are unable to cyclize unreduced polyketides. A different set of CYCs is present in PKSs lacking KR. The octaketide-specific ZhuI and ZhuJ from the R1128 biosynthetic pathway can be combined with the *act* minimal PKS to produce the unreduced anthraquinone 3,6,8-dihydroxy-1-methylanthraquinone-2-carboxylic acid (TMAC).[100] TcmN found in the decaketide-specific *tcm* PKS can catalyze the regioselective cyclization between C9 and C14 of unreduced (1) octaketides to yield RM77,[101] (2) nonaketides to yield PK8[102] and (3) decaketides to yield RM80.[101] Similarly, equipping the *whiE, sch* and *pms* PKSs with dodecaketide-specific cyclases yields the compounds TW95a and TW95b (C9–C14).[91] Additional cyclases can transform dodecaketides into the highly elaborate ring systems observed in pradimycin-related compounds.

3.2.3   Primer Unit Modifications

The primer unit represents an attractive site for introducing alternative chemical functionalities into aromatic polyketides, largely due to the exceptionally high substrate specificity of the minimal PKS toward malonyl-ACP. Alternative extender units have not been observed in the backbones of aromatic polyketides, leaving the starter unit as the only accessible position for structural alteration. Heterologous recombination of initiation and elongation modules has been employed successfully to alter the primer unit in aromatic polyketides.[100] Functional assembly of these modules has facilitated the biosynthesis of completely new aromatic polyketide scaffolds as well as the regioselective modification of known polyketides (Figure 9, bottom).

In the presence of the R1128 initiation module, the *act* KS–CLF produced alkylacyl-primed hexaketides instead of acetyl-primed octaketides.[100] The reduced (when KR is included) and unreduced hexaketides cyclized to form YT46 and YT84,[103] respectively. The bicyclic structures of YT46 and YT84 have not been observed in polyketides before. The *tcm* minimal PKS produced alkylacyl-primed octaketides instead of acetyl-primed decaketide when outfitted with the R1128 initiation module.

Reduced and unreduced alkylacyl octaketides spontaneously rearrange to yield YT85 and YT82/82C, which are alkyl-primed versions of mutactin and SEK4/4B, respectively.[103] Furthermore, equipping the bimodular *tcm* PKS with tailoring enzymes specific for octaketides resulted in the transformation of the nascent alkylacyl octaketide into the expected anthraquinones YT127 and YT128.[100] Both YT127 and YT128 displayed superior activity than R1128 against the breast cancer cell line MCF-7, and possessed inhibitory activity against glucose-6-phosphate translocase, an attractive target for the treatment of Type II diabetes.[100]

The *fren* minimal PKS was also able to accept the starter unit furnished by the R1128 initiation module.[103] The novel compound YT87 was biosynthesized by an *S. coelicolor* strain that expressed both modules. YT87 is an alkylacyl-primed heptaketide that is reduced at C7 (as observed in RM18b). Similarly, overexpressing the R1128 initiation module along with the dodecaketide-specific *pms* minimal PKS resulted in the biosynthesis of alkylacyl-primed decaketides.[78] Alkylacyl-primed analogs of SEK15, RM20b were detected in these bimodular constructs. Coexpression of additional decaketide-specific tailing enzymes, including the *gris* ARO, *dps* CYC, produced alkylacyl aklanonic acid YT296, an analog of a key intermediate in the biosynthesis of daunorubicin.

From the above studies, it is evident that the minimal PKS controls polyketide chain length by counting the number of atoms incorporated into the backbone rather than the number of elongation cycles. For example, when the decaketide (C_{20})-specific *tcm* PKS is primed with a C_6 alkyl starter unit, the minimal PKS behaves as an octaketide synthase by performing seven additional iteration cycles as compared to nine iterations following acetate priming.[100] The final number of carbons in the backbone remains at C_{20}. In contrast, auxiliary PKS enzymes such as KR, ARO and CYC recognize specific functional groups in the polyketide backbone rather than overall chain length and the identities of the starter unit. The *act* KR, ARO and CYC all efficiently transformed an alkylacyl-primed octaketide into a DMAC analog, YT127. It is also apparent that most of the complex aromatic polyketides, such as daunorubicin and tetracycline, can be outfitted with new starter units using the bimodular PKSs with the same tailoring enzymes, which might improve their therapeutic properties.

The alkylacyl starter units incorporated into aromatic polyketides are chemically inert and cannot easily be utilized as orthogonal reactive handles for semisynthetic modifications. The substrate specificity of the *fren*[104] and R1128 initiation modules are largely determined by the KSIII enzymes,[83] which are inflexible toward alternative acyl-CoA substrates. To expand the functionality that can be realized with the bimodular PKSs, initiation modules that can synthesize alternative starter units are required. This goal can be accomplished through the characterization of new initiation modules, such as the malonamate-specific tetracycline initiation module. The recently sequenced hedamycin gene cluster encodes an iterative Type I PKS that synthesizes a diene starter unit for its Type II minimal PKS.[105] Hedamycin therefore serves as an ideal model for studying starter unit channeling between Type I and Type II PKSs. Alternatively, the substrate specificities of known initiation modules can be expanded through protein engineering efforts. The crystal structure of the R1128 KSIII has recently been elucidated,[106] hence allowing rational manipulation of the gatekeeping enzyme.

When the decaketide-specific *tcm* minimal PKS was reconstituted *in vitro*, alkylacyl-CoAs were examined as sources of primer units.[84] The acyl-CoA concentrations were elevated in the millimolar range to compensate for low rates of activation. In the presence of the R1128 AAT, the *tcm* minimal PKS was readily primed by propionyl-CoA to yield propionyl-primed version SEK15 and SEK15B; by butyryl-CoA to yield the butyryl-primed nonaketides; by hexanoyl-CoA to yield hexanoyl-primed versions of SEK4 and SEK4B; by octanoyl-CoA to yield octanoyl-primed heptaketides; and by decanoyl-CoA to yield decanoyl-primed hexaketides. Each of the unnatural compounds contains a total of 20 carbons in the polyketide backbone. An incremental increase in the starter unit led to a corresponding decrease in the number of decarboxylative condensations. These *in vitro* results suggested that unnatural starter units can be efficiently incorporated into polyketide backbones using engineered biosynthesis if a metabolic pathway for accumulation of the starter unit can be engineered *in vivo*.

Precursor-directed biosynthesis has also been employed successfully in engineered biosynthesis of Type II polyketides. Moore and co-workers[107] inactivated phenylalanine ammonia lyase in the enterocin PKS that is required for the synthesis of the benzoate starter unit. The resulting *Streptomyces maritimus* strain was unable to synthesize enterocin in the absence of exogenously supplied benzoic acid. Enterocin analogs containing the unnatural aryl starter units were recovered when novel aryl acids, including *p*-fluorobenzoic acid, 2-thiophenecarboxylic acid and cyclohex-1-enecarboxylic acid, were administered to the mutant strain. The aryl acids were converted to the corresponding aryl-CoAs and were used as starter units in the absence of benzoyl-CoA.

3.2.4   *Reshuffling of Downstream Tailoring Enzymes*

Manipulation of further downstream enzymes through gene inactivation and insertion can afford additional unnatural aromatic compounds.[27] The most frequently targeted downstream enzymes for engineered biosynthesis are the different transferases and oxygenases. The results of reshuffling these enzymes are considerably less predictable than the recombination of enzymes from initiation, elongation and immediate tailoring modules. Due to both the number of tailoring enzymes available and the vast variance in substrate structures, no clear programing rule has emerged regarding these enzymes.

One common approach to produce hybrid aromatic compounds involves moving an entire biosynthetic cluster into a heterologous host containing a different set of PKS genes. Heterologous, compatible interactions may yield hybrid products that exhibit features of both parent compounds. Since a large number of enzymes are involved, this approach may actually yield numerous unnatural compounds produced at yields too low for characterization. When the entire tetracenomycin gene cluster was introduced into the urdamycin A producer of *S. fradiae* Tü2717, the novel antibiotic 6-hydroxytetracenomycin was produced.[108] The hydroxylation of tetracenomycin had apparently been the action of an urdamycin oxygenase, which was believed to catalyze the hydroxylation of C12 during urdamycin biosynthesis. Clearly, cross reactivity of the oxygenase toward the unnatural tetracenomycin scaffold could not be predicted *a priori*. Similarly, when a fragment of the mithramycin (*mtm*) PKS was introduced into the tetracenomycin producer *Streptomyces glaucescens*, the fortuitous action of a putative mithramycin fourth-ring cyclase led to the synthesis of novel tetracenomycin M.[109] The apparent compatibility between components of the *mtm* and *tcm* biosynthetic pathways led to the further experiment in which *tcm* oxygenases were expressed in the mithramycin producer *Streptomyces argillaceus*. The heterologous interactions resulted in the biosynthesis of novel decaketide premithramycinone H.[110]

Rational recombination of heterologous tailoring enzymes has been successful in the engineered biosynthesis of natural product analogs. Parts of the mithramycin intermediate methylpremithramycinone and the urdamycin aglycon aquayamicin share a similar aromatic scaffold. It was hypothesized that the urdamycin glycotransferase, UrdGT2, which naturally transfers D-olivose to C9 of aquayamicin, could also act on an analogous positon of methylpremithramycinone. Indeed, replacement of the *mtm* glycotransferases and methyltransferases in *S. argillaceus* with UrdGT2 yields the predicted compound 9-*C*-olivosylpremithamycinone, as well as the unanticipated 9-*C*-mycarosylpremithramycinone.[111] Addition of a second heterologous glycotransferase LanGT1 from the landomycin A PKS inserts a second D-olivose sugar to yield 9-*C*-diolivosylpremithramycinone. These studies showed that selected glycotransferases show substantial substrate tolerance toward both the sugar moiety and the polyketide aglycon.

The sugar components are important for the biological activities of polyketides. The mono-, di- and oligosaccharides are generally involved in molecular recognition of the cellular target and are therefore candidates for combinatorial biosynthesis. For example, biosynthesis of the clinically important antitumor drug daunorubicin requires the addition of daunosamine to the ε-rhodomycinone aglycon. Epirubicin is a semisynthetic analog of daunorubicin and has higher antitumor activity and lower cardiac cytotoxicity. Epirubicin has an epimerized 4-epidaunosamine sugar moiety in place of daunosamine. To produce epirubicin biosynthetically, 4-ketohexose reductases from avermectin and erythromycin PKSs that reduce the 4-keto functionality with opposite stereospecificity were introduced into the daunorubicin producer. The hybrid sugar biosynthetic pathways correctly generated the desired 4-epidaunosamine and the downstream daunorubicin enzymes were able to utilize the unnatural sugar and synthesize the targeted 4′-epirubicin.[112]

4 Type III Polyketide Synthase

4.1 Type III PKS Consists of a Homodimeric Ketosynthase

Type III PKSs are found widely among plants and bacteria.[12] They consist of a homodimer ketosynthase that iteratively condenses malonyl-CoA to give relatively smaller aromatic polyketides than those discussed in the last section (Figure 10). Type III PKS differs from Type II PKS in two important aspects: (1) the KS directly recruits malonyl-CoA in the absence of an acyltransferase or an ACP; and (2) the KS active site catalyzes decarboxylative condensation and chain elongation, *and also* defines the regioselectivity of intramolecular cyclization. The relatively simple mode of chain assembly has made Type III PKSs attractive targets for engineered biosynthesis.

Polyketide synthesis by a Type III PKS involves three steps. First, an appropriate primer unit in the form of acyl-CoA is recruited. The acyl group is transferred to the

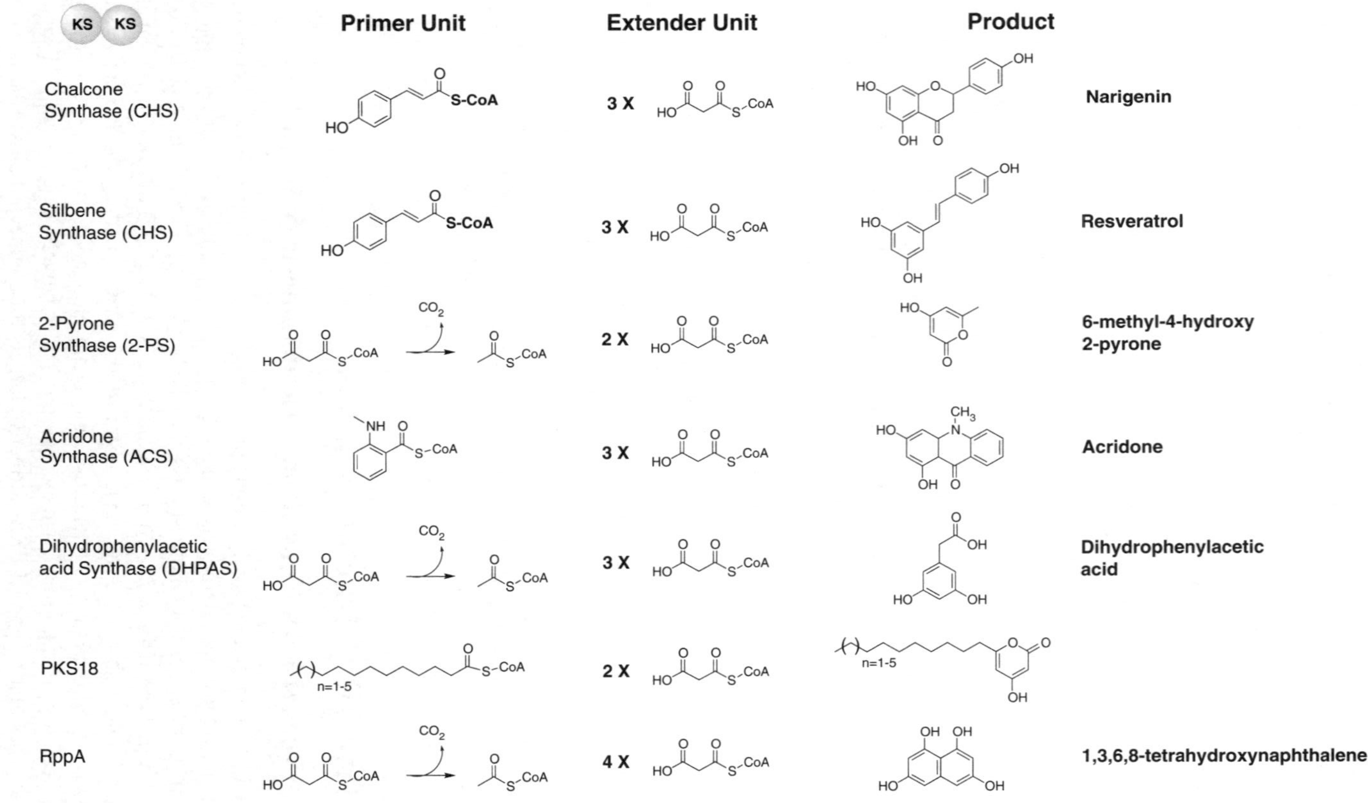

Figure 10 *Type III PKSs and their products*

active-site cysteine of KS and primes the synthase for condensation. Second, the KS iteratively condenses malonyl-CoA to yield a full-length polyketide.[113] The number of chain extending cycles can range from two[114] to six,[115] depending on both the size of the starter unit and the size of the KS active site. Lastly, the polyketide is cyclized within the KS and released. The regiospecificity of the cyclization reactions are largely determined by both the electrostatic environment of the active site as well as the orientation of the substrate.[116] Tremendous insights into the mechanism of Type III PKS have been gathered through the structural elucidation of several KS dimers.[114,116–120] These insights have allowed rational mutagenesis of the KS to afford unnatural compounds.

The best-characterized Type III PKSs are the chalcone synthase (CHS)[116] and stilbene synthase (STS)[120] found in higher plants (Figure 10). Chalcone is a key metabolic precursor that can be converted to important plant metabolites such as flavanone, flavanol and anthocyanin. STS synthesizes the antifungal and antioxidant compound resveratrol. Both CHS and STS use *p*-coumaroyl-CoA as the primer unit and condense three malonyl-CoAs to yield an identical tetraketide thioester bound to the active-site cysteine. The two synthases cyclize the tetraketide with different regioselectivity: CHS catalyzes the C6–C1 Claisen condensation to yield naringenin chalcone, while STS catalyzes the C2–C7 intramolecular aldol condensation to yield resveratrol. Comparing the crystal structures of the alfalfa CHS[116] and Scots pine STS[120] revealed subtle structural differences between the two enzymes. The CHS was converted into an efficient STS when STS-specific residues were introduced at its active site. The difference in cyclization patterns between the two enzymes was attributed to a unique "aldol switch" mechanism that allows the STS to cleave the enzyme-bound thioester prior to cyclization. The resulting free tetraketide with a terminal acid group preferentially undergoes an intramolecular aldol reaction instead of the Claisen condensation observed in CHS.

The 2-pyrone synthase (2-PS) found in several plant species is the simplest Type III PKS.[114] 2-PS condenses three malonyl-CoAs and yields a spontaneously cyclized triketide lactone (TKL). The first malonyl-CoA is decarboxylated and serves as the acetate primer, analogous to the decarboxylative mechanism observed in Type II PKSs. The 2-PS is unable to accept larger starter units due to a considerable smaller active site than that of CHS. Mutating the residues that surround the CHS-binding pocket to those found in 2-PS yielded a mutant CHS that is biochemically indistinguishable from 2-PS.[114]

Several important chalcone synthase-related PKSs have recently been discovered in bacteria.[121] RppA is a Type III PKS that synthesizes 1,3,6,8-tetrahydroxynaphthalene (THN) from five malonyl-CoAs in *Streptomyces griseus*.[122] RppA specifically cyclizes the pentaketide intermediate via two intramolecular condensations to yield the bicyclic THN. The crystal structure of a closely related THN synthase (THNS) from *S. coelicolor* was recently determined, uncovering the presence of a second active-site cysteine that stablizes the growing polyketide and prevents aberrant cyclization of the highly reactive pentaketide.[119]

PKS18 and PKS11 are Type III PKSs found in *Mycobacterium tuberculosis* and catalyze two rounds of chain elongation using long-chain acyl-CoAs such as lauroyl- and palmitoyl-CoAs as starter units.[123] The products of these novel PKSs are long-chain α-pyrones that may contribute to the virulence properties of *M. tuberculosis*. The crystal structure of PKS18 has been resolved as well, highlighting an unusually long active-site tunnel that accommodates the fatty acyl CoAs.[124] Rational mutagenesis to introduce larger residues around the pocket, such as phenylalanine, prohibited PKS18 from utilizing lauroyl and palmitoyl acyl-CoAs, while allowing efficient extension with medium-chain-length hexanoyl and octanoyl-CoAs as starting units.

Another important Type III PKS is the dihydroxyphenylacetic acid synthase (DHPAS) encoded by the balhimycin gene cluster in *Amycolatopsis mediterranei*.[125] DHPA is the precursor of the nonproteinogenic amino acid *S*-3,5-dihydroxyphenylglycine (DHPG), which is found in vancomycin-related antibiotics. DHPAS condenses four malonyl-CoAs to yield an acetate-primed tetraketide, and catalyze the C6–C1 Claisen condensation observed in CHS.

4.2   Engineered Biosynthesis of Type III Polyketides

Unnatural polyketides have been synthesized using techniques such as precursor-directed synthesis and rational mutagenesis. Most of the experiments performed to date have been *in vitro* using purified enzymes (sometimes at millimolar concentrations) and substrates (at concentrations unobtainable under *in vivo* conditions). Results from *in vitro* experiments, while providing snapshots of the substrate tolerance and catalytic turnover of the (mutant) synthases, are not accurate representation of the biosynthetic potential of Type III PKSs *in vivo*. This is largely due to the limiting concentration of both the enzyme and the necessary unnatural acyl substrates *in vivo*.

The Type III PKS can be engineered to produce shunt products. Shunt products are aberrantly cyclized polyketides. The wild-type CHS synthesize coumaroyltriacetic acid lactone (CTAL) as a minor shunt product, which is a result of spontaneous α-pyrone formation instead of the KS-controlled C6–C1 Claisen condensation. The ratio of CTAL to naringenin can be increased up to threefold by a single amino acid substitution (G256A or G256V) at the active site of CHS.[118] The mutation alters the orientation of the tetraketide intermediate at the active site and favors α-pyrone formation. Shunt products were also produced by the THNS containing a single mutation at the second active-site cysteine (C106S). The cysteine was hypothesized to stabilize the polyketide intermediate prior to controlled cyclization. Loss of stabilization led to the exclusive synthesis of triacetic acid lactone (TAL) instead of the wild-type product flaviolin.[119]

Altering the starter unit can also change the orientation and size of the polyketide and lead to alternatively cyclized compounds. Type III PKSs incorporate an array of alkyl- and aryl-CoA starter units. The CHS KS displays intrinsic promiscuity toward various starter units *in vitro*. In addition to the cognate *p*-coumaroyl-CoA, additional aryl-CoAs, such as *m*-coumaroyl-, 3-phenylpropionyl-, phenylacetyl-CoA are able to prime the KS and yield products.[126] CHS,[127] STS[128] and RppA[122] are able to accept alkylacyl-CoA ranging from acetate (C2) to tetradecanoate (C14) to initiation polyketide elongation. Depending on the size of the starter unit, shunt products derived from two, three and five rounds of chain elongation can be obtained. Analogs of the triketide lactone and tetraketide CTAL are detected when unnatural starter units are supplied. *Para*-substituted courmaroyl-CoAs, including *p*-F-coumaroyl-, *p*-Cl-coumaroyl, *p*-Br-coumaroyl and *p*-methoxycoumaroyl-CoA, were all able to prime the *Scutellaria baicalensis* CHS *in vitro*.[127] While only the fluorinated coumaroyl-CoA yielded an analog of naringenin, the bulkier substitutions yielded only analogs of bis-noryangonin and CTAL, presumably due to altered orientation of the substrate in the CHS active site. Interestingly, the same CHS was able to accept an unnatural starter unit in which the coumaroyl ring was replaced by thiophene or furan, and catalyze the synthesis of the corresponding naringenin analogs.[127]

The alfalfa CHS has been rationally engineered to accept larger substrates.[126] The wild-type CHS cannot be primed by the bulkier N-methylanthraniloyl-CoA, which is the primer unit used by the acridone synthase (ACS) in the biosynthesis of 1,3-dihydroxy-N-methylacridone. To compensate for the increase in substrate size, two phenylalanine residues (Phe265 and Phe215) located at the entrance of the CHS active site were mutated to smaller residues. The single mutant CHS (F215S) was able to accept N-methylanthraniloyl-CoA with a k_{cat}/K_m similar to that of ACS, and turned over naringenin with ~200-fold lower efficiency than that of wild-type CHS. The CHS (F215S) synthesized the N-methylanthranilonate-primed unnatural product N-methylanthraniloyltriacetic acid lactone.

Methylmalonyl-CoA has been examined as an extender unit for Type III PKSs.[122,129] In the absence of the cognate substrate malonyl-CoA, p-coumaroyl-primed CHS, STS and benzalacetone synthase (BAS) were able to perform one round of chain extension using methylmalonyl-CoA to yield the unnatural polyketide 1-(4-hydroxyphenyl)pent-1-en-3-one. CHS and STS was also able to perform up to two additional rounds of elongation using methylmalonyl-CoA, although the turnover is extremely slow.[129] The RppA KS catalyzed a one-step decarboxylative condensation using methylmalonyl-CoA when primed with a diketide starter unit acetoacetyl-CoA.[122] The ability of Type III PKSs to utilize C_α-substituted extender units, albeit with poor efficiency, is in contrast to the stringent substrate specificity toward malonate displayed by Type II minimal PKSs. Succinyl-CoA was also investigated as a potential extender unit. No product formation was observed.[129]

5   Conclusions

Polyketides are the most valuable natural products in the treatment of many human diseases. Engineering polyketide diversity through rational and combinatorial biosynthesis has therefore been, and will continue to be, an important technological advance in the development of new therapeutics, including antibiotics that can combat current modes of bacterial resistance and cancer drugs that have novel antitumor properties. The last decade has seen the accumulation of a large knowledge base for polyketide biosynthesis, aided by the sequencing of a vast array of gene clusters and the development of versatile genetic tools. Design rules have been devised enabling regioselective and stereoselective modification of all three major types of polyketides. The fundamental mechanisms (repeated Claisen-like condensation) and stages (initiation, elongation and termination) of polyketide assembly are analogous across different synthase platforms. An attractive route to additional diversity in polyketide structures and biological activities is through reprograming of functional, hybrid synthases that adopt biochemical features from all three different types of polyketide synthases. In addition, synergistic collaboration between chemical synthesis and biological synthesis will allow access to novel areas of chemical space and generate hitherto unanticipated "unnatural" natural products.

Acknowledgments

Research on polyketide biosynthesis in the author's laboratories is supported by grants from the National Institutes of Health.

References

1. R.L. Monaghan and J.S. Tkacz, *Annu. Rev. Microbiol.*, 1990, **44**, 271.
2. S. Jennewein and R. Croteau, *Appl. Microbiol. Biotechnol.*, 2001, **57**, 13.
3. C.R. Hutchinson, *Curr. Opin. Microbiol.*, 1998, **1**, 319.
4. D.E. Cane, C.T. Walsh and C. Khosla, *Science*, 1998, **282**, 63.
5. D.O'Hagan, *The Polyketide Metabolites*, Ellis Howard, Chichester, UK, 1991.
6. D.A. Hopwood, *Chem. Rev.*, 1997, **97**, 2465.
7. R.J. Heath and C.O. Rock, *Nat. Prod. Rep.*, 2002, **19**, 581.
8. D.L. Stassi, S.J. Kakavas, K.A. Reynolds, G. Gunawardana, S. Swanson, D. Zeidner, M. Jackson, H. Liu, A. Buko and L. Katz, *Proc. Natl. Acad. Sci. USA*, 1998, **95**, 7305.
9. Y. Kato, L. Bai, Q. Xue, W.P. Revill, T.W. Yu and H.G. Floss, *J. Am. Chem. Soc.*, 2002, **124**, 5268.
10. B.J. Rawlings, *Nat. Prod. Rep.*, 1997, **14**, 523.
11. B.J. Rawlings, *Nat. Prod. Rep.*, 1999, **16**, 425.
12. M.B. Austin and J.P. Noel, *Nat. Prod. Rep.*, 2003, **20**, 79.
13. J. Cortes, K.E. Wiesmann, G.A. Roberts, M.J. Brown, J. Staunton and P.F. Leadlay, *Science*, 1995, **268**, 1487.
14. J.M. Weber, J.O. Leung, S.J. Swanson, K.B. Idler and J.B. McAlpine, *Science*, 1991, **252**, 114.
15. M. Oliynyk, M.J. Brown, J. Cortes, J. Staunton and P.F. Leadlay, *Chem. Biol.*, 1996, **3**, 833.
16. L. Tang, S. Shah, L. Chung, J. Carney, L. Katz, C. Khosla and B. Julien, *Science*, 2000, **287**, 640.
17. R. Zirkle, J.M. Ligon and I. Molnar, *Microbiology*, 2004, **150**, 2761.
18. R. McDaniel, S. Ebert-Khosla, D.A. Hopwood and C. Khosla, *Science*, 1993, **262**, 1546.
19. B.A. Pfeifer, S.J. Admiraal, H. Gramajo, D.E. Cane and C. Khosla, *Science*, 2001, **291**, 1790.
20. C.M. Starks, Y. Zhou, F. Liu and P.J. Licari, *J. Nat. Prod.*, 2003, **66**, 1313.
21. C. Khosla, *Chem. Rev.*, 1997, **97**, 2577.
22. S. Donadio, M.J. Staver, J.B. McAlpine, S.J. Swanson and L. Katz, *Science*, 1991, **252**, 675.
23. R.S. Gokhale, D. Hunziker, D.E. Cane and C. Khosla, *Chem. Biol.*, 1999, **6**, 117.
24. Z. Hu, B.A. Pfeifer, E. Chao, S. Murli, J. Kealey, J.R. Carney, G. Ashley, C. Khosla and C.R. Hutchinson, *Microbiology*, 2003, **149**, 2213.
25. Y. Xue, L. Zhao, H.W. Liu and D.H. Sherman, *Proc. Natl. Acad. Sci. USA*, 1998, **95**, 12111.
26. Y.Q. Cheng, G.L. Tang and B. Shen, *Proc. Natl. Acad. Sci. USA*, 2003, **100**, 3149.

27. U. Rix, C. Fischer, L.L. Remsing and J. Rohr, *Nat. Prod. Rep.*, 2002, **19**, 542.
28. R.S. Gokhale, S.Y. Tsuji, D.E. Cane and C. Khosla, *Science*, 1999, **284**, 482.
29. R.S. Gokhale and C. Khosla, *Curr. Opin. Chem. Biol.*, 2000, **4**, 22.
30. J. Cortes, S.F. Haydock, G.A. Roberts, D.J. Bevitt and P.F. Leadlay, *Nature*, 1990, **348**, 176.
31. J. Lau, D.E. Cane and C. Khosla, *Biochemistry*, 2000, **39**, 10514.
32. G.F. Liou, J. Lau, D.E. Cane and C. Khosla, *Biochemistry*, 2003, **42**, 200.
33. S.C. Tsai, L.J. Miercke, J. Krucinski, R. Gokhale, J.C. Chen, P.G. Foster, D.E. Cane, C. Khosla and R.M. Stroud, *Proc. Natl. Acad. Sci. USA*, 2001, **98**, 14808.
34. S. Donadio and L. Katz, *Gene*, 1992, **111**, 51.
35. R. Pieper, S. Ebert-Khosla, D. Cane and C. Khosla, *Biochemistry*, 1996, **35**, 2054.
36. R. McDaniel, C.M. Kao, H. Fu, P. Hevezi, C. Gustafsson, M. Betlach, G. Ashley, D.E. Cane and C. Khosla, *J. Am. Chem. Soc.*, 1997, **119**, 4309.
37. R. McDaniel, C.M. Kao, S.J. Hwang and C. Khosla, *Chem. Biol.*, 1997, **4**, 667.
38. C.M. Kao, M. McPherson, R. McDaniel, H. Fu, D.E. Cane and C. Khosla, *J. Am. Chem. Soc.*, 1998, **120**, 2478.
39. C.M. Kao, R. McDaniel, R. McDaniel, H. Fu, D. Cane and C. Khosla, *J. Am. Chem. Soc.*, 1997, **119**, 11339.
40. D. Bedford, J.R. Jacobsen, G. Luo, D.E. Cane and C. Khosla, *Chem. Biol.*, 1996, **3**, 827.
41. L. Liu, A. Thamchaipenet, H. Fu, M. Betlach and G. Ashley, *J. Am. Chem. Soc.*, 1997, **119**, 10553.
42. C. Khosla, R.S. Gokhale, J.R. Jacobsen and D.E. Cane, *Annu. Rev. Biochem.*, 1999, **68**, 219.
43. C.D. Reeves, S. Murli, G.W. Ashley, M. Piagentini, C.R. Hutchinson and R. McDaniel, *Biochemistry*, 2001, **40**, 15464.
44. R. Regentin, L. Cadapan, S. Ou, S. Zavala and P. Licari, *J. Ind. Microbiol. Biotechnol.*, 2002, **28**, 12.
45. R. McDaniel, A. Thamchaipenet, C. Gustafsson, H. Fu, M. Betlach and G. Ashley, *Proc. Natl. Acad. Sci. USA*, 1999, **96**, 1846.
46. P. Kumar, A.T. Koppisch, D.E. Cane and C. Khosla, *J. Am. Chem. Soc.*, 2003, **125**, 14307.
47. C.M. Kao, L. Katz and C. Khosla, *Science*, 1994, **265**, 509.
48. C.M. Kao, G. Luo, C.M. Kao, D.E. Cane and C. Khosla, *J. Am. Chem. Soc.*, 1995, **117**, 9105.
49. C.M. Kao, G. Luo, L. Katz, D.E. Cane and C. Khosla, *J. Am. Chem. Soc.*, 1994, **116**, 11612.
50. C.D. Reeves, S.L. Ward, W.P. Revill, H. Suzuki, M. Marcus, O.V. Petrakovsky, S. Marquez, H. Fu, S.D. Dong and L. Katz, *Chem. Biol.*, 2004, **11**, 1465.
51. K. Watanabe, C.C. Wang, C.N. Boddy, D.E. Cane and C. Khosla, *J. Biol. Chem.*, 2003, **278**, 42020.
52. G.F. Liou and C. Khosla, *Curr. Opin. Chem. Biol.*, 2003, **7**, 279.
53. S. Kuhstoss, M. Huber, J.R. Turner, J.W. Paschal and R.N. Rao, *Gene*, 1996, **183**, 231.
54. A.F. Marsden, B. Wilkinson, J. Cortes, N.J. Dunster, J. Staunton and P.F. Leadlay, *Science*, 1998, **279**, 199.

55. J.R. Jacobsen, C.R. Hutchinson, D.E. Cane and C. Khosla, *Science*, 1997, **277**, 367.

56. C. Carreras, S. Frykman, S. Ou, L. Cadapan, S. Zavala, E. Woo, T. Leaf, J. Carney, M. Burlingame, S. Patel, G. Ashley and P. Licari, *J. Biotechnol.*, 2002, **92**, 217.

57. R.P. Desai, E. Rodriguez, J.L. Galazzo and P. Licari, *Biotechnol. Prog.*, 2004, **20**, 1660.

58. K. Kinoshita, P.G. Williard, C. Khosla and D.E. Cane, *J. Am. Chem. Soc.*, 2001, **123**, 2495.

59. C.N. Boddy, K. Hotta, M.L. Tse, R.E. Watts and C. Khosla, *J. Am. Chem. Soc.*, 2004, **126**, 7436.

60. C. Olano, B. Wilkinson, C. Sanchez, S.J. Moss, R. Sheridan, V. Math, A.J. Weston, A.F. Brana, C.J. Martin, M. Oliynyk, C. Mendez, P.F. Leadlay and J.A. Salas, *Chem. Biol.*, 2004, **11**, 87.

61. C. Olano, B. Wilkinson, S.J. Moss, A.F. Brana, C. Mendez, P.F. Leadlay and J.A. Salas, *Chem. Commun.*, 2003, **22**, 2780.

62. J. Piel, *Proc. Natl. Acad. Sci. USA*, 2002, **99**, 14002.

63. L. Hendrickson, C.R. Davis, C. Roach, D.K. Nguyen, T. Aldrich, P.C. McAda and C.D. Reeves, *Chem. Biol.*, 1999, **6**, 429.

64. Y. Abe, T. Suzuki, C. Ono, K. Iwamoto, M. Hosobuchi and H. Yoshikawa, *Mol. Genet. Genomics*, 2002, **267**, 636.

65. J. Kennedy, K. Auclair, S.G. Kendrew, C. Park, J.C. Vederas and C.R. Hutchinson, *Science*, 1999, **284**, 1368.

66. B.J. Rawlings, *Nat. Prod. Rep.*, 1999, **16**, 425.

67. M.J. Butler, C. Binnie, I.S. Hunter, D.A. Sugden and M. Warren, *Dev. Ind. Microbiol.*, 1990, **31**, 41.

68. L. Han, K. Yang, E. Ramalingam, R.H. Mosher and L.C. Vining, *Microbiology*, 1994, **140** (12), 3379.

69. R. McDaniel, S. Ebert-Khosla, D.A. Hopwood and C. Khosla, *Nature*, 1995, **375**, 549.

70. J. Dreier, A.N. Shah and C. Khosla, *J. Biol. Chem.*, 1999, **274**, 25108.

71. R.G. Summers, A. Ali, B. Shen, W.A. Wessel and C.R. Hutchinson, *Biochemistry*, 1995, **34**, 9389.

72. A.T. Keatinge-Clay, D.A. Maltby, K.F. Medzihradszky, C. Khosla and R.M. Stroud, *Nat. Struct. Mol. Biol.*, 2004, **11**, 888.

73. C. Bisang, P.F. Long, J. Cortes, J. Westcott, J. Crosby, A.L. Matharu, R.J. Cox, T.J. Simpson, J. Staunton and P.F. Leadlay, *Nature*, 1999, **401**, 502.

74. Y. Tang, S.C. Tsai and C. Khosla, *J. Am. Chem. Soc.*, 2003, **125**, 12708.

75. A.T. Koppisch and C. Khosla, *Biochemistry*, 2003, **42**, 11057.

76. C.W. Carreras and C. Khosla, *Biochemistry*, 1998, **37**, 2084.

77. B. Shen and C.R. Hutchinson, *Science*, 1993, **262**, 1535.

78. T.S. Lee, C. Khosla and Y. Tang, *J. Am. Chem. Soc.*, 2005, **127**, 12254.

79. C.R. Hutchinson, *Chem. Rev.*, 1997, **97**, 2525.

80. I. Chopra, P.M. Hawkey and M. Hinton, *J. Antimicrob. Chemother.*, 1992, **29**, 245.

81. B.S. Moore, C. Hertweck, J.N. Hopke, M. Izumikawa, J.A. Kalaitzis, G. Nilsen, T.O'Hare, J. Piel, P.R. Shipley, L. Xiang, M.B. Austin and J.P. Noel, *J. Nat. Prod.*, 2002, **65**, 1956.

82. T. Marti, Z.H. Hu, N.L. Pohl, A.N. Shah and C. Khosla, *J. Biol. Chem.*, 2000, **275**, 33443.

83. E.S. Meadows and C. Khosla, *Biochemistry*, 2001, **40**, 14855.

84. Y. Tang, A.T. Koppisch and C. Khosla, *Biochemistry*, 2004, **43**, 9546.

85. R. McDaniel, S. Ebert-Khosla, H. Fu, D.A. Hopwood and C. Khosla, *Proc. Natl. Acad. Sci. USA*, 1994, **91**, 11542.

86. R. McDaniel, S. Ebert-Khosla, D.A. Hopwood and C. Khosla, *J. Am. Chem. Soc.*, 1994, **116**, 10855.

87. L. Prado, E. Fernandez, U. Weissbach, G. Blanco, L.M. Quiros, A.F. Brana, C. Mendez, J. Rohr and J.A. Salas, *Chem. Biol.*, 1999, **6**, 19.

88. H. Fu, D.A. Hopwood and C. Khosla, *Chem. Biol.*, 1994, **1**, 205.

89. B. Shen and C.R. Hutchinson, *Proc. Natl. Acad. Sci. USA*, 1996, **93**, 6600.

90. R. McDaniel, S. Ebert-Khosla, D.A. Hopwood and C. Khosla, *J. Am. Chem. Soc.*, 1993, **115**, 11671.

91. T.W. Yu, Y. Shen, R. McDaniel, H.G. Floss, C. Khosla, D. Hopwood and B.S. Moore, *J. Am. Chem. Soc.*, 1998, **120**, 7749.

92. Y. Shen, P. Yoon, T.W. Yu, H.G. Floss, D. Hopwood and B.S. Moore, *Proc. Natl. Acad. Sci. USA*, 1999, **96**, 3622.

93. M. Metsa-Ketela, K. Palmu, T. Kunnari, K. Ylihonko and P. Mantsala, *Antimicrob. Agents. Chemother.*, 2003, **47**, 1291.

94. G. Blanco, H. Fu, C. Mendez, C. Khosla and J.A. Salas, *Chem. Biol.*, 1996, **3**, 193.

95. H. Fu, S. Ebertkhosla, D.A. Hopwood and C. Khosla, *J. Am. Chem. Soc.*, 1994, **116**, 4166.

96. H. Fu, R. McDaniel, D.A. Hopwood and C. Khosla, *Biochemistry*, 1994, **33**, 9321.

97. N. Lomovskaya, Y. Doi-Katayama, S. Filippini, C. Nastro, L. Fonstein, M. Gallo, A.L. Colombo and C.R. Hutchinson, *J. Bacteriol.*, 1998, **180**, 2379.

98. V.B. Rajgarhia, N.D. Priestley and W.R. Strohl, *Metab. Eng.*, 2001, **3**, 49.

99. J. Kantola, T. Kunnari, A. Hautala, J. Hakala, K. Ylihonko and P. Mantsala, *Microbiology*, 2000, **146** (1), 155.

100. Y. Tang, T.S. Lee and C. Khosla, *PLoS Biol.*, 2004, **2**, 227.

101. R. McDaniel, C.R. Hutchinson and C. Khosla, *J. Am. Chem. Soc.*, 1995, **117**, 6805.

102. P.J. Kramer, R.J. Zawada, R. McDaniel, C.R. Hutchinson, D. Hopwood and C. Khosla, *J. Am. Chem. Soc.*, 1997, **119**, 635.

103. Y. Tang, T.S. Lee, H.Y. Lee and C. Khosla, *Tetrahedron*, 2004, **60**, 7659.

104. M.J. Bibb, D.H. Sherman, S. Omura and D.A. Hopwood, *Gene*, 1994, **142**, 31.

105. T. Bililign, C.G. Hyun, J.S. Williams, A.M. Czisny and J.S. Thorson, *Chem. Biol.*, 2004, **11**, 959.

106. H. Pan, S. Tsai, E.S. Meadows, L.J. Miercke, A.T. Keatinge-Clay, J.O'Connell, C. Khosla and R.M. Stroud, *Structure*, 2002, **10**, 1559.

107. J.A. Kalaitzis, M. Izumikawa, L. Xiang, C. Hertweck and B.S. Moore, *J. Am. Chem. Soc.*, 2003, **125**, 9290.

108. H. Decker, S. Haag, G. Udvarnoki and J. Rohr, *Angew. Chem. Int. Ed. Engl.*, 1995, **34**, 1107.

109. E. Kunzel, S.-E. Wohlert, C. Beninga, S. Haag, H. Decker, C.R. Hutchinson, G. Blanco, C. Mendez, J.A. Salas and J. Rohr, *Chem. Eur. J.*, 1997, **3**, 1675.

110. F. Lombo, E. Kunzel, L. Prado, A.F. Brana, K.U. Bindseil, J. Frevert, D. Bearden, C. Mendez, J.A. Salas and J. Rohr, *Angew. Chem. Int. Ed. Engl.*, 2000, **39**, 796.

111. A. Trefzer, G. Blanco, L. Remsing, E. Kunzel, U. Rix, F. Lipata, A.F. Brana, C. Mendez, J. Rohr, A. Bechthold and J.A. Salas, *J. Am. Chem. Soc.*, 2002, **124**, 6056.

112. K. Madduri, J. Kennedy, G. Rivola, A. Inventi-Solari, S. Filippini, G. Zanuso, A.L. Colombo, K.M. Gewain, J.L. Occi, D.J. MacNeil and C.R. Hutchinson, *Nat. Biotechnol.*, 1998, **16**, 69.

113. J.M. Jez, J.L. Ferrer, M.E. Bowman, R.A. Dixon and J.P. Noel, *Biochemistry*, 2000, **39**, 890.

114. J.M. Jez, M.B. Austin, J. Ferrer, M.E. Bowman, J. Schroder and J.P. Noel, *Chem. Biol.*, 2000, **7**, 919.

115. I. Abe, Y. Utsumi, S. Oguro and H. Noguchi, *FEBS Lett.*, 2004, **562**, 171.

116. J.L. Ferrer, J.M. Jez, M.E. Bowman, R.A. Dixon and J.P. Noel, *Nat. Struct. Biol.*, 1999, **6**, 775.

117. J.M. Jez, M.E. Bowman, R.A. Dixon and J.P. Noel, *Nat. Struct. Biol.*, 2000, **7**, 786.

118. J.M. Jez, M.E. Bowman and J.P. Noel, *Biochemistry*, 2001, **40**, 14829.

119. M.B. Austin, M. Izumikawa, M.E. Bowman, D.W. Udwary, J.L. Ferrer, B.S. Moore and J.P. Noel, *J. Biol. Chem.*, 2004, **279**, 45162.

120. M.B. Austin, M.E. Bowman, J.L. Ferrer, J. Schroder and J.P. Noel, *Chem. Biol.*, 2004, **11**, 1179.

121. N. Funa, Y. Ohnishi, I. Fujii, M. Shibuya, Y. Ebizuka and S. Horinouchi, *Nature*, 1999, **400**, 897.

122. N. Funa, Y. Ohnishi, Y. Ebizuka and S. Horinouchi, *J. Biol. Chem.*, 2002, **277**, 4628.

123. O.A. Trivedi, P. Arora, V. Sridharan, R. Tickoo, D. Mohanty and R.S. Gokhale, *Nature*, 2004, **428**, 441.

124. R. Sankaranarayanan, P. Saxena, U.B. Marathe, R.S. Gokhale, V.M. Shanmugam and R. Rukmini, *Nat. Struct. Mol. Biol.*, 2004, **11**, 894.

125. V. Pfeifer, G.J. Nicholson, J. Ries, J. Recktenwald, A.B. Schefer, R.M. Shawky, J. Schroder, W. Wohlleben and S. Pelzer, *J. Biol. Chem.*, 2001, **276**, 38370.

126. J.M. Jez, M.E. Bowman and J.P. Noel, *Proc. Natl. Acad. Sci. USA*, 2002, **99**, 5319.

127. I. Abe, H. Morita, A. Nomura and H. Noguchi, *J. Am. Chem. Soc.*, 2000, **122**, 11242.

128. I. Abe, T. Watanabe and H. Noguchi, *Phytochemistry*, 2004, **65**, 2447.

129. I. Abe, Y. Takahashi and H. Noguchi, *Org. Lett.*, 2002, **4**, 3623.

Combinatorial Synthetic Design: The Balance of Novelty and Familiarity

A. GANESAN

School of Chemistry, University of Southampton, Southampton SO17 1BJ, United Kingdom

Organic chemists love to make molecules. During much of our history, this activity was target-driven and focused on making compounds one at a time with exact precision. In the last 20 years, combinatorial chemistry has irreversibly altered the traditional paradigm. For the first time, we can rapidly generate thousands to millions of compounds using unsophisticated equipment within the reach of an academic laboratory or small startup company. With this power comes a responsibility: how do we select which compounds we should be making? A number of strategies for library design[*][†] in drug discovery, and their evolution over time, are discussed in this chapter.

1   Biological Macromolecules – Strength in Numbers

Geysen's pioneering peptides on pins in Australia were the first[‡] truly combinatorial libraries to be made. An early example[4] was the identification of an immunogenic epitope for the coat protein VP1 of the foot and mouth disease virus, outbreaks of which can have disastrous effects on a country's livestock. The VP1 protein has 213 amino acid residues, and it is known that epitopes are between five to eight residues in length. Geysen prepared all the possible hexapeptides embedded within VP1 (amino acids 1–6, 2–7, 3–8, *etc.*), and identified an epitope by testing their relative binding to antisera. Epitopes may not be contiguous but comprise amino acids in

---

[*] For recent general reviews on combinatorial libraries and design strategies, see Ref. 1.
[†] For exhaustive compilations on combinatorial libraries, the surveys (now annual) by Roland Dolle are invaluable, see Ref. 2.
[‡] For an engaging personal account from pioneers of combinatorial chemistry including Geysen, see Ref. 3.

close proximity in 3-D space rather than primary sequence. By testing the peptide library against a discontinuous antigen, Geysen was able to identify a linear mimic ("mimotope") of the same.

The next stage of evolution was to prepare all possible peptides of a given length. For a hexapeptide, this represents $20^6 = 64$ million sequences. Obviously, it is logistically impossible to synthesize each of these individually. Even if this were done, nobody would be able to perform 64 million assays. A simple yet ingenious solution, whereby the peptides were prepared as complex mixtures, was devised. The library comprised 400 samples, each containing 20^4 individual peptide sequences in a defined format. A sample that contained a hit could then be resynthesized and tested in cycles of successively narrower complexity in order to yield a single active sequence. Geysen applied this to the discovery[5] of a known epitope for myohemerythrin (Figure 1), using mixtures of amino acids in his library synthesis. Meanwhile, the groups of Houghten and Lam prepared[6] similar hexapeptide libraries

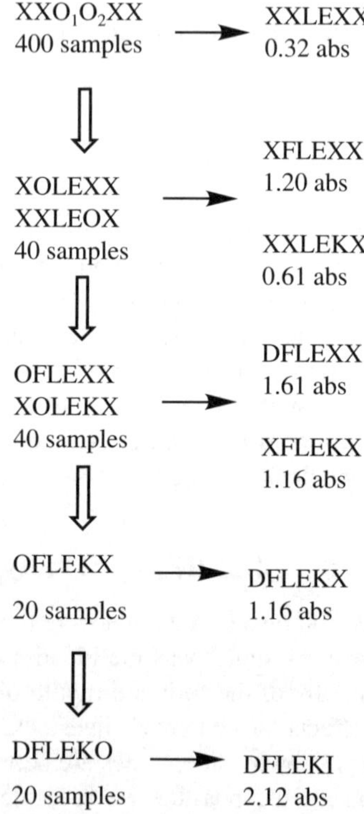

Figure 1 *Identification of a linear hexapeptide epitope by Geysen. In the first cycle, all possible hexapeptides are prepared as sample pools where four positions are randomized mixtures of 20 amino acids (X), and O_1 and O_2 are defined amino acids, leading to 20x20 = 400 samples. The best sequence, XXLEXX, had an ELISA absorbance of 0.32. In the second cycle, the LE was fixed and two other positions varied, leading to the new best sequences XFLEXX and XXLEKX. Further cycles of synthesis and screening led to the epitope as a single sequence*

by using Furka's theoretical split-and-mix concept for combinatorial synthesis, in which each solid-phase bead is treated as an individual microreactor producing a single compound. In screening such mixture libraries, some important assumptions are made, such as the lack of non-additive effects, the independence of one substituent upon another, and the likelihood that the hit will be a local minimum rather than the best molecule present. Even so, the method has proven remarkably successful in finding hits in a variety of drug-discovery applications.

Molecular biologists were already familiar with combinatorial concepts. In applications like error-prone polymerized chain reaction (PCR), oligonucleotides are prepared that contain all four possible bases at a given position. It was a relatively short step from such syntheses to truly combinatorial oligonucleotide libraries. Since the chemistry of oligonucleotide synthesis is simpler than peptide synthesis and less sequence dependent, fairly large oligomers (tens of residues) could be assembled, leading to billion-molecule libraries. These peptide and oligonucleotide libraries, which marked the beginning of combinatorial chemistry, have some unusual features that are worth emphasizing.

1.1　Congruence between Biological and Chemical Space

The goal of the early peptide libraries was to seek epitopes, which are peptides. Similarly, oligonucleotide libraries were meant to identify oligonucleotide sequences. This may seem like restating the obvious, but it means there was no disconnect between the molecules tested and the types of compounds sought after. Nowadays, combinatorial libraries are often used in more abstract ways for drug discovery, without a guarantee that a successful solution exists within the composition of the library. Indeed, a large number of structurally diverse compounds may be screened without ever finding a hit. This is not necessarily a failure of combinatorial chemistry, but a reflection of the imprecision of our knowledge of the biologically active chemical space for a given target.

1.2　The Libraries are Exhaustive within the Defined Boundaries

If a linear hexapeptide solution existed, it must have been present in Geysen's library (provided the synthesis worked perfectly, which is a separate issue). There are no additional unrepresented sequences, and mathematically such libraries are called NP-complete (NP = non-deterministic polynomial time). Such NP-complete libraries are actually quite rare in combinatorial chemistry. Although there are infinite numbers of peptides, for Geysen's epitope mapping, he needed to examine only natural amino acids. This reduces the complexity for a sequence of length n to 20^n, a number that increases exponentially with n but nevertheless remains finite.

Among peptide libraries today, such NP-complete sets are uncommon. It is often profitable to expand the monomer set to include unnatural amino acids, thus once again leading to infinite possibilities. Even within the narrower context of natural peptides, the complete set of 20 amino acids is seldom employed in synthetic libraries. In Houghten's example above, both cysteine and tryptophan were omitted to avoid problems due to oxidative side reactions, while others often substitute methionine by the more inert norvaline. Besides peptides, oligonucleotides are the other main class of NP-complete combinatorial libraries. There are two main reasons for this. Firstly, the diversity of nucleotide building blocks is limited, unlike the case with amino acids. Secondly, although the introduction of unnatural nucleotides increases chemical diversity, this is outweighed by the inability to use PCR for the amplification of hits from complex mixtures.

1.3 Highly Optimized Synthesis Procedures were Available

The preparation of peptides and nucleotides involves the iterative repetition of three simple steps: monomer activation, coupling to the growing chain, and deprotection to unmask a new reactive terminus. The arithmetical demon of linear syntheses, whereby the overall yield decreases exponentially with any loss in yield for each cycle, had forced chemists to perfect these reactions to the point of being practically quantitative. The availability of such robust protocols was a necessary prerequisite for combinatorial chemistry. Outside the peptide and nucleotide areas, synthesis of a library is often far less time consuming than the development of optimized reaction conditions.

2 Oligomer Synthesis – Improving on Mother Nature

Most targets in drug discovery are proteins. It is not surprising that their functions can be modulated by smaller peptide fragments, nor that such compounds can display exceptional affinity and selectivity. However, peptides tend to suffer from disadvantages such as high molecular weight, large numbers of H-bond donors and acceptors, and many rotatable bonds, all of which are factors that are likely to result in poor oral bioavailability. In addition, they are metabolically unstable to hydrolysis by acid or proteases. Nevertheless, for combinatorial chemistry, peptides have obvious advantages in their ease of synthesis. The question then arose if other monomers could be designed that can be easily transformed into oligomers with improved drug-like properties over peptides. This approach has been termed "unnatural biopolymers", which is somewhat of an oxymoron.

A classic example is the peptoid backbone introduced[7] by workers at Chiron. In its later version, the reaction cycle involved[8] amide-bond formation with bromoacetic acid, after which the halide is displaced by a primary amine. The secondary amine is then ready for addition of the next monomer (Figure 2). Alignment of the peptoid backbone against peptides shows how the two differ in their repeating unit. The overall synthetic procedure is straightforward, and many groups have employed peptoid combinatorial libraries. Among leads discovered from such efforts are nanomolar ligands[9] for the α1-adrenergic receptor from Chiron, and recent examples[10] such as a transactivation domain mimic and multidrug resistance reversal agents (Figure 3).

A host of other oligomer-based scaffolds have been described. Often, the scope and limitations are difficult to gauge, as their use is largely restricted to the academic groups that first devised the methodology. Schultz,[11] for example, has reported on the solid-phase synthesis of oligocarbamates. In one application, a library of

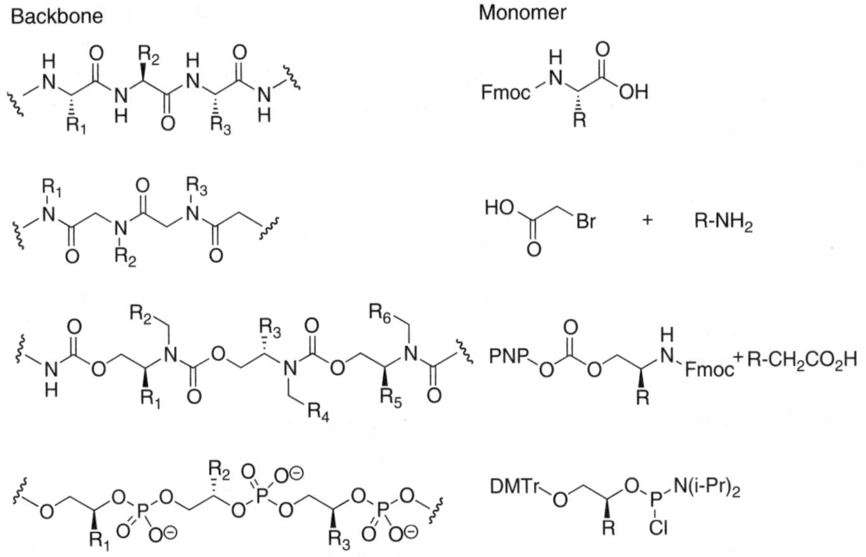

Backbone

Monomer

Figure 2 *The structure of a tripeptide and the corresponding monomer for solid-phase synthesis, contrasted with trimers and monomers for three other oligomer libraries prepared from bifunctional monomers*

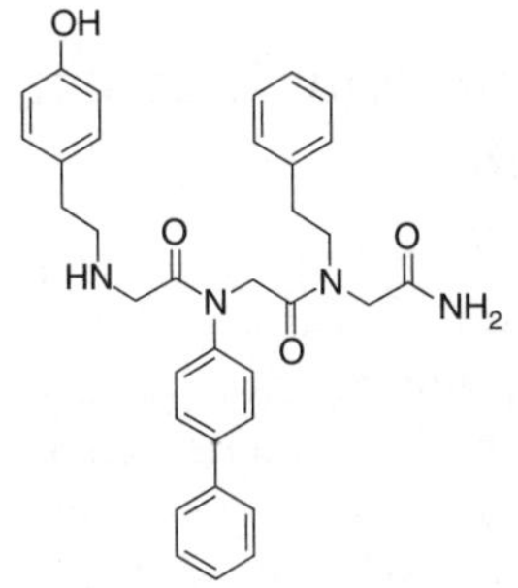

peptoid adrenergic receptor antagonist (Chiron) 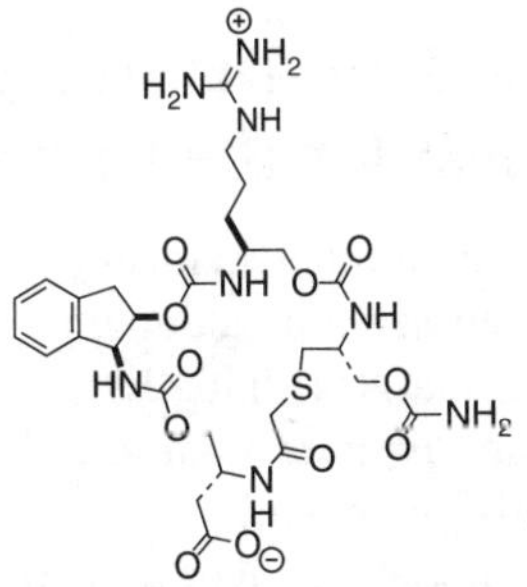　　　oligocarbamate RGD mimic (Schultz)

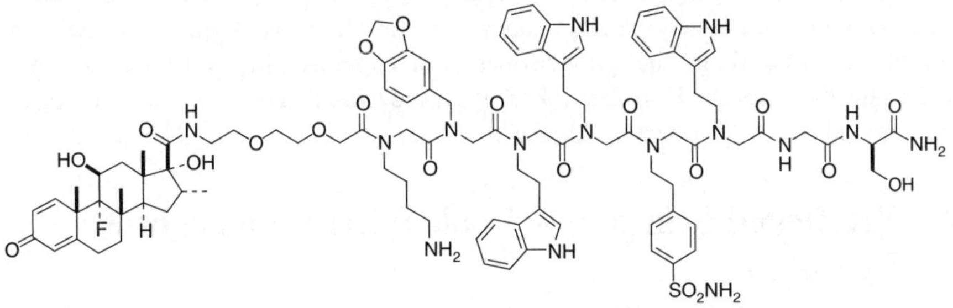

peptoid transactivation domain (Kodadek)

Figure 3 *Three examples of biologically active oligomers discovered from screening peptoid and oligocarbamate combinatorial libraries*

>500,000 cyclic trimer and tetramer oligocarbamates was screened[12] for binding to GPIIb/IIIa. This integrin binds to fibrinogen in a critical step of platelet formation, an interaction mediated by an Arg–Gly–Asp (RGD) sequence. The Schultz library was screened on beads, with positives cleaved and sequenced by mass spectrometry. The best ligand identified predictably contains both the Arg and Asp equivalents of the natural RGD peptide. Nevertheless, this functionality is now displayed on the unnatural oligocarbamate scaffold with a respectable affinity of 3.9 nM, not far off from 1.5 nM for the control ligand G4120 or 2 nM for kistrin, a protein antagonist from snake venom.

We described[13] an unusual oligomer whereby α-amino acids were converted into α-hydroxy acids by oxidative deamination with retention of stereochemistry, followed by reduction to 1,2-diols. The primary alcohol was protected by the dimethoxytrityl group, and the remaining secondary alcohol converted to a phosphitylating agent. These monomers could then be incorporated into oligomers under the standard conditions of automated DNA synthesis. The resulting molecules retain amino acid side-chains, but have the phosphodiester backbone more familiar to nucleic acids.

It is not always clear if any of the new scaffolds are significantly superior to nature's peptides or nucleotides. Just like the latter, all oligomers inherently suffer from the disadvantage of increasingly likely violation of Lipinski's "Rule of Five" or other descriptors of drug-like properties as the number of residues in the sequence increases.

3   Random, Discovery, or Prospecting Libraries – the Quest for the Universal Scaffold

Early combinatorial chemistry was driven by the belief that large compound collections would serve as an engine for lead discovery across a spectrum of unrelated therapeutic targets. Initially, the term "random libraries" was used to distinguish these from "focused libraries" designed for a specific purpose. Random, however, has negative connotations as it implies that the generation of large numbers, regardless of compound type or properties, would suffice. While some might actually have believed this at the outset of combinatorial chemistry, it rapidly became clear that random libraries required as much thought and planning as their focused counterparts. Other terminologies began to appear, such as "universal library", although the composition of a library that provides useful hits against all targets remains as elusive as the quest for the Holy Grail. Perhaps more realistic are "discovery library" or Bartlett's phrase[14] "prospecting library".

4   Privileged Scaffolds – Look Where the Light is Brightest

As is true when panning the world's rivers for gold, only a minute fraction of prospecting libraries can be expected to be successful. For drug-discovery applications, one strategy is to learn from the lessons of the past. Certain ring skeletons are

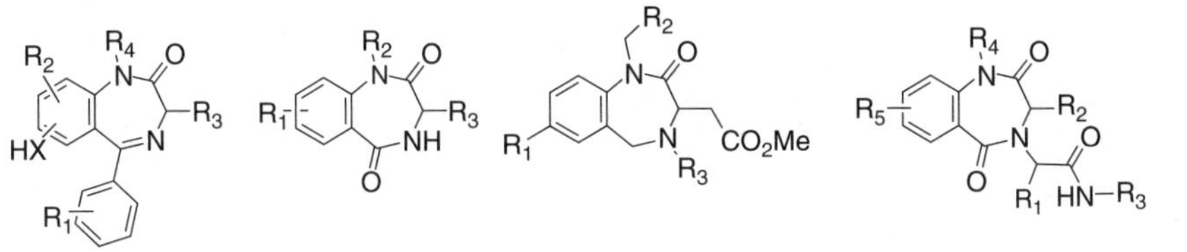

Figure 4 *Four examples of large (>100 member) benzodiazepine combinatorial libraries prepared in academia or industry, by solid- or solution-phase parallel synthesis*

recurrent in drugs, and such privileged scaffolds were originally defined[15] as "… a single molecular framework able to provide ligands for diverse receptors …". Benzodiazepines are the quintessential privileged scaffold, and were the target of the first[16] parallel solid-phase synthesis of a small-molecule non-oligomer library. Although only eight examples were reported, this first publication by Ellman was instrumental in alerting the pharmaceutical industry to the potential of combinatorial chemistry, and numerous benzodiazepine libraries have since appeared[17] (Figure 4). These are undoubtedly the tip of the iceberg, as many others remain unpublished in corporate collections. In the same manner, many other privileged scaffolds are popular targets for discovery libraries. Our group, for example, has prepared libraries based on well-known structural motifs such as β-amino alcohols,[18] hydantoins and thiohydantoins,[19] quinolinones,[20] biaryls,[21] and acyclic guanidines.[22] While it is easy to rationalize the value of such scaffolds based on proven success in past drugs, this is akin to shining a torchlight on the ground, and exploring only the narrow patch that is then illuminated.

5   The Decoration or Synthesis of Novel Scaffolds – Aid for the Underprivileged

Are certain scaffolds privileged because they are heavily explored, or do they inherently possess favourable characteristics for the discovery of new drugs? The two explanations appear equally likely. One of the greatest achievements of combinatorial chemistry is the testing of this hypothesis by the exploration of relatively uncharted chemical space. Thanks to parallel synthesis, enough library members can be made to evaluate the new scaffold with reasonable confidence. In addition to this curiosity-driven search of new scaffolds, another important motivation is the pressing need to source novel and patentable compound classes that do not overlap with those of competitors.

A major constraint in the design of novel scaffolds is synthetic accessibility. Ideally, the library should be assembled in a few robust steps, and two approaches are generally followed. Firstly, a specific core molecule can be chosen as starting point, followed by its decoration by variable groups. Secondly, the core is assembled during the course of the synthesis. Regardless of which strategy is followed, given

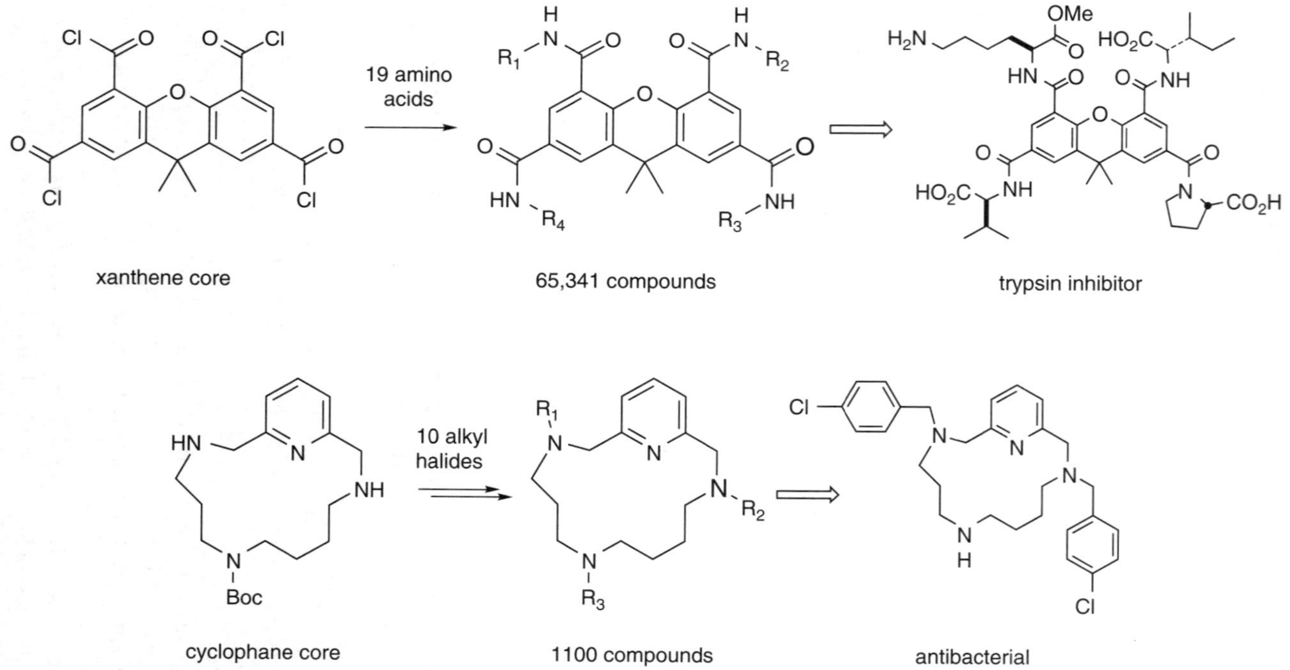

Figure 5 *Examples of scaffold decoration for the solution-phase synthesis of pooled combinatorial libraries from Rebek and Isis. In both cases, biologically active compounds were identified*

the size of the global combinatorial chemistry community, many groups have converged on the same scaffold either independently or following up on results from one another.

A classic illustration of scaffold decoration is the trisubstituted 1,3,5-triazine. The starting material trichloro-1,3,5-triazine is inexpensive, and the halogens can be displaced by nucleophilic aromatic substitutions one by one. Such chemistry was well precedented in pre-combinatorial days, and used on a large scale for the synthesis of "colour-fast" reactive dyes. The overall reaction sequence has an appeal in its simplicity, and both academic and industrial practitioners have reported a steady trickle of such triazine-based libraries over the last 20 years. Novelty will come either from the particular set of nucleophiles employed or the assay targets.

More unique are the xanthene and cubane cores (Figure 5) chosen[23] by Rebek for the solution-phase synthesis of highly pooled combinatorial libraries in a single step. By iterative deconvolution, a thrombin inhibitor was identified from this complex collection. Similarly, workers at Isis reported[24] the functionalization of a novel polyazacyclophane core, which led to the discovery of antibacterial leads. Clearly, in both these cases, the hits are structurally unrelated to common drug-discovery scaffolds.

As an example of scaffold synthesis rather than decoration, we adapted the Lacey–Dieckmann synthesis of tetramic acids to a combinatorial strategy (Figure 6). The overall sequence is only three steps, and all the three reaction inputs involve readily available building blocks: amino acids, aldehydes, and carboxylic acids. We published[25] both solution- and solid-phase protocols, and this was followed soon thereafter[26] by similar reports from industrial groups. Clearly, the ability to access this heterocycle with ample room for diversification was highly attractive, and each of us had independently developed this idea.

Early combinatorial libraries steered towards high molecular weight and lipophilic tendencies. A recent example[27] from my group illustrates a deliberate attempt to avoid such problem structures (Figure 7). The triazinedione skeleton has

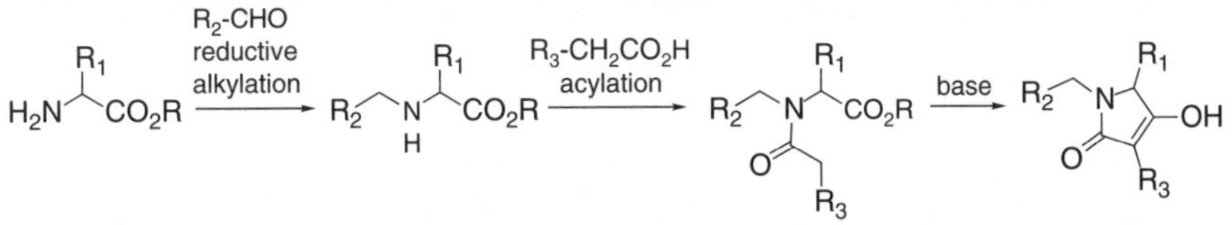

Figure 6 *A three-step sequence leading to tetramic acid libraries. Both parallel solution-phase (R = Me, Et) and solid-phase (R = polystyrene resin) have been reported*

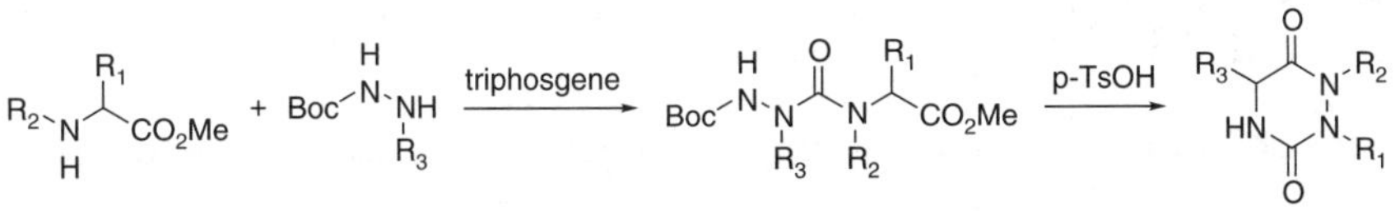

Figure 7 *Synthesis of a triazinedione "underprivileged" scaffold*

similarities to a diketopiperazine, but is significantly less common, with only 13 examples in the SciFinder database. The parent heterocycle has a molecular weight of only 115, giving plenty of room for introducing three variable substituents without exceeding a molecular weight of 500. Should one or more of these be hydrophobic, it is partially compensated by making the core heteroatom rich with three nitrogens, two oxygens, and containing H-bond donors and acceptors.

6 Target Class Libraries – Diversity with a Purpose

Perhaps the most successful lead-discovery libraries are those aimed at a specific target class. Rather than treat targets as a black box at which to throw diverse molecules, it is more profitable to bin them into families according to structural or functional homology. Combinatorial libraries can then be specifically designed to interact with the target class, in an approach that yields a significantly higher hit rate than purely "random" discovery libraries. Kinase inhibitors are a prime example, and there are many array syntheses of heterocyclic libraries designed to bind to the cofactor ATP-binding pocket in the enzyme. Schultz has been particularly active[28] in this area, preparing a series of substituted purines by solid- and solution-phase. Once these compounds are in hand, there may be additional bonuses (Figure 8). The sulfotransferases have a similar cofactor, and screening the kinase library by Bertozzi's group led to the identification[29] of micromolar inhibitors. More recently, a purine was shown[30] to cause the dedifferentiation of a committed mouse myogenic cell lineage.

Given their importance as drug-discovery targets, G-protein coupled receptors (GPCRs) are another target class that is intensively investigated[31] by combinatorial libraries. Among the privileged motifs for GPCRs are 2-arylindoles, and Merck has reported[32] the solid-phase synthesis of a 128,000-member mixture library, as 320 samples containing 400 compounds each. Screening against 16 GPCRs followed by deconvolution led to the discovery of a number of selective ligands (Figure 9). Another library from Roche,[33] albeit on a smaller scale and involving discrete compounds, explored the spiropiperidine as a privileged GPCR motif in combination with the 3,5-bis(trifluoromethyl)phenyl moiety for selective interaction with NK-1 receptors.

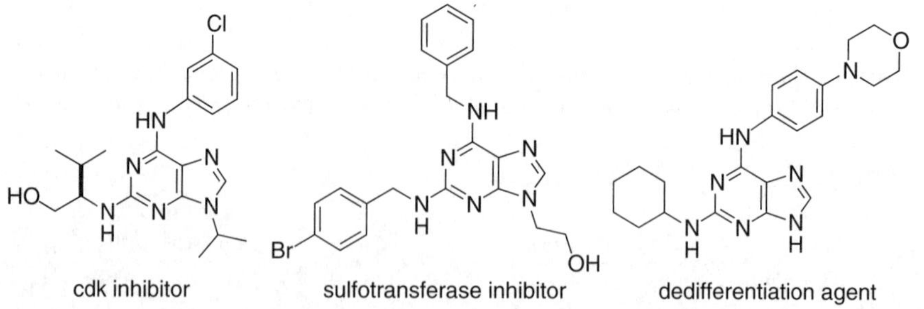

Figure 8 *Three biologically active compounds from target class purine libraries directed at protein kinases*

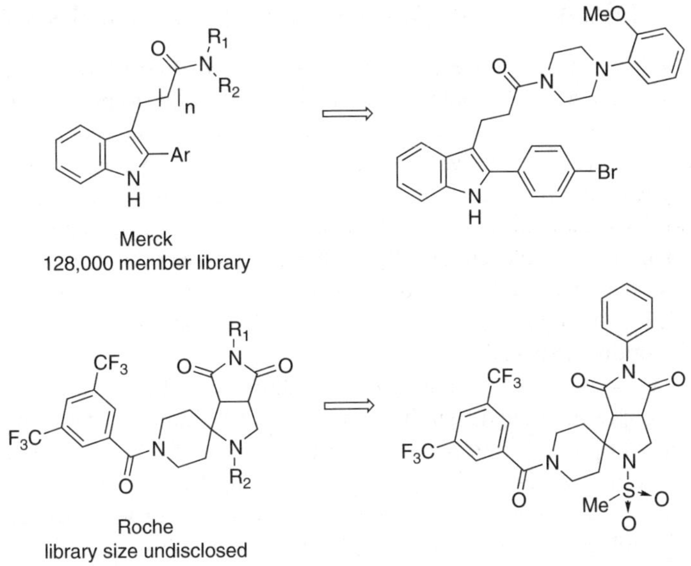

Merck
128,000 member library

Roche
library size undisclosed

Figure 9 *Two examples of GPCR target class libraries prepared by solid-phase synthesis reported by industrial groups. In both cases, an example of a neurokinin 1 receptor ligand discovered is shown*

7   Peptide and Nucleotide Libraries Redux

We began by describing peptides and nucleotides, the first classes of compounds to be prepared combinatorially. Both of these suffer from some disadvantages for classical drug discovery, as alluded to earlier. Nevertheless, when used creatively, peptide libraries can be a valuable resource for medicinal chemistry and the discovery of novel leads. An impressive recent development[§] is the synthesis of mixture-based peptide libraries containing motifs for protease inhibition. These libraries can then be employed to profile the detailed substrate specificity of a given protease. Undoubtedly, a major reason for their success is that the peptide libraries seek to answer a biological question in peptide chemical space, rather than discover leads that still need to be translated to non-peptidic compounds for further development as therapeutic agents.

Meanwhile, combinatorial oligonucleotide libraries have become established as a powerful method for the identification of high-affinity ligands for a given target. This "aptamer" technology[35] benefits from the ability to routinely make very large (10^9) mixture-based oligonucleotide libraries, and yet apply stringent screening criteria to pan for binding members in a series of iterative cycles. Biological amplification of sequences by PCR is a tremendous boon that facilitates the identification of active components from a complex mixture, a step that remains the bottleneck in non-nucleotide combinatorial libraries. While aptamers have proven highly successful as high-affinity ligands for their target, perhaps their likely role is as biological research tools rather than directly for drug discovery.

[§] For reviews and recent examples from the Ellman group, see Ref. 34.

8   Lead Discovery or Drug Discovery – Size does Matter

The discovery of a lead and the discovery of a drug are almost always two separate processes. Outside natural products, it is rare for compounds to progress all the way from initial identification as leads to clinical candidates. This being the case, an important issue for library design is whether combinatorial chemists should focus on maximizing lead- or drug-like features in their libraries. Historical analyses[36] indicate that the two have some differences. As a lead is optimized by medicinal chemists, it tends to grow in size and complexity with respect to various descriptors. Astex, for example, have described[37] a "Rule of Three" for lead discovery: MW < 300, $ClogP < 3$, H-bond donors <3, rotatable bonds <3, as opposed to Lipinski's "Rule of Five" for orally bioavailable drugs.

How does one go about designing lead-like libraries? The constraints imply that only fairly simple compounds will be suitable. By definition, the initial lead may be relatively low in potency, and biological testing will need less stringent conditions than High-Throughput Screening (HTS) (close to mM rather than μM concentrations). A number of groups have implemented[38] methods for the discovery of such "lead-like" fragments. Ellman, for example, has screened a set of 305 aldehydes to identify potential weak binders or inhibitors, followed by their linkage to derive higher-affinity leads. The method was applied[39] to the discovery of inhibitors for targets including tyrosine kinases and sulfotransferases (Figure 10). Abbott and others

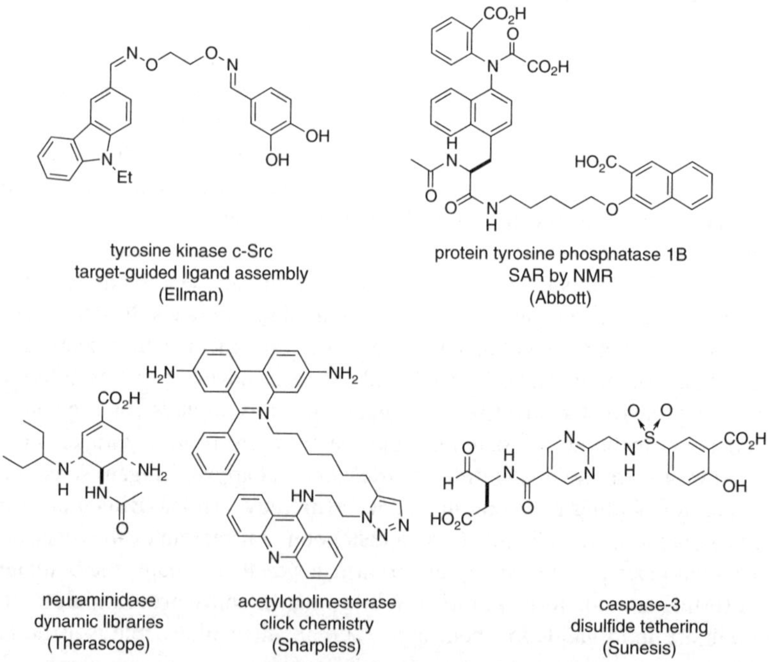

tyrosine kinase c-Src
target-guided ligand assembly
(Ellman)

protein tyrosine phosphatase 1B
SAR by NMR
(Abbott)

neuraminidase
dynamic libraries
(Therascope)

acetylcholinesterase
click chemistry
(Sharpless)

caspase-3
disulfide tethering
(Sunesis)

Figure 10 *Five examples of compounds discovered by fragment assembly combinatorial approaches. In each case, the target and methodology are indicated*

have employed "SAR by NMR" to screen sets of compounds for their binding. In one example,[40] screening 10,000 compounds led to a nearly millimolar tyrosine phosphatase inhibitor. Following optimization and rescreening of a second fragment library, a nanomolar compound was identified. The *in situ* combination of reactive building blocks in the presence of the biological target is a third and popular approach, using techniques such as dynamic combinatorial chemistry,[41] click chemistry,[42] and disulfide tethering.[43]

9 Natural Product Scaffolds for Combinatorial Chemistry – Why Reinvent the Wheel?

Natural products, either directly or as the original lead, are the source[44] for over 50% of our current drugs. That suggests that natural product scaffolds are a highly fruitful starting point for combinatorial chemistry. In the past, the relationship between these two fields has been somewhat adversarial. Combinatorial libraries were partially responsible for the downsizing or abandonment of natural product screening by the pharmaceutical industry. More recently, it has come to be realized that this was a premature decision. Natural products and combinatorial chemistry have much to offer each other,[45] and in the future a more symbiotic relationship between the two is likely.

There are two basic postulates that underpin the exploitation of natural products combinatorially. Firstly, the actual natural product may not be the optimized compound of its class for the activity detected. Thus, further explorations around the scaffold may produce improved therapeutic candidates; this idea is akin to the target class libraries described earlier. Secondly, the scaffolds chosen by nature have evolved as privileged motifs for the display of chemical information in 3D space. Thus, they may be valuable starting points for general-purpose discovery libraries, above and beyond any particular biological activities associated with the original natural product. Both of these postulates imply that natural products possess some unusual features complementary to synthetic scaffolds favoured by medicinal chemists. Recent computational studies[46] do provide supporting evidence for this belief. On average, natural products have more chiral centres and bridgehead atoms than synthetic trade drugs, while their elemental composition is overweighted in oxygen at the expense of nitrogen, sulfur, and halogens.

The combinatorial decoration or derivatization of existing natural products is a popular method for library creation. Among targets for such exercises are the alkaloids yohimbine[47] and scopolamine,[48] terpenoids and steroids,[49] and the heterocyclic peptide antibiotic GE-2270A (Figure 11). In the latter case, combinatorial variation[50] was instrumental in the discovery of a developmental candidate now in clinical trials at Vicuron. Pfizer, the world's largest pharmaceutical company, recently acquired Vicuron for the sum of $1.9 billion, a mark of approval for natural product-based combinatorial chemistry.

The above examples were focused exercises in which the libraries were designed for evaluation against one or a limited set of biological targets, and took advantage of existing scaffolds. If some time and resource is invested, an alternative approach

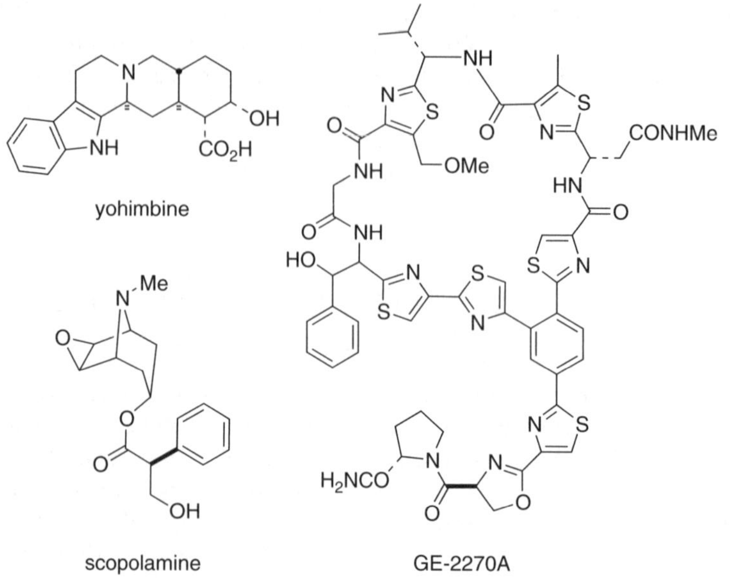

Figure 11 *Three examples of natural product scaffolds that have been diversified by decoration*

is the *de novo* synthesis of the natural product, by a combinatorial strategy that facilitates analogue preparation. There are many examples[51] of such projects, particularly from academic laboratories. In our group, we recently accomplished[52] the solid-phase total synthesis of kahalalide A (Figure 12), a cyclic depsipeptide with antimycobacterial activity. The synthesis was then applied to the preparation of analogues, and a compound with higher activity than the natural product was identified.

Besides focused natural product libraries that are essentially used in lead optimization, an increasing number of general discovery libraries incorporate privileged scaffolds present in natural products. One with which we have personal experience[53] is the tetrahydro-β-carboline skeleton, which occurs in numerous indole alkaloids as well as medicinal chemistry leads and trade drugs. The scaffold is readily prepared by the Pictet–Spengler reaction of tryptophan or tryptamine derivatives, and this accessibility has led to the preparation of combinatorial libraries by many groups. The versatility of the scaffold can be seen from the variety of targets for which leads have been identified[54] (Figure 13).

10   From Natural Products to Natural Product-Like Libraries – Hubris or Progress?

In the preceding section, we have seen the value of natural products in drug discovery and development. Now that we have a better understanding of the complementarity between natural products and synthetic chemistry, can we design new scaffolds that lean towards the former? Compared to classical skeletons favoured by medicinal chemists, the newer "natural product-like" scaffolds[55] incorporate a high degree of

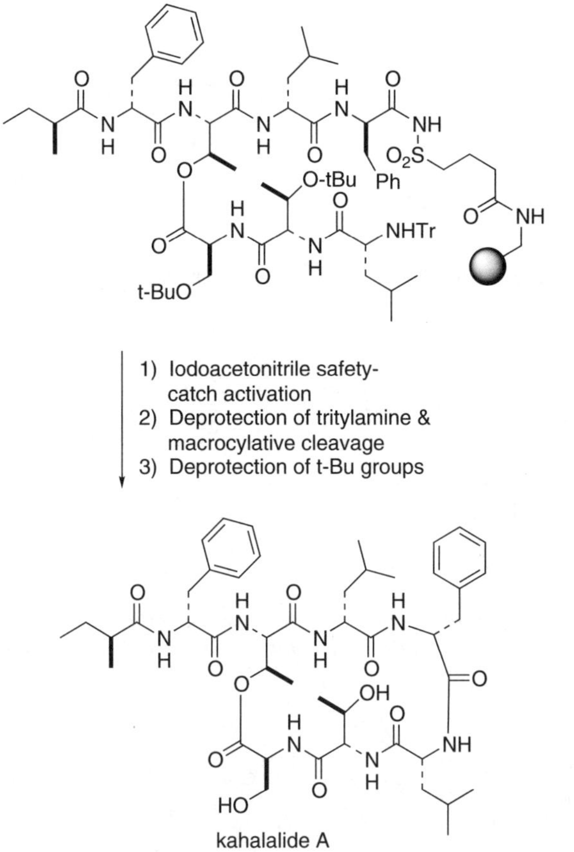

Figure 12 *A solid-phase total synthesis of the antimycobacterial cyclodepsipeptide kahalalide A. The synthesis relies on the Kenner "safety-catch" linker for attachment to the peptide backbone, followed by macrocyclative cleavage of the linear depsipeptide*

stereochemical information and fused, angular, or polycyclic ring systems (Figure 14). This is an ongoing global experiment, primarily by academic practitioners,[56] and it will be some time before this approach can be realistically assessed. The underlying hypotheses for exploring natural product-like space are certainly appealing, but the successful mimicry of what nature has achieved over millions of years is unlikely to be straightforward.

11   Lead Discovery and Combinatorial Chemistry – What have We Learned?

In this chapter, we have surveyed the most popular strategies for the design of combinatorial libraries. We can summarize the experiences over the last two decades as follows, some of these lessons having been learnt the hard way.

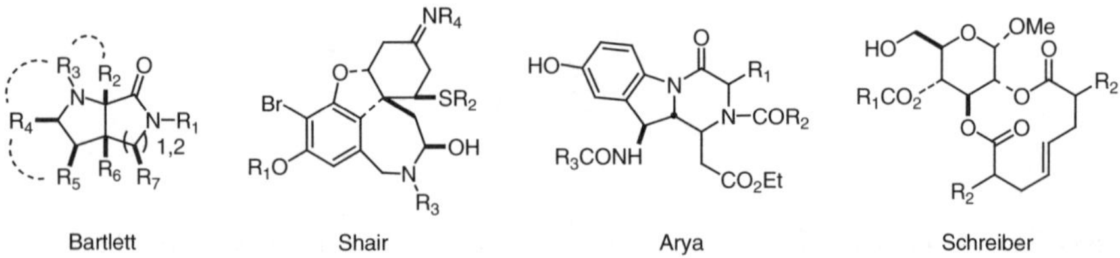

Figure 13 *Four examples of biologically active compounds discovered from libraries based on the tetrahydro-β-carboline natural product privileged scaffold*

Figure 14 *Some examples of "natural product-like" libraries*

11.1 The Drug-Discovery Process cannot be Simplified to a Single Blueprint

There are drugs on the market that look ridiculously simple in structure, and others that are horrendously complex, with everything in between. Similarly, the development of these drugs did not follow a single tried and tested pathway. For this reason, no single technology, whether it is structure-based design, combinatorial chemistry, or proteomics, to pick three recently fashionable developments, will by itself solve the bottlenecks of drug discovery. All of them need to be integrated as part of the medicinal chemist's toolkit, to be used judiciously when appropriate.

11.2 Combinatorial Chemistry is an Extremely Powerful Technology

The ability to generate large collections of compounds and evaluate them for activity is a positive feature that cannot be underestimated. Instead of collecting data on single compounds, the medicinal chemist can now examine an array of compounds in evaluating structure–activity relationships. This increase in information content is of obvious value, provided it is used wisely.

11.3 Combinatorial Chemistry is at its Best in Lead Optimization

Once a lead compound is found, optimization by parallel synthesis will clearly result in an increased productivity and accelerated timelines compared to the low throughput of traditional medicinal chemistry. It is much less clear that combinatorial chemistry, at least in its early manifestations, was efficient in discovering the initial leads. In Golebiowski *et al.*'s reviews[57] on this subject, it is practically impossible to find convincing examples of compounds that have fallen directly out of a discovery library and progressed to preclinical or clinical development. Partly, this is due to the slowness of the drug-discovery process. Programmes at an advanced stage right now would have started at about the same time as combinatorial chemistry, when libraries were less sophisticated than present. Another reason is that the combinatorial chemistry origins of compounds now in late stage development may not be publicly disclosed.

11.4 Combinatorial Chemistry is about Making the Compounds that Fit Your Needs, not How They are Made

In the synthesis of combinatorial libraries, there is a raft of tactical issues that need to be tackled. Will the library be made of mixtures or discrete compounds? Prepared by solid-phase or solution-phase? Screened in solution or attached to beads? What level of purification and characterization is needed? Will hits be identified by deconvolution, encoding techniques, or other means? These are all crucial operational aspects of combinatorial chemistry but it is equally important not to concentrate on them to the extent of missing the big picture. At the end of the day, neither a biological assay nor a medicinal chemist care how a compound was made. It is vital, though, that the tactical decisions do not prevent one from making the right compounds. Combinatorial synthesis is a means to an end, not an end in itself.

References

1. (a) D.W. Hobbs and T. Guo, in *Combinatorial Library Design and Evaluation*, A.K. Ghose and V.N. Viswanadhan (eds), Marcel Dekker, New York, 2001, 1–49. (b) A. Ganesan, *Drug Discov. Today*, 2002, **7**, 47–55.
2. (a) R.E. Dolle, *Mol. Divers.*, 1998, **3**, 233–256. (b) R.E. Dolle and K.H. Nelson Jr., *J. Comb. Chem.*, 1999, **1**, 235–282. (c) R.E. Dolle and K.H. Nelson Jr., *J. Comb. Chem.*, 2000, **2**, 383–433. (d) R.E. Dolle, *J. Comb. Chem.*, 2001, **3**, 477–517. (e) R.E. Dolle, *J. Comb. Chem.*, 2002, **4**, 369–418. (f) R.E. Dolle, *J. Comb. Chem.*, 2003, **5**, 693–753. (g) R.E. Dolle, *J. Comb. Chem.*, 2004, **6**, 623–679.
3. M. Lebl, *J. Comb. Chem.*, 1999, **1**, 3–24.
4. (a) H.M. Geysen, R.H. Meloen and S.J. Barteling, *Proc. Natl. Acad. Sci.*, 1984, **81**, 3998–4002. (b) H.M. Geysen, S.J. Rodda and T.J. Mason, *Mol. Immunol.*, 1986, **23**, 709–715.

5. H.M. Geysen and T.J. Mason, *Bioorg. Med. Chem. Lett.*, 1993, **3**, 397–404.
6. (a) K.S. Lam, S.E. Salmon, E.M. Hersh, V.J. Hruby, W.M. Kazmierski and R.J. Knapp, *Nature*, 1991, **354**, 82–84. (b) R.A. Houghten, C. Pinilla, S.E. Blondelle, J.R. Appel, C.T. Dooley and J.H. Cuervo, *Nature*, 1991, **354**, 84–86.
7. R.J. Simon, R.S. Kania, R.N. Zuckermann, V.D. Huebner, D.A. Jewell, S. Banville, S. Ng, L. Wang, S. Rosenberg, C.K. Marlowe, D.C. Spellmeyer, R. Tan, A.D. Frankel, D.V. Santi, F.E. Cohen and P.A. Bartlett, *Proc. Natl. Acad. Sci.*, 1992, **89**, 9367–9371.
8. R.N. Zuckermann, J.M. Kerr, S.B.H. Kent and W.H. Moos, *J. Am. Chem. Soc.*, 1992, **114**, 10646–10647.
9. R.N. Zuckermann, E.J. Martin, D.C. Spellmeyer, G.B. Stauber, K.R. Shoemaker, J.M. Kerr, G.M. Figliozzi, D.A. Goff, M.A. Siani, R.J. Simon, S.C. Banville, E.G. Brown, L. Wang, L.S. Richter and W.H. Moos, *J. Med. Chem.*, 1994, **37**, 2678–2685.
10. (a) B. Liu, P.G. Alluri, P. Yu and T. Kodadek, *J. Am. Chem. Soc.*, 2005, **127**, 8254–8255. (b) I. Masip, N. Cortes, M.-J. Abad, M. Guardiola, E. Perez-Paya, J. Ferragut, A. Ferrer-Montiel and A. Messeguer, *Bioorg. Med. Chem. Lett.*, 2005, **13**, 1923–1929.
11. C.Y. Cho, E.J. Moran, S.R. Cherry, J.C. Stephans, S.P.A. Fodor, C.L. Adams, A. Sundaram, J.W. Jacobs and P.G. Schultz, *Science*, 1993, **261**, 1303–1305.
12. C.Y. Cho, R.S. Youngquist, S.J. Paikoff, M.H. Beresini, A.R. Hebert, L.T. Berleau, C.W. Liu, D.E. Wemmer, T. Keough and P.G. Schultz, *J. Am. Chem. Soc.*, 1998, **120**, 7706–7718.
13. P. Lin and A. Ganesan, *Bioorg. Med. Chem. Lett.*, 1998, **8**, 511–514.
14. M.R. Spaller, M.T. Burger, M. Fardis and P.A. Bartlett, *Curr. Opin. Chem. Biol.*, 1997, **1**, 47–53.
15. (a) A.A. Patchett and R.P. Nargund, in *Annual Reports in Medicinal Chemistry*, Vol 35, A.M. Doherty (ed), Academic Press, San Diego, 2000, 289–298. (b) D.A. Horton, G.T. Bourne and M.L. Smythe, *Chem. Rev.*, 2003, **103**, 893–930.
16. B.A. Bunin and J.A. Ellman, *J. Am. Chem. Soc.*, 1992, **114**, 10997–10998.
17. (a) B.A. Bunin, M.J. Plunkett and J.A. Ellman, *New J. Chem.*, 1997, **21**, 125–130. (b) C.G. Boojamra, K.M. Burow and J.A. Ellman, *J. Org. Chem.*, 1995, **60**, 1240–1256. (c) G. Bhalay, P. Blaney, V.H. Palmer and A.D. Baxter, *Bioorg. Med. Chem. Lett.*, 1996, **6**, 2441–2444. (d) C. Hulme, L. Ma, M.P. Cherrier, J.J. Romano, G. Morton, C. Duquenne, J. Salvino and R. Labaudiniere, *Tetrahedron Lett.*, 2000, **41**, 1883–1887.
18. B.L. Chng and A. Ganesan, *Bioorg. Med. Chem. Lett.*, 1997, **7**, 1511–1514.
19. M.M. Sim and A. Ganesan, *J. Org. Chem.*, 1997, **62**, 3230–3235.
20. (a) B.A. Kulkarni and A. Ganesan, *Chem. Commun.*, 1998, 785–786. (b) M.M. Sim, C.L. Lee and A. Ganesan, *Tetrahedron Lett.*, 1998, **39**, 6399–6402.
21. J.D. Revell and A. Ganesan, *Chem. Commun.*, 2004, 1916–1917.
22. P.A. Boguszewski, S.S. Rahman and A. Ganesan, *J. Comb. Chem.*, 2004, **6**, 32–34.

23. (a) T. Carell, E.A. Wintner, A. Bashir-Hashemi and J. Rebek Jr., *Angew. Chem. Int. Ed. Engl.*, 1994, **33**, 2059–2061.(b) T. Carell, E.A. Wintner, A. Bashir-Hashemi and J. Rebek Jr., *Angew. Chem. Int. Ed. Engl.*, 1994, **33**, 2061–2064. (c) T. Carell, E.A. Wintner, A.J. Sutherland, J. Rebek Jr., Y.M. Dunayevskiy and P. Vouros, *Chem. Biol.*, 1995, **2**, 171–183.

24. H. An, L.L. Cummins, R.H. Griffey, R. Bharadwaj, B.D. Haly, A.S. Fraser, L. Wilson-Lingardo, L.M. Risen, J.R. Wyatt and P.D. Cook, *J. Am. Chem. Soc.*, 1997, **119**, 3696–3708.

25. (a) B.A. Kulkarni and A. Ganesan, *Angew. Chem. Int. Ed. Engl.*, 1997, **36**, 2454–2455. (b) B.A. Kulkarni and A. Ganesan, *Tetrahedron Lett.*, 1998, **39**, 4369–4372.

26. (a) J. Matthews and R.A. Rivero, *J. Org. Chem.*, 1998, **63**, 4808–4810. (b) L. Weber, P. Iaiza, G. Biringer and P. Barbier, *Synlett*, 1998, 1156–1158. (c) T.T. Romoff, L. Ma, Y. Wang and D.A. Campbell, *Synlett*, 1998, 1341–1342.

27. S.K. Radford, P. Doman and A. Ganesan, unpublished results.

28. (a) Y.-T. Chang, N.S. Gray, G.R. Rosania, D.P. Sutherlin, S. Kwon, T.C. Norman, R. Sarohia, M. Leost, L. Meijer and P.G. Schultz, *Chem. Biol.*, 1999, **6**, 361–375. (b) S. Ding, N.S. Gray, X. Wu, Q. Ding and P.G. Schultz, *J. Am. Chem. Soc.*, 2002, **124**, 1594–1596.

29. (a) J.I. Armstrong, A.R. Portley, Y.-T. Chang, D.M. Nierengarten, B.N. Cook, K.G. Bowman, A. Bishop, N.S. Gray, K.M. Shokat, P.G. Schultz and C.R. Bertozzi, *Angew. Chem. Int. Ed.*, 2000, **39**, 1303–1306. (b) D.E. Verdugo, M.T. Cancilla, X. Ge, N.S. Gray, Y.-T. Chang, P.G. Schultz, M. Negishi, J.A. Leary and C.R. Bertozzi, *J. Med. Chem.*, 2001, **44**, 2683–2686.

30. S. Chen, Q. Zhang, X. Wu, P.G. Schultz and S. Ding, *J. Am. Chem. Soc.*, 2004, **126**, 410–411.

31. K.H. Bleicher, L.G. Green, R.E. Martin and M. Roger-Evans, *Curr. Opin. Chem. Biol.*, 2004, **8**, 287–296.

32. C.A. Willoughby, S.M. Hutchins, K.G. Rosauer, M.J. Dhar, K.T. Chapman, G.G. Chicchi, S. Sadowski, D.H. Weinberg, S. Patel, L. Malkowitz, J. Di Salvo, S.G. Pacholok and K. Cheng, *Bioorg. Med. Chem. Lett.*, 2002, **12**, 959–962.

33. K.H. Bleicher, Y. Wüthrich, G. Adam, T. Hoggmann and A.J. Sleight, *Bioorg. Med. Chem. Lett.*, 2002, **12**, 3073–3076.

34. (a) D.J. Maly, L. Huang and J.A. Ellman, *ChemBiochem*, 2002, **3**, 16–37. (b) C.M. Salisbury, D.J. Maly and J.A. Ellman, *J. Am. Chem. Soc.*, 2002, **124**, 13964–13965. (c) B. Lehting, K.D. Pryor, J.K. Wu, F. Marsilio, R.A. Patel, C.S. Craik and J.A. Ellman, *Biochem. J.*, 2003, **371**, 525–532. (d) D.N. Gosalia, C.M. Salisbury, D.J. Maly, J.A. Ellman and S.L. Diamond, *Proteomics*, 2005, **5**, 1292–1298.

35. (a) M. Rimmele, *ChemBiochem*, 2003, **4**, 963–971. (b) S.M. Nimjee, C.P. Rusconi and B.A. Sullenger, *Ann. Rev. Med.*, 2005, **56**, 555–583.

36. (a) T.I. Oprea, A.M. Davis, S.J. Teague and P.D. Leeson, *J. Chem. Inf. Comput. Sci.*, 2001, **41**, 1308–1315. (b) M.M. Hann and T.I. Oprea, *Curr. Opin. Chem. Biol.*, 2004, **8**, 255–263.

37. M. Congreve, R. Carr, C. Murray and H. Jhoti, *Drug Discov. Today*, 2003, **8**, 876–877.

38. D.A. Erlanson, R.S. McDowell and T. O'Brien, *J. Med. Chem.*, 2004, **47**, 3463–3482.

39. (a) D.J. Maly, I.C. Choong and J.A. Ellman, *Proc. Natl. Acad. Sci.*, 2000, **97**, 2419–2424. (b) J.W. Kehoe, D.J. Maly, D.E. Verdugo, J.I. Armstrong, B.N. Cook, Y.-B. Ouyang, K.L. Moore, J.A. Ellman and C.R. Bertozzi, *Biorg. Med. Chem. Lett.*, 2002, **12**, 329–332.

40. B.G. Szczepankiewicz, G. Liu, P.J. Hajduk, C. Abad-Zapatero, Z.H. Pei, Z.L. Xin, T.H. Lubben, J.M. Trevillyan, M.A. Stashko, S.J. Ballaron, H. Liang, F. Huang, C.W. Hutchins, S.W. Fesik and M.R. Jirousek, *J. Am. Chem. Soc.*, 2003, **125**, 4087–4096.

41. M. Hochgurtel, R. Biesinger, H. Kroth, D. Piecha, M.W. Hofmann, S. Krause, O. Schaaf, C. Nicolau and A.V. Eliseev, *J. Med. Chem.*, 2003, **46**, 356–358.

42. W.G. Lewis, L.G. Green, F. Grynszpan, Z. Radic, P.R. Carlier, P. Taylor, M.G. Finn and K.B. Sharpless, *Angew. Chem. Int. Ed.*, 2002, **41**, 1053–1057.

43. I.C. Choong, W. Lew, D. Lee, P. Pham, M.T. Burdett, J.W. Lam, C. Wiesmann, T.N. Luong, B. Fahr, W.L. DeLano, R.S. McDowell, D.A. Allen, D.A. Erlanson, E.M. Gordon and T. O'Brien, *J. Med. Chem.*, 2002, **45**, 5005–5022.

44. (a) D.J. Newman, G.M. Cragg and K.M. Snader, *Nat. Prod. Rep.*, 2000, **17**, 215–234. (b) D.J. Newman, G.M. Cragg and K.M. Snader, *J. Nat. Prod.*, 2003, **66**, 1022–1037.

45. (a) J.-Y. Ortholand and A. Ganesan, *Curr. Opin. Chem. Biol.*, 2004, **8**, 271–280. (b) A. Ganesan, *Curr. Opin. Biotech.*, 2004, **15**, 584–590. (c) A. Ganesan, in *Combinatorial Synthesis of Natural Product-Based Libraries*, A. Boldi (ed), CRC Press, Boca Raton, FL, 2006.

46. (a) T. Henkel, R.M. Brunne, H. Müller and F. Reichel, *Angew. Chem. Int. Ed.*, 1999, **38**, 643–647. (b) M.L. Lee and G. Schneider, *J. Comb. Chem.*, 2001, **3**, 284–289. (c) M. Feher and J.M. Schmidt, *J. Chem. Inf. Comput. Sci.*, 2003, **43**, 218–227.

47. A. Atuegbu, D. Maclean, C. Nguyen, E.M. Gordon and J.W. Jacobs, *Bioorg. Med. Chem.*, 1996, **4**, 1097–1106.

48. N.S. Aberle, A. Ganesan, J.N. Lambert, S. Saubern and R. Smith, *Tetrahedron Lett.*, 2001, **42**, 1975–1977.

49. L.C. Ciobanu and D. Poirier, *J. Comb. Chem.*, 2003, **5**, 429–440.

50. J. Clough, S.Q. Chen, E.M. Gordon, C. Hackbarth, S. Lam, J. Trias, R.J. White, G. Candiani, S. Donadio, G. Romano, R. Ciabatti and J.W. Jacobs, *Bioorg. Med. Chem. Lett.*, 2003, **13**, 3409–3414.

51. (a) P.M. Abreu and P.S. Branco, *J. Braz. Chem. Soc.*, 2003, **14**, 675–712. (b) A.M. Boldi, *Curr. Opin. Chem. Biol.*, 2004, **8**, 281–286.

52. L. Bourel, K.V. Rao, M.T. Hamann and A. Ganesan, *J. Med. Chem.*, 2005, **48**, 1330–1335.

53. (a) H. Wang and A. Ganesan, *Tetrahedron Lett.*, 1997, **38**, 4327–4328. (b) H. Wang and A. Ganesan, *Org. Lett.*, 1999, **1**, 1647–1649. (c) H. Wang, T. Usui, H. Osada and A. Ganesan, *J. Med. Chem.*, 2000, **43**, 1577–1585. (d) H. Wang and A. Ganesan, *J. Org. Chem.*, 2000, **65**, 4685–4693. (e) D. Bonnet and A.

Ganesan, *J. Comb. Chem.*, 2002, **4**, 546–548. (f) N. Srinivasan and A. Ganesan, *Chem. Commun.*, 2003, 916–917. (g) J.D. Revell, N. Srinivasan and A. Ganesan, *Synlett*, 2004, 1428–1430.

54. (a) A. van Loevezijn, J.D. Allen, A.H. Schinkel and G.-J. Koomen, *Bioorg. Med. Chem. Lett.*, 2001, **11**, 29–32. (b) T. Redemann, M.E. Brooks and G. Jung, cited in ref. 2c. (c) S. Hotha, J.C. Yarrow, J.G. Yang, S. Garrett, K.V. Renduchintala, T.U. Mayer and T.M. Kapoor, *Angew. Chem. Int. Ed.*, 2003, **42**, 2379–2382.

55. (a) S.L. Schreiber, *Science*, 2000, **287**, 1964–1969. (b) M.D. Burke and S.L. Schreiber, *Angew. Chem. Int. Ed.*, 2004, **43**, 46–58. (c) A. Reayi and P. Arya, *Curr. Opin. Chem. Biol.*, 2005, **9**, 240–247. (d) S. Shang and D.S. Tan, *Curr. Opin. Chem. Biol.*, 2005, **9**, 248–258.

56. (a) M.A. Marx, A.-L. Grillot, C.T. Louer, K.A. Beaver and P.A. Bartlett, *J. Am. Chem. Soc.*, 1997, **119**, 6153–6167. (b) H.E. Pelish, N.J. Westwood, Y. Feng, T. Kirchhausen and M.D. Shair, *J. Am. Chem. Soc.*, 2001, **123**, 6740–6741. (c) Z. Gan, P.T. Reddy, S. Quevillon, S. Couve-Bonnaire and P. Arya, *Angew. Chem. Int. Ed.*, 2005, **44**, 1366–1368. (d) Y. Kim, M.A. Arai, T. Arai, J.A. Lamenzo, E.F. Dean III, N. Patterson, P.A. Clemons and S.L. Schreiber, *J. Am. Chem. Soc.*, 2004, **126**, 14740–14745.

57. (a) A. Golebiowski, S.R. Klopfenstein and D.E. Portlock, *Curr. Opin. Chem. Biol.*, 2001, **3**, 273–284. (b) A. Golebiowski, S.R. Klopfenstein and D.E. Portlock, *Curr. Opin. Chem. Biol.*, 2003, **7**, 308–325.

Compound Collections: Acquisition, Annotation, and Access

REG RICHARDSON

ChemBridge Europe, 4 Clarks Hill Rise, Evesham, Worcs WR11 2FW, UK

1   Introduction

The juggernaut that is drug discovery continues to move forward at a relentless pace, imposing ever-changing requirements on those trying to create the drugs of the future. This chapter reviews the information and chemical libraries that are available for use in the drug-discovery environment and for the sourcing, selection and annotation of compounds destined to be screened as those potential new drugs.

While the cost of getting new chemical entities to the market continues to grow, it is now estimated to cost $900 M (Figure 1). As a consequence, the scientists within the drug-discovery teams need to make more informed decisions regarding what to make, what to buy/acquire, what to test and what to designate as viable hits to take forward as candidates for lead optimisation. What products and technologies are out there to help with this process?

The basic premise of pharmaceutical R&D has not changed for decades; in order to discover new drugs, new chemical entities (NCEs) are needed. These compounds need to be designed or sourced and acquired, and then screened. Traditionally, these NCEs have been synthesised by pharmaceutical companies internally. However, outsourcing has become an increasingly attractive alternative, as library providers offer some real competition to the internal generation of screening candidates. With funding for external R&D seemingly on the increase, the use of these external library and service providers is growing accordingly.

Where do the NCEs come from, how good are the companies that provide them and how are molecules themselves being chosen to increase the chances of success?

Commercial sources can be mercurial institutions, at times growing with the trend for external acquisition, at other times fading as the interest in commercial offerings

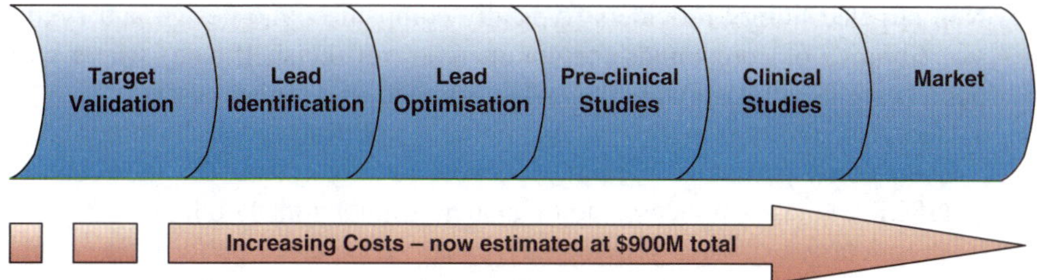

Figure 1 *The cost of advancing a new chemical entity (NCE) to the marketplace*

declines or companies are acquired, often by a large pharma collaborator, to supplement internal activities. Several providers have a shared, non-exclusive, pool of virtual as well as stocked molecules; others ensure exclusivity by in-house design and synthesis or by isolation of previously unreported products from natural sources. Several providers approach the business from multiple angles.

Internal compound collections within the pharma industry reflect the history of that company's clinical focus. As a result legacy collections are often narrow in diversity and even narrower in sub-structure. In order to increase diversity in terms of both structure and pharmacophore profiles, commercial library providers are able to supply millions of previously unreported compounds that have novel structures and that explore new regions of chemical and pharmacophore space. With the universe of theoretically possible small molecules conservatively estimated to be 10^{60} compounds[1] and with on the order of 60,000,000–120,000,000 being available on the shelves of screening facilities worldwide[2] it is easy to see why R&D organisations are attracted to external sources for novel chemical samples – there are plenty to choose from. However, it is obvious that even the commercially available compounds, are just a very small drop in a very large ocean of potential small molecules.

2 Commercial Offerings

There are many sources of commercially available small molecules. Some of those sources would acknowledge that not all of the molecules they offer are suitable as potential pharmaceutical materials and that many more could only be considered as reagents or intermediates. However, there are still a very large number of providers giving access to high quality small molecules which may be considered good candidates for hit-to-lead or lead optimisation chemistry, should they be identified as a hit in a biological screen.

The commercial suppliers fall into four broad categories:

1. Shared-pool, non-exclusive, providers (typically companies sourcing compounds from the former Russian states and other eastern bloc countries as well as from India, China and elsewhere) where the compounds are available in very large numbers at a relatively low cost and each compound is available from one or more providers.

2. Non-targeted library providers (typically companies who design and synthe-sise libraries of compounds for commercial, off-the-shelf sale) where compounds are usually available in smaller numbers at a higher cost.
   a. Targeted library providers whose libraries are designed and synthesised to act against particular biological targets or target families, most notably GPCRs, kinases, proteases, ion-channels – among many others.
   b. Proprietary, exclusive providers (typically companies performing contract or custom synthesis) offering compound design and/or synthesis services on a fee-for-service or FTE basis. Such arrangements tend to come with the highest cost and are often associated with downstream milestones and/or royalties payable on a success-related basis.

Table 1 lists a selection of companies recognised as providers of high-quality collections of small molecules that are available off-the-shelf and which may be assembled by collection methods or by using parallel synthesis techniques.

The involvement of the library providers, regardless of their business model, is generally restricted to the early stages of the drug-discovery process (Figure 2).

Table 1 *Providers of stock libraries acquired from academic communities or syn-thesised using in-house resources*

#	Company name	Web
1	ASDI Biosciences	www.asdibiosciences.com
2	Albany Molecular Research Inc	www.albmolecular.com
3	ArQule	www.arqule.com
4	Asinex	www.asinex.com
5	Biofocus	www.biofocus.com
6	Bionet	www.keyorganics.ltd.uk
7	ChemBridge Corporation	www.chembridge.com
8	ChemDiv	www.chemdiv.com
9	ChemStar	www.chemstar.ru
10	Chem T&I	www.chemti.com
11	Comgenex	www.comgenex.hu
12	EMC MicroCollections	www.microcollections.de
13	Enamine	www.enamine.relc.com
14	IF Lab	www.iflab.kiev.ua
15	InterBioScreen	www.ibscreen.com
16	LaboTest	www.labotest.de
17	MicroChemistry	www.mch.ru
18	Moscow MedChem Labs Ltd.	www.mosmedchemlabs.com
19	Nanosyn	www.nanosyn.com
20	Otava	www.otava.com.ua
21	Peakdale	www.peakdale.co.uk
22	Pharmacopeia	www.pharmacop.com
23	Polyphor	www.polyphor.com
24	SPECS	www.specs.com
25	Timtec	www.timtec.com
26	TOSLab	www.toslab.com
27	VITAS-M	www.vitasmlab.com

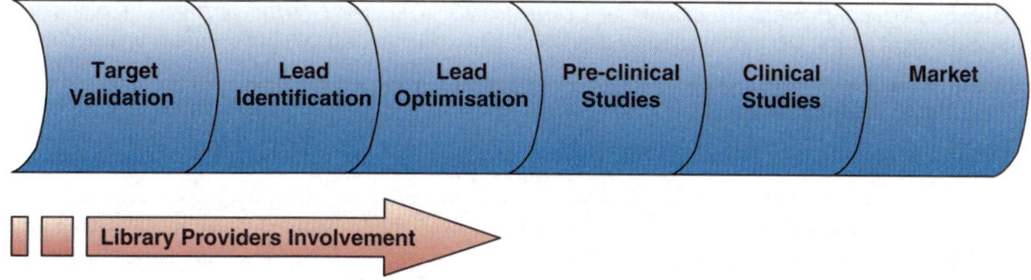

Figure 2 *The involvement of library providers in the drug discovery pipeline*

3 Companies Providing Non-Proprietary, Non-Parallel Synthesised Libraries (Shared-Pool/'Collected Collections')

The shared-pool providers number well over 50, although the better recognised companies are limited to less than 25. These companies combined are capable of supplying several million compounds between them; some are unique to individual providers, while many are offered by multiple providers. In the past, these compound collections have been viewed with some suspicion with respect to quality of the structures and the purity of the samples themselves. However, in recent years the improvement in quality from almost all providers has meant that valuable leads are now being found in these inexpensive collections.

These 'collected collections' often originate from academic institutes throughout the former Russian states or from other eastern bloc countries as well as India, China and other far eastern countries. The compounds are generally classically synthesised, often as part of academic training, and are acquired by providers who put the samples through rigorous quality control models to prove structure and purity. Many providers now use multiple structural and cheminformatic "filters" to remove molecules that are unsuitable for drug discovery. The profile of these compound collections has, in most instances, improved dramatically as the pharmaceutical industry has made it clear what types of compounds it considers useful (as discussed below). The sheer numbers in these collections, however, ensure that a large number of diverse and interesting compounds are available, from which researchers are usually able to 'cherry-pick' according to their selection protocols. The re-emergence and progress of chemical genetics as a screening methodology has also allowed researchers to screen many tens or hundreds of thousands of diverse compounds sufficiently quickly and inexpensively that the need for cherry-picking of individual samples is reduced.

The shared-pool suppliers, as seen in Table 2 below, are capable of providing 70,000–450,000+ compounds each from stock, generally as dry powder/dry film or as solutions in DMSO (from cherry-picked selections or pre-plated sets). The cost of these libraries is relatively low, allowing discovery scientists the opportunity to screen a large number of structurally and pharmacophorically diverse compounds easily and cheaply. By screening a relevant subset from the compounds offered, the user often can use the hits as the basis for structure queries in sourcing analogues for

Table 2 *Shared-pool collections – compounds acquired from primarily academic sources*

#	Company name	'Collected collection' size (as of October 2004)*	Collection name (where applicable)
1	ASDI Biosciences	~130,000	Premier powders collection
2	Albany Molecular Research	~100,000	Chemical sample collection
3	Asinex	230,000	Gold collection
4	ChemBridge Corporation	445,000	EXPRESS-pick
5	ChemStar	60,000	ChemStar library
6	Chem T&I	165,000	n/a
7	Enamine	350,000	Enamine screening collection
8	IF Lab	149,000	n/a
9	InterBioScreen	370,000	Synthetic compounds collection
10	LaboTest	70,000	n/a
11	MicroChemistry	106,000	n/a
12	Moscow MedChem Labs Ltd.	200,000	SHARK database
13	Otava	45,5000	Supplier stock database
14	SPECS	220,000	n/a
15	Timtec	130,000	General screening collection
16	TOSLab	22,000	
17	VITAS-M	150,000	STK

* Numbers are approximate. Information taken from company web sites where available.

rapid SAR generation, without the need for any synthetic chemistry at all. One potential drawback of these 'collected collections' is that they do contain singletons (*i.e.*, compounds for which no close analogues are present); while many assay managers try to avoid screening singletons, many others see them as valid members of a diverse collection of compounds. The inherent problem with singletons is that analogues may be difficult or impossible to obtain, thus requiring synthesis before an SAR can be generated. It takes time and effort to develop a synthetic route to such analogues, although it often does clear the way to claim valuable, additional IP at a very early stage.

At face value, it would appear that these providers could supply a total number of compounds well in excess of 3,000,000; however, in reality there is a significant amount of overlap between these simple, and essentially design-free, collections. In fact the total number of unique, structurally different compounds is something less than 2,000,000 once the overlap is taken into account and, of these, just 700,000–750,000 might be considered 'drug-like' once a set of suitable 'medchem' filters have been applied. (The broad outlines of the cheminformatics methods being used to analyse diversity and to apply such filters to these commercial compound collections will be discussed below.)

There has been considerable interest in collections of natural products in recent years, although at present the use of such libraries has declined for a number of reasons. The general preference for screening individual compounds means that isolation of pure components from natural extracts is necessary. The purification procedures are often complex, especially if the number of components in a mixture is high. Furthermore, the structural complexity of many natural products is proving unattractive for medicinal chemists trying to convert a complicated hit structure into a drug-like small molecule. Finally, the situation with respect to intellectual property is fast becoming unclear as territories try to protect the so-called biodiversity rights, covering all natural products found within their boundaries. As with all small molecule drug-discovery programs, it is rarely the 'hit' that proves to be the final drug. While the same can be said for natural products, in many cases purified (although not necessarily single agent) natural extracts are an acceptable form of treatment. Thus, there is concern that a strong IP position on these extracts may become more difficult to obtain.

4   Companies Providing In-House Designed, Parallel Synthesised Libraries

While there are a significant number of providers positioning themselves as providers of both 'collected collections' *and* in-house designed libraries, the number of companies specialising only in the latter is growing rapidly. The number of companies that provide general and focused screening libraries from parallel synthesis continues to increase, despite the tendency for pharma companies themselves to capture their technology or IP by acquisition.

In the late 1980s and early 1990s the advent and rapid development of combinatorial chemistry and parallel synthesis methods, along with the associated but absolutely necessary high throughput screening techniques, promised to revolutionise drug discovery. The vast numbers of small molecules that could be made quickly and cheaply were expected to provide a huge increase in the number of hits and useful leads, resulting in the maintenance of the double-digit percentage growth that pharma was projecting – and in many cases required. Although the results fell short of these promises, many of the concepts of combinatorial chemistry, in particular strategies for parallel synthesis driven by cheminformatics concepts, are being used by most drug-discovery chemistry groups.

The shortfall of many of the initial combinatorial chemistry approaches can be traced to two problems. First, the early production of complex compound mixtures by the so-called split-and-mix method[3] led to considerable difficulties in the screening step, necessitating resynthesis or other "deconvolution" methods to identify the active components. Worse, in many cases an active sample simply proved to be combination of weakly active components in the mixture. Second, in many cases the design methods employed resulted in libraries that were very similar in structure, with little diversity in substructures or pharmacophores. Thus, while the number of hits may have increased, all that was ultimately being gained was a faster SAR, not necessarily more good lead molecules. Current library providers and drug-discovery research teams now use advanced cheminformatics methodology to assist in library design for almost all applications. These cheminformatics strategies will be discussed below in describing approaches to compound selection.

The so-called 'sparse' libraries are now the norm, designed by creating larger virtual combinatorial libraries and then removing the unattractive or undesirable components with structural, pharmacophore and ADMET filters; this method is applicable to both diverse and targeted/focused libraries. Although the manipulation of reagents is a little more complex in synthesis of a "partially combinatorial" library, the resulting collection is better at providing molecules that can meet all the typical requirements of drug-like leads.

Companies that provide these libraries can be split roughly into two overlapping groups: those that provide diverse libraries and those that provide targeted/focused libraries; either type of library may be 'off-the-shelf' or offered through custom synthesis. Table 3 lists a small selection of the companies using parallel-synthesis techniques to produce libraries for general hit identification or for targeting a specific receptor or enzyme family.

A much more comprehensive compendium of molecular diversity and chemistry service providers has recently been published, although details regarding the libraries that are actually stocked are somewhat limited.[4]

In addition to the databases made available by individual library providers (either freely or under protection of Non-Disclosure Agreement), a number of other, accumulated databases can be accessed. These datasets include the ACD (Available Chemicals Directory) from MDL (although this is more of a directory of available reagents), the SCD (Screening Compounds Directory), also from MDL, and DiscoveryGate (www.discoverygate.com), the online extension to the SCD. The SCD claims to have over 2 million structures from 49 suppliers of screening compounds. ChemNavigator's iResearch Library (www.chemnavigator.com) is another source of multiple databases as well as procurement services; listing claims to have access to 8 million unique structures from almost 150 compound providers. However, both DiscoveryGate and ChemNavigator are subscription services that require users to pay for access to the databases and associated services, such as synthesis pathways and

Table 3 *Providers of off-the-shelf libraries assembled using parallel-synthesis techniques*

#	Company name	Library size*	Product name(s)
1	Albany Molecular Research	~100,000	n/a
2	Asinex	120,000	Platinum collection
3	Biofocus	25,000	Focused/targeted libraries
4	Bionet	35,000	n/a
5	ChemBridge Corporation	160,000	PHARMACophore: NovaCore: GPCR: Kinase targeted
6	ChemDiv	~600,000	n/a
7	Comgenex	~200,000	n/a
8	EMC Microcollections	30,000	n/a
9	Otava	25,500	n/a
10	Peakdale	8,000	n/a
11	Polyphor	30,000	Polyphor 1, Polyphor 2
12	SPECS	~10,000	Targeted libraries
13	TOSLab	1,000	n/a
14	Tripos	85,000	Diverse library

* Information taken from company web sites where available.

literature citations. Of course, these costs have to be offset against the time and expense that would be required to assemble (and format and update) individual databases in-house.

5   Compound Selection and Database Filtering

It is one thing to have millions of individual small molecules at one's disposal, but quite another to select the compounds you want from this growing pool of diversity! The first question to be answered is, of course, 'do we need to select compounds at all?', because the answer is not always 'yes, we do'! Very often a random selection will perform as well as any pre-selected set, particularly when assaying orphan targets or those for which little or no information is available regarding endogenous ligands or 3D structures.

Companies planning to assemble a new, non-focused corporate screening collection from ground level have several options available to them. The most straightforward may be to take a diverse collection of compounds as offered by the provider. Although a logical assumption is that the companies with the most compounds would be able to offer the widest diversity, this is not necessarily the case. A well designed, small collection of compounds may out-perform a selection from a larger set of compounds if the diversity within the large collection is limited. Indeed, more effective approaches may be to combine supplier databases and analyse the collection as a whole for diversity, or to analyse each supplier's offering individually to build the desired diversity.

Such selections can be based, for example, on sub-structure similarity/dissimilarity calculations, or pharmacophore analyses; each approach will generally lead to very different selections, even within a single set of compounds. The detailed concept of generating diversity has already been covered in this book, but it is valuable to review the basic concepts.

6   Sub-structure Similarity/Dissimilarity

Sub-structure diversity is most easily defined using metrics such as the Tanimoto Dissimilarity Index. These metrics are based on linear bitmaps (fingerprints) generated from the molecular fragments or compound sub-structures (Figure 3). This approach has been developed extensively by Daylight Chemical Information Systems.[5]

For each molecular fragment that is present in the molecule, 1 'bit' is set within the linear table. For example, the fingerprint above can be written digitally as:

$$01000011000101000000001001100100000000111010000$$

with each occupied 'bin' being assigned a value of 1 and each empty 'bin' assigned as 0. This analysis is carried out independently for all compounds in the collection to generate a complete set of molecular 'fingerprints' for the library.

Comparison of these fingerprints to a reference compound is useful in searching for either similarity or dissimilarity. The reference compound itself can be chosen at

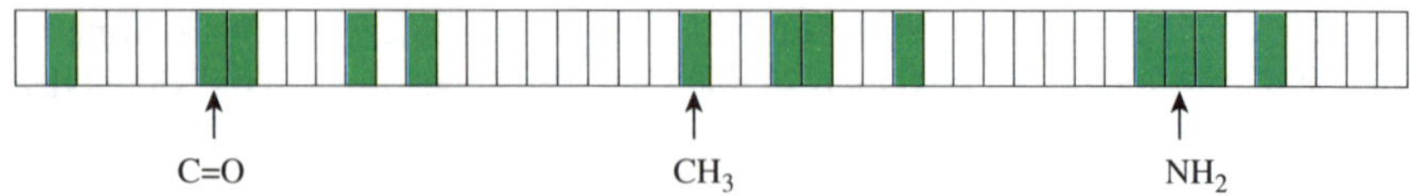

Figure 3 *Example of a fingerprint useful in determining similarity*

random or predetermined. The comparison involves the calculation of the so-called Tanimoto coefficients (Equation 1):

$$\text{Tanimoto Coefficient} = T(x,y) = \frac{N(x,y)}{N(x)+N(y)-N(x,y)} = \text{a measure of similarity}$$

$$\text{Tanimoto Dissimilarity Index} = 1 - T(x,y)$$

Equation 1 *Calculation of Tanimoto coefficients for similarity and dissimilarity*

The basis of the calculation is defined as the number of bits set to 1 that are common to both structures, $N(x,y)$ versus the number of bits set as 1 in either set, $N(x)$, $N(y)$. The selection process then follows the following general path:

With the reference compound as the starting point, pair-wise comparison of every compound with the reference structure identifies the most dissimilar compound, which is added to the collection. The software calculates the mean Tanimoto coefficient from the growing diverse collection ($\bar{x} = \sum x/n$: The sum of all coefficients divided by the number of compounds in the growing collection), and the compound with the coefficient furthest from that mean is added. This process is repeated until the collection either reaches the desired size or no further compounds can be added without, for example, exceeding some lower threshold for the difference in coefficients.

A similar pair-wise comparison can be used to evaluate the self-similarity of a database of structures. This approach also allows direct visual comparison of databases or database subsets if the coefficient distributions are plotted as a graph or histogram and this approach can be used for either self-similarity or for database comparison (Chart 1).

Many other approaches have been developed for measuring similarity and dissimilarity, including a multitude of variations on clustering or partitioning strategies. It is not possible to summarise here the wide variety of structure-based methods for diversity analysis; however, a number of detailed papers and reviews are available, particularly recommended for advanced theory is that of Agrafiotis.[6,7]

7   Pharmacophore Analysis

The generation of a diverse set of compounds based on analysis of the pharmacophores they contain is rather more complex, requiring the generation of a pharmacophore fingerprint for all conformations of every molecule in the database. The pharmacophore concept discards the traditional perceptions of chemical structure,

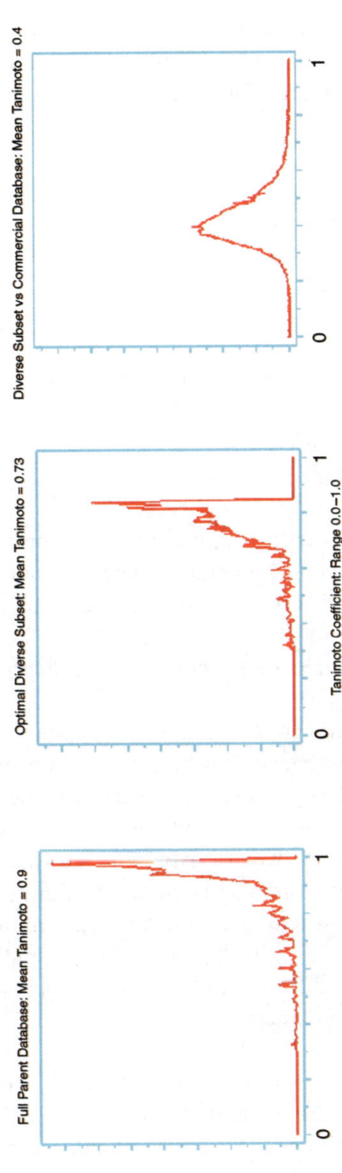

Chart 1 *Graphical example of self-similarity within a database and comparison of a diverse subset with an external commercial database*

focusing instead on a series of descriptors that represent functional groups and that could interact with a receptor. The obvious assumption is that a molecule must have valid pharmacophores to offer to a receptor in order to be considered drug-like; logic then dictates that compounds that have no valid pharmacophores should be removed from a database.

The number of descriptors commonly used in a pharmacophore analysis is refreshingly small; however, this advantage can be offset by the huge numbers of conformations that flexible molecules are able to adopt. The commonly used descriptors include:

- Hydrogen-bond donor,
- Hydrogen-bond acceptor,
- Charged centre,
- Aromatic ring,
- Hydrophobic group.

In addition to these, the following customisable descriptors may be added:

- Acid (including Lewis acid)
- Base (including Lewis base)

Generation of every three-centre pharmacophore is the method most commonly used; however, since three-centre pharmacophores do not represent three-dimensional information such as volume or chirality, four-centre pharmacophores are becoming increasingly common.

Figure 4 shows how a single small molecule can display a variety of pharmacophores.

Using all available structural conformations, the pharmacophores represented in a single molecule can also be encoded in much the same way as the fingerprints described above for dissimilarity searching (Figure 5) using distance metrics.[8]

Pairwise comparison of these fingerprints, in the same way as the dissimilarity analysis described above, is one way of using this information to generate a pharmacophorically diverse set of compounds. Since the Tanimoto and pharmacophore analyses measure different structural characteristics, a library that is "diverse" by one criterion may not be diverse by the other. For example, the decoration of a common core structure with different functional groups could lead to a wide variation in pharmacophore combinations without greatly changing the Tanimoto fingerprints.

Yet another approach was proposed by Muegge *et al.*, who suggested that molecules that are not "drug-like" are often 'under functionalised' and, as a consequence, do not offer valid pharmacophores to the target. Muegge also took the rather unusual viewpoint that the only pharmacophore elements that should be considered are those that might distinguish drug-like from non drug-like compounds. In this analysis, aromatic and other lipophilic groups are considered irrelevant; instead a series of 12 key functional groups are used. These include amine, amide, alcohol, ketone, sulphone,

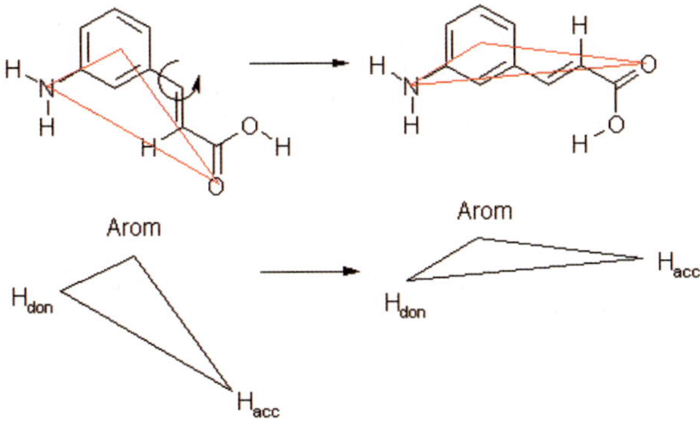

Figure 4 *The multiple pharmacophore concept from a single molecule*

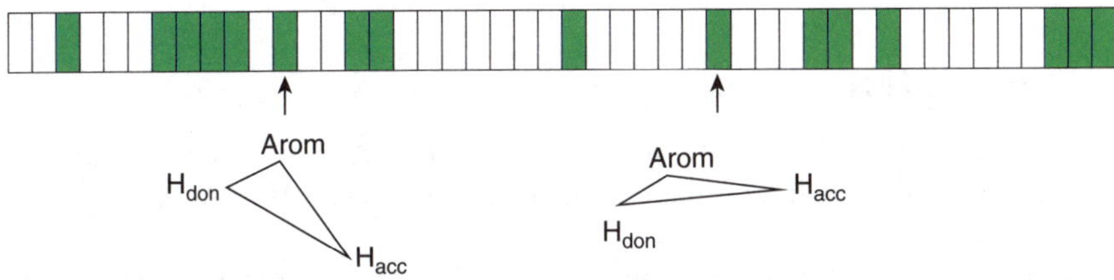

Figure 5 *Pharmacophore fingerprints*

sulphonamide, carboxylate, ester, carbamate, guanidine, urea and amidine moieties, which are, according to Muegge, the pharmacophore elements responsible for key interactions between ligands and targets. Since any compound exhibiting two or more independent (non-fused) pharmacophore elements may be considered drug-like in the absence of any other structural filters, it is essential to use this approach in parallel with additional structural filters that remove unwanted functional groups, such as reactive centres.

According to the more traditional pharmacophore approach, the molecule illustrated in Figure 6 could be a useful addition to a drug-like library. However, according to Muegge this compound only has a single pharmacophoric element, the amide. Aromatic groups and additional hydrogen-bonding options provided by the thiophene and morpholino rings, respectively, are not taken into consideration.

Sub-structure-, similarity- and pharmacophore-based search queries can also be used to generate targeted or focused compound sets, with pharmacophore-based methods also enabling 'lead-hopping',[9] a technique that can identify molecules with diverse core sub-structures which have the potential to match the target interactions of a lead ligand.[10] The use of such structural and pharmacophore-based analysis is currently gaining support in the form of library annotation, a process being adopted for the generation of collections of compounds having improved lead-like or drug-like properties.

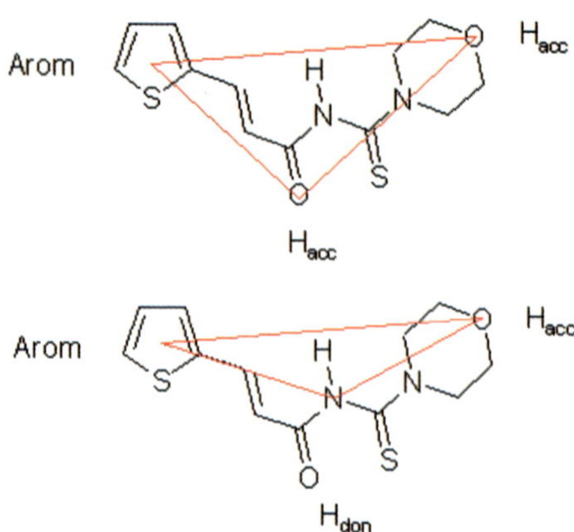

Figure 6 *Example of a molecule failing the Muegge analysis for drug-likeness*

8   Annotation

\An`no*ta"tion\, n. [L. annotatio: cf. F. annotation.] A note, added by way of comment, or explanation; usually in the plural (Webster's 1913 dictionary).

Annotation can take one of two main forms – biological annotation, which is the assignment of specific biological properties to known active compounds,[11] and chemical/structural annotation, which will be the only focus of this chapter.

The definition of annotation above illustrates just what is involved in this process – for each of a series of molecules in a database, notes, in this instance physicochemical data, are added. Comparison or filtering of the individual compounds according to these 'notes' can be used to determine which may be most lead- or drug-like.

Although the practice of using more lead- or drug-like compounds in primary biological screens has taken a while to emerge, the advancement in computational capabilities makes it increasingly practical to use such characteristics in selecting high-quality compounds for screening. Many authors have described methods for the evaluation of 'lead-likeness' and 'drug-likeness' within chemical libraries and compound collections; indeed, the approaches outlined are often as diverse as the compounds being analysed.[12–17]

Recent publications by Mozziconacci *et al.*,[18] Baurin *et al.*,[19] and Muegge *et al.*,[12] all demonstrate how the process of annotation can be used in manipulating multiple databases containing many hundreds of thousands or even millions of potential screening compounds. The detailed analyses are not so general that the described methods are universally applicable, since the filters and limits applied are often tailored to the target being considered. However, the methodology is adaptable. The general principles will be described here with frequent reference to the papers cited above. It is important to note that the various annotation methods are not necessarily right or wrong, or quantitatively accurate; rather, they represent more rational ways to select compounds from large databases.

The various authors describe how, from a series of compound databases offered by library providers, compounds were compared, filtered and annotated to yield relatively small subsets that might be considered lead-like. Although most research organisations are reluctant to reveal the specific metrics and filters used in their analysis procedures, it is likely that most companies follow similar approaches in assessing libraries from different vendors.

Most of the suppliers shown above in Table 1 are represented in the databases used by Baurin *et al.*.[19] The analysis that they reported is contained in Table 5, although it is now outdated. Their objective was to identify all of compounds available in commercial databases that may be considered drug-like, according to their criteria. The principle of their analysis is quite general, it is only the parameters that are likely to change from problem to problem, not the comparative operations themselves. Thus, different parameters and metrics may be employed by different research groups and in different disciplines (*e.g.* pharma, crop science, animal health, *etc.*), but the overall principle is the same. Indeed, the operations described below are applied, with many variants, by most of the major pharma research organizations, in many cases using proprietary metrics developed in-house.

The first question addressed by Baurin *et al.* was 'how many unique compounds can be identified from all of the databases?'. Based on data provided by 23 suppliers, the total number of compound entries was over 2.67 million. Removing duplicates from this list, mainly from the 'shared-pool' providers, left more than 1.62 million 'unique' structures. The next consideration was to select the appropriate filters to apply to determine drug-likeness, with the caveat that all of these filters would likely be subject to different interpretations by different research groups.

Nevertheless, there is a fairly general consensus within the pharma industry regarding functional groups that render a compound unacceptable as a drug candidate. Table 4 below lists groups that may lead to rejection of a compound for screening; this list is not necessarily complete, and some groups in the list may not be rejected

Table 4 *Functional groups that may lead to rejection of compounds as hits or candidates*

Alkyl halide (not F)	*Anhydride*	*Acylating agent*	*Polyfluoro alkyl*
Thiol	Disulphide	Peroxy	Sulphonyl/sulphinyl halide
Epoxide	Aziridine	Phosphoryl/phosphonyl/ phosphinyl halide	Azo
Diazo	Diazonium	Schiff base/imine	Ketene
Carbodiimide	Nitroso	Alkyl ketone	Alkyl aldehyde
Sulphate	Isonitrile	Isocyanide	Isocyanate
Isothiocyanate	N-halo	N-Hydroxy amine (oximes/oxime ether)	N-oxide
Open chain thiourea	Thiocarbonyl	Hydrazone	Non-metals (Si, B, P, As)
Organometallic	Adamantyl	Polynitrophenyl (nitro>=2)	Polyhalophenyl (halo >2)
Orthoquinone	Benzidine	β-Naphthylamine	Polyaromatics
Michael acceptor	Dye (extended chromophore)	Aminals	Ketal/acetal
Alpha-halo ketone	Alpha-dicarbonyl	Sulphonate ester	Phosphonate ester

by all researchers. However, since these groups are often associated with generation of false-positives because of generalised cytotoxicity or unwanted ligand receptor interactions such as irreversible binding,[20] it is usually desirable to remove compounds that contain them from screening collections.

There are, in addition to these simple functional group filters, a number of property-based filters that may be applied. These filters take the form of calculated metrics, such as the Lipinski Rule of Five (LRoF; Hydrogen-bond donors, Hydrogen-bond acceptors, Lipophilicity, Molecular weight), Solubility, total Polar Surface Area (tPSA), Blood–brain-barrier (BBB) Permeability, calculated metabolic filters (cADMET; Absorption-Distribution-Metabolism-Excretion-Toxicity) and Bioavailability.

9   Lipinski Rule-of-Five (LRoF)

The Lipinski Rule-of-Five almost certainly represents the first real attempt to identify a drug-like filter for molecular databases. Lipinski analysed the compounds within the World Drug Index that had progressed to Phase II clinical trials or better and discovered a simple set of rules that have been used extensively since their publication in 1997.[21] The LRoF states that a compound *may* have lower membrane permeability or poorer absorption if two or more of the rules are violated. These rules are: a compound should have no more than 5 hydrogen-bond donors, no more than 10 hydrogen bond acceptors, a calculated log P of 5 or less, and a molecular weight of 500 or less. Log P is the logarithm of the octanol-water partition coefficient, and thus is a measure of the lipophilicity of a molecule. The rule derives its name from the fact that all of the individual requirements are divisible by five (although there are only four rules!).

Not all drugs need to have a high level of permeability, nor do all drugs obey the rules, so the pass rate for marketed drugs is not 100% (compounds with active transport mechanisms of delivery are also excluded from the rules). Therefore, while the LRoF is a very useful decision-support tool for drug-likeness, it is not really a stand-alone function and, like many other tools, it should be used in conjunction with other filters, especially those based on structural requirements.

10   Topological Polar Surface Area (tPSA) and Blood–Brain-Barrier Permeability (Log BB)

These two functions are not truly independent parameters, since both correlate with the compound's log P value. Indeed, Log BB has also been calculated using PSA alone, as well as several other solvation and free energy methods.[22–24] After Young showed that experimental Log P values could be correlated with experimental Log BB data[25] algorithms were developed to calculate Log BB based on tPSA and c Log P (c Log P is a computed Log P value, based on a calculation of the contributions of the individual atoms and fragments in a molecule to its overall partition coefficient).[26,27]

The PSA is defined as the sum of the surfaces of the polar atoms in a molecule. Because this value can vary significantly with conformation, different approaches have been devised in order to use it practically. One approach has been to calculate the PSA of the molecule in a single energy minimised conformation. Another has been to calculate a topological PSA (tPSA), which is based on average contributions that the polar functional groups make in a variety of molecules. The calculation of tPSA is much faster, since energy minimization and/or a search for a low-energy conformation is not required. Of course, these different approaches do not all yield the same results, so PSA values, even for a given conformation, can vary widely!

One correlation between PSA and Log BB was provided by Clark *et al.*[27] who proposed the following, extremely simplistic, calculation:

$$\text{Log BB} = -0.0148\,\text{PSA} + 0.152c\,\text{Log}\,P + 0.139$$

This equation relies on a PSA value generated from a single conformer. More recent methods include variations from Fehér *et al.*, and Kelder *et al.*, [28,29] the latter group proposed a more direct correlation between Log BB and a so-called dynamic PSA (dPSA). The dPSA is challenging to obtain, since it is determined from a Boltzmann-weighted ensemble of molecular conformations. Note that, in Kelder *et al.*'s formulation of Log BB, *c* Log *P* is not considered at all:

$$\text{Log BB} = -0.032\,\text{dPSA} + 1.33$$

Moreover, these authors were able to simplify their calculation by correlating dPSA with static PSA values:

$$\text{dPSA} = 0.97\,\text{PSA} + 1.97$$

to give as a final equation:

$$\text{Log BB} = -0.031\,\text{PSA} + 1.267$$

Although Log BB is not relevant if blood–brain-barrier penetration is not an issue, the value of tPSA is of general importance, since it can be used as a secondary filter to discriminate between CNS, peripheral and gastrointestinal (Gi) targets. For example, van de Waterbeemd has shown that tPSA should ideally be below 90Å^2, with an acceptable working range of $20–120\text{Å}^2$, if compounds are going to penetrate the CNS. In contrast, for peripheral or Gi applications, tPSA can be as high as 190Å^2 for high molecular weight compounds (actual working values are dependent on the specific algorithm employed).[30]

Baurin employed tPSA as part of his 'Veber Filter';[31] a 'rule of two' involving tPSA and the number of rotatable bonds in a molecule. This filter allows an upper limit of 140Å^2 for tPSA and less than 10 rotatable bonds. In retrospect, these values are quite generous in light of van de Waterbeemd's data; a more generally used limit is 8 rotatable bonds, which limits the flexibility of a potential ligand.

11 Solubility

The experimental solubility of any given compound is quite straightforward to determine, given sufficient material. However, with a large number of samples, experimental determination of solubility becomes time consuming and tedious, and for databases encompassing millions of compounds, probably none of which are in hand, the measurement is impossible. While poor solubility does not necessarily rule the compound out as a marketable pharmaceutical, molecules with reasonable solubility are by far preferred over insoluble compounds for optimisation and development.

Although, a number of computational models for calculating solubility have been published, this filter remains one of the least used in selecting compounds from the larger commercial databases. Since the quality of many of the 'historical' compound collections was assessed by NMR, there is the reasonable expectation that these compounds are sufficiently soluble to screen. The observed retention time of a sample analysed by reversed-phase LC also gives an indication that a compound ought to have *some* aqueous solubility. As a consequence, many people rely on the simple fact that if a sample is sufficiently soluble in DMSO to give a good NMR spectrum or provides an LC trace, it will have enough solubility for screening

Nonetheless, although the details will not be presented here, it is possible to compute the solubility of a compound with sufficient confidence to be useful. For example, Baurin *et al.*[19] developed a proprietary algorithm based on 40 group-count descriptors and 5 general descriptors: molecular weight, polar surface area, $S \log P$ (a value of $\log P$ derived using atomic property contributions),[32] and two parameters, PEOE_RPC$^-$ and PEOE_RPC$^+$, that reflect polarisation of positive and negative charges on the molecule. The 40 group-count descriptors are based on SMILES representations generated within the Molecular Operating Environment (MOE) software from the Chemical Computing Group and represent the presence or absence of specific functional groups. Although the details are not presented here, the authors found that this combination of computed properties and functional descriptors provide solubility predictions that are sufficiently accurate to be useful in database filtering.

In addition to these fairly straightforward methods, several newer, more complex, analytical approaches have been developed. Among these are the multidimensional "BCUT" parameters developed by Pearlman[33] or adaptations of the genetic algorithms and neural networks originally used for assessment of combinatorial library diversity.[34–36]

Another important subset of current strategies are the virtual screening methods, which can be used as standalone applications or in conjunction with the other filters just described. Virtual or *"in silico"* screening has gained prominence in recent years as a method for identifying molecules that may interact favourably with a biological target prior to actual acquisition and screening.[37] From the point of view of the library vendors, this approach reduces the volume of their business, since virtual screening is employed to limit the number of compounds tested. Often, the actual wet screen will then focus on small selections of closely related compounds, potentially affording useful SAR information from the first screening round.

The starting requirements for a virtual screen are (1) structural information for the biological target and (2) a database ("virtual library") of 3D structures as the candidate ligands. Two different approaches have been adopted for addressing the conformational flexibility of most small molecule ligands: multiple 3D conformations may be generated and stored ahead of time for each molecule, or all the conformers may be generated 'on-the-fly' during the virtual screen, which reduces the storage requirements for the database.[38] The virtual screen itself involves docking the molecules within the target receptor/enzyme computationally and "scoring" the resulting complex by considering favourable and unfavourable interactions. The better the score, the better the binding energy between the ligand and receptor is expected to be. With the improvement in computer hardware, the virtual screening process has increased in speed and efficiency, and, of course, complexity, leading to better modelling of ligand-receptor interactions and thus more effective selection of compounds for use in the 'wet' screening process.

A further evolutionary step based on virtual screening is the so-called 'design in receptor' or fragment screening approaches, which take as their starting point experimental observations of weakly bound, generally quite small molecules that bind to the receptor site. The basic strategy involves the selection of small, drug-like molecules that are co-crystallised with the target receptor. X-ray crystallography determines which molecules interact with the receptor in an appropriate way, such that they can serve as a "core seed" that can be elaborated and grown into a more potent compound. This approach borrows from both focused screening and structure-based design concepts.

12   Examples of the Use of Chemical Annotation and Pharmacophore-Based Lead-Hopping

It is instructive to show real examples of how database management and analysis techniques have been used to interrogate chemical databases in order to retrieve compounds useful for screening. Publications by Baurin and Webb illustrate how commercial databases can be analysed to identify drug-like compounds, and how assembly and screening of a diverse 'internal' collection can generate potential new lead- or drug-like compounds for a target based on pharmacophore information.

As described above, Baurin *et al.*,[19] analysed the commercial databases provided by a number of compound library vendors. Table 5, below, shows the results of their analysis, which, although now out of date, illustrates how millions of compounds can be condensed to a more manageable number of molecules more likely to be useful as leads. Some explanation of the individual filters is given in the footnote to the table, and elaborated in the original publication. This table is not included as an endorsement for either the methods or the algorithms used in Baurin *et al.*'s calculations; we reproduce the data simply to illustrate the operations that can be applied to a master database in attempting to enhance the properties of a subset of compounds prior to the application of further selection criteria.

In an earlier study carried out at the University of Orleans, also involving Baurin,[18] an analysis of 15 databases from commercial and other sources showed that these offerings can be of very acceptable quality. Their findings are reproduced in Figure 8, which shows the percentage distribution of compounds from each supplier that fail one or more filters applied to the individual collections.

These analytical filters reduce an otherwise overpowering number of compounds to a manageable number of higher quality structures primed for more individualised selection processes. As mentioned above, one of the individualised approaches for

Table 5 *Analysis of drug-like properties computed for compounds from 23 suppliers[19] (Reproduced with permission of the American Chemical Society from Ref. 19)*

Provider	Library size	Lip-3	Lip-4	Veber	Medchem-1	Medchem-2	Medchem-1+ Medchem-2	Solubility	Caco-2	All-filtered drug-like
AnalytiCon	765	572/75%	208/27%	148/19%	640/84%	745/97%	620/81%	650/85%	509/67%	76/10%
ChemOvation	1049	1016/97%	767/73%	1034/99%	1044/100%	1035/99%	1030/98%	824/79%	1044/100%	721/69%
Aurora	2245	2186/97%	1974/88%	2135/95%	1735/77%	2063/92%	1553/69%	1966/88%	2210/98%	1325/59%
Menai	3777	3709/98%	3239/86%	3725/99%	3121/83%	2636/70%	1980/52%	3079/82%	3771/100%	1518/40%
TOSLab	6669	5788/87%	4422/66%	6325/95%	5853/88%	4770/72%	3954/59%	4181/63%	6607/99%	2402/36%
MDPI	8734	8295/95%	6947/80%	8135/93%	6625/76%	6840/78%	4731/54%	6658/76%	8632/99%	3744/43%
Biofocus	10486	10227/98%	8755/83%	10415/99%	10253/98%	9987/95%	9754/93%	8733/83%	10219/97%	7635/73%
Bionet	34699	33870/98%	26317/76%	34449/99%	31944/92%	28982/84%	26227/76%	25541/74%	34598/100%	18410/53%
Maybridge	56405	55237/98%	45868/81%	55422/98%	49252/87%	44708/79%	37555/67%	43463/77%	56031/99%	27736/49%
Tripos	92920	89400/96%	74685/80%	89242/96%	79467/86%	82624/89%	69171/74%	74873/81%	92207/99%	50870/55%
MCL	105868	88674/84%	57397/54%	100209/95%	82922/78%	77783/73%	54837/52%	45531/43%	105589/100%	24693/23%
Vitas-M	113669	105944/93%	84035/74%	108939/96%	94494/83%	86634/76%	67459/59%	71909/63%	113027/99%	42402/37%
UKR	118426	99313/84%	70645/60%	106954/90%	106226/90%	95351/81%	83151/70%	59050/50%	118184/100%	33179/28%
Enamine	122859	112516/92%	85808/70%	117107/95%	117200/95%	112702/92%	107043/87%	75610/62%	122189/99%	59618/49%
TimTec	128181	119889/94%	96230/75%	120582/94%	105343/82%	92364/72%	69526/54%	84588/66%	126803/99%	45500/35%
IFLab	149862	140446/94%	111746/75%	143309/96%	134705/90%	119526/80%	104369/70%	94610/63%	148815/99%	62225/42%
Comgenex	164663	120301/73%	63807/39%	129847/79%	142080/86%	155855/95%	133272/81%	66409/40%	164478/100%	37396/23%
Chem T&I	165054	144080/87%	111375/67%	159219/96%	147320/89%	140990/85%	123256/75%	89638/54%	164663/100%	60489/37%
Asinex	202542	191589/95%	158556/78%	193433/96%	173521/86%	163233/81%	134212/66%	142074/70%	200517/99%	94092/46%
Specs	220558	204541/93%	157936/72%	212515/96%	190465/86%	177899/81%	147806/67%	135105/61%	219307/99%	88776/40%
IBS	283825	262741/93%	200287/71%	269413/95%	254317/90%	223889/79%	194381/68%	174145/61%	280122/99%	108803/38%
ChemBridge	319537	314309/98%	262965/82%	311830/98%	275327/86%	264474/83%	220264/69%	229916/72%	317508/99%	154800/48%
ChemDiv	362229	336308/93%	261553/72%	343289/95%	315420/87%	280319/77%	233510/64%	224476/62%	359038/99%	138647/38%
Sum	2675022									
Unique	1622763	1450529/89%	1077731/66%	1511941/93%	1423152/88%	1341231/83%	1141620/70%	949652/59%	1611920/99%	607223/37%
WDI	1141	1128/99%	1037/91%	1043/91%	1052/92%	930/82%	841/74%	1062/93%	1073/94%	708/62%

Note: From Baurin *et al.*,[19]: "To facilitate reading of the table, the suppliers are sorted according to the size of their library. Lip-3 is the number of molecules fulfilling 3 out of 4 Lipinski rules, and Lip4 is the number of molecules fulfilling all 4 Lipinski rules. MedChem-1 and MedChem-2 columns contain the number of compounds passing these MedChem tractability filters. MedChem-1+MedChem-2 compounds pass MedChem-1 and MedChem-2 filters. The all-filtered drug-like column totals the compounds passing Lip-4 + Veber + MedChem-1 + MedChem-2 + Solubility + Caco-2 filters".
From the total of 2.67 million compounds available from the 23 suppliers, 1.62 million unique structures were identified, of which just over 607,000 (37%) pass all the applied filters for drug-likeness. The databases for all 23 suppliers would need to be analysed to gain access to all 607,000 structures. The question thus arises: could that operation be made more manageable if only a subset of suppliers were to be considered; in other words, how few suppliers would be able to provide, between themselves, a workable proportion, say 90%, of those unique compounds? The graph in Figure 7 shows that, by using approximately half of the suppliers, 90% of the unique, drug-like compounds would be accessible.

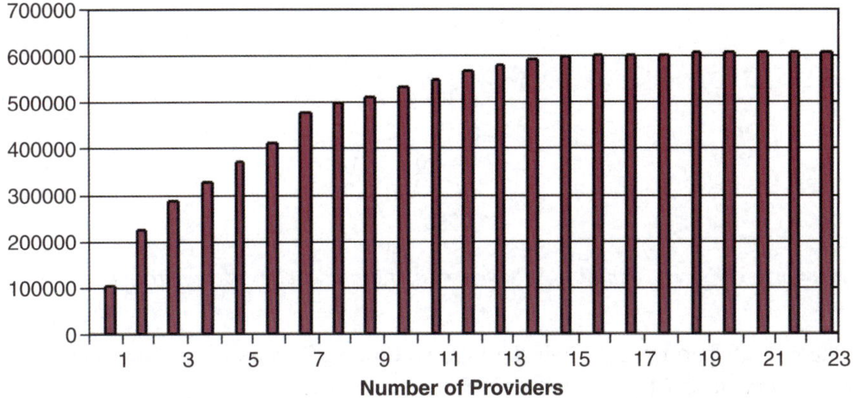

Figure 7 *Plot of the cumulative number of all unique structures passing all filters (Reproduced with permission of the American Chemical Society from Ref. 19)*

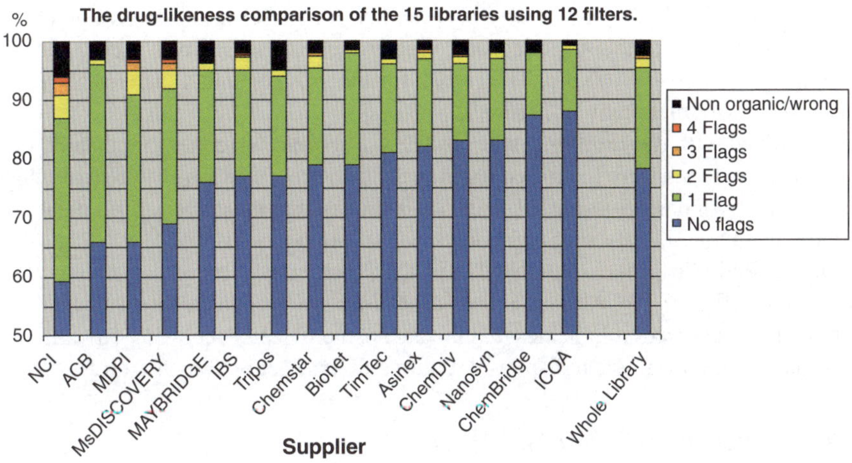

Figure 8 *Drug-likeness comparisons from 15 supplier databases (Reproduced with permission of the University of Orleans)*

further compound selection is lead-hopping, which involves identifying structures that match pharmacophoric models generated from known ligands. An illustration of this approach was reported by Webb *et al.*,[10] who sought novel ligands that would bind selectively to one of the adenosine receptors, specifically sub-type A_3. The pharmacophore query used to drive the lead-hopping process was generated from 1 of 4 compounds previously reported to have A_3 receptor selectivity.[39–41] Using a commercial software package, Chem-X, as the pharmacophore generator and search engine and compound 1 as the starting point, 9 distinct and viable pharmacophores were generated using the default software settings. These 9 pharmacophores were then used as search queries within a refined version of a set of compounds biased

Compound 1 *Known adenosine A$_3$ ligand used to generate pharmacophore queries*

towards CNS-permeability and the hits were ranked according to availability and promiscuity (number of rotatable bonds).

Hit compounds that exhibited at least 4 of the 9 available pharmacophores and had the lowest promiscuity were selected. Using these simple criteria, 186 compounds were chosen for screening, yielding 16 compounds that were specific for the A$_3$ adenosine receptor, 3 more that were specific for the A$_1$ receptor and a further 6 that were specific for the A$_{2A}$ receptor. Subsequently, 6 more examples were found that bound specifically to the A$_{2B}$ adenosine receptor, one of which was highly specific with a K_i value below 100 nm. The results for the A$_1$, A$_{2A}$ and A$_3$ adenosine receptors are shown graphically in Figure 9.

Combined with the additional data for the A$_{2B}$ receptor screen, more than 30 new sub-structure types were identified for the adenosine receptor family, none of which had been reported previously. This analysis demonstrated the power of this database-mining technique in selecting compounds for screening.

Interestingly 20 of the 26 active hits in the graph above shared a common pharmacophore (Figure 10). Although, this result does not account for the sub-type differentiation observed, it indicates that the adenosine receptor recognises a particular pharmacophore, even when displayed within a diverse set of sub-structures.

13 Compound Acquisition

The majority of the companies listed in Tables 1 and 2 store their available compounds in the form of dry powders (or whatever the natural state is), although samples dissolved in DMSO are becoming more available. For dry, solvent-free compounds selected from a supplier database, the samples will be weighed, usually manually, into cluster tubes or vials before shipping to the end user. This mode of operation may seem tedious, however, it does keep costs down for the preparation of samples for which automated weighing, *e.g.* of resinous compounds, is profoundly difficult.

Handling samples obtained in solution is more straightforward. The solutions are usually in DMSO, stored frozen between +4 and –20 °C. Automated plate duplication and cherry-picking of individual wells are now commonplace procedures. Thus, end users are not forced to accept all compounds from a specific plate, since the ability to generate daughter plates from parents is a simple matter. Samples can be dispensed into any compatible well-based system, such as monoblocks or cluster tubes, or 96-well plates can be combined to create higher density plates. The use of solution-phase samples also allows a significant reduction in deliverable sample sizes, often

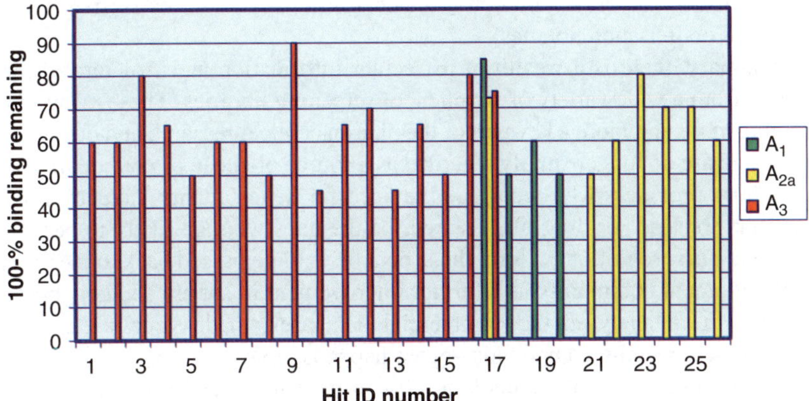

Figure 9 *Distribution of active adenosine ligands identified in in vitro assays (Reproduced with permission of the authors from Ref. 6)*

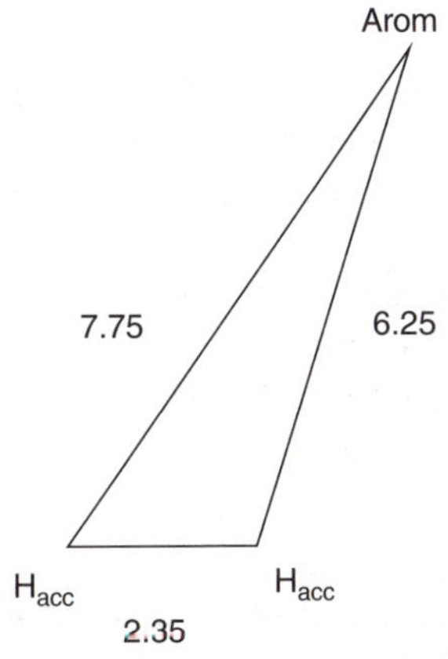

Figure 10 *Pharmacophore present in 20 of 26 active hits (Reproduced with permission of the author of Ref. 10)*

well below the limits of accurate dry powder weighing, and it is of course much faster. If delivery of the compound in solution is not desirable (because DMSO solutions should ideally be transported in the frozen state), solvent removal is always an option.

Plate formats are flexible, with more library suppliers accepting requests for the higher density screening plates, such as 384- and 1536-well formats, with a concomitant reduction in sample size delivered. Of course, the reduction in sample size however does not necessarily result in a pro-rata reduction in cost!

The vast majority of compounds are provided with no intellectual property rights retained by the supplier. As a consequence, the end user is able to capture IP for use of the compounds immediately on making a discovery. Composition of matter rights is another issue, however, since in almost all cases, the compounds are already in the public domain. Thus, in these situations, the end user only creates novel compositions of matter through chemical modifications to the initial hits as part of a lead-optimisation effort. The exceptions are libraries provided on a custom-synthesis basis (frequently hit-to-lead/lead optimisation libraries).

In summary, there is a wealth of molecular information and structural diversity available from a wide variety of suppliers of screening libraries. These resources are becoming more and more a key part of the pharmaceutical research and development process and are no longer simply regarded as a source of cheap or poor quality structures. Those providers who recognise the need for better quality molecules that fit descriptors for lead- or drug-likeness will be the ones who are likely to benefit the most from this recognition. In turn, those benefits will be passed back to the research and development community as a further improvement in supply and collaboration.

So, you have analysed the commercial databases, made your selections and ordered your screening compounds – what happens next?

Subsequent chapters in this book describe further methods for measuring library diversity, the design of focused libraries, and compound library management and screening.

Acknowledgments

At the time of writing the group at the University of Orleans have published new data on an expanded selection of chemical library providers. This information was part of the 10th Electronic Computational Chemistry Conference (ECCC10) and is available only online:

A. Monge, A. Arrault, C. Marot and L. Morin-Allory, *Analysis of a set of 2.6 million unique compounds gathered from the libraries of 32 chemical providers*
http://www.univ-orleans.fr/icoa/eposter/eccc10/monge/.

References

1. R.S. Bohacek, C. McMartin and W.C. Guida, The art and practise of structure based drug design, *Med. Res. Rev.*, 1996, **16**(1), 3–50.
2. M.M. Hann and T.I. Oprea, Pursuing the leadlikeness concept in pharmaceutical research, *Current Opinion in Chemical Biology*, 2004, **8**, 255–263.
3. K.S. Lam, S.E. Salmon, E.M. Hersh, V.J. Hruby, W.M. Kazmierski and R.J. Knapp, A new type of synthetic peptide library for identifying ligand-binding activity, *Nature*, 1991, **354**(6348), 82–84.
4. W.A. Warr, *The Directory of Molecular Diversity Suppliers*, Publ., Select Biosciences Ltd., Crestland House, Sudbury, UK, 2004.
5. Website created on July 28–30, 1998. http://www.daylight.com/meetings/summerschool98/course/basics/fp.html.
6. J.W. Raymond, C.J. Blankley and P. Willett, Comparison of chemical clustering methods using graph-based and fingerprint-based similarity measures, *J. Mol. Graphics and Modelling*, 2003, **21**(5), 421–433.
7. D.K. Agrafiotis, Diversity of chemical libraries, in *The Encyclopedia of Computational Chemistry*, P.v.R. Schleyer, N.L. Allinger, T. Clark, J. Gasteiger, P.A. Kollman, H.F. Schaefer III and P.R. Schreiner (eds), Wiley, Chichester, 1998, **1**, 742–761.

8. J.S. Mason and D.L. Cheney, *Library Design and Virtual Screening Using Multiple 4-Point Pharmacophore Fingerprints*, Pacific Symposium on Biocomputing, 2000, **5**, 573–584.
(also at http://helix-web.stanford.edu/psb00/mason.pdf).

9. R.D. Cramer, M.A. Poss, M.A. Hermsmeier, T.J. Caulfield, M.C. Kowala and M.T. Valentine, Prospective identification of biologically active structures by topomer shape similarity searching, *J. Med. Chem.*, 1999, **42**, 3919–3933.

10. T.R. Webb, N. Melman, D. Lvovskiy, X-D. Ji and K.A. Jacobson, The utilization of a unified pharmacophore query in the discovery of new antagonists of the adenosine receptor family, *Bioorg. Med. Chem. Lett.*, 2000, **10**, 31–34.

11. D.E. Root, S.P. Flaherty, B.P. Kelley and B.R. Stockwell, Biological mechanism profiling using an annotated compound library, *Chem. Biol.*, 2003, **10**, 881–892.

12. I. Muegge, Selection criteria for drug-like compounds, *Med. Res. Rev.*, 2003, **23**, 302–321.

13. M. Brustle, B. Beck, T. Schindler, W. King, T. Mitchell and T. Clark, Descriptors, physical properties & drug-likeness, *J. Med. Chem.*, 2002, **45**, 3345–3355.

14. G.M. Rishton, Nonleadlikeness and leadlikeness in biochemical screening, *Drug Discovery Today*, 2003, **8**, 86–96.

15. W.P. Walters and M.A. Murcko, Prediction of drug likeness, *Adv. Drug Delivery Rev.*, 2002, **54**, 255–271.

16. M.A. Miller, Chemical database techniques in drug discovery, *Nat. Rev.*, 2002, **1**, 220–227.

17. V.V. Zernov, K.V. Balakin, A.A. Ivaschenko, N.P. Savchuk and I.V. Pletnev, Drug discovery using support vector machines. The case studies of drug-likeness, agrochemical-likeness and enzyme inhibition predictions, *J. Chem. Inf. Comput. Sci.*, 2003, **43**, 2048–2056.

18. J-C. Mozziconacci, E. Arnoult, N. Baurin, C. Marot and L. Morin-Allory, *Preparation of a Molecular Database From a Set of 2 Million Compounds for Virtual Screening Applications: Gathering, Structural Analysis and Filtering*, http://www.univ orleans.fr/SCIENCES/ICOA/eposter/eccc9/ECCC9i.htm.

19. N. Baurin, R. Baker, C. Richardson, I. Chen, N. Foloppe, A. Potter, A. Jordan, S. Roughley, M. Parratt, D. Greaney, D. Morley and R.E. Hubbard, Drug-like annotation and duplicate analysis of a 23-Supplier chemical databasetotalling 2.7 million compounds, *J. Chem. Inf. Comput. Sci.*, 2004, **44**, 643–651.

20. G.M. Rishton, Reactive compounds and *in vitro* false positives in HTS, *Drug Discovery Today*, 1997, **2**, 382–384.

21. C.A. Lipinsky, F. Lombardo, B.W. Dominy and P.J. Feeney, Experimental and computational approaches to estimate solubility and permeability in drug discovery and development settings, *Adv. Drug Delivery Rev.*, 1997, **23**, 3–25.

22. X.C. Fu, C.X. Chen, W.Q. Liang and Q.S. Yu, Predicting blood-brain barrier penetration of drugs by polar molecular surface area and molecular volume, *Acta Pharmacol. Sin.*, 2001, **22**(7), 663–668.

23. Y.N. Kaznessis, M.E. Snow and C.J. Blankley, Prediction of blood-brain partitioning using monte carlo simulations of molecules in water, *J. Comp. Aided Mol. Des.*, 2001, **15**, 697–708.

24. C-L. Ma, C. Chen and J. Yang. Predictive model of blood-brain barrier penetration of organic compounds, *Acta Pharm. Sin.*, 2005, **26**, 500–512.

25. R.C. Young, R.C. Mitchell, T.H. Brown, C.R. Ganellin, R. Griffiths, M. Jones, K.K. Rana, D. Saunders, I.R. Smith, N.E. Sore and T.J. Wilks, Development of a new physicochemical model for brain penetration and its application to the design of centrally acting H_2 receptor histamine antagonists, *J. Med. Chem.*, 1988, **31**, 656–671.

26. R.F. Rekker, *The Hydrophobic Fragmental Constant: It's Derivation and Application With a Means of Characterizing Membrane Systems*, Elsevier Scientific Publishing Company, The Netherlands, 1977.

27. D.E. Clark, Rapid calculation of polar molecular surface area and its application to the prediction of transport phenomena. 2. Prediction of blood-brain barrier penetration, *J. Pharm. Sci.*, 1999, **88**, 815–821.

28. M. Fehér, E. Sourial and J.M. Schmidt, A simple model for the prediction of blood-brain partitioning, *Int. J. Pharm.*, 2000, **201**, 239–247.

29. J. Kelder, P.D.J. Grootenhuis, D.M. Bayada, L.P.C. Delbressine and J.P. Ploemen, Polar molecular surface as a dominating determinant for oral absorption and brain penetration of drugs, *Pharm. Res.*, 1999, **16**, 1514–1519.

30. H. van de Waterbeemd, D.A. Smith and B.C. Jones, Lipophilicity in PK design: methyl, ethyl, Futile, *J. Comp. Aided Mol. Design*, 2001, **15**, 273–286.

31. D.F. Veber, S.R. Johnson, H-Y. Cheng, B.R. Smith, K.W. Ward and K.D Kopple, Molecular properties that influence the oral bioavailability of drug candidates, *J. Med. Chem.*, 2002, **45**, 2615–2623.

32. S.A. Wildman and G.M. Crippen, Prediction of physiochemical parameters by atomic contributions, *J. Chem. Inf. Comput. Sci.*, 1999, **39**(5), 868–873.

33. R.S. Pearlman and K.M. Smith, *3D-QSAR in Drug Design*, Vol. 2, H. Kubinyi *et al.*, (ed), Kluwer-Escom, The Netherlands, 1998, 339–353.

34. J. Sadowski, M. Wagener and J. Gasteiger, Assessing similarity and diversity of combinatorial libraries by spatial autocorrelation functions and neural networks, *Angew. Chem. Intl. Ed. Engl.*, 1995, **34**, 2674–2677.

35. R.P. Sheridan and S.K. Kearsley, Using a genetic algorithm to suggest combinatorial libraries, *J. Chem. Inf. Comput. Sci.*, 1995, **35**, 310–320.

36. R.D. Brown and Y.C. Martin, Use of structure-activity data to compare structure-based clustering methods and descriptors for use in compound selection, *J. Chem. Inf. Comput. Sci.*, 1996, **36**, 572–584.

37. W.P. Walters, M.T. Stahl and M.A. Murcko, Virtual screening – an overview, *Drug Discovery Today*, 1998, **3**, 160–178.

38. B. Waszkowycz, T.D.J. Perkins, R.A. Sykes and J. Li, Large scale virtual screening for discovering leads in the postgenomic Era, *IBM Systems J.*, 2001, **40**(2), 360–376.

39. Y. Karton, J-I. Jiang, X-D. Ji, N. Melman, M.E. Olah, G.L. Stiles and K.A. Jacobson, Synthesis and biological activities of flavonoid derivatives as A_3 adenosine receptor antagonists, *J. Med. Chem.*, 1996, **39**, 2293–2301.

40. Y-C. Kim, X-D. Ji and K.A. Jacobson, Derivatives of the triazoloquinazoline adenosine antagonist (CGS15943) Are selective for the human A_3 receptor subtype, *J. Med. Chem.*, 1996, **39**, 4142–4148.

41. M.A. Jacobson, P.K. Chakravarty, R.G. Johnson and R. Norton, Novel, selective, non-xanthine A_3 adenosine receptor antagonists, *Drug Devel. Res.*, 1996, **37**, 131.

Chemical Diversity: Definition and Quantification

ALAN C. GIBBS AND DIMITRIS K. AGRAFIOTIS

Johnson & Johnson Pharmaceutical Research and Development, LLC,
665 Stockton Drive Exton, PA 19431, USA

1 Introduction

The advent of chemogenomics, combinatorial chemistry, and high-throughput screening has necessitated the development of efficient means to characterize the structural and functional diversity of compound collections. While functional diversity implies biological heterogeneity, it is implicitly related to structural diversity – compound constitution, configuration, and conformation. The following chapter, though not exhaustive, describes the quantitative characterization of chemical diversity, a means to measure the extent of differential features and properties within a compound collection.

The measurement of molecular diversity requires the definition of a chemical space. This N-dimensional chemical space is represented by a group of selected molecular descriptors. Each compound in a collection can be assigned coordinates based on the measurement of its descriptor values. Increasing distance, within the dimensions of the assigned coordinate space, should correlate with increasing diversity (or decreasing similarity) between compounds.

The dimensionality of chemical structure space exceeds that of known biological functional space. The dimensionality of biological functional space has increased in recent years due to the discovery of a multitude of genes, largely from the Human Genome Project.[1] This chapter, however, will focus on chemical diversity rather than functional diversity. Quantification of chemical diversity involves two areas: first, the predefinition of a chemical space, accomplished by selection of a diversity metric and a compound representation (*i.e.*, molecular descriptors); and second, a rational subset selection, or classification, method dependent on efficient dimensionality reduction. Here, we describe these methods, prerequisites for a definition of chemical diversity. For further details on this subject, the reader can refer to a number of excellent reviews.[2–6]

2   Diversity Metrics

A diversity metric is a function to aid the quantification of the diversity of a set of compounds in some predefined chemical space. Diversity metrics fall into three main classes: (1) Distance-based methods, which express diversity as a function of the pairwise molecular dissimilarities defined through measurement. (2) Cell-based methods, which define diversity in terms of occupancy of a finite number of cells that represent disjoint regions of chemical space. (3) Variance-based methods, which quantify diversity based on the degree of correlation between a compound's important features.

2.1   Distance-Based Metrics

Distance-based metrics quantify the diversity of a set of compounds as a function of their pairwise (dis)similarities in a descriptor space.[7,8] It is important to mention that distance coefficients are analogous to distances in multidimensional geometric space, although they are usually not equivalent to such distances. For a distance coefficient to be described as a metric, it must possess the following four properties[9]: (1) Distance values must be nonzero and the distance from an object to itself must be zero. (2) Distance values must be symmetric. (3) Distance values must obey the triangular inequality. (4) Distances between nonidentical objects must be greater than zero. A coefficient containing only the first three properties is dubbed a pseudo-metric, and one without the third property is a nonmetric.

The descriptors used for pairwise distance measurements can be continuous, as in a physicochemical property, or binary (*e.g.*, the presence or absence of a specific substructure). For continuous chemical spaces, nearly all metrics are based on the generalized Minkowski metric given in (1), where x_{ik} represents the kth feature of the ith molecule, k is the total number of features, and r the order of the metric.

$$d_{ij} = \left[\sum_{k=1}^{K} \left| x_{ik} - x_{jk} \right|^r \right]^{1/r} \tag{1}$$

The most common example, the Euclidean distance, where $r = 2$ is shown in (2).

$$d_{ij} = \sqrt{ \sum_{k=1}^{K} \left| x_{ik} - x_{jk} \right|^2 } \tag{2}$$

Two popular distance metrics are the Manhattan distance metric ($r = 1$), which represents the sum of the absolute descriptor differences, and the ultrametric ($r = \infty$), which represents the maximum absolute descriptor difference. Both the Manhattan and Euclidean distance metrics obey all four metric properties.

When the descriptors represent counts or sums, the metrics in (3) and (4) have been used, where x_{ik} is (as it is in the generalized Minkowski metric) the value of the kth descriptor in the ith compound, and K the total number of descriptors.[10]

$$d_{ij} = 1 - \frac{2.\sum\limits_{k=1}^{K} \min(x_{ik}, x_{jk})}{\sum\limits_{k=1}^{K} x_{ik} + \sum\limits_{k=1}^{K} x_{jk}} \tag{3}$$

$$d_{ij} = 1 - \frac{\sum\limits_{k=1}^{K} \min(x_{ik}, x_{jk})}{\sum\limits_{k=1}^{K} \max(x_{ik}, x_{jk})} \tag{4}$$

Both indices range from zero to one, zero indicating complete identity and one indicating that the two structures are most dissimilar.

For binary descriptors, the most commonly used distance function is the Tanimoto (or Jaccard) coefficient given in (5). Here x and y are two binary sets (encoded molecules). *AND* is the bitwise "and" operation (a bit in the result is set if both corresponding bits in the two operands are set), and *IOR* the bitwise "inclusive or" operation (a bit in the result is set if either corresponding bits in the two operands is set). The result, T, is a measure of the number of features shared by the two molecules relative to the ones they could have in common.

$$T = \frac{|AND(x, y)|}{|IOR(x, y)|} \tag{5}$$

Other metrics include the Hamming distance metric, given by (6). *XOR* is the bitwise "exclusive or" operation (a bit in the result is set if the corresponding bits in the two operands are different), and N the number of bits in each set. The Dice coefficient is defined by (7).

$$H = \frac{|XOR(x, y)|}{N} \tag{6}$$

$$D = \frac{2|AND(x, y)|}{|x| + |y|} \tag{7}$$

With respect to metric properties, the Tanimoto coefficient obeys all four properties if dichotomous variables are used. Complement does not obey the triangular inequality in the Dice coefficient.

When using binary descriptors, the Euclidean distance can be reformed to the form given in (8). Where $NOT(x)$ denotes the binary complement of x, and the expression $|XOR(x, NOT(y))|$ represents the number of bits that are identical in x and y (either ones or zeroes).

$$E = \sqrt{N - |XOR(x, NOT(y))|} \tag{8}$$

The Euclidean distance is a good measure of similarity when the binary sets are relatively rich, and is mostly used in situations in which similarity is measured in a *relative* sense (the distance of two compounds to the same target). James and co-workers[11] prefer the Tanimoto coefficient when *absolute* comparisons (between two independent pairs of molecules) are made.

After a distance function is defined, the diversity of a compound collection can be measured in a number of ways. Minimum intermolecular dissimilarity (9) (where d_{ij} is the distance between the *i*th and *j*th compounds in the collection C), and average nearest neighbor distance, (10), are two common examples of distance-based diversity measures. Figure 1 illustrates examples of compound subsets using a nearest-neighbor design metric.

$$D(C) = \min_{i<j} d_{ij} = \min_i \min_{j \neq i} d_{ij} \qquad (9)$$

$$D(C) = \frac{1}{N} \sum_i \min_{j \neq i} d_{ij} \qquad (10)$$

The calculation time for all these functions (and related functions such as the power-sum, product, minimum-spanning tree, *etc.*) depends on the size of the compound collection according to a square-law – their quadratic dependence on the number of compounds in C.[12] When the dimensionality of the space is relatively small (<10), the nearest neighbor computation in (9) and (10) can be carried out in an efficient manner using a combinatorial data structure known as a *k*-dimensional tree. However, this algorithm becomes less efficient as the dimensionality of the space increases and eventually degrades to a quadratic-order algorithm, with the additional overhead of constructing and traversing the tree.

An alternative algorithm, the Kolmogorov–Smirnov statistic has recently been proposed, which is not effected by the dimensionality problem.[13] The Kolmogorov–Smirnov statistic does not depend on the size of the compound collection, and can be used to perform diversity analysis on both global and local scale. Thus, the main advantage of the Kolmogorov–Smirnov statistic is that it allows meaningful comparison of data sets of different cardinality. Higgs and co-workers,[14] illustrate three examples of distance-based designs. The authors describe edge designs, spread designs, and coverage designs. Edge designs attempt to identify molecules at the edge of descriptor space that produce minimum variance estimates for linear-model parameters. Spread designs are used to identify subsets of molecules that are maximally dissimilar with respect to each other. Coverage designs identify a subset of molecules that are maximally similar to the ca ndidate set of molecules. These designs use algorithms with linear-time complexity; hence, they may also be applied to large chemical libraries.

2.2 Cell-Base Diversity Metrics

Cell-base diversity attempts to quantify diversity by dividing chemical space into hyper-rectangular regions and measuring the occupancy of the resulting cells. One advantage of these methods is that, unlike distance-based techniques, they can

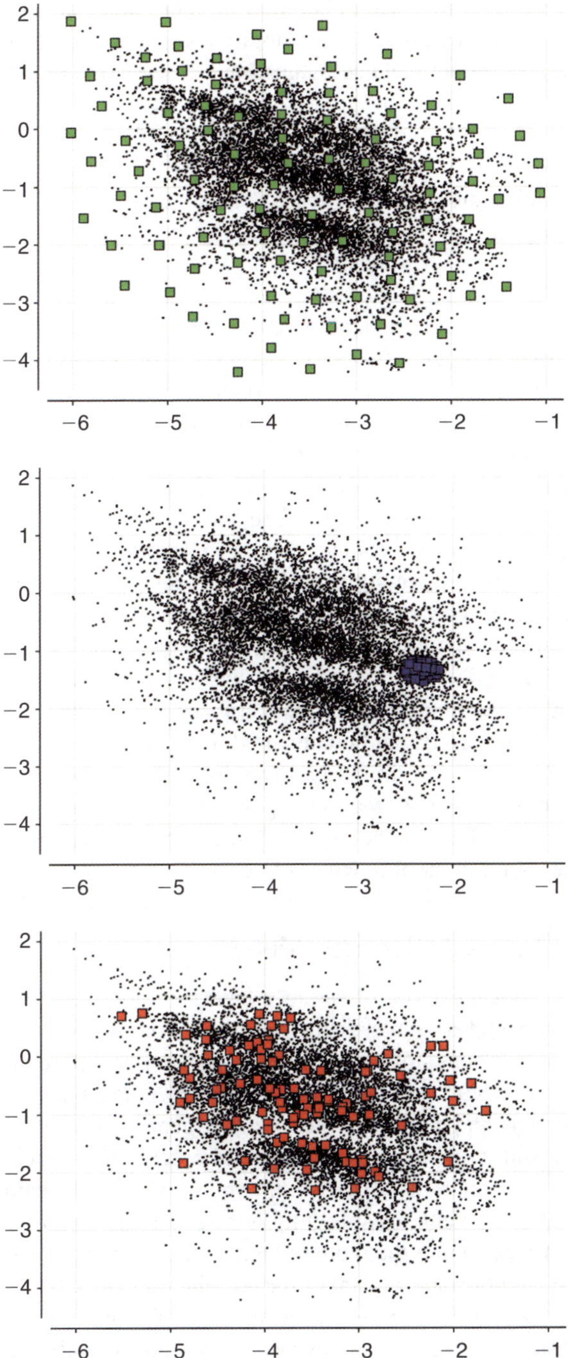

Figure 1 *SPE maps (see Section 4) showing subset selections using the nearest-neighbor distance metric for a diverse selection (top), and a similar selection (middle). A random selection is shown (bottom) for comparison. All selections are 100 compounds from a 10,000-member library*

encode absolute position in space in addition to intermolecular distance.[15] Another advantage is that diversity estimation and library comparisons can be carried out significantly faster, and are not plagued by quadratic complexity, as are most distance-based metrics.

The most intuitive cell-based diversity metric is simply the number of cells occupied by a design defined in (11) where δ_i is one if the ith cell is occupied and zero if it is not, M the total number of cells.

$$D(C) = \sum_{i=1}^{M} \delta_i \qquad (11)$$

This metric is a measure of absolute diversity, and can be used to determine the diversity of any library regardless of size. However, it does not take into account the clustering of the data set, and can be very poor at discriminating collections with similar spans but very different distributions. To address this potential issue, Cummins and co-workers[16] have described a cell-based "binning" method using a Riemann gridding scheme to subdivide the chemical space, followed by diversity comparison and diverse subset selection. The authors give a thorough discussion of the effect of bin resolution and detail an elegant approach for the removal of outliers. In addition, Agrafiotis and Rassokhin[17] have reported a rigorous method for the selection of optimal bin size, by examining the discriminating power of the diversity score at several grid resolutions by using a recursive partitioning (RP) scheme.

A special case of cell-based methods is a diversity measure proposed for binary fingerprints. Unlike continuous descriptors, binary descriptors such as structural keys and hashed fingerprints can be compared using fast binary operations to give rapid estimates of molecular similarity, diversity, and complementarity. The most common example of a diversity measure applied to binary descriptors is the binary union (inclusive "or"). This can be exploited in a number of different ways; elegant examples can be found in the following references.[18,19]

2.3   Variance-Based Diversity Metrics

In this section, we detail a series of variance-based methods rooted in the principles of optimal experimental design.[20] Using such methods, a selection of compounds can be thought of as a series of experiments to probe a predefined molecular property space. The data from these experiments are represented as matrices – rows representing compounds and columns descriptors. A general method to determine the redundancy of a square matrix is to compute the determinant of the matrix. The determinant of the correlation matrix is null when variables are entirely redundant, or correlated, and unity when they are fully orthogonal. This relationship extends to larger design matrices so that the more redundant the columns of the design matrix are, the closer to zero the determinant of the correlation matrix is, bringing us to the definition of the D-optimal criterion as a measure of molecular diversity in (12). Here X is the design matrix, X' the transpose of that matrix, and $|X'X|$ represents the determinant.

$$D(C) = |X'X| \qquad (12)$$

For a collection of N compounds, the design matrix is a N × M matrix where the M columns represent a distinct molecular attribute or descriptor. A maximally diverse collection of compounds, therefore, is one that maximizes the value of D.

Martin and co-workers[21] have adopted the D-optimal criterion in the design of combinatorial libraries (see Section 5).

3 Molecular Description

There is a wide variety of descriptors available to describe molecules; the molecular representation they encode is key to the measurement of diversity. Descriptors directly influence the metrics and algorithms used in the design or analysis, the nature of the chemical space in question, and the location of molecules within the chemical space. Therefore, it is important to select descriptors most appropriate to the problem at hand. Such a selection of the appropriate descriptors is nontrivial, the requirement of compound representation is that it contains enough information to incorporate structure and, in some investigations, function.

These descriptors can vary in complexity from simple atom counts to electronic properties derived from high-level *ab initio* calculations. Available molecular descriptors fall into three broad categories: (1) Two-dimensional, encoding the topology of a molecule. (2) Three-dimensional, based on the 3D structure of a molecule. (3) Physicochemical (and electronic) representing whole molecule properties.

The following subsection highlights common examples of molecular descriptors, and descriptor selection. A more thorough discussion of chemical descriptors can be found in the following reference.[22]

3.1 Two-Dimensional Descriptors

Topological indices are numerical values calculated from characteristics of a molecular structure, represented as a molecular graph. These indices, including molecular connectivity, make up the majority of two-dimensional descriptors. Two-dimensional indices efficiently capture many aspects of compound constitution including ring count, ring size, branching, bond order, *etc.* Topological indices may be classified into four main groups: (1) Adjacency matrices (*e.g.*, Zagreb, Randic connectivity, Platt index, *etc.*). (2) Topological distance matrices (*e.g.*, Wiener, polarity number, distance sum, Altenburg polynomial, *etc.*). (3) Centric. (4) Information-theoretic indices (*e.g.*, Shannon, chromatic, orbital and topological information, *etc.*).

Two-dimensional fragment descriptors are common in substructure-searching applications. Examples include: atom pairs (APs), topological torsions, augmented atoms (AAs), and atom sequence. APs are patterns of the form $A_i\text{-}d\text{-}A_j$, where A_i and A_j define the AP (nonhydrogen atoms) and d the interatomic distance measured in bonds along the shortest path connecting the two atoms.[23] This descriptor includes the number of nonhydrogen atom connections, as well as the number of π-electron pairs on each atom. Topological torsions were designed to complement the predictive ability of AP descriptors.[24] They take the form $A_i\text{-}A_j\text{-}A_k\text{-}A_m$ representing sequentially bonded, nonhydrogen, atoms. Each atom is described by its atom type, which includes the element type, the number of nonhydrogen branches attached (excluding the branches that make up the torsion itself), and the number of π-electron pairs. Interestingly, topological torsions were conceived from the basic conformational element, the torsion angle. The AA descriptor consists of a central atom with bonded nonhydrogen atoms and a description of the associated bonds. Developed shortly after the AA descriptor, ganglia-augmented atom (gAA) descriptors are extensions of AA descriptors.[24] Each gAA descriptor consists of an AA descriptor with additional bonds on all the atoms. Each structure, therefore, contains equal numbers of AA and gAA descriptors. It is possible, however, to have several more distinct gAA keys than there are distinct AA keys. Hence, gAA keys are more discriminating. A less complex descriptor is the Atom sequence. These descriptors represent linear sequences of a given set of connected atoms, with the corresponding bonding pattern.

Moreau[25] has proposed the autocorrelation of topological molecular structure (ATS) value. The function is defined in (13).

$$A(d) = \sum P_i P_j \tag{13}$$

where P_i and P_j represent values of an atomic property of the ith and jth atoms and d is the interatomic distance measured in bonds along the shortest path connecting the two atoms. ATS values are attractive because they can be computed readily from the connection table and compounds encoded using a fixed-length vector of small rank. Molecular properties encoded by this method include volume, electronegativity, hydrophobicity, and hydrogen-bonding potential.

Descriptors binary in nature have found wide use in substructure keys and hashed fingerprints. Such substructure keys encode a molecular structure as a bit-string, where each bit in the string indicates the presence (1) or absence (0) of a particular pattern or structural feature. Structural keys rely on predefined lists of structural fragments. Fragment lists obey two principles: requirement of nonredundant information, and a fragment frequency close to equal. Features encoded may include the number of occurrences of a particular element (*e.g.*, the presence of one, two, or three carbon atoms), hybridization or atom types (*e.g.*, sp^2 carbon or aromatic carbon), functional groups (alcohols, carbonyls, amines, *etc.*), and ring systems. Hashed fingerprints are bit-strings derived directly from the connection table. Fingerprints are different from structural keys because they do not depend on preselected structural fragments to perform bit assignment and are therefore very general. Every compound pattern, up to a predefined path length, in contrast to structural keys, is systematically enumerated, serving as input for a hashing algorithm. Both substructure keys and hashed fingerprints were developed for rapid database searching and proven equally effective for similarity and diversity profiling.

Recently, a group at Novartis has described a binary-descriptor method for representing compounds that not only attempts to identify those that bind to the same target but also that bind to other homologous targets.[26] This "homology-based similarity searching" requires a compound description that reflects the ability of a particular molecule to interact with other pertinent target proteins. The description uses the so-called Similog keys based on molecular constitution and number of atom triplets. Atom triplets represent the graph distance between atom types that include hydrogen-bond donors/acceptors, electronegativity, and bulkiness. A comparison using Similog keys with other two-dimensional descriptors reveals that Similog keys are more effective in the identification of ligands that recognize homologous targets.

3.2 Three-Dimensional Descriptors

The two-dimensional descriptors previously discussed are based solely on the molecular graph. Three-dimensional descriptors, in contrast, use distance information derived from spatial (usually Euclidean) arrangement of atoms or atom groups. The most common three-dimensional descriptors are of the fragment type. It should be noted that most two-dimensional fragment descriptors have a three-dimensional counterpart, and both types are usually encoded using binary bit-strings. Kearsley and co-workers[27] at Merck have described a three-dimensional version of the AP. These three-dimensional descriptors, the geometric AP and the geometric-binding property pair, differ from APs in that the distance between the atoms is through space rather than through bonds. In the case of the geometric-binding property pair, the atom type is grouped into one of seven classes (cation, anion, H-bond donor, H-bond acceptor, polar, hydrophobic, and other).

Two popular three-dimensional fragment-based descriptors are the potential pharmacophore point (PPP) pairs and the potential pharmacophore point triangles.[28] The PPP-pair encodes distances between all pairs of potential pharmacophoric points. These points maybe one of five different types: H-bond donor, H-bond acceptor, positive charged, negative charged, or hydrophobic. Similar to the geometric APs, PPP-pairs use through space-atom distances, but differ by using these five predefined points, or atom types. PPP-triangles encode all potential three-point pharmacophores available in a molecule. Essentially, PPP-triangles are triplets of PPP-pairs with their associated distances. The triplet thus encodes, for example, circumference and wing-area information in addition to distance information implicit in the formed triangle.

Other three-dimensional nonfragment-based descriptors are the so-called field descriptors, which are generated from the superimposition of molecules on a rectangular grid or lattice. Steric, electrostatic, and hydrophobic are among types of fields that can be compared via partial least-square analysis to identify and compare regions of molecules that are important for biological activity. Comparitive molecular field analysis (CoMFA) is one example of this type of analysis.[29] Since the basis of this technique is the accurate alignment of molecules when comparing molecules, there is much interest in the literature regarding this key element of the method. While the assumption that scaffold-based, combinatorial-derived compounds can be superimposed with respect to the common scaffold, it is not clear how this method can be used to compare compounds that do not contain a common core. The alignment of other properties, rather than the superimposition of a common core, such as molecular moments of inertia or quadrupole moments, have been suggested as possibilities to circumvent this problem.[30] Arguments have been made that in some cases a common alignment is undesirable due to poor model predictability, since it has been shown that alternative binding modes exist for similar compounds.[31] It has been suggested that an alignment from a docking simulation (if receptor coordinates are available), of each compound in the training set may help this problem.[32] Interestingly, CoMFA has also been used as a diversity-analysis application, called the topomeric CoMFA analysis.[33] In this type of analysis, similar field descriptors among structurally diverse compounds can be illuminated with the idea that they may have similar biological activity.

Another common class of three-dimensional descriptors are the weighted holistic invariant molecular (WHIM) descriptors.[34] These descriptors are based on the principle components analysis of the weighted covariance matrix obtained from the atomic coordinates of the compound. They differ from field descriptors in that they are invariant to rotation and translation. Since the coordinates are centered, the resulting descriptors are invariant to translation. Moreover, the principle component analysis (PCA) step also gives a unique solution that is invariant to rotation. This characteristic is a distinct advantage over other three-dimensional descriptors that rely heavily on superimposition, rotation, and translation. This methodology has been applied to properties other than atomic coordinates, such as molecular surfaces[35] and a grid-based analog of interaction fields.[36]

3.3   Physicochemical and Electronic Descriptors

Empirical, semiempirical, and *ab initio* methods have been used extensively to calculate molecular descriptors. These molecular property descriptors help capture important characteristics of compounds such as bioavailability and receptor affinity. Descriptors such as octanol–water partition coefficient (log P), HOMO/LUMO energies, hammett σ, total energy, heats of formation, ionization potential, atomic charges, electron densities, dipole/quadrupole moments, volume, and polar surface area are common examples. For an excellent review of physicochemical descriptors, the reader is directed to the following reference.[37]

3.4   Descriptor Selection

Selecting suitable descriptors is as much art as science, requiring extensive chemical and problem-specific expertise. One of the challenges in finding a 'best' set of descriptors is that different classes of descriptors encode different types of information. Two-dimensional structural fragments are the most widely used, and exceed in most cases descriptors of other types in their ability to recognize biologically active and inactive compounds.[38] Traditional whole-molecule descriptors such as log P, pK_a, HOMO/LUMO gap, *etc.*, have been inappropriate for this task. Pearlman and Smith[15] have suggested three reasons why these types of descriptors make rather poor chemistry space metrics: (1) The correlation problem: the axes of a vector space should be orthogonal or uncorrelated. Despite their diverse origin, many of the traditional descriptors are highly correlated. Redundant variables affect distance in descriptor space and have the tendency to overemphasize certain molecular characteristics at the expense of others (see Section 4). (2) Some descriptors are related to drug transport or pharmacokinetics, but are weakly related to receptor affinity. (3) Many of the traditional, whole-molecule descriptors possess little information about substructural differences within compounds, which is the basis of structural diversity. Therefore, physicochemical descriptor space, usually, does not effectively relate structure activity relationship (SAR) information.

A common objective in descriptor selection is to define a valid descriptor set, which shows an SAR for compounds with affinity for a given biological target. It has been noted that valid descriptors should be expected to illuminate SAR around a series of compounds, but should not be expected to support a qualitative structure activity relationship (QSAR) for the same set of compounds.[39] In other words, good chemistry space descriptors may not be good QSAR descriptors. The following example illustrates the so-called "similarity paradox", where a small change in chemical structure can often lead to a pronounced change in biological activity. Many chiral drugs differ in configuration at only one center and would be near each other in many chemical spaces. However, the enantiomers often have very different biological properties. In these cases, QSAR-related descriptor validation would be weak with respect to SAR descriptor validation and vice versa.

Randic[39] points out a few potential pitfalls regarding descriptor selection. The selection of descriptors from a large descriptor set may often be biased, when excluding descriptors that show limited correlation with the desired property when used alone. However, these descriptors, when combined with others, may generate a high-quality regression. Likewise, descriptors that show high correlation with previously selected descriptors are often removed from SAR analyses. It is suggested that they should not be removed, since the only useful criterion for the removal of a descriptor should be its inability to reduce the standard error of the regression.

Recently, an entropy-based approach has been introduced to compare the intrinsic and extrinsic variability of different descriptors, independent of their units and value ranges. The method was originally introduced in communication theory and is based on Shannon entropy, which calculates descriptor-entropy values using histogram representations. Shannon entropy is defined as:

$$SE = -\sum p_i \log_2 p_i \qquad (14)$$

where p is the sample probability of a data point, count c, to be within a specified data range i. This method has been shown to identify compound descriptors that are most sensitive to systematic differences in databases consisting of synthetic compounds, drug-like compounds, and natural products. Descriptors with consistently high-information content were detected, and database-specific differences were quantified. Different sets of only very few descriptors were found to be most responsive to principal differences between the different databases.[40]

Descriptor selection is an important step in diversity analysis and depends on the nature of the compound collections (number of compounds, heterogeneity, homogeneity, combinatorial, *etc.*), the technique being used, and the biological target (in SAR/QSAR analyses). A descriptor set valid for a library of natural products may be poor for a combinatorial-library analysis. Descriptors able to discriminate between high- and low-affinity compounds for one target maybe poor for another target, even within the same receptor family.

4   Dimensionality Reduction

As the previous section illustrates, descriptor selection is an important concept. The problem is amplified due to inherently high-dimensional representations (due to large numbers of descriptors), which have a significant impact on the speed of computation and can limit the number of available analysis options (*e.g.*, poor *k*-dimensional trees and increased complexity of cell-based analysis). Therefore, it is necessary to reduce the dimensionality of the chemical space by removing dimensions that do not contribute to the overall picture. Four common methods of dimensionality reduction are outlined below: (1) Principle component analysis. (2) Singular-value decomposition. (3) Factor analysis (FA). (4) Multidimensional scaling. (5) Stochastic Proximity Embedding

4.1   Principle Component Analysis

PCA is a statistical technique that has been used ubiquitously in multivariate data analysis.[41] Given a set of input vectors described by partially cross-correlated variables, the PCA will transform them into a set that is described by a smaller number of orthogonal variables, the principle components, without a significant loss in the variance of the data. The principle components correspond to the eigenvectors of the covariance matrix, m_{ij}, a symmetric matrix that contains the variances of the variables in its diagonal elements and the covariances in its off-diagonal elements (15):

$$m_{ji} = m_{ji} = \frac{1}{N} \sum_{k=1}^{N} (x_{ki} - u_i)(x_{kj} - u_j) \tag{15}$$

where u_i is the mean of variable *i*th variable, and N is the number of compounds in the collection, (16):

$$u_i = \frac{1}{N} \sum_{j=1}^{N} x_{ij} \tag{16}$$

The eigenvalues of this matrix represent the variances of the principal components. PCA reduces the dimensionality by eliminating variables that contribute the least to the variance of the data, *i.e.*, those with the smallest eigenvalues. After diagonalization of the covariance matrix, the original data can be transformed by, in (17).

$$x' = V^T x \tag{17}$$

where V^T is the transpose of the filtered eigenvector matrix, x the original input vector in the original coordinate frame, and x' the coordinates of that sample in the transformed frame. Thus, the components of x' are linear combinations of the original, cross-correlated variables. The method has been employed by a number of research groups; examples include Martin and co-workers,[21] and Gibson and co-workers.[41]

4.2   Singular-Value Decomposition

Singular value decomposition (SVD) is a method similar to PCA in that it allows for the determination of the rank of a matrix. An M × N matrix, M, can be expressed as the product of U, an orthogonal M × M matrix, S, an M × N diagonal matrix with real, nonnegative elements, and V^T, the transpose of an orthogonal N × N matrix Eq(18).

$$M = USV^T \tag{18}$$

If the rank of M is less than M and N, then some elements of the matrix S will be zero, and the number of nonnegative elements will equal the rank of M. SVD has been used successfully to reduce the dimensionality of descriptor space of chemical libraries.[42] Xie and co-workers report mapping chemicals with 10% of high-dimensional distances into two dimensions ranging from 30 to 100%, depending on the types of compounds and descriptors used.

4.3 Factor Analysis (FA)

FA attempts to extract coherent subsets of variables that are relatively independent of one another, not directly observable, and are thought to be representative of the underlying process that has created the correlations.[43] Both PCA and FA rely on eigenvalue analysis of the covariance matrix as well as linear combinations of variables to explain a set of observations. The two techniques differ in that the quantities of interest in PCA are the observed variables themselves. The observed variables are of little intrinsic value in FA; what is of interest is the underlying factors. Cummins and co-workers[16] used FA to reduce a set of 61 molecular descriptors to four factors. The remaining four factors were then used to compare the diversity of five chemical databases using a cell-based approach. In a similar study, however, Gibson and co-workers[41] compared 100 different heterocyclic aromatic ring fragments from known HIV-1 reverse transcriptase inhibitors and concluded that FA did not significantly reduce the complexity of the analysis.

4.4 MultiDimensional Scaling

While PCA and FA attempt to decorrelate data, multidimensional scaling (MDS) on the other hand, attempts to construct a configuration of points in a low-dimensional space based on distance information between data points. In particular, given a set of k data points in the input space $\{x_i, I = 1, 2, \dots k\}$, a symmetric matrix, r_{ij}, of the observed dissimilarities between these points, and a set of images of x_i on a d-dimensional display plane $\{y_i, I = 1, 2, \dots k; y \in \Re_d\}$, the objective is to place y_i on the plane in such a way that their Euclidean distances $d_{ij} = \|y_i - y_j\|$ approximate, as closely as possible, the corresponding values r_{ij}. A sum-of-squares error function can be used to measure the quality of the embedding. A common sum-of-squares error function is Kruskal's stress given in (19).

$$S = \sqrt{\frac{\sum_{i<j}(d_{ij} - r_{ij})^2}{\sum_{i<j} d_{ij}^2}} \tag{19}$$

The actual embedding is carried out in an iterative fashion. The process starts by: (1) generating an initial set of coordinates y_i., (2) computing the distances d_{ij}, (3) finding a new set of coordinates, y_i ,using a steepest-descent algorithm (Kruskal's linear regression), and (4) repeating steps (2) and (3) until the change in the stress function falls below some predefined threshold.

The first application using MDS in molecular diversity analysis was introduced by a group at Chiron as a means of reducing the enormous dimensionality of binary chemical descriptors.[21] They found that 2048-bit Daylight fingerprints associated with 721 commercially available primary amines could be reduced to only five dimensions that reproduced all 260,000 original dissimilarities with a standard deviation of only 10%. Similarly, only seven dimensions were required to reduce the 642,000 pairwise similarities among a set of 1133 carboxylic acids and acid chlorides to the same standard deviation.

4.5 Stochastic Proximity Embedding

MDS has certain disadvantages. Although it works well with linear or quasi-linear subspaces, it fails to detect nonlinear structures, curved manifolds, and arbitrarily shaped clusters. Another disadvantage is that MDS attempts to preserve all pairwise distances in the data sample, both local and remote. It has been known for some time that conventional similarity measures, such as the Euclidean distance, tend to underestimate the proximity of points on a nonlinear manifold and lead to erroneous embeddings.[44,45] Sammons's nonlinear mapping (NLM) algorithm alleviates this problem somewhat by introducing a normalization factor in the error function to give increasing weight to short-range versus long-range distances (20).

$$S = \frac{\displaystyle\sum_{i<j} \frac{(d_{ij} - r_{ij})^2}{r_{ij}}}{\displaystyle\sum_{i<j} r_{ij}} \tag{20}$$

This scheme, however, is arbitrary and fails with highly folded topologies. Tenenbaum[46] introduced the Isomap (isometric feature mapping) method to help alleviate this problem. The Isomap method uses an estimated geodesic distance instead of the conventional Euclidean distance, as would be input to MDS. The geodesic distances are estimated by connecting each point to its nearest neighbors, then tracing the shortest path between all pairs of points on the resulting graph. It was shown that, Isomap recovers the true dimensionality and geometric structure of the data if it belongs to a certain class of Euclidean manifolds; however, this advantage is of little practical use since the quadratic complexity of the embedding algorithm precludes its use with large data sets. Locally linear embedding (LLE) is a related approach that produces globally ordered maps by constructing locally linear relationships between data points;[47] however, it has a similar scaling problem.

A recently introduced algorithm, called stochastic proximity embedding (SPE), is a novel self-organizing scheme that addresses the key limitations of Isomap (isometric feature mapping) and LLE.[48–51] SPE builds on the same geodesic principle first proposed and exploited in Isomap, but introduces two algorithmic advances: SPE circumvents the calculation of estimated geodesic distances, and uses a pairwise refinement scheme that does not require the complete distance, d_{ij}, or proximity, r_{ij}, matrix. Due to these advances, the method scales linearly with the number of points.

SPE minimizes the stress function given in (21).

$$S = \frac{\sum_{i<j} \dfrac{f(d_{ij}, r_{ij})^2}{r_{ij}}}{\sum_{i<j} r_{ij}} \tag{21}$$

where $f(d_{ij}, r_{ij})$ is the pairwise stress defined as $f(d_{ij}, r_{ij}) = (d_{ij} - r_{ij})^2$ if $r_{ij} \leq rc$ or $d_{ij} < r_{ij}$, and $f(d_{ij}, r_{ij}) = 0$ if $r_{ij} > rc$ and $d_{ij} \geq r_{ij}$, and rc is a predefined neighborhood radius. This calculation is accomplished using a stochastic approximation of steepest descent that attempts to bring each individual term $f(d_{ij}, r_{ij})$ rapidly to zero. The method starts with an initial configuration and iteratively refines it by repeatedly selecting two points at random, and adjusting their coordinates so that their Euclidean distance on the map, d_{ij}, matches more closely their corresponding proximity, r_{ij}. The correction is proportional to the disparity $\lambda(|r_{ij} - d_{ij}|/d_{ij})$, where λ is a learning-rate parameter that decreases during the course of the refinement in order to avoid oscillatory behavior. If $r_{ij} > rc$ and $d_{ij} \geq r_{ij}$, *i.e.*, if the points are nonlocal and their distance on the map is already greater than their proximity, r_{ij}, their coordinates remain unchanged. Similar to Isomap and LLE, SPE depends highly on the choice of the neighborhood radius, rc. If rc is too large, the local neighborhoods will include data points from other branches of the manifold, shortcutting them, and leading to substantial errors in the final embedding. If rc is too small, it will lead to discontinuities, causing the manifold to fragment into a large number of disconnected clusters.

In contrast to MDS, SPE preserves exact distances between neighboring points and lower bounds between remote points, thus allowing the manifold to unfold and reveal its true intrinsic dimensionality. Essentially, the method views the input distances between remote points as lower bounds of their true geodesic distances, and uses them as a means to impose global structure.

5   Subset Selection and Classification

Exhaustive assaying of all compounds in a large collection is usually impossible due to restrictions on time and resources. It is desirable, if not necessary, for subset selection to be made rationally. A subset that maximizes the information content of the entire library and increases chance of finding an active compound or series is clearly preferred. The selection, or library design, can be carried out at the reagent or product level. In either case, an efficient algorithm is required to sort through the large number of possibilities, often simultaneously attempting the optimization of frequently conflicting objectives. The identification of the N most diverse compounds in a collection containing C compounds is generally unfeasible for large numbers of N and C; this conundrum is known as the combinatorial optimization problem.[52] Depending on the scale of the problem and central processing unit (CPU) expense, these algorithms can be prohibitively slow and often converge to suboptimal local minima. Stochastic algorithms attempt to circumvent the multiple-minima problem by allowing the generation of successor states that might be inferior to their predecessors. Several methods have been investigated, including simulated annealing,[53,54] genetic algorithms,[55] and particle swarms.[56,57]

In essence, the selection problem can be viewed as a heuristic search in which each state of the search space represents a particular subset of the virtual library. This section highlights a few common methods of subset selection.

5.1 Clustering

Clustering was one of the first methods to be applied to diversity analysis.[8] In general, clustering is a multivariate analysis technique that attempts to organize information about variables in a set of relatively homogeneous groups called 'clusters'. These clusters must be internally homogenous (compounds in the same cluster must be structurally similar) and externally heterogeneous (compounds in different clusters must be dissimilar to each other).

There are four major steps in cluster analysis. (1) An appropriate set of features must be selected, and scaled in a meaningful way. (2) A similarity matrix must be constructed as a record of the distances between each pair of objects in the collection. (3) A decision must be made about the number and interpretation of the clusters. (4) The cluster solution must be validated by visual or statistical means. Figure 2 illustrates one possible clustering of a 10,000-member library using SPE.

Clustering methods fall into two main families based on how they form clusters. The first method, nonhierarchical clustering (k-nearest-neighbor clustering), which produces a single set of clusters based on some user-defined criteria. This technique constructs the clusters by computing the k-nearest neighbors of each compound, then clustering compounds together if they are on each other's nearest neighbor lists and share some minimum number of nearest neighbors. A popular member of the nonhierarchical clustering family is the Jarvis–Patrick algorithm.[58] The main advantage of this technique is speed; the main disadvantage is its tendency to generate too

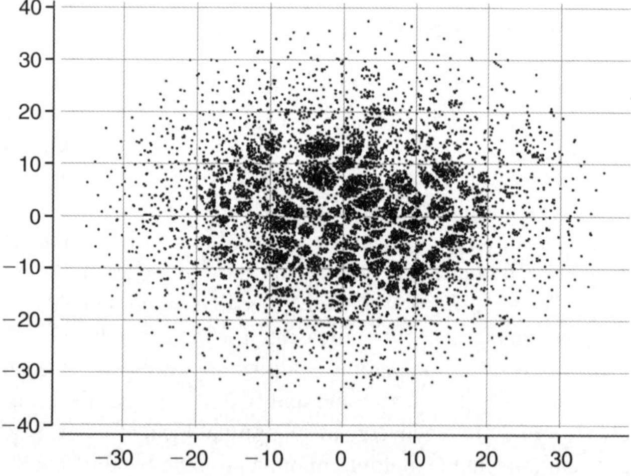

Figure 2 *SPE map showing a clustering of a 10,000-member library*

many singletons (clusters with a single compound) or too few large clusters depending on the stringency of the clustering criteria. The second clustering method, hierarchical clustering, produces output in the form of a dendrogram or tree. Hierarchical clustering may be top–down and employ division (divisive clustering), or bottom–up and employ aggregation (agglomerative clustering). In top–down clustering, the process starts with a single cluster of compounds that is subsequently subdivided into smaller and smaller groups until each compound is a member of its own cluster, a singleton. A common example of a top–down clustering method is the Guenoche algorithm.[59] Conversely, bottom–up clustering approaches start with singletons and work their way to the top by combining clusters together to form larger clusters. The most popular agglomerative clustering method is described by Ward,[60] where clusters are merged together to minimize intra-cluster variance and maximize inter-cluster variance.

Generally speaking, cluster analysis requires the user to make certain decisions, which may have a profound influence on the constitutions and populations of the clusters. In order to design diverse subsets, the user typically selects a representative from each cluster (usually the centroid), and then optionally examines the resulting subset for possible multicolinearities. If the selection is nonorthogonal, the suspect compounds are replaced with other members of the same cluster and the new solution is re-evaluated. This cycle is iterated until a quasi-orthogonal subset is accepted. In several studies, hierarchical methods, in particular Ward's clustering, give superior results compared to nonhierarchical methods.[28,61]

5.2 Partitioning Methods

Partitioning methods are useful techniques for compound subset selection. Presently, cell-based methods and RP are the most popular partitioning methods. Cell-based methods, originally proposed by Cummins and co-workers[16] and later by Pearlman,[62] partition the property space or functional space into a set of multidimensional cells by dividing each axis into a finite number of equally sized bins. Following cell formation, the compounds are assigned to particular cells, based on the presence or absence of a particular property or set of properties. The method is flexible, intuitive, direct, and is ideally suited to a wide variety of diversity-related tasks. Cell-based methods are advantageous because certain tasks, such as filling diversity voids (unoccupied cells or fractions of cells) and selecting diverse subsets for diversity analysis, are usually straightforward. Their main disadvantage is that in order to truncate the combinatorially explosive enumeration of discrete cells, the method requires a low-dimensional representation of the chemical space.

RP refers to a class of decision-tree methods in which the tree is grown using a greedy iterative procedure. RP attempts to divide compound collections along decision trees and identifies combinations of binary-compound descriptors that ultimately copartition compounds into learning sets that share the same activity.[63,64] RP has the advantage of speed and ease of implementation, and has been applied to a variety of chemical design problems.[63–65] A possible limitation of RP is that the method constructs the decision regions in a global fashion; as a result, there is a higher probability for many data samples to be near the region boundaries. It can be argued that such a construction causes some samples to be poorly represented by their local regions and therefore classified with less accuracy.

Bajorath and co-workers[65] have recently proposed the median partitioning (MP) method for diverse subset selection and compound classification. Similar to RP, the MP method is an iterative and statistically based approach for compound classification. In contrast to RP, MP does not use a decision-tree structure to subdivide a compound collection based on the presence or absence of particular properties. Moreover, the method does not depend on the processing of learning sets for the derivation of rules and predictive models. MP uses one property descriptor for each partitioning step to iteratively divide the entire compound collection into subsets above and below the median value of each selected descriptor. The median is characterized as the value within a distribution that divides the collection into two equal subsets above and below the median. Depending on how many descriptors, N, are chosen for the analysis, a total of 2^N different partitions will be generated. Due to the nature of the method, binary descriptors (see Section 3) cannot be used; the method is amenable to continuous descriptors only. Like RP, MP does not depend on pairwise compound comparisons and is feasible for partitioning of large compound collections.

5.3   Experimental Design

Martin and co-workers[21] introduced a selection technique with the objective of providing a rational method for the selection of a representative set of amines and carboxylic acids to be used as reagents in an N-substituted, peptoid combinatorial library. D-optimal design was used for the selection following the definition of appropriate chemical space. D-optimal design takes as input a design matrix, X, and selects a subset of points from a larger pool of candidates that maximize the determinant of the "information matrix", X^TX. The rows of X represent monomers or compounds, and the columns represent the original features, or higher-order terms such as their squares, cubes, or cross terms. This approach, in effect, minimizes the determinant of the inverse, which is the variance of the parameter estimates for the model that is encoded by the design matrix. The algorithm builds up the design in a stepwise manner, starting with some preselected set of compounds, and gradually augments that set by including the reagent that best complements the existing solution until the maximum number is reached.

The objective of D-optimal design is to identify points that are not only spread out in property space, but are also orthogonal. This result is achieved by maximizing volume in covariance space. The determinant is large when the variances are large (the points are spread out); when the covariances are small, the properties are orthogonal. D-optimal design biases the selection of compounds toward the extreme edges of property space and tends to ignore the central region. Other related criteria include, A-, G-, and I-optimality, which, is model-based like D-optimality. S- and U-optimality are distance based and are designed to fill space and maximize spread.

In contrast to optimal design, Hamprecht and co-workers[66] recently introduced a space-filling design technique for compound selection. This stochastic method uses the best linear unbiased estimator, in the form of Kriging,[67] to construct selection designs that optimize the integrated mean-square prediction error, or entropy. This space-filling design attempts to sample chemical space more evenly than *D*-optimal design, when the response surface is not uniform but irregular, as is often the case with heterogeneous compound collections.

5.4 Reagent-Based Versus Product-Based Design

The aforementioned selection algorithms can be very computationally demanding, especially those that scale to the square of the number of molecules in the collection. Divide-and-conquer approaches, designed specifically for combinatorial library analysis, have been developed to help alleviate the computational problem by dealing with each substitution site independently of the others. This approach is based on the assumption that a diverse set of reagents would lead to a diverse set of compounds. The validity of this assumption was challenged by Gillet and co-workers.[68,69] This challenge consisted of generating virtual libraries of three published combinatorial libraries, by both product- and reagent-based design. The diversity of all the libraries was then measured using the diversity index described by Turner and co-workers.[70] In all the cases, the product-based libraries showed a diversity that was greater than that of the reagent-based libraries. Jamois[71] has come to a similar conclusion, but found that the benefit depends to a large extent on the types of descriptors used and the nature of the library. These authors also reported that the efficiency of reagent-based selection is compromised when there is degeneracy in the reagent lists. With reagent-based approaches, the selections of R_1, R_2, and R_n are made independently, which introduces the possibility of significant overlap between selected R groups.

5.5 Random Versus Rational Design

Subset selection has traditionally been classified into two general categories, selections that are representative of the collection and selections that are diverse. It has been suggested that representative sets attempt to represent the density of compounds in a collection, whereas diverse sets represent the volume occupied by a compound collection.[66] Random selection, due to its stochastic nature, leads to subsets that represent the density of a compound collection. Hamprecht has pointed out that volumes and densities in chemistry space are continuous notions, and that the compounds in that space are just a collection of points with zero total volume.

Potter and Matter[72] compared random versus diverse subset selection in the context of maximizing structural diversity and compound biological activity. Using an optimal diverse subset, the authors were able to select members from all the biological active classes in a public database, containing 1283 compounds in 55 biological activity classes, selecting no more than 38% of all the compounds. A random selection, however, required 3.5–3.7 times as many compounds selected in order to span 90% of the entire biological active classes. Additionally, a CoMFA experiment was used to evaluate designed and randomly selected subsets for a database containing inhibitors of the angiotensin-converting enzyme. The authors detail the superiority of dissimilar designed subsets, which lead to robust QSAR models with higher predictive power, compared to randomly chosen compounds.

6   Conclusion

In this chapter, we have attempted to provide the reader with descriptions of a few methods that define and quantify chemical diversity. The basic elements involved in chemical diversity definition include: a measure of distance between molecules, a measure of molecular diversity that realistically quantifies the diversity of a set of compounds, and a sampling or selection method for identifying a diverse subset of compounds from large libraries. Although conceptually simple, molecular diversity is a concept that is difficult to give a single hard-coded definition, due to the range of metrics and molecular representations possible for a given analysis. The choice of metrics and molecular representation can only be validated to the extent that they meet the goal of the diversity analysis at hand, *e.g.*, compound selection to enrich a high-throughput screening campaign. Proper validation will only come from comparison with appropriate control experiments, but, regrettably, these are often not feasible due to difficulties in the design and expense of the experiments.

Abbreviations

AA	augmented atom
gAA	ganglia-augmented atom
AP	atom pair
CoMFA	comparative molecular field analysis
CPU	central processing unit
FA	factor analysis
LLE	local linear embedding
MDS	multidimensional scaling
MP	median partitioning
NLM	nonlinear mapping
PCA	principle component analysis
PPP	potential pharmicophore point
QSAR	qualitative structure activity relationship
RP	recursive partitioning
SAR	structure activity relationship
SPE	stochastic proximity embedding
SVD	singular value decomposition
WHIM	weighted holistic invariant molecular

References

1. Human Genome Project Information, U.S., website:http://www.ornl.gov/sci/techresources/Human_Genome/home.shtml, *Department of Energy Office of Science.*
2. J.M. Blaney and E.J. Martin, *Curr. Opin. Chem. Biol.*, 1997, **1**, 54–59.
3. P. Willett, *Curr. Opin. Biotechnol.*, 2000, **11**, 85–88.
4. M.G. Bures and Y.C. Martin, *Curr. Opin. Chem. Biol.*, 1998, **2**, 376–380.

5. D.K. Agrafiotis, V.S. Lobanov and F.R. Salemme, *Nat. Rev. Drug Discov.*, 2002, **1**, 337–346.
6. M. Farnum, R. DesJarlais and D.K. Agrafiotis, in *Chemoinformatics - From Data to Knowledge,* J. Gasteiger (ed), Wiley-VCH, Weinheim, 2003.
7. M.A. Johnson and G.N. Maggiora, eds. *Concepts and Applications of Molecular Similarity,* Wiley, New York, 1990.
8. P. Willett, *Similarity and Clustering in Chemical Information Systems,* Wiley, New York, 1987.
9. P. Willett, J.M. Barnard and G.M. Downs, *J. Chem. Inf. Comput. Sci.*, 1998, **38**, 983–996.
10. S.K. Kearsley, S. Sallamack, E.M. Fluder, J.D. Andose, R.T. Mosley and R.P. Sheridan, *J. Chem. Inf. Comput. Sci.*, 1996, **36**, 118–127.
11. C.A. James, D. Weininger and J. Delany, *Daylight Chemical Information System, Inc.* 2004.
12. S.V. Trepalin, V.A. Gerasimenko, A.V. Kozyukov, N.P. Savchuk and A.A. Ivaschenko, *J. Chem. Inf. Comput. Sci.*, 2002, **42**, 249–258.
13. D.N. Rassokhin and D.K. Agrafiotis, *J. Mol. Graph. Model.*, 2000, **18**, 368–382.
14. R.E. Higgs, K.G. Bemis, I.A. Watson and J.H. Wikel, *J. Chem. Inf. Comput. Sci.*, 1997, **37**, 861–870.
15. R.S. Pearlman and K.M. Smith, *J. Chem. Inf. Comput. Sci.*, 1999, **39**, 28–35.
16. D.J. Cummins, C.W. Andrews, J.A. Bentley and M. Cory, *J. Chem. Inf. Comput. Sci.*, 1996, **36**, 750–763.
17. D.K. Agrafiotis and D.N. Rassokhin, *J. Chem. Inf. Comput. Sci.*, 2002, **42**, 117–122.
18. S.D. Pickett, J.S. Mason and I.M. McLay, *J. Chem. Inf. Comput. Sci.*, 1996, **36**, 1214–1223.
19. N.E. Shemetulskis, D. Weininger, C.J. Blankley, J.J. Yang and C. Humblet, *J. Chem. Inf. Comput. Sci.*, 1996, **36**, 862–871.
20. D.C. Montgomery, *Design and Analysis of Experiments,* 4th. Ed., Wiley, New York, 1996.
21. E.J. Martin, J.M. Blaney, M.A. Siani, D.C. Spellmeyer, A.K. Wong and W.H. Moos, *J. Med. Chem.*, 1995, **38**, 1431–1436.
22. D.K. Agrafiotis, J.C. Myslik and F.R. Salemme, *Annu. Rep. Combinator. Chem. Mol. Divers.*, 1999, **2**, 71–92.
23. R.E. Carhart, D.H. Smith and R. Venkataraghavan, *J. Chem. Inf. Comput. Sci.*, 1985, **25**, 64–73.
24. R. Nilakantan, N. Bauman, J.S. Dixon and R. Venkataraghavan, *J. Chem. Inf. Comput. Sci.*, 1987, **27**, 82–85.
25. G. Moreau and P. Broto, *Nouv. J. Chim.*, 1980, **4**, 359–360.
26. A. Schuffenhauer, P. Floersheim, P. Acklin and E. Jacoby, *J. Chem. Inf. Comput. Sci.*, 2003, **43**, 391–405.
27. R.P. Sheridan, M.D. Miller, D.J. Underwood and S.K. Kearsley, *J. Chem. Inf. Comput. Sci.*, 1996, **36**, 128–136.
28. R.D. Brown and Y.C. Martin, *J. Chem. Inf. Comput. Sci.*, 1996, **36**, 572–584.
29. R.D., Cramer III and J.D. Bunce, *Pharmacochem. Libr.*, 1987, **10**, 3–12.
30. J.J. Perez, *Chem. Soc. Rev.*, 2005, **34**, 143–152.

31. C. Mattos and D. Ringe, in *3D QSAR in Drug Design: Theory, Methods and Applications*, H. Kubinyi (ed), ESCOM, Leiden, 1993, 226–254.

32. G. Jones and P. Willett, *Curr. Opin. Biotechnol.*, 1995, **6**, 652–656.

33. R.D. Clark, A.M. Ferguson and R.D. Cramer, *Perspect. Drug Discov. Des.*, 1998, **9/10/11**, 213–224.

34. R. Todeschini, R. Lasagni and E. Marengo, *J. Chemometr.*, 1994, **8**, 263–272.

35. G. Bravi, E. Gancia, P. Mascagni, M. Pegna, R. Todeschini and A. Zaliani, *J. Comp.-Aid. Molec. Des.*, 1997, **11**, 79–92.

36. R. Todeschini, G. Moro, R. Boggia, L. Bonati, U. Cosentino, M. Lasagni and D. Pitea, *Chemometr. Intell. Lab. Sys.*, 1997, **36**, 65–73.

37. D.J. Livingstone, *J. Chem. Inf. Comput. Sci.*, 2000, **40**, 195–209.

38. R.D. Brown and Y.C. Martin, *J. Chem. Inf. Comput. Sci.*, 1997, **37**, 1–9.

39. M. Randic, *J. Molec. Graph. Model.*, 2001, **20**, 19–35.

40. J.W. Godden and J. Bajorath, *J. Chem. Inf. Comput. Sci.*, 2002, **42**, 87–93.

41. S. Gibson, R. McGuire and D.C. Rees, *J. Med. Chem.*, 1996, **39**, 4065–4072.

42. D. Xie, A. Tropsha and T. Schlick, *J. Chem. Inf. Comput. Sci.*, 2000, **40**, 167–177.

43. H.H. Harmon, *Modern Factor Analysis,* 2nd. Ed. University of Chicago Press, Chicago, 1967.

44. R.N. Shepard and J.D. Carroll, in *Multivariate Analysis: Proceedings of an international symphosium*, P.R. Krishnaiah (ed), Academic Press, New York, 1966, 561–592.

45. T. Martinetz and K. Schulten, *Neural Networks*, 1994, **7**, 507–522.

46. J.B. Tenenbaum, V. de Silva and J.C. Langford, *Science*, 2000, **290**, 2319–2323.

47. S.T. Roweis and L.K. Saul, *Science*, 2000, **290**, 2323–2326.

48. D.K. Agrafiotis and H. Xu, *Proc. Natl. Acad. Sci. USA*, 2002, **99**, 15869–15872.

49. D.K. Agrafiotis and H. Xu, *J. Chem. Inf. Comput. Sci.*, 2003, **43**, 475–484.

50. D.K. Agrafiotis, *J. Comput. Chem.*, 2003, **24**, 1215–1221.

51. D.N. Rassokhin and D.K. Agrafiotis, *J. Mol. Graph. Model.*, 2003, **22**, 133–140.

52. B. Korte and J. Vygen, *Combinatorial Optimization: Theory and Algorithms,* 2nd Ed., Springer, Berlin, 2002.

53. D.K. Agrafiotis, *J. Chem. Inf. Comput. Sci.*, 1997, **37**, 841–851.

54. V.S. Lobanov and D.K. Agrafiotis, *J. Chem. Inf. Comput. Sci.*, 2000, **40**, 460–470.

55. V.J. Gillet, W. Khatib, P. Willett, P.F. Fleming and D.V.S. Green, *J. Chem. Inf. Comput. Sci.*, 2002, **42**, 375–385.

56. D.K. Agrafiotis and W. Cedeno, *J. Med. Chem.*, 2002, **45**, 1098–1107.

57. W. Cedeno and D.K. Agrafiotis, *J. Comp.-Aid. Molec. Des.*, 2003, **17**, 255–263.

58. R.A. Jarvis and E.A. Patrick, *IEEE Trans. Comput.*, 1973, **C22**, 1025–1034.

59. A. Guenoche, P. Hansen and B. Jaumard, *J. Classif.*, 1991, **8**, 5–30.

60. J. Ward, *J. Am. Stat. Assoc.*, 1963, **58**, 236.

61. G.M. Downs, P. Willett and W. Fisanick, *J. Chem. Inf. Comput. Sci.*, 1994, **34**, 1094–1102.

62. R.S. Pearlman, Website created on June 1996, http://www.netsci.org/Science/Combichem/feature08.html.

63. A. Rusinko III, S.S. Young, D.H. Drewry and S.W. Gerritz, *Combinator. Chem. High T. Scr.*, 2002, **5**, 125–133.

64. A. Rusinko III, M.W. Farmen, C.G. Lambert, P.L. Brown and S.S. Young, *J. Chem. Inf. Comput. Sci.*, 1999, **39**, 1017–1026.

65. J.W. Godden, J.R. Furr and J. Bajorath, *J. Chem. Inf. Comput. Sci.*, 2003, **43**, 182–188.

66. F.A. Hamprecht, W. Thiel and W.F. van Gunsteren, *J. Chem. Inf. Comput. Sci.*, 2002, **42**, 414–428.

67. N.A.C. Cressi, *Math. Geol.*, 1990, **22**, 239–252.

68. V.J. Gillet, P. Willett and J. Bradshaw, *J. Chem. Inf. Comput. Sci.*, 1997, **37**, 731–740.

69. V.J. Gillet, *Mol. Divers.*, 2002, **5**, 245–254.

70. D.B. Turner, S.M. Tyrrell and P. Willett, *J. Chem. Inf. Comput. Sci.*, 1997, **37**, 18–22.

71. E.A. Jamois, *Curr. Opin. Chem. Biol.*, 2003, **7**, 326–330.

72. T. Potter and H. Matter, *J. Med. Chem.*, 1998, **41**, 478–488.

Section 3

Conceptual Advances in Synthesis: "Mining"–Turning a Hit into a Lead

Focused Libraries: The Evolution in Strategy from Large-Diversity Libraries to the Focused Library Approach

RUBEN TOMMASI AND IVAN CORNELLA

Lead Synthesis & Chemogenetics – US, Novartis Institutes for Biomedical Research, Inc., 250 Massachusetts Avenue, Cambridge, MA 02139, USA

1 Introduction

One of the main challenges in the pharmaceutical industry has been the identification of suitable starting points for drug-discovery programs. The perception that hit rates from high-throughput screening (HTS) are low has driven a search for more innovative starting points for new lead-optimization efforts. In the early 1990s, many pharmaceutical companies invested heavily to build up combinatorial chemistry groups whose task was to synthesize libraries to supplement traditional compound collections for lead-finding efforts. At the time, there was tremendous enthusiasm for the idea that drugs could be discovered more rapidly with the use of combinatorial chemistry strategies. The initial thought was that large-mixture libraries would revitalize the discovery of new drugs by providing suitable hits and leads for any target of interest. The philosophy was relatively simple: Make as many diverse compounds as possible and the lead (or drug!) would be in the library, provided it was sufficiently large and explored enough diversity. Moreover, it was also felt that subsequent optimization of a hit compound into a drug candidate would be much easier, since the synthetic methods for analog synthesis would already have been developed, and, moreover, enough structure-activity relationship (SAR) information would be revealed by the initial library screen.

However, experience soon showed that success in this arena would not be so simple. Foremost among the problems were variable hit rates: either too many hits for a target were identified or none at all. Moreover, validation of the hits obtained from screening these libraries was an enormous challenge; in addition to the tedious deconvolution steps required to identify the active components in large-mixture libraries, the high rate of false positives, likely as a result of pleotropic effects, resulted in large expenditures of time on resynthesis of compounds that gave no fruitful results in the end. The pressure to make this technology useful for the drug-discovery process led to the development of numerous orthogonal tools to assist in improving library design. In addition to the evolution of many new technologies, combinatorial chemistry has expanded our thinking about synthetic methods, purification strategies, automation and also various abstract subjects such as chemistry space, chemical diversity and their relation to biological space. The strategies arising from combinatorial chemistry concepts have certainly not been static, and the manner in which combinatorial libraries are designed and prepared has advanced considerably since the first split-mix libraries of the early 1990s. The focus of this chapter is to present an overview of this evolution and the resulting shift in strategy toward smaller targeted libraries of purified compounds.

2   A Synergistic, Multidisciplinary Approach to Library Conception

The evolutionary pressure over library design and development for the identification of novel lead compounds has boosted the development of technological advances both in organic chemistry and adjacent scientific areas. The need to adapt our synthetic processes and approaches to produce large numbers of compounds in a timely fashion, while maintaining satisfactory standards of purity and increasing their relevance for biological and pharmacological interaction, has provoked synergistic-enhancing effects in many disciplines.

2.1   Improvements in Synthetic Methods

Novel approaches to synthetic organic chemistry are indeed among the core processes for new library developments, as addressed in Chapters 3 and 4 in this volume. Significant advances in the way that chemical reactions are carried out, whether in solution or on solid phase,[1] in traditional glassware or state-of-the-art reactors,[2] in step-wise fashion or as multicomponent reactions[1,3] are continuously being developed, improved and adapted to the production of libraries. The expansion of synthetic tools, in turn, has lead to the development of new ways to purify, quantitate and identify products in a high throughput and reliable fashion.[4] Automated chromatographic purification systems, together with extraction and scavenging methods[5] are routine approaches replacing traditional single-reaction purification techniques. Analytical methods have been adapted for characterization and quality control of the large number of compounds prepared. Mass Spectrometry (MS)[6] and HPLC,[7] capable of handling small amounts of material and large collections of molecules, have replaced NMR as the primary method of choice for quality control of libraries. The shift in strategy from large-mixture libraries to collections of purified individual compounds ("singletons") could not have been accomplished without these recent advances in analytical methods. The focused libraries that are being developed today are possible only as a result of these advances.

2.2 Impact of *In Silico* Tools for Library Design

In addition to the advances in synthesis, purification and analysis, the production of focused libraries is also benefiting greatly from recent developments in *in silico* tools for library design and analysis. Many cheminformatics and bioinformatics approaches have been developed recently to provide us with a collection of software applications to assist in the design and synthesis of relevant compounds. Lead-finding libraries designed for HTS and hit-to-lead campaigns are benefiting from the advances in these computer-assisted drug design (CADD) programs.

Computer-based lead finding and feature-based pharmacophore algorithms[*,8] are used for designing and optimizing as many chemical scaffold properties as possible. From single, small molecules to large virtual libraries, structures are designed to have drug-like features and screened with *in silico* docking experiments to identify those that have the greatest potential to interact with a target active site.[9] Quantitative structure-activity relationship (QSAR) analysis,[10] clustering[11] and principal component analysis (PCA)[†] are mathematical tools which are routinely utilized to optimize, render and cluster the 2D or 3D structures of chemical entities and correlate them with biological activity. The calculated physicochemical information for the compounds of interest is evaluated and correlated with what is known about the structure of the target itself. All of these insights are fed back into the design of a set of compounds that are optimized for the desired target and appropriate physical characteristics (natural product-likeness, drug-likeness) within a diversity set. Even when the structure of the biological target is not readily available, cheminformatics approaches can still be useful in library design based on known pharmacophoric features of the chemical scaffolds.

An example of how this approach is utilized is provided by a recent report by Krier *et al.*[12] on the synthesis of a focused library of phosphodiesterase-4 (PDE-4) inhibitors. Using a method, which they coined as "Scaffold-Linker-Functional Group" (SLF), they aimed to maximize the diversity in their library design, while minimizing the size of the requisite library. A virtual library based on zardarvarine (Scheme 1) was generated and compounds were selected for synthesis by docking into the X-ray crystal structure of the human PDE-4 catalytic domain. Nine of the compounds which achieved the highest docking scores were synthesized and screened for their ability to inhibit PDE-4, with the most active having subnanomolar activity.

---

[*] The number of published and patented computational methods and software packages is so vast that a lengthy enumeration and description, even of the better-validated tools, falls beyond the scope of this chapter. For detailed reviews and other recent papers relating CADD with medicinal chemistry see Ref. 8.

[†] PCA is a mathematical method for data analysis that identifies patterns and highlights similarities and differences. It is usually combined with clustering algorithms to provide information on the relationships between the calculated descriptors and the biological activity studied.

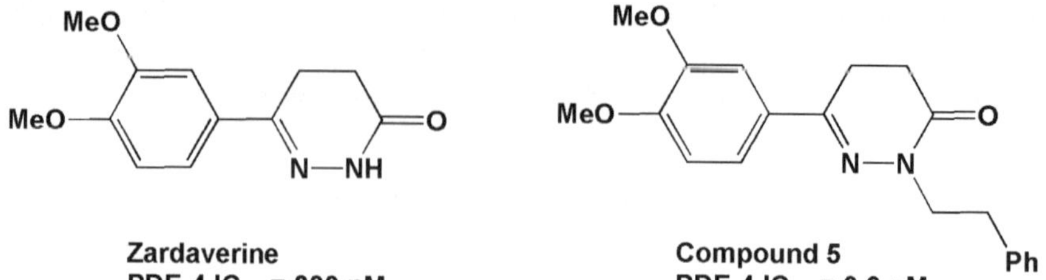

MeO

MeO

Zardaverine
PDE-4 IC$_{50}$ = 800 nM

MeO

MeO

Compound 5
PDE-4 IC$_{50}$ = 0.9 nM

Ph

Scheme 1 *Zardaverine-based analogs as inhibitors of PDE-4*

Computational science is not only relevant for *a priori* rational design of compound libraries but also plays an important role during and after execution. When the time comes to step into the lab, software tools are playing an increasing role for building block and reagent selection.[13] Compound enumeration[14] and prioritization for screening are also improved through software for lead optimization, including analysis of potency, selectivity and synthetic feasibility. These computational experiments can improve the speed and accuracy with which homology models are correlated with experimental observations. Nevertheless, although state-of-the-art tools are available for the rational design and preparation of chemical libraries, improvements are still being made and we can look forward to evermore sophisticated ways of integrating predictive and experimental information.

2.3   Influence of Biology in Library Design

One of the most significant synergistic advances in library design toward relevant chemical space has come from biology. The completion of the genome opened the gate to the "omics" age, providing a wealth of information and triggering a race to annotate the proteome,[‡,15] in a quest for the identification of new protein targets for drug discovery.

Continuous advances in cellular systems and molecular biology,[16] crystallography,[17] improved affinity purification methods,[18] bioinformatics tools for protein structure clustering[19] and automated microscopy,[20] just to cite a few, are providing a better understanding of biological systems and more information about potential targets than ever before. The information from structural analysis of potential targets and from high-throughput bioassays are having a profound effect on library design, in the same way as assays of focused libraries comprised of more biologically relevant compounds are contributing to a comprehensive exploration of biological space. It is not the aim of this chapter to review these contributions in detail, but the influence of biology (*e.g.* target information) will be exemplified below in describing efforts toward target-focused libraries.

---

‡ The proteome can be defined as the complete collection of proteins expressed by a given genome, and proteomics the branch of knowledge that studies it. Proteomics and chemical biology have become common disciplines at the interface between chemistry and biology. For a global perspective of the synergy between chemistry and biology in drug discovery see Ref. 15.

In summary, the synergistic combination of advances in chemical synthesis, automation and purification, coupled with *in silico* tools and biological target information, has paved the way for the design of more efficiently focused libraries.

3 Library Design Concepts

3.1 Impact of Diversity on Library Design

The design strategies employed to improve combinatorial chemistry have evolved considerably since the early days of peptide and peptidomimetic libraries. The main concern early was on the availability of suitable synthetic methods that could be applied to the synthesis of libraries of small molecules; however, this early obstacle has been intensively addressed and at this point can be considered overcome (for examples of new methodology developed for library production see Ref. 21). With the ability in hand to prepare many different types of molecules in a variety of formats, the current challenge is to decide what compounds to make. As a consequence, much attention is now focused on the definition and analysis of chemical diversity.

The term chemical diversity is used collectively to describe structural (scaffold) or functionality differences within a collection of compounds. There are numerous scientific publications and reviews on methods for understanding chemical diversity and its application to drug discovery.

An early report by Martin and co-workers[22] illustrated the use of "Chemical Functionality" descriptors to generate molecular 'fingerprints', which were compared on the basis of their Tanimoto coefficients[23] (a metric for the degree of similarity between two bit strings). This approach has evolved into more sophisticated analyses of chemical diversity, even extending to 3-dimensional (3D) pharmacophoric fingerprints.[24] It is important to distinguish the difference in design strategies between the large-diversity libraries of the early 1990s and the more recent focused library approach. As noted above, the large-diversity libraries sought to cover as much chemical space as possible, but typically resulted in either too many low-quality hits or no hits at all. This experience suggested that there is a 'limit' to the diversity which can be attained with a given scaffold or chemical template. In hindsight, it seems obvious that a given scaffold is going to provide only a limited range of orientations that the substituents can adopt, unless it is extremely flexible (which is itself undesirable). Thus, within a given library there are functional-group relationships that can be easily reached and others that are unattainable due to geometric and steric constraints. This realization has helped to drive a shift in combinatorial design from large libraries (>10,000 members) built on a limited number of scaffolds to smaller, more-focused libraries (<1000 members) as a way to diversify scaffolds and increase diversity overall.

A recent review by Sauer and Schwarz[25] illustrate how a "high degree of shape (hence scaffold) diversity in screening collections will increase the odds of addressing a broad range of biological targets". Using a computational method to assess molecular shape, they show that individual compound classes tend to generate a limited set of molecular shapes, and, furthermore, that increasing the number of substitutions does not significantly improve the space coverage. Their analysis highlights the importance of scaffold assessment in library design and corroborates the notion underlying the industry trend toward smaller library size with novel scaffolds.

However, this shift in strategy has not come without a price. The early promise of combinatorial chemistry was that once the initial (large) investment in methods development and library synthesis was complete, one could benefit immediately from the easy generation of large numbers of compounds. However, optimizing scaffold diversity requires more effort in methods development to generate the diverse scaffolds, and results in smaller numbers of compounds for a similar investment. Although this tactic may appear to be more expensive, there are other considerations beyond the cost of library generation. For example, the synthetic methods developed for the library synthesis can still be applied in following up the hits that are found, and, most importantly, there is a greater likelihood of finding a useful lead from a number of smaller libraries than from a single large library.

3.2   Diversity-Oriented Synthesis in Prospecting Library Design

The concept of increasing scaffold diversity in the design of libraries that are not derived from a target model or lead series (the so-called "prospecting libraries"[26]) has stimulated the development of "combinatorial synthetic design"[27] and "diversity-oriented synthesis" (DOS)[28] as new aspects of traditional organic synthesis. The aim of the DOS library approach is to generate structurally complex compounds with higher diversity than would arise from designs based on a target focus. In addition to the diversity achieved from incorporation of a variety of building blocks, further diversity can be obtained from stereochemical complexity and the use of different "branching pathways" which, after a split step, lead to different backbones. Although the synthetic investment in generating such libraries is much higher than alternative strategies, the expectation is that more complex, natural product-like structures will result in the identification of novel leads for difficult biological targets such as protein–protein interactions. In this light, the DOS strategy can be viewed as combining the concepts of large, natural product-like libraries with the increased scaffold diversity of the focused library approach.

3.3   Target-Oriented Library Design

The design of target-oriented libraries differs from that leading to large diverse combinatorial libraries, in that the former are directed toward a particular target or family of targets, based upon existing medicinal chemical knowledge and biological principles relevant to the target class. Including a known pharmacophore in designing a library against a particular target would be an example of this strategy.

It is interesting to analyze the impact that this approach has had on the use of "diversity" in the various design strategies (for a definition of chemical diversity related to biological assays see Ref. 29). A graphical method for understanding these differences is elaborated from a recent analysis by Lipinski and Hopkins,[30] as illustrated in Figure 1. Consistent with this work (Figure 1A), the blue cube is

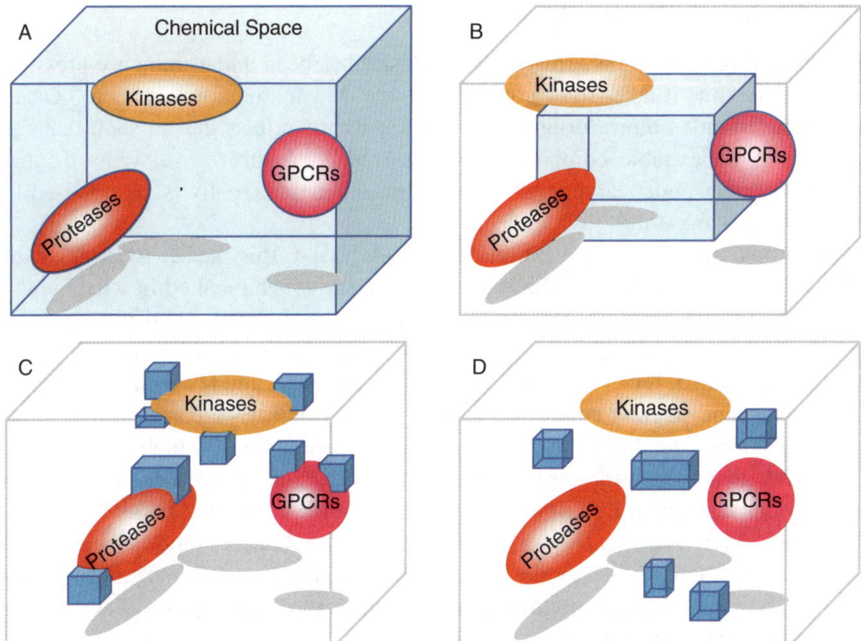

Figure 1 *Representation of some of the interfaces between biology and chemistry space: (a) the continuum of chemistry space continuum with representative regions of specific biological activity highlighted; (b) large combinatorial libraries seek to cover as much space as possible across several biological families but with one core scaffold; (c) smaller focused libraries, shown as the small blue cubes, are designed with relevant biological targets in mind; and (d) libraries that do not overlap with relevant biological space are undesirable*

meant to represent the continuum of chemical space, which contains discrete areas of biologically relevant compounds depicted as globular shapes. The differences in approaches between large diversity and the smaller focused approaches can be envisaged by analyzing the differences between the panels in this figure. The goal of large-diversity libraries is to use a single scaffold and synthetic route to cover as much of this space as possible, without concentrating in any one region (Figure 1B). In contrast, the realization that large areas of biological space cannot be covered with one scaffold underlies the design of smaller libraries that cover different regions of chemical space and increase the possibility of overlap with a particular target family (Figure 1C). A goal of focused library design is also to avoid libraries that do not fall into an area of chemical space with no known biological interest (Figure 1D). Of course, the underlying assumption is that the relevant and irrelevant regions of chemical space, both structural and functional, are known. The fact that there are still biological systems and types of targets, such as protein–protein interactions, for which no general modulation strategy has emerged, suggests that additional regions of chemical space remain to be identified.

3.4 Focus on Drug-Like Libraries

Over the last decade, the pharmaceutical industry has been under immense pressure
to reduce the time it takes to discover new drugs. Much attention has been focused
on making libraries more 'drug-like', trying to improve their design such that the
synthesis of 'undesirable' compounds is minimized.[31,32] In this regard, we have come
to a much better understanding as to what drug-likeness actually is, as opposed to
just 'looking like a drug'.

Numerous scientific publications have addressed this issue, starting with
Lipinski's seminal analysis of the World Drug Index which resulted in what is now
commonly known as Lipinski's 'Rule of 5'.[33] His analysis revealed that the vast
majority (90%) of orally absorbed drugs have molecular weights less than 500, log
P values less than 5, fewer than five hydrogen-bond donors and fewer than 10 hydrogen-bond acceptors. Based on these guidelines, many libraries are analyzed *in silico*
prior to synthesis to avoid the combination of functionalities, which may lead to
compounds that violate these principles. Veber *et al.*[34] expanded these concepts to
include other molecular descriptors such as polar surface area and number of rotatable bonds. More recently, Lipinski[35] has also highlighted the importance of solubility considerations in library design. The modern day task of library design thus
involves the evaluation and optimization of all these parameters prior to any significant investment in synthesis.

4 Focused Libraries

4.1 Libraries Focused on Pharmacophore Models

A pharmacophore is defined as the ensemble of steric and electronic features that
ensure the optimal interactions with a specific biological target to modulate its biological response.[36] A pharmacophore is not an actual molecule or collection of
specific functional groups; rather, it is an abstract representation that summarizes
what is known about the way a group of molecules interacts with the target.
Incorporating a particular pharmacophore in the design of a focused library may
increase the likelihood that the library will contain compounds with activity against
the target family.

An example of a pharmacophore-focused library from Krier *et al.*[12] is presented
above to illustrate the impact of *in silico* tools in library design (see Scheme 1).
Another example is shown in Scheme 2. In 2003, Chen and co-workers[37] described
the design and parallel synthesis of piperidine libraries targeting the nociceptin
receptor (NOP).[§] Based on three scaffolds known from the literature as having ago-
nistic or antagonistic properties against NOP, a new piperidine pharmacophore was
designed. After library design and optimization, 18 confirmed hits with moderate
potency and selectivity were obtained. Moreover, the initial hits served as useful
starting points for an optimization library that resulted in significant improvements
in both selectivity and potency.

---

§ NOP receptor belongs to the G-protein-coupled receptors (GPCR) superfamily (see Ref. 36 for more
  information).

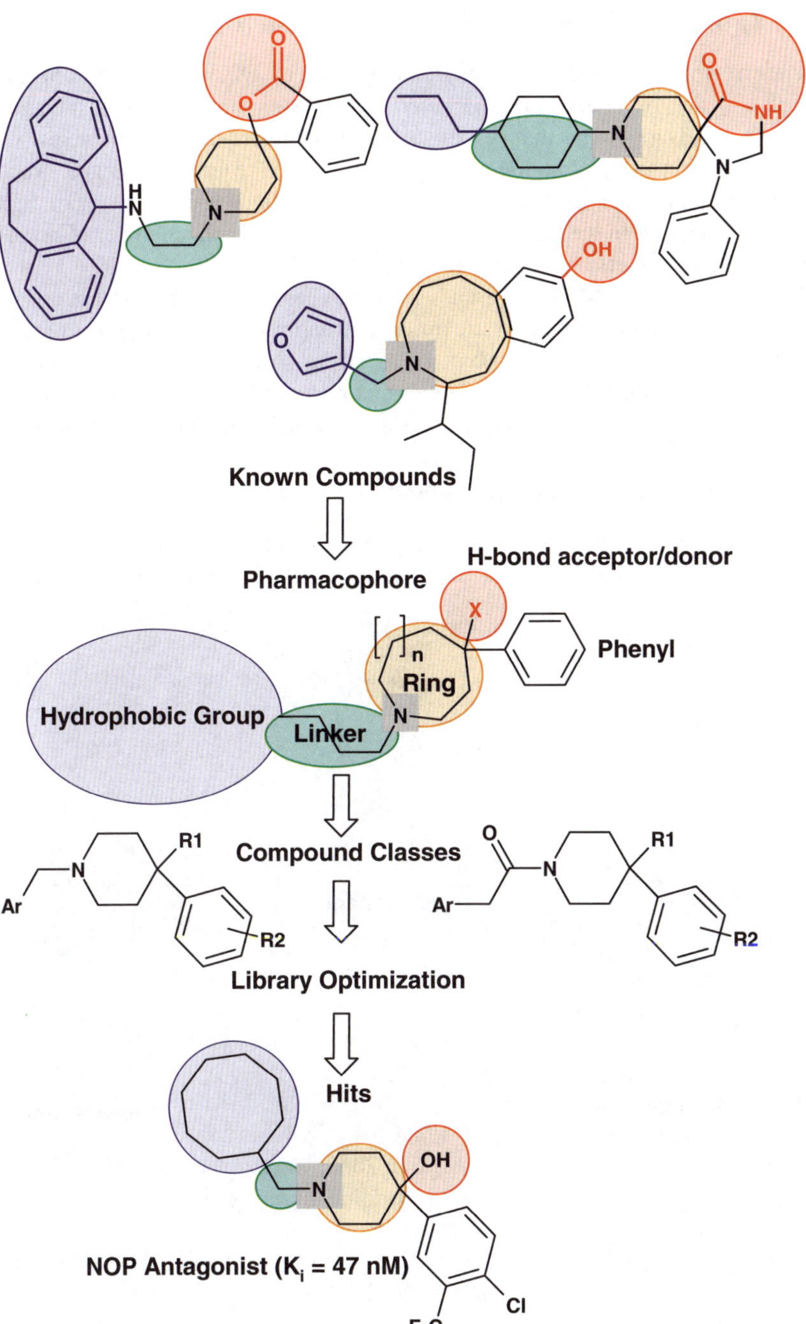

Scheme 2 *A pharmacophore-focused piperidine library designed from known NOP receptor antagonists. Blue, hydrophobic group; Green, linker; Orange Core Ring; Red H-bond donor/acceptor*

4.2 Libraries Focused on Privileged Structures

The term "privileged structure" was first described by Evans[38] in 1988 in his analysis of a series of benzodiazepine-based cholecystokinin-1 (CCK-1) antagonists (Scheme 3). Evans found that certain chemical scaffolds afford derivatives, which bind to a diverse group of biological targets; he coined the term 'privileged structure' for the compound classes that appear to have this characteristic. Benzodiazepines have been reported in the literature as privileged scaffolds on a number of occasions because of their diverse biological activities.[39]

In a recent review, Muller[40] discusses the privileged structure concept and highlights several paradigms of these structures. The definition of privileged structure can be easily confused with the concept of recurrent structural element. While some recurrent elements can be desirable, those that result in hits across many screens can generate irrelevant compounds or false positives, which should obviously be avoided.

4.3 Libraries Focused on Target Classes

Many advances have been made in biological screening, including higher throughput assay formats which provide rapid turnaround for optimization efforts. This

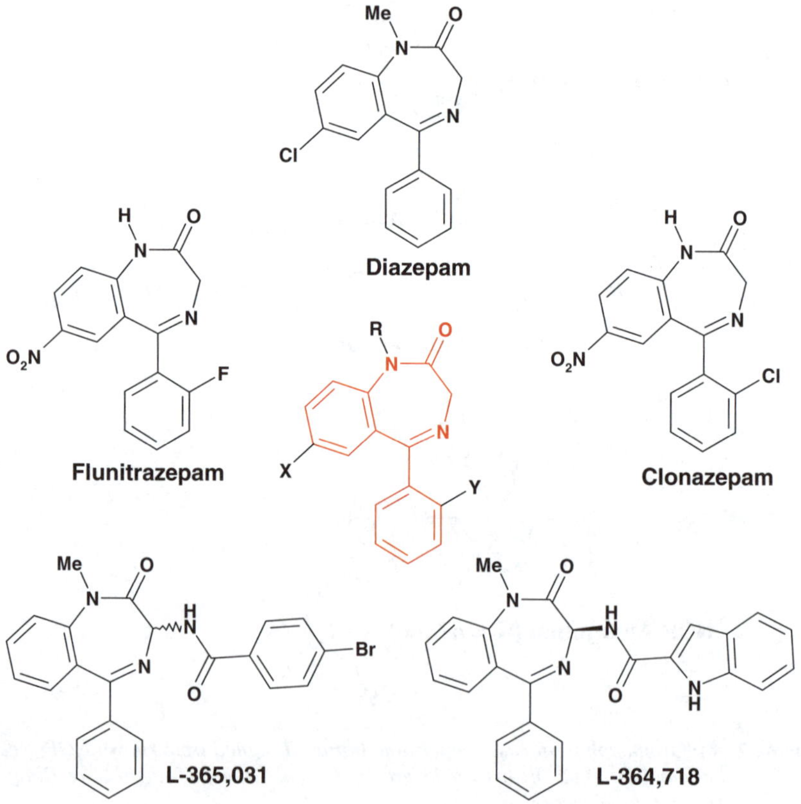

Scheme 3 *Examples of drugs derived from benzodiazepine-privileged structure*

wealth of data can provide the combinatorial and medicinal chemist with useful information to help narrow down the number of structures/scaffolds that need to be synthesized in order to identify and optimize biologically active molecules for drug discovery (for a review of target family-directed libraries, see Ref. 41). Thus, in the modern approach to library design, high-throughput chemistry efforts utilize as much of the known target information as possible to ensure that both novel and biologically relevant compounds are generated. With the higher output in both chemistry and screening, the focused library benefits from an iterative optimization approach in which the design of subsequent compound collections takes advantage of the biological results produced by the previous collection. In the following sections, we describe several examples of biologically targeted focused libraries.

4.3.1   GPCR-Targeted Libraries

The GPCRs are a large and widely distributed super-family of membrane-bound proteins that are involved in signal transduction in many physiological and pathological processes. Examples of GPCRs include the receptors for the neurotransmitters acetylcholine, muscarine, dopamine and opioids. GPCRs are one of the most relevant targets for small molecules, representing the site of action of *ca.* 40% of the marketed drugs. Even considering their relevance and popularity as targets for drug discovery, GPCRs can still be challenging targets. Because they are integral membrane proteins, little structural information is available. For some members of this super family, the so-called orphan GPCRs, the identity of their natural ligands is not even known. As a consequence, focused libraries that target GPCRs have mainly come from a privileged-structure approach. A recent review by Crossley[42] on GPCR-targeted libraries highlights several other design principles, including the use of available in-house historical knowledge (*e.g.* the historical compound archive of a large pharmaceutical company) as starting point. Another excellent review on libraries for GPCRs has been recently published by Guo and co-workers.[43] An example of a GPCR library by Bleicher and co-workers[44] is illustrated in Scheme 4.

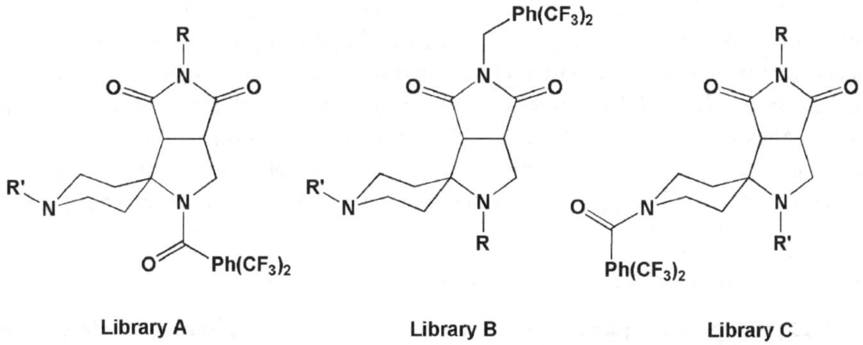

Library A   **Library B**   **Library C**

Scheme 4 *Human NK-1 libraries based on a spiropyrrolo–pyrrole template with the inclusion of the bis-trifluoromethyl benzoyl moiety to provide the desired selectivity*

Bleicher reports the generation of human NK-1 ligands based on the spiropy-rrolo–pyrrole scaffold, which is considered a promiscuous privileged structure. In order to overcome the liability associated with lack of selectivity of this pharma-cophore, the bis-trifluorobenzoyl group (a pharmacophore which provides selectiv-ity for the NK-1 receptor) was utilized as a 'needle' to generate specific compounds for the NK-1 receptor. This example highlights an efficient parallel library approach in which the 'needle' group was placed in several places on the scaffold, and small libraries were prepared at each position to thoroughly explore the SAR of the series. These efforts lead to ligands with nanomolar affinities to the receptor.

4.3.2 Kinase-Targeted Libraries

The protein kinases constitute a family of approximately 500 known enzymes that act as mediators in regulatory processes ranging from cell proliferation and differentiation, to transcription, metabolism and apoptosis, among others. Their mode of action involves phosphorylation of certain amino acid residues (most often Ser, Thr, or Tyr) in specific proteins by transfer of a phosphate group from a donor molecule (usually ATP).

In a 2002 paper, Manning *et al.* provided an overview of the "kinome" (*e.g.* the complete protein kinase set expressed by the genome) and reviewed its relevance to drug discovery.[45] They emphasize that the many roles that kinases play and their potential tractability as drug targets have made them an attractive target for developing new therapies against a diverse array of diseases. However, the vast majority of kinase inhibitors interact with the ATP-binding site, which can lead to low specificity for the desired target and therefore undesirable side effects. Thus, much of the effort directed toward the discovery of potent and selective kinase inhibitors has involved the design of libraries to overcome these challenges.

This approach is exemplified in the work of Rich and collaborators[46] who designed a library targeted against the janus family of kinases (JAK). These kinases mediate cellular responses to cytokines in many developmental and immunological processes; hence, they are key points for intervention in pathological and malignant cellular behavior. It is also known that JAK kinases contribute to tumor progression. Tyrphostin and related compounds are one of the best-studied scaffolds for JAK kinase inhibition. The authors describe the synthesis of a focused family of 599 compounds around the scaffold of Tyrphostin AG490 (see Scheme 5). Tyrphostin was selected based on its known activity against JAK2 and JAK3 as well as other kinases; hence, compounds with increased potency and selectivity were sought. Two members of the library, 5H4 and 8B9, show a significantly improved inhibition of proliferation in cell lines dependent on JAK2 or JAK3 kinases. It is worth noting that potency increase resulted from the incorporation of hydrophobic trifluoromethyl groups, which can enhance hydrophobic interactions as well as cell-membrane penetration.

4.3.3 Natural Product-Based Focused Libraries

Natural products have played a disproportionate role in the discovery of leads and the development of successful drugs. Close to 50% of the marketed drugs based on small molecules for the last 25 years have come from natural products, and the

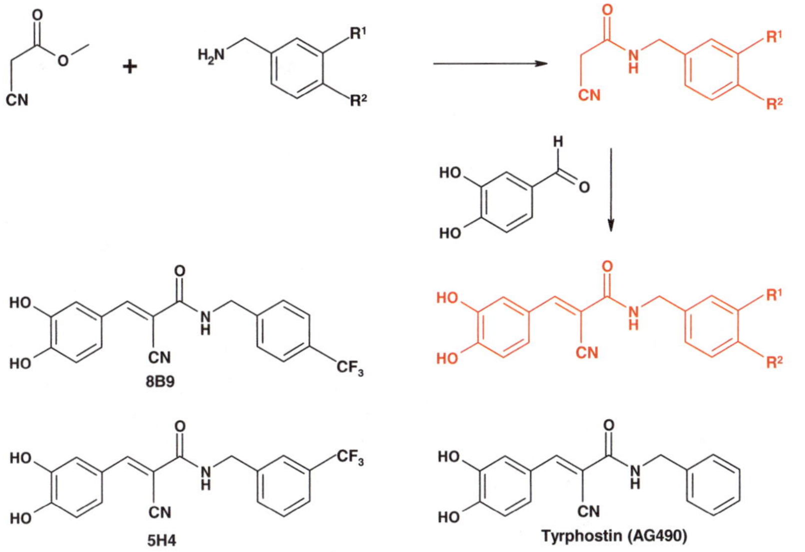

Scheme 5 *Library of JAK-2 and JAK-3 inhibitors based on the tyrphostin (AG490) scaffold*

percentage is even higher for therapeutic uses in cancer (*ca.* 60%) or infectious diseases (*ca.* 75%) (for more recent reviews on the role of natural products in drug discovery see Ref. 47). In light of these figures, the appeal of these evolution-privileged zones in the continuum of chemical space for the design of focused libraries is understandable. The discovery of an interesting biological activity for a naturally occurring scaffold attracts great attention, both to study the SAR of the chemical series and to design focused libraries around the novel structure. The epothilones, taxol, rapamycin, vancomycin and cyclosporines are just a few examples of novel structural classes that nature has given to drug discovery. Trying to take advantage of this evolutionary "know-how" underlies the intensive efforts in the quest for more novel structures. Recently, Waldmann and co-workers[48] have reviewed the ways that knowledge gleaned from naturally relevant chemical scaffolds, combined with protein structure similarity clustering (PSSC), can yield improved results, illustrating the advantages to be gained by exploring nature's pre-validated starting points.

Natural products have provided a challenge to chemists since the early times of total synthesis, hence it is not surprising to find a number of reports of natural product-based focused libraries in the literature. An example from Fürstner's group[49] illustrates a successful approach to the design and synthesis of a library based on the natural product latrunculin (Scheme 6). Latrunculins are marine macrolides with cytotoxic and antiviral activity that act by binding to the actin cytoskeleton without disrupting microtubules. The desire to understand their SAR, combined with a difficult and low-yielding extraction and isolation from the natural source, fostered the synthesis of a small library of scaffold-related compounds. The synthesis of the macrolide skeleton was accomplished via ring-closing alkyne metathesis and Lindlar hydrogenation,[50] to yield both latrunculin A and B as well as 10 other analogs.

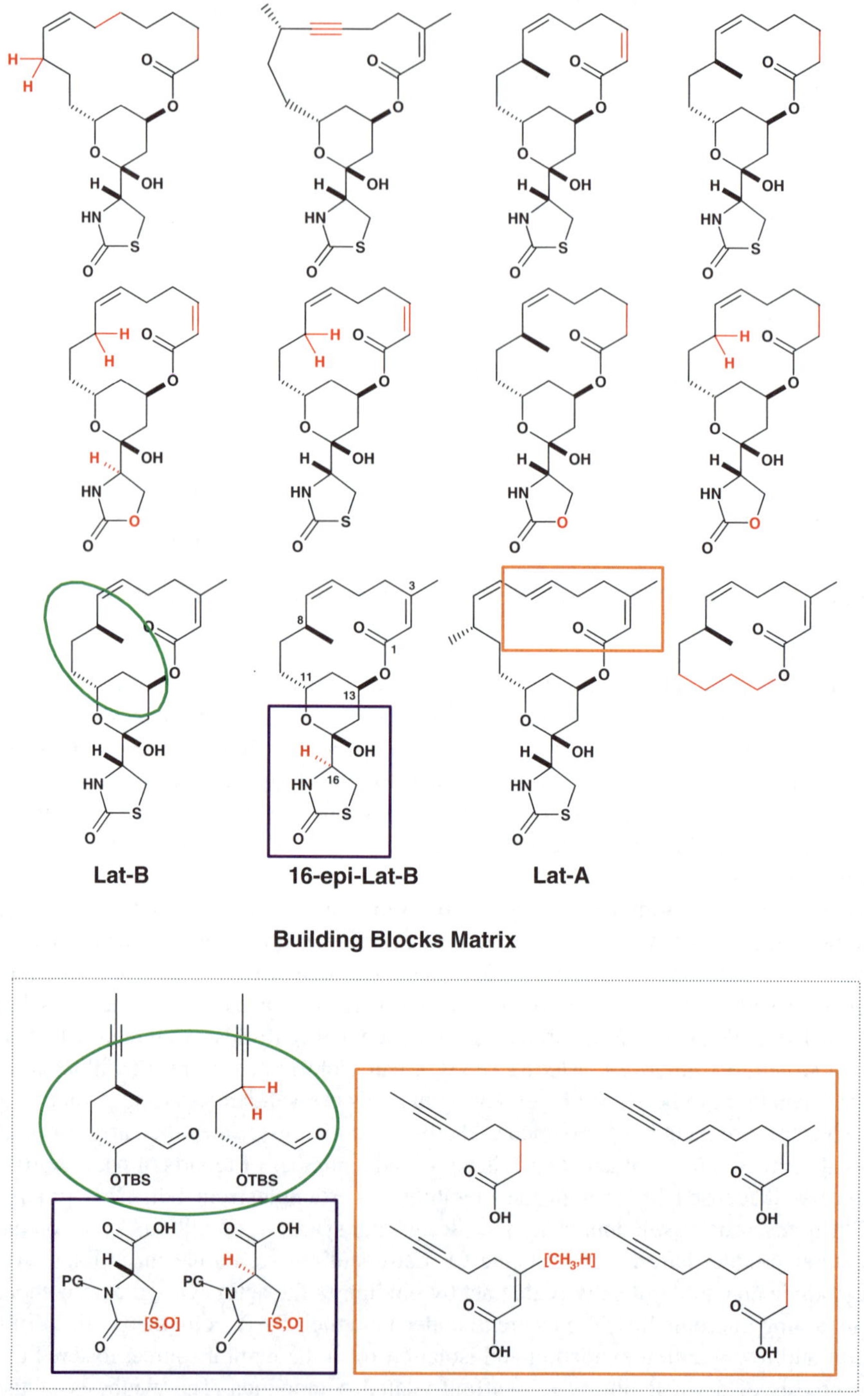

Lat-B

16-epi-Lat-B

Lat-A

Building Blocks Matrix

Scheme 6 *Focused library of latrunculin analogs*

4.4 Early Optimization or Hit-to-Lead Libraries

There has been much recent attention devoted to the subject of hit-to-lead advancement, and in most large pharmaceutical companies it is not uncommon to find chemistry groups that are dedicated to this early optimization effort. The focus of these efforts is to optimize hit structures into advanced compounds appropriate for full lead optimization, typically through the use of high-throughput chemistry methods. Several chemical scaffolds, typically identified from high-throughput screens, are evaluated in parallel through the design and synthesis of focused libraries, which are used to establish an SAR for each series. All of the considerations of drug-likeness mentioned above are relevant to this approach as well. However, at this stage, the goal is to determine for each series if there is an adequate SAR on which to base a full-scale lead optimization effort. As several scaffolds are typically evaluated in parallel, HTS methods play an important role although, instead of the rather large libraries generated for an HTS screen, this phase usually involves parallel synthesis of tens to hundreds of compounds. Since reliable SAR data is the desired product of these efforts, the quality of the individual compounds is very important. For this reason, all compounds in the collection are usually purified by HPLC. A practical strategy for preparing such hit-to-lead libraries is to generate on a large scale an advanced intermediate that then serves as the starting point for preparation of the final compounds with different points of diversity.

Raynaud and co-workers[51] recently reported the optimization of inhibitors of cyclin-dependent kinase-2 (CDK-2), based on a purine scaffold (Scheme 7). The synthesis

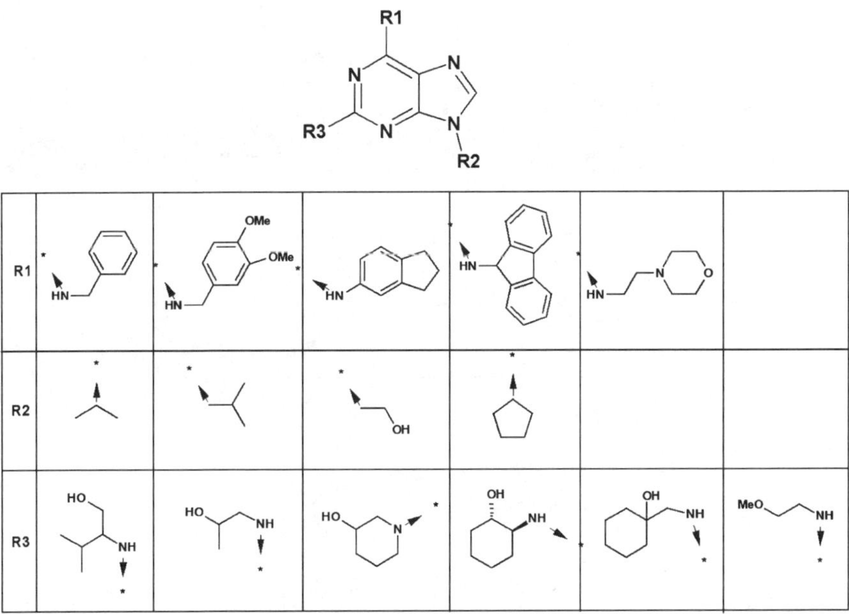

Scheme 7 *Structure of 2,6,9-trisubstituted purines analogs prepared by Raynaud and co-workers as CDK-2 inhibitors for PK profiling*

was carried out in parallel to generate a focused library, which was used to improve the pharmacokinetics properties of the leads. This work highlights the early improvements made possible with the marriage of modern chemistry techniques and those in other disciplines (in this case cassette dosing for rapid pharmacokinetic analysis).

In another recent effort, Poulain and co-workers[52] illustrate the improvements in efficiency that focused libraries can bring to early lead optimization. The authors describe the hit-to-lead optimization of μ opiate ligands, starting with a micromolar hit from HTS. Using a pharmacophore-based approach, several focused libraries were designed and synthesized, as illustrated in Scheme 8. Combining the results of these libraries suggested a series of chimeric compounds, which exhibited sub-nanomolar activity and improved physico-chemical properties.

These, among the increasing number of reported examples, demonstrate how the focused library approach extends further along the drug-discovery pipeline than the initial HTS screen.

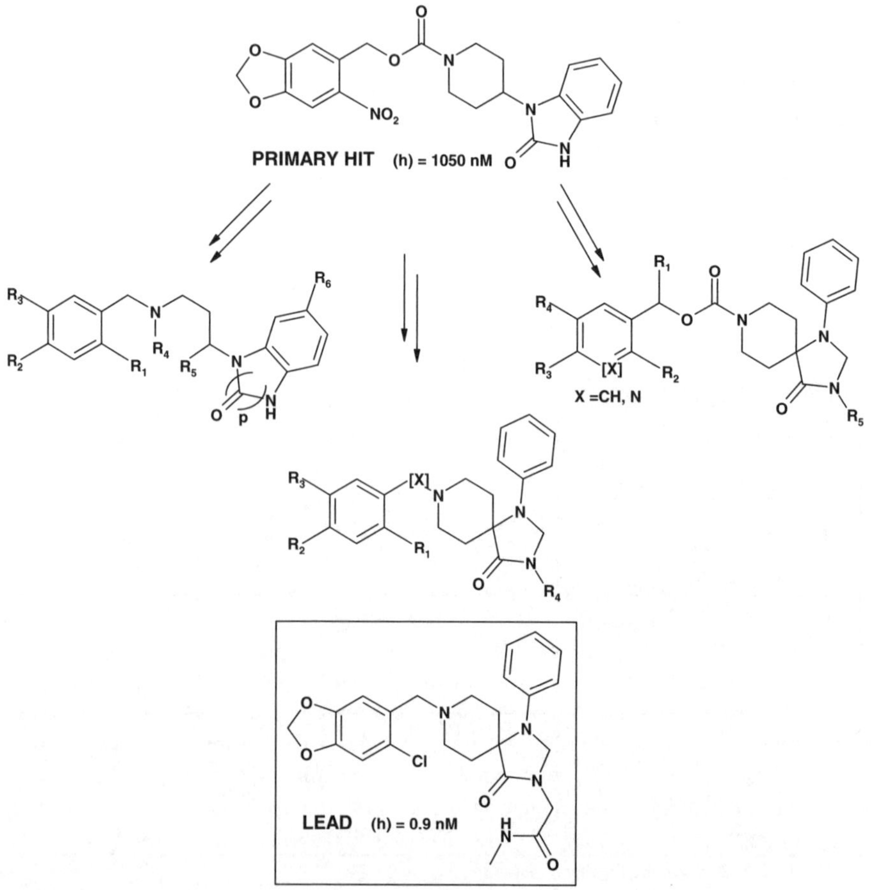

Scheme 8 *Hit-to-lead-focused library for early optimization against the μ opiate receptor*

5 Summary

The evolution of HTS methods from the early combinatorial libraries to the more recent focused approaches has proceeded very rapidly. These efforts have benefited from developments in a number of related technologies that together have shaped the way in which the modern combinatorial and medicinal chemist views library design. Technological advances in the areas of automated synthesis, cheminformatics and purification techniques are being made every day and it is evident that these advances will continue to accelerate the drug-discovery process. Today, the use of privileged structures, pharmacophore models and biological target information to assist in the design of focused libraries is improving the way in which we approach lead finding and hit-to-lead efforts. As focused library approaches reach ever further into the lead optimization process, it is evident that the full extent to which these high-throughput methods can contribute to the drug-discovery process is yet to be reached.

References

1. (i) M. Pulici, G. Cervi, K. Martina and F. Quartieri, Use of multicomponent, domino, and other one-pot syntheses on solid phase: powerful tools for the generation of libraries of diverse and complex compounds, *Comb. Chem. High Throughput Screening*, 2003, **6**(7), 693. (ii) N.A. Boyle and K.J. Janda, Formats for combinatorial synthesis: solid-phase, liquid-phase and surface, *Curr. Opin. Chem. Bio.*, 2002, **6**(3), 339. (iii) T. Carell, E.A. Wintner, A.J. Sutherland, J. Rebek Jr., Y.M. Dunayevskiy and P. Vouros, New promise in combinatorial chemistry: synthesis, characterization, and screening of small-molecule libraries in solution, *Chem. Bio.*, 1995, **2**(3), 171.

2. (i) C.O. Kappe, Synthetic methods, Controlled microwave heating in modern organic synthesis, *Angew. Chem. Int. Ed.*, 2004, **43**(46), 6250. (ii) C.J. Cullen, R.C.R. Wootton and A.J. de Mello, Microfluidic systems for high-throughput and combinatorial chemistry, *Curr. Opin. Drug Disc. Dev.*, 2004, **7**(6), 798. (iii) S.H. DeWitt, Microreactors for chemical synthesis, *Curr. Opin. Chem. Bio.*, 1999, **3**(3), 350.

3. (i) D. Tejedor, D. Gonzalez-Cruz, A. Santos-Exposito, J.J. Marrero-Tellado, P. de Armas and F. Garcia-Tellado, Multicomponent domino processes based on the organocatalytic generation of conjugated acetylides: efficient synthetic manifolds for diversity-oriented molecular construction, *Chem. Eur. J.*, 2005, **11**(12), 3502. (ii) A. Ulaczyk-Lesanko and D.G. Hall, Wanted: new multicomponent reactions for generating libraries of polycyclic natural products, *Curr. Opin. Chem. Bio.*, 2005, **9**(3), 266. (iii) L. Weber, The application of multicomponent reactions in drug discovery, *Curr. Med. Chem.*, 2002, **9**(23), 2085.

4. E. Letot, G. Koch, R. Falchetto, G. Bovermann, L. Oberer and H.-J. Roth, Quality control in combinatorial chemistry: determinations of amounts and comparison of the "purity" of LC-MS-purified samples by NMR, LC-UV and CLND, *J. Comb. Chem.*, 2005, **7**(3), 364.

5. (i) C. Edwards and D.J. Hunter, High-throughput purification of combinatorial arrays, *J. Comb. Chem.*, 2003, **5**(1), 61. (ii) M.S. Congreve, S.V. Ley and

J.J. Scicinski, Analytical construct resins for analysis of solid-phase chemistry, *Chem. Eur. J.*, 2002, **8**(8), 1768. (iii) D.B. Kassel and P.L. Myers, The analytical challenge: keeping pace with combinatorial chemistry, *Pharm. News*, 2002, **9**(3), 171. (iv) J. Eames and M. Watkinson, Polymeric scavenger reagents in organic synthesis, *Eur. J. Org. Chem.*, 2001, **7**, 1213. (v) W.C. Ripka, G. Barker and J. Krakover, High-throughput purification of compound libraries, *Drug Disc. Today*, 2001, **6**(9), 471. (vi) H.N. Weller, Purification of combinatorial libraries, *Mol. Diver.*, 1998, **4**, 47.

6. (i) P.S. Marshall, B. Leavens, O. Heudi and C. Ramirez-Molina, Liquid chromatography coupled with inductively coupled plasma mass spectrometry in the pharmaceutical industry: selected examples, *J. Chromatogr. A*, 2004, **1056**(1–2), 3. (ii) L. Prokai, A. Zharikova, T. Janaky, X. Li, A.C. Braddy, P. Perjesi, L. Matveeva, D.H. Powell and K. Prokai-Tatrai, Integration of mass spectrometry into early-phase discovery and development of central nervous system agents, *J. Mass Spec.*, 2001, **36**(11), 1211.

7. Y. Shintani, K. Hirako, M. Motokawa, T. Iwano, X. Zhou, Y. Takano, M. Furuno, H. Minakuchi and M. Ueda, Development of miniaturized multi-channel high-performance liquid chromatography for high-throughput analysis, *J. Chromatogr. A*, 2005, **1073**(1–2), 17.

8. (i) O. Dror, A. Shulman-Peleg, R.H. Nussinov and H.J. Wolfson, Predicting molecular interactions in silico: I.A guide to pharmacophore identification and its applications to drug design, *Curr. Med. Chem.*, 2004, **11**(1), 71. (ii) J.H. Van Drie, Pharmacophore discovery: A critical review, in *Computational Medicinal Chemistry for Drug Discovery*, P. Bultinck (ed), Marcel Dekker, Inc., New York, 2004, 437. (iii) J.L. Jenkins, M. Glick and J.W. Davies, A 3D similarity method for scaffold hopping from known drugs or natural ligands to new chemotypes, *J. Med. Chem.*, 2004, **47**(25), 6144. (iv) E.M. Krovat and T. Langer, Non-peptide angiotensin II receptor antagonists: chemical feature based pharmacophore identification, *J. Med. Chem.*, 2003, **46**(5), 716.

9. H. Kubinyi, Combinatorial and computational approaches in structure-based drug design, *Curr. Opin. Drug Disc. Dev.*, 1998, **1**(1), 16.

10. R.A. Lewis, A general method for exploiting QSAR models in lead optimization, *J. Med. Chem.*, 2005, **48**(5), 1638.

11. (i) S.J. Wilkens, J. Janes and A.I. Su, HierS: hierarchical scaffold clustering using topological chemical graphs, *J. Med. Chem.*, 2005, **48**(9), 3182. (ii) M. Pintore, O. Taboureau, F. Ros and J.R. Chretien, Database mining applied to central nervous system (CNS) activity, *Eur. J. Med. Chem.*, 2001, **36**(4), 349.

12. M. Krier, J.X. Araujo-Junior, M. Schmitt, J. Duranton, H. Justiano-Basaran, C. Lugnier, J.J. Bourguignon and D. Rognan, Design of small-sized libraries by combinatorial assembly of linkers and functional groups to a given scaffold: application to the structure-based optimization of a phosphodiesterase 4 inhibitor, *J. Med. Chem.*, 2005, **48**, 3816.

13. M. Braban, L. Pop, X. Willard and D. Horvath, Reactivity prediction models applied to the selection of novel candidate building blocks for high-throughput organic synthesis of combinatorial libraries, *J. Chem. Inf. Comp. Sci.*, 1999, **39**(6), 1119.

14. (i) P. Watson, M. Verdonk and M.J. Hartshorn, A web-based platform for virtual screening, *J. Mol. Graph Mod.*, 2003, **22**(1), 71. (ii) J.B. Bremner, K. Castle, R. Griffith, P.A. Keller and D.D. Ridley, Mining the chemical abstracts database with pharmacophore-based queries, *J. Mol. Graph. Mod.*, 2002, **21**(3), 185.

15. (i) J. Drews, Drug discovery: a historical perspective, *Science*, 2000, **287**(5460), 1960. (ii) F. Darvas, G. Dorman, L. Urge, I. Szabo, Z. Ronai and M. Sasvari-Szekely, Combinatorial chemistry. Facing the challenge of chemical genomics, ©2001 IUPAC, *Pure App. Chem.*, 2001, **73**(9), 1487. (iii) N.S. Gray, L. Wodicka, A.-M.W.H. Thunnissen, T.C. Norman, S. Kwon, F.H. Espinoza, D.O. Morgan, G. Barnes, S. LeClerc, L. Meijer, S.-H. Kim, D.J. Lockhart and P.G. Schultz, Exploiting chemical libraries, structure, and genomics in the search for kinase inhibitors, *Science*, 1998, **281**(5376), 533.

16. (i) M.W. Kirschner, The meaning of systems biology, *Cell*, 2005, **121**(4), 503. (ii) E.C. Butcher, Innovation: can cell systems biology rescue drug discovery? *Nat. Rev. Drug Disc.*, 2005, **4**(6), 461.

17. (i) T.L. Blundell and S. Patel, High-throughput X-ray crystallography for drug discovery, *Curr. Opin. Pharmacol.*, 2004, **4**(5), 490. (ii) T.L. Blundell, H. Jhoti and C. Abell, High-throughput crystallography for lead discovery in drug design, *Nat. Rev. Drug Disc.*, 2002, **1**(1), 45.

18. A. Bauer and B. Kuster, Affinity purification-mass spectrometry. Powerful tools for the characterization of protein complexes, *Eur. J. Biochem., /FEBS*, 2003, **270**(4), 570.

19. M.A. Koch and H. Waldmann, Protein structure similarity clustering and natural product structure as guiding principles in drug discovery, *Drug Disc. Today*, 2005, **10**(7), 471.

20. T.J. Mitchison, Small-molecule screening and profiling by using automated microscopy, *Chem. Bio. Chem.*, 2005, **6**(1), 33.

21. (i) A.B. Beeler, S.E. Schaus and J.A. Porco, Chemical library synthesis using convergent approaches, *Curr. Opin. Chem. Bio.*, 2005, **9**(3), 277. (ii) A. Ganesan, Recent developments in combinatorial organic synthesis, *Drug Disc. Today*, 2002, **7**(1), 47. (iii) T.S. Haque, A.P. Combs and L.A. Thompson, New synthetic methodologies, in *Integrated Drug Discovery Technologies*, H-Y. Mei and A.W. Czarnik (eds), Marcel Dekker, Inc., New York, 2002, 407.

22. E.J. Martin, J.M. Blaney, M.A. Siani, D.C. Spellmeyer, A.K. Wong and W.H. Moos, Measuring diversity: experimental design of combinatorial libraries for drug discovery, *J. Med. Chem.*, 1995, **38**, 1431.

23. P. Willet, *Similarity and clustering in chemical information systems*, Wiley, New York, NY, 1987, 54.

24. B.R. Beno and J.S. Mason, The design of combinatorial libraries using properties and 3D pharmacophore fingerprints, *Drug Disc. Today*, 2001, **6**(5), 251.

25. W.H.B. Sauer and M.K. Schwarz, Size doesn't matter: scaffold diversity, shape diversity and biological activity of combinatorial libraries, *Chimia*, 2003, **57**(5), 276.

26. M.R. Spaller, M.T. Burger, M. Fardis and P.A. Bartlett, Synthetic strategies in combinatorial chemistry, *Curr. Opin. Chem. Bio.*, 1997, **1**, 47.

27. M.A. Marx, A.-L. Grillot, C.T. Louer, K.A. Beaver and P.A. Bartlett, Synthetic design for combinatorial chemistry. Solution and polymer-supported synthesis of polycyclic lactams by intramolecular cyclization of azomethine ylides, *J. Am. Chem. Soc.*, 1997, **119**, 6153.

28. (i) D. Lee, J.K. Sello and S.L. Schreiber, Pairwise use of complexity-generating reactions in diversity-oriented organic synthesis, *Org. Lett.*, 2000, **2**, 709. (ii) M.D. Burke and S.L. Schreiber, A planning strategy for diversity-oriented synthesis, *Angew. Chem. Int. Ed.*, 2004, **43**(1), 46.

29. H.-J. Roth, There is no such thing as 'diversity'!, *Curr. Opin. Chem. Bio.*, 2005, **9**(3), 293.

30. C. Lipinski and A. Hopkins, Navigating chemical space for biology and medicine, *Nature*, 2004, **432**, 855.

31. L. Weber, Molecular diversity analysis and combinatorial library design, *Methods Principles Med. Chem.*, 2000, **8**, 137.

32. R.A. Lewis, The design of small- and medium-sized focused combinatorial libraries: design of focused combinatorial libraries, in *Molecular Diversity in Drug Design*, P.M. Dean and R.A. Lewis (eds), Kluwer Academic Publishers, Dordrecht, 1999, 221.

33. C.A. Lipinski, F. Lombardo, B.W. Dominy and P.J. Feeney, Experimental and computational approaches to estimate solubility and permeability in drug discovery and development settings, *Adv. Drug Deliv. Rev.*, 1997, **23**, 3.

34. D.F. Veber, S.R. Johnson, H.-Y. Cheng, B.R. Smith, K.W. Ward and K.D. Kopple, Molecular properties that influence the oral bioavailability of drug candidates, *J. Med. Chem.*, 2002, **45**(12), 2615.

35. C.A. Lipinski, Solubility in the design of combinatorial libraries, *Chem. Anal.*, 2004, **163**, 407.

36. C.G. Wermuth, C.R. Ganellin, P. Lindberg and L.A. Mitscher, ©1998 IUPAC, *Pure Appl. Chem.*, 1998, **70**(5), 1129.

37. Z. Chen, W.S. Miller, S. Shan and K.J. Valenzano, Design and parallel synthesis of piperidine libraries targeting the nociceptin (N/OFQ) receptor, *Bioorg. Med. Chem. Lett.*, 2003, **13**(19), 3247.

38. B.E. Evans, K.E. Rittle, M.G. Bock, R.M. DiPardo, R.M. Freidinger, W.L. Whitter, G.F. Lundell, D.F. Veber and P.S. Anderson, Methods for drug discovery: development of potent, selective, orally effective cholecystokinin antagonists, *J. Med. Chem.*, 1988, **31**, 2235.

39. (i) D.J. Parks, L.V. LaFrance, R.R. Calvo, K.L. Milkiewicz, V. Gupta, J. Lattanze, K. Ramachandren, T.E. Carver, E.C. Petrella, M.D. Cummings, D. Maguire, B.L. Grasberger and T. Lu, 1,4-Benzodiazepine-2,5-diones as small molecule antagonists of the HDM2-p53 interaction: discovery and SAR, *Bioorg. Med. Chem. Lett.*, 2005, **15**(3), 765. (ii) B.A. Bunin, M.J. Plunkett and J.A. Ellman, The combinatorial synthesis and chemical and biological evaluation of a 1,4-benzodiazepine library, *Proc. Natl. Acad. Sci.*, 1994, **91**(11), 4708.

40. G. Müller, Medicinal chemistry of target family-directed masterkeys, *Drug Disc. Today*, 2003, **8**(15), 681.

41. S.J. Shuttleworth, R.V. Connors, J. Fu, J. Liu, M.E. Lizarzaburu, W. Qiu, R. Sharma, M. Wanska and A.J. Zhang, Design and synthesis of protein

superfamily-targeted chemical libraries for lead identification and optimization, *Curr. Med. Chem.*, 2005, **12**(11), 1239.

42. R. Crossley, The design of screening libraries targeted at G-protein coupled receptors, *Curr. Topics Med. Chem.*, 2004, **4**, 581.

43. T. Guo and D.W. Hobbs, Privileged structure-based combinatorial libraries targeting G protein-coupled receptors, *Drug Dev. Tech.*, 2003, **1**(4), 579.

44. K.H. Bleicher, A.G. Wüthrich, T. Hoffmann and A.J. Sleight, Parallel solution- and solid-phase synthesis of spiropyrrolo–pyrroles as novel neurokinin receptor ligands, *Bioorg. Med. Chem. Lett.*, 2002, **12**, 3073.

45. G. Manning, D.B. Whyte, R. Martinez, T. Hunter and S. Sudarsanam, The protein kinase complement of the human genome, *Science*, 2002, **298**, 1912.

46. L. Gu, H. Zhuang, B. Safina, X.-y. Xiao, W.W. Bradford and B.E. Rich, Combinatorial approach to identification of tyrphostin inhibitors of cytokine signaling, *Bioorg. Med. Chem.*, 2005, **13**(13), 4269.

47. (i) D.J. Newman, G.M. Cragg and K.M. Snader, Natural products as sources of new drugs over the period 1981–2002, *J. Nat. Prod.*, 2003, **66**(7), 1022. (ii) F.E. Koehn and G.T. Carter, The evolving role of natural products in drug discovery, *Nat. Rev. Drug Disc.*, 2005, **4**(3), 206. (iii) J. Clardy and C. Walsh, Lessons from natural molecules, *Nature*, 2004, **432**, 829.

48. R. Balamurugan, F.J. Dekker and H. Waldmann, Design of compound libraries based on natural product scaffolds and protein structure similarity clustering (PSSC), *Mol. Bio. Syst.*, 2005, **1**(1), 36.

49. A. Fürstner, D. Kirk, M.D.B. Fenster, C. Aissa, D. De Souza and O. Mueller Diverted total synthesis: preparation of a focused library of latrunculin analogues and evaluation of their actin-binding properties, *Proc. Natl. Acad. Sci.*, 2005, **102**(23), 8103.

50. A. Fürstner and L.Turet, Concise and practical synthesis of latrunculin A by ring-closing Enyne-Yne metathesis, *Angew. Chem. Int. Ed.*, 2005, **44**(22), 3462.

51. F.I. Raynaud, P.M. Fischer, B.P. Nutley, P.M. Goddard, D.P. Lane and P. Workman, Cassette dosing pharmacokinetics of a library of 2,6,9-trisubstituted purine cyclin-dependent kinase 2 inhibitors prepared by parallel synthesis, *Mol. Cancer Ther.*, 2004, **3**, 353.

52. R. Poulain, D. Horvath, B. Bonnet, C. Eckhoff, B. Chapelain, M.-C. Bodinier and B. Deprez, From hit to lead. Combining two complementary methods for focused library design. Application to opiate ligands, *J. Med. Chem.*, 2001, **44**(21), 3378.

Translating Peptides into Small Molecules

GERD HUMMEL, ULRICH REINEKE AND ULF REIMER

JERINI AG, Invalidenstraße 130, 10115 Berlin, Germany

1   Peptides as Drugs: The Good, the Bad and the Ugly

Nature entrusts peptides with numerous tasks ranging from passing on messages as peptide hormones to priming the immune system, tuning metabolism and protein degradation or defending microorganisms as antibiotic agents. In addition, researchers have identified a variety of peptides with artificial biological functions that are not represented by naturally occurring molecules, *e.g.* receptor antagonists or inhibitors of protein–protein interactions. These peptides or their derivatives are valuable starting points for drug development and, indeed, a number of peptides or peptide-derived drugs have already made their way into the clinic.[1] Compared to the standard properties of drug-like small molecules matching the "rule of five", which are today's preferred formats, peptide-based drugs sometimes overcome the barrier of "undrugable targets" and are often extremely potent. Furthermore, many peptides are highly specific and do not accumulate in the body since they have short half-lives. As a consequence, peptides often show less systemic toxicity compared to small molecules.[1] In addition, peptides are ideal, rapidly accessible lead structures for drug-target validation as well as further drug development due to their modular structure, variable presentation of different functional groups and ease of preparation, either chemically or biologically.

Despite their favourable properties, peptide-based drugs are under-represented in the pharmaceutical market. This discrimination is usually due to their poor bioavailability, which sometimes necessitates non-oral administration or even special medical devices such as inhalers. Another related major disadvantage of peptides is their low metabolic stability due to proteolytic degradation. In addition, costs of goods for the drug substance are sometimes tremendous. Therefore, there is considerable interest to transform the "active principle" of biologically active peptides into small molecules with improved pharmacokinetic properties. In this chapter, we present an overview of ways to identify active peptides and how to tailor them for transformation into peptidomimetics or small molecules. This introduction is followed by descriptions of classical transformation approaches using stepwise substitutions of peptidic features. Conceptually different is the strategy in which peptides are used to generate pharmacophore models representing all necessary functional properties in the appropriate spacing and 3-D orientation. These pharmacophore models are then applied to *in silico* screens of small-molecule libraries to identify compounds that resemble the "active principle" of the starting peptides.

2   Origin of Biologically Active Peptides

Organisms from all species produce a huge variety of peptides in order to respond to certain physiological or pathophysiological stimuli. Reflecting their many different functions, their molecular structures are extremely diverse, ranging from very short peptides such as the enkephalins to complex peptide hormones like insulin with 51 amino acids in two chains connected by two disulfide bonds.

Also very similar to peptides occurring in nature are active peptides identified from the primary structure of larger proteins (knowledge- or sequence-based approach). This approach involves scanning the entire sequence of the protein with overlapping peptides, usually not longer than 15 amino acids (peptide scan), which are screened for interactions with a binding partner. The sequence common to the interacting peptides is the binding site[2] and such peptides often inhibit the respective protein–protein interaction.

As an alternative to the knowledge or sequence-based approaches, many methods to identify bioactive peptides *de novo* have been developed. Biological display techniques such as phage display were described[3] and widely applied very early on. Contemporary standard libraries of linear or cyclic peptides have a diversity of approximately 10^9 independent clones, meaning libraries with up to seven randomised positions can theoretically guarantee comprehensive coverage of the potential sequence repertoire. However, *in vitro* translation systems result in peptide libraries with even higher diversity, since coupling of the peptide with its mRNA is achieved in a cell-free system involving small particles of mRNA/peptide/ribosome or only mRNA–peptide complexes. In addition, no diversity-limiting transformation steps are required. Libraries with diversities from 10^{10} up to 10^{14} different peptides can be prepared using polysome or ribosome display[4] or the PROfusion™ technology.[5]

Chemically prepared peptide libraries can be classified into three different types: (1) Multiple peptide synthesis techniques permit the preparation of hundreds or even thousands of individual peptides. Each sequence is known from its position during the, usually automated, synthesis process and large amounts of comparatively pure compounds are obtained. However, the feasible number of peptides is rather limited compared to the diversity achieved with biological libraries. In principle, all peptide-synthesis strategies can be applied to generate these libraries, including conventional solid or solution-phase procedures as well as array-based syntheses like the SPOT™ method.[6] (2) Combinatorial library techniques generating mixtures of beads with one individual peptide each[7] use a process called portion mixing or the "one-bead-one-peptide" approach.[8] Depending on the bead size and reactor volume, up to 10^7

or even 10^8 peptides with natural as well as non-natural building blocks can be generated.[7] The disadvantage of this technique is the need to identify the structures of the active compounds after screening using sophisticated but rather tedious coding and decoding or sequencing processes. (3) The sequence identification step can be circumvented by using combinatorial peptide libraries with randomised as well as defined positions.[9] In this approach, the entire library is subdivided into a small number of peptide mixtures that have individual amino acids at certain positions: O_1XXXXX, XO_2XXXX, XXO_3XXX, $XXXO_4XX$, $XXXXO_5X$ and $XXXXXO_6$ (O = position with a defined amino acid, X = position with a mixture of amino acids). If the 20 naturally encoded amino acids are each used for the defined positions (O), this library comprises 120 separate mixtures that are screened for binding to the target. Subsequently, individual peptides representing all possible combinations of the most active amino acids at each position (positional scanning approach) are synthesised and screened. All randomised positions must be deconvoluted by an iterative process based on the results obtained with the starting library.[10]

3 General Strategy for Translating Peptides into Small Molecules

The most commonly used strategy to transform peptides into non-peptides is outlined in Figure 1.[11]

Once the primary structure of the biologically active peptide has been determined, the first step is to identify the smallest active fragment required for biological activity. This step involves preparing truncated peptides in which amino acids from the amino and carboxyl termini have been removed, one at a time. Subsequently, the influence of each individual amino acid on the biological activity is determined by systematically replacing each residue in the peptide with specific amino acids, such as alanine or D-amino acids. After the structure–activity relationship (SAR) of each amino acid in the peptide has been explored, the bioactive conformation is investigated by introducing constraints at various positions in the peptide to reduce its conformational flexibility. Whenever possible, all the effects of the introduced constraints should be analysed by biophysical methods. In the final step, the essential amino acid side chains are positioned on carefully selected non-peptide scaffolds to correspond with the derived model of the bioactive conformation. Additionally, the 3-D pharmacophore model can be used for virtual screening of compound libraries (see Section 5).

Although several examples exist in which peptides have been converted successfully into non-peptidic drug candidates, the process is difficult and there is no guarantee of success. The most common methods for the stepwise transformation of peptides into small molecules are presented in more detail below.

4 Tailoring Peptide Sequences for their Translation into Small Molecules

Peptides to be transformed into small molecules should be as small and as rigid as possible. There is no rule to say which molecular weight leads to a successful

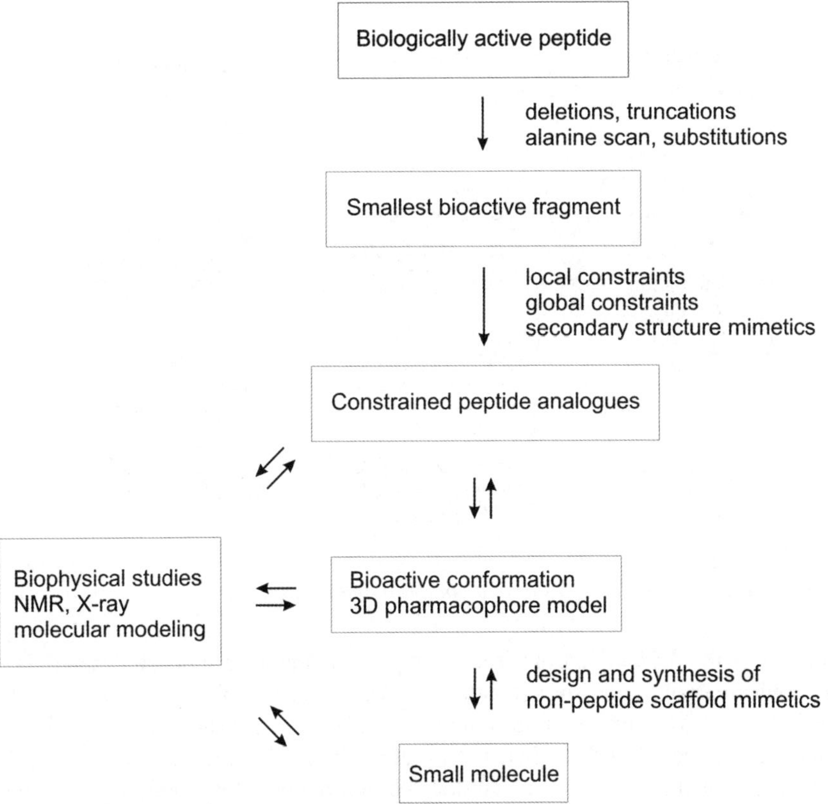

Figure 1 *Strategy for transforming peptides into small molecules*

transformation process, but as a rule of thumb, peptides with fewer than six residues are very good starting points, whereas peptides between 6–15 residues are challenging.

Biologically active peptides often contain a well-defined core of key residues. In addition, these peptides include other dispensable positions, resulting from either the predefined peptide length used in the library design or evolutionary processes. In order to narrow down the peptide to the "active principle" or to minimise the molecular weight to facilitate peptide-based drug design, three different types of libraries are useful: (1) truncation libraries (synonyms: size scan, window scan) comprise peptides omitting one or more *N*-, *C*- or *N*- and *C*-terminal amino acids (Figure 2A). (2) Peptides from libraries of deletion analogues (Figure 2B) have one or more consecutive amino acid(s) deleted at all possible positions. (3) Compared to deletion libraries, combinatorial deletion libraries additionally cover peptides with two or more positions omitted independently throughout the sequence (Figure 2C). It should be noted that the number of peptide analogues covered by a combinatorial deletion library rapidly increases, depending on the number of deleted positions and the peptide length.

Peptides are typically highly flexible, but should be as rigid as possible for the transformation process, for two reasons: (1) the binding-free energy of a peptide

A	B	C
Truncation	**Deletion**	**Combinatorial Deletion**

Figure 2 — Library design diagram

```
        A                          B                          C
    Truncation                 Deletion            Combinatorial Deletion

  OOOOOOOOOO               OOOOOOOOOO                 OOOOOOOOOO
   OOOOOOOOO              -OOOOOOOOO                 --OOOOOOOO
    OOOOOOOO              O-OOOOOOOO                 -O-OOOOOOO
     OOOOOOO              OO-OOOOOOO                 -OO-OOOOOO
    . . . .               OOO-OOOOOO                 -OOO-OOOOO
                          OOOO-OOOOO                 -OOOO-OOOO
  OOOOOOOOO               OOOOO-OOOO                 -OOOOO-OOO
   OOOOOOOO               OOOOOO-OOO                 . . . .
    OOOOOOO               OOOOOOO-OO
     OOOOOO               OOOOOOOO-O                 O--OOOOOOO
                          OOOOOOOOO-                 O-O-OOOOOO
    . . . .                                          O-OO-OOOOO
                          --OOOOOOOO                 O-OOO-OOOO
   OOOOOOOO               O--OOOOOOO                 O-OOOO-OOO
    OOOOOO                OO--OOOOOO                 O-OOOOO-OO
     OOOO                 OOO--OOOOO                 . . . .

    . . . .                . . . .
```

Figure 2 *Library design for analysis and optimisation of peptide length*

interacting with a binding partner can be improved, and even more important (2) a rigid conformation helps in generating a peptide-based pharmacophore model because functional groups that are important for activity are already prepositioned and can be assigned in three dimensions, preferably by NMR techniques.

Conformational flexibility can be reduced by introducing local and/or global constraints at various positions in the peptide. Local conformational constraints can be achieved by incorporating modified amino acids (D-, *N*-methyl, α-methyl, cyclic, α,β-dehydro, β-substituted amino acids), replacing the amide moiety by isosteres ($CH=CH$, CH_2CH_2, CH_2NH, $NHCO$, $CSNH$, $COCH_2$, CH_2O, CH_2S, $SOCH_2$, $CH(OH)CH_2$, *etc.*) and short-range cyclisations, either within a single amino acid (proline or proline mimetics) or between adjacent residues (Figure 3).

Global constraints are achieved by medium- or long-range cyclisations involving *C*- and *N*-termini, backbone-to-backbone, side chain-to-side chain or side chain–to-backbone cyclisations. Cyclisation can impose significant conformational restrictions on the peptide backbone and the location of attached side chains. The most common examples of side chain-to-side chain cyclisation include the formation of disulfide bridges between cysteine residues and the formation of lactam bridges between glutamic/aspartic acid and lysine residues (Figure 4).

The binding conformation is usually unknown since structure determination of peptide–protein complexes by X-ray crystallography or NMR is time consuming and laborious. Furthermore, docking of peptides to binding partners *in silico* is one of the most complex modelling problems due to the tremendous intrinsic flexibility of the ligand. Therefore, a large number of cyclic peptide analogues have to be synthesised and screened to seek out the proper conformation of a biologically active peptide. A systematic approach is the "cyclisation scan" comprising all possible combinations of two cysteine residues within the starting peptide (Figure 5). A cyclisation scan by

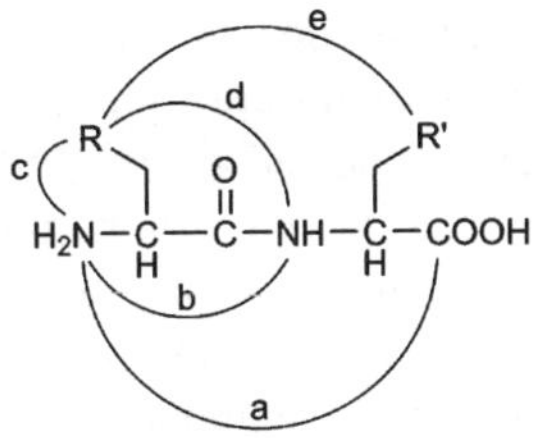

a = end to end (head to tail)
b = *N*-backbone to *N*-backbone / end to *N*-backbone
c = end to side-chain
d = side-chain to *N*-backbone
e = side-chain to side-chain

Figure 3 *Short-range cyclisations between adjacent amino acids*

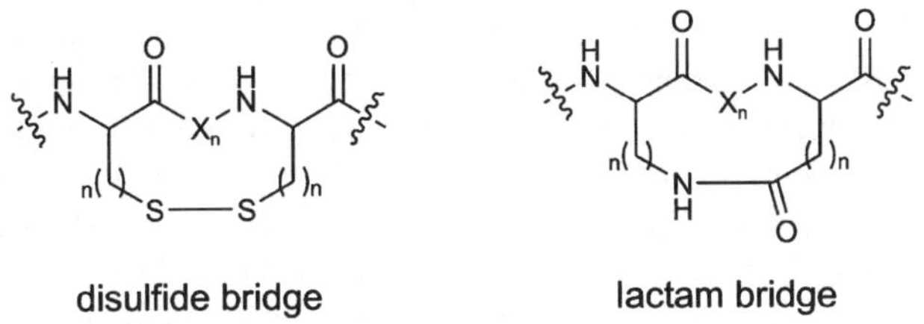

Figure 4 *Examples of long-range cyclisations (X: amino acid)*

Cyclization Scan

OOOOOOOOOO
CCOOOOOOOO
CO**C**OOOOOOO
COO**C**OOOOOO
COOO**C**OOOOO
COOOO**C**OOOO
COOOOO**C**OOO
• • • •

O**CC**OOOOOOO
O**C**O**C**OOOOOO
O**C**OO**C**OOOOO
O**C**OOO**C**OOOO
O**C**OOOO**C**OOO
O**C**OOOOO**C**OO
• • • •

Figure 5 *Cyclisation scan*

disulfide-bond formation via cysteine residues is the most easily implemented strategy. However, several other chemical cyclisation strategies can be similarly applied.

Additionally, secondary structure mimetics can be used to introduce constraints that may either have local or global effects. The secondary structure of a peptide is dominated by energetically favoured torsion angels ϕ, φ and ω, together with additional stabilizing factors such as hydrogen bonds and hydrophobic contacts. The most common secondary structures found in peptides and proteins are α-helices, β-sheets and reverse turns. Reverse turns have often been implicated as recognition elements for peptide–receptor interactions, and therefore introducing turn mimetics into peptides is an important strategy for probing bioactive conformations.[12] Reverse turns are classified according to the number of amino-acid residues involved as γ-turns (three amino acids), β-turns (four amino acids), α-turns (five amino acids) or π-turns (six amino acids).

The most common naturally occurring β-turn is usually defined as any tetrapeptide sequence, occurring in a non-helical region, in which the distance between $C_\alpha(i)$ and $C_\alpha(i + 3)$ is less than 7 Å.[13] A β-turn is often stabilised by a hydrogen bond between the carbonyl function of residue i and the NH-group of residue i + 3 to give a 10-membered ring (Figure 6). The majority of the β-turn mimetics synthesised are dipeptide replacements for residues i + 1 and i + 2. A large number of β-turn mimetics have been reported in the literature;[14] two examples are shown in Figure 6.

A γ-turn, which is a more rare reverse turn, is defined by a three-residue turn forming a seven-membered hydrogen bonded-ring between the carbonyl of the i residue and the amide NH of the i + 2 residue.[17] Most of the γ-turn mimetics described in the literature consist of a six-or seven-membered heterocyclic-or carbocyclic-ring structure (Figure 7).

The introduction of conformational constraints should influence the backbone conformation without compromising any crucial side chain interaction with the receptor.

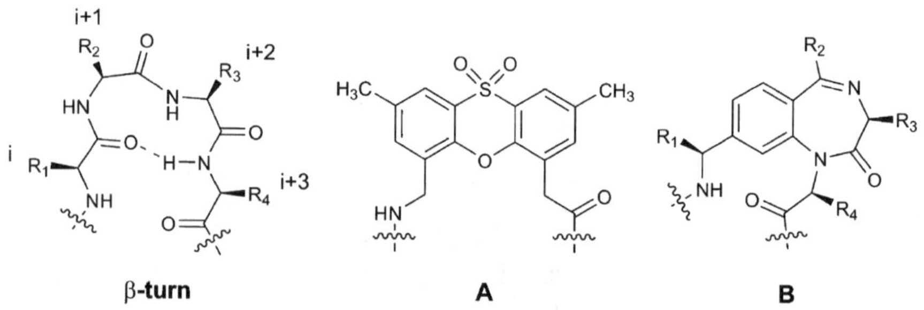

Figure 6 *β-Turn and β-turn mimetics (A,[15] B[16])*

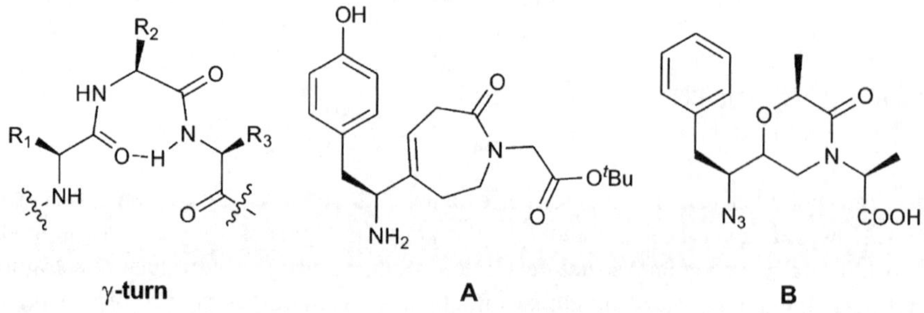

Figure 7 *γ-Turn and γ-turn mimetics (A,[18] B[19])*

Amino Acid Substitution Scan

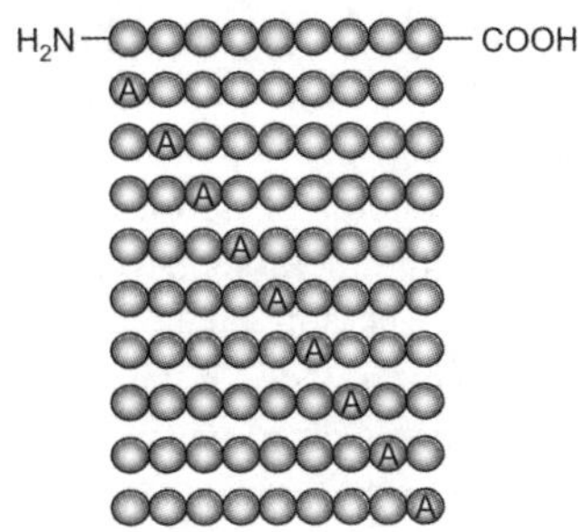

Complete Substitutional Analysis

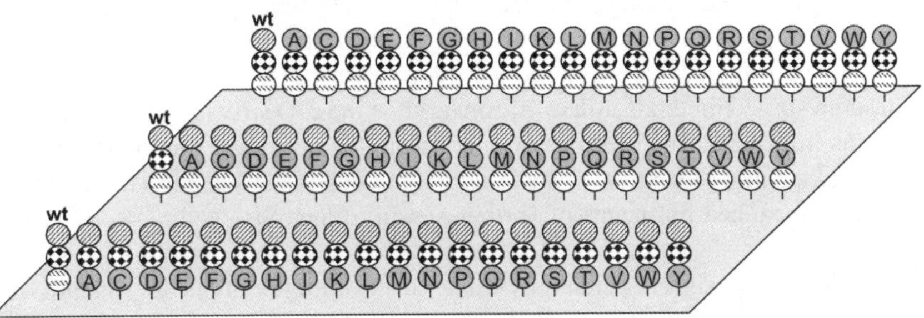

Figure 8 *Substitution scan and substitutional analysis*

A loss in activity after the introduction of a conformational constraint may either be caused by steric hindrance between the ligand and the receptor, due to the added constraining atoms, or by inability of the ligand to adopt the proper conformation. If activity is retained in the constrained analogue, the analogue is able to adopt the proper conformation, and binding is allowed in spite of the added restricting atoms.

Throughout the minimisation and conformational stabilisation process, the amino acid composition may have to be reoptimised to maintain activity and selectivity. This process is typically achieved by extensive synthesis and testing of amino acid substitution analogues. Systematic approaches are (1) amino-acid-substitution scans in which all positions of a peptide are substituted by one amino acid, *e.g.* alanine or proline and (2) complete substitutional analyses in which each position is substituted by a larger set of building blocks, such as all genetically encoded amino acids (Figure 8). Unnatural building blocks significantly increase the chemist's options for peptide optimisation and tailoring.

5 Transformation of Peptide Ligands into Small Molecules using Computational Approaches

The SAR of the constrained analogues together with the information obtained from biophysical studies (X-ray, NMR) and computational methods can be used, in an iterative process, to provide information about the receptor-bound and/or biologically active conformation. This model in turn can serve as the basis for a 3-D pharmacophore model. In the last step, topographic information from the 3-D pharmacophore model is used to position the amino acid side chains in the correct spatial arrangement on selected non-peptidic scaffolds or templates, such as small (five- to seven-membered) ring systems of defined stereochemistry. This process is well supported by the computational methods described in this section.

Rational approaches take into account all available information about the target–ligand system in question to narrow down the chemical space of potential small molecule ligands. In the following section, we focus on some computational techniques that can support the transformation of peptidic ligands into small molecule ligands. Computational chemistry provides a number of techniques for this transformation, which can be divided into structure- and ligand-based techniques. Both categories rely on information about the structure of the target or the bound ligand. The so-called 2-D methods, which only take into account the topology of ligand molecules, have not proven to be very successful for the transformation of peptidic structures into small molecules.

Structure-based methods utilise information on the 3-D structure of the target protein. This information can be used as a template for docking experiments in which libraries of small molecules are fitted into the binding site of the target. The docking solutions are ranked by means of various scoring functions, leading to a hit list of potential binders.

Ligand-based design depends on structural information about the ligands. This ligand structure is either known from experimental techniques such as X-ray crystallography or NMR spectroscopy or can be deduced from other ligands with a known structure, such as proteins, or from SAR data and sequence information about the ligands. The ligand structure can be translated into a pharmacophore model, which is a 3-D representation of the ligand's functionalities or features important for the target–ligand interaction. This pharmacophore can be used for the virtual screening of huge libraries of small molecules. This virtual screening process leads to a relatively small number of hits, which are tested for activity. Pharmacophore-based virtual screening methods are standard tools in computational chemistry. Implementation of such algorithms is achieved using a number of commercially available programs such as Catalyst, Cerius[II] (both Accelrys), Unity (Tripos) or MOE (CCG), to name but a few. The key to making this approach feasible is by determining the structure of a bound peptide. Three starting situations are conceivable: detailed information on the active complex comprising the protein target and its peptidic ligand is available; the structure of the target protein and the sequence of one or more active peptidic ligands are known; or ligand sequences and detailed structure–activity data for these ligands are known.

An alternative approach in ligand-based design is to take the functional groups of the peptide that interact with the target and graft them onto a different, non-peptidic scaffold. The scaffold is chosen such that the groups adopt the appropriate 3-D relationship. The program CAVEAT was developed for this purpose, enabling the rapid search of 3-D molecular databases to identify templates that have bonds that adopt the correct orientation to serve as attachment points.[20] This approach has been used in devising inhibitors of Factor Xa[21] and antagonists of the bradykinin B2 receptor.[22]

A pharmacophore model can be constructed directly from the information available from the structure of a complex between a peptidic ligand and its target protein, which can reveal which interactions are productive and necessary for binding the ligand. Functionalities involved in the interaction pattern between the target and the ligand are called features. The spatial arrangement and the properties of these features are known from the structure of the complex. Typical pharmacophore features include: hydrophobes, ring centres, hydrogen-bond donors and acceptors and positive and negative ionisable functional groups. Mapping these features onto the ligand structure in the complex results in a pharmacophore model that can be readily used for virtual screening.

As an example, we describe the pharmacophore generation and virtual screening for the model-case thrombin. Due to the wealth of structural information about thrombin–inhibitor complexes, and the considerable number of known inhibitors, this protease is frequently used as a test case in computational chemistry. We use the coordinates of the bound inhibitor from the high-resolution crystal structure of a thrombin–D-Phe-Pro-Arg-chloromethyl ketone complex (pdb code 1ppb, resolution 1.92 Å, Figure 9A) as a template to generate a pharmacophore model. The four backbone atoms of the peptidic ligand involved in hydrogen bonds to thrombin are translated into one hydrogen-bond donor or three acceptor features in this model (Figure 9B). The phenylalanine aromatic ring in the peptide is overlaid with an aromatic ring feature and the peptide's proline ring is reflected by a hydrophobic feature. Finally, the guanidinium group of the peptidic arginine is represented as a positively ionisable feature in the pharmacophore model.

To test the performance of this rapidly generated pharmacophore model, we compiled a dataset of 10,160 randomly chosen molecules from the catalogue of a commercial supplier of screening compounds (Maybridge). This set was spiked with 100 molecules described as thrombin inhibitors. In the subsequent virtual screening process, a conformational model of each compound in the screening library was fitted to the pharmacophore model. The resulting score is a measure of the quality of the overlap between the pharmacophore model in question and the respective library molecule. In this case, we used the Catalyst software package (Accelrys Inc.). After the screening procedure, the library was ranked according to the fitting scores. With our simple pharmacophore model, we were able to enrich almost 60% of the spiked thrombin inhibitors in the first 10% of the entire test library (Figure 9C). The performance of such pharmacophore models can be improved by including excluded volumes or shape information from the target molecule. Information about where a potential hit molecule might clash with the target protein and the shape of an interacting molecule can both be determined from the intricate structure of a target–peptide complex.

The thrombin-inhibitor example demonstrates a path from the bioactive 3-D structure of a peptide to small molecules. However, unambiguous structural information on the bioactive ligand conformation needed for such a transformation is only available in a limited number of cases. In most cases, additional complementary information has to be used to deduce the bioactive conformation of a bound ligand, and thus a pharmacophore model for virtual screening or design.

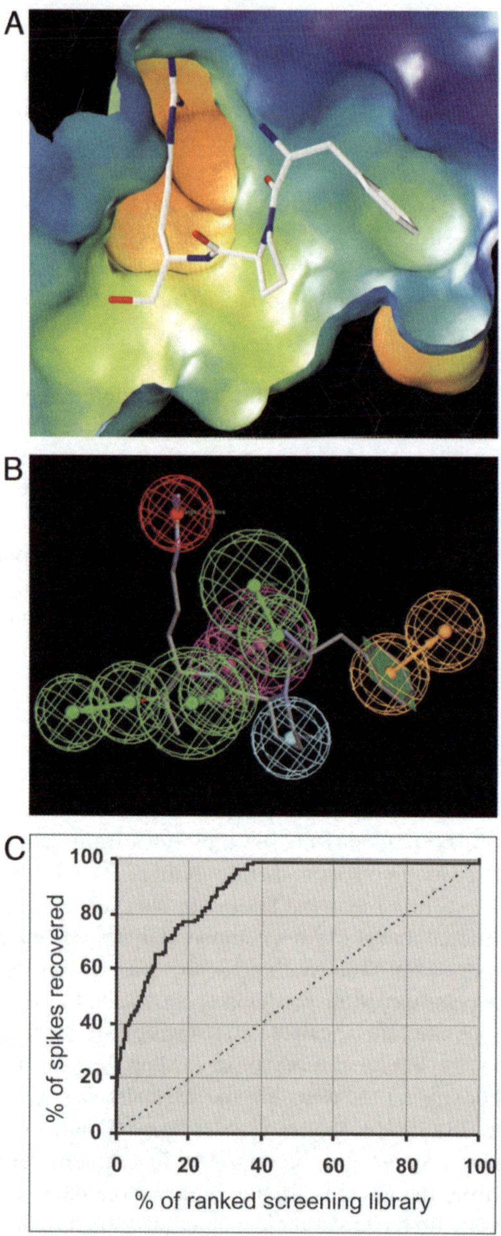

Figure 9 *Virtual screening procedure based on the high-resolution (1.92 Å) complex crystal structure of thrombin and an inhibitory peptide. (A) Surface representation of the binding pocket of thrombin complexed with D-Phe-Pro-Arg-chloromethylketone. (B) Overlay of the bound structure of the thrombin ligand and the resulting pharmacophore (green: hydrogen-bond acceptors; violet: hydrogen-bond donor; brown: ring aromatic; blue: hydrophobic; red: positively ionizable). (C) Enrichment curve for 100 spikes of known thrombin inhibitors in a library of 10,260 molecules after virtual screening using the pharmacophore model shown in (B)*

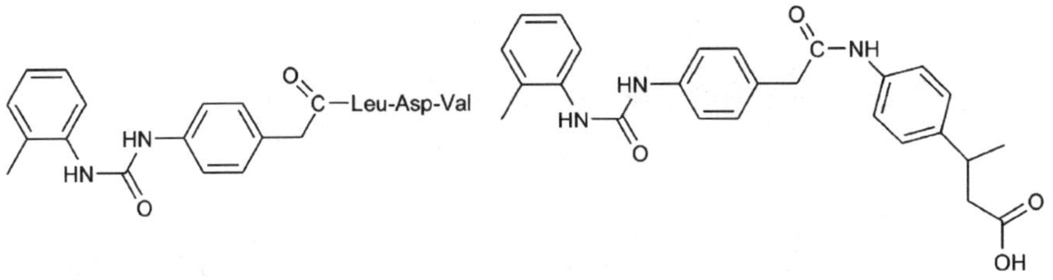

Figure 10 *Starting structure of the tripeptide derivative (left, IC_{50} = 0.6 nM) and resulting virtual screening hit (right, IC_{50} = 1.3 nM)*

If a peptidic ligand presented to a target is a fragment of a protein with a known 3-D structure, or the structure of a different binding protein is known, these structures can help deduce the binding conformation of the peptidic ligand. For instance, this strategy was successful in screening for novel $\alpha_4\beta_1$ (very late antigen-4, VLA-4) antagonists. A derivative of the tripeptide Leu-Asp-Val is known to bind tightly to $\alpha_4\beta_1$. However, the bioactive conformation of this peptide is unknown. To deduce this conformation, the X-ray structure of the vascular cell adhesion molecule-1 (VCAM-1) was used.[23] VCAM-1 is a natural ligand of $\alpha_4\beta_1$. The tripeptide derivative was based on the X-ray crystal structure of the integrin-binding region of VCAM-1. After virtual design and screening of a library of 8,624 molecules, 12 molecules were identified and synthesised as potential $\alpha_4\beta_1$ antagonists. All of these molecules were active, with the most potent compound having an IC_{50} of 1 nM (Figure 10).

Other approaches were successful even in the absence of any structural information about a complex or a ligand protein that could be used as a template. The main challenge in such cases is to cope with the inherent flexibility of peptidic molecules. A rapid exchange between different conformations is observed in the solution. However, depending on the sequence of the particular peptide, certain conformations can be stabilised. Such preferred conformations might be similar to the binding conformation in target–peptide complexes. NMR spectroscopy is an ideal technique for determining peptide conformation in solution. In addition, energetically favoured peptide conformations can also be calculated using different computational methods, and both NMR and computational methods have been used to estimate binding conformations of peptides. The probability of success increases when all available data concerning the SAR of the peptide family under investigation are taken into account. All possible introduced conformational constraints such as cyclisations or the incorporation of restraining amino acids such as imino acids or D-amino acids help to limit the vast conformational space of peptidic structures. Library strategies for the optimisation of peptide rigidity are described above.

The strategy is well illustrated by an example in which non-peptidic urotensin II receptor antagonists were identified based on the 11-amino acid cyclic peptide urotensin-II.[24] The SAR of urotensin-II was evaluated by truncation libraries and Ala and D-amino acid scans. These libraries revealed that the residues WKY are key for peptide function and lead to a 200-fold less active cyclic hexapeptide Ac-CFwKYC-NH$_2$ containing a D-amino acid. D-configuration amino acids stabilise the formation

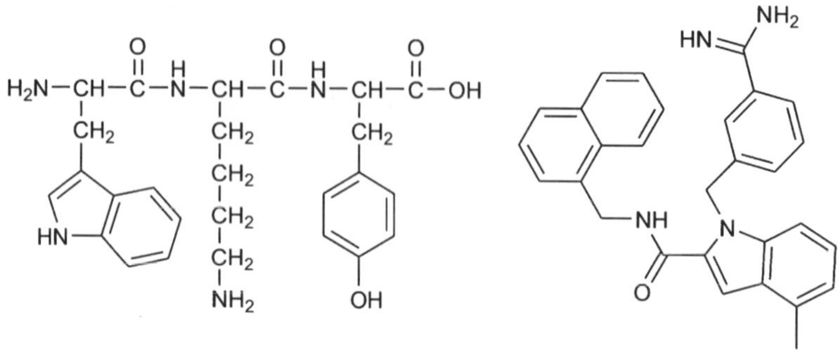

Figure 11 *Starting fragment of the peptidic structure of urotensin II (left, EC_{50} of 2.5 nM for urotensin II) and hit structure from virtual screening (right, EC_{50} = 400 nM)*

of β-turn structures and thus introduce a conformational constraint into the flexible peptide. Consequently, although the solution structure of the full-length urotensin-II as measured by the NMR spectroscopy shows considerable conformational fluctuations, the conformation of Ac-CFwKYC-NH$_2$ is very well defined. Both structures were used as templates for generating two pharmacophore models, each containing a positive ionisable and two hydrophobic aromatic features, reflecting the spatial arrangement of the side chains of the key residues wKC and WKC, respectively. The pharmacophore models were used for a virtual screen of the Aventis compound repository. Hit compounds were screened against the urotensin II receptor using a functional fluorometric imaging plate reader assay. For the urotensin II-derived pharmacophore, a hit rate of 2% was achieved with IC_{50} values ranging from 400 nM to 7 μM (Figure 11). The verified hits belong to six different scaffold classes.

This success contrasts with the screening results for the second pharmacophore model generated from the structurally better-defined cyclic hexapeptide. The hit rate of 0.2% in this screen is barely higher than the hit rates seen in high-throughput screens for G-protein-coupled receptor (GPCR) antagonists. This fact illustrates one problem with the ligand-based approach: the solution structure or a calculated low-energy structure does not necessarily reflect the bound conformation of the active peptide and may be a misleading starting point for pharmacophore generation and virtual screening. Therefore, selection of the "right" conformation is a critical step. This feature is further complicated by the fact that the binding conformation of active ligands is often not a low-energy conformation. An investigation of the conformational energies of complex crystal structures revealed an energetically non-favourable conformation for over 60% of the bound ligands.[25] This factor is one of the reasons why computational assessment of binding conformations using straightforward techniques of simulated annealing or molecular dynamics is critical. When sampling the conformational space, the most important parameter for the simulation is the energy of the conformations, and normally, low energy conformations are selected. Moreover, due to the huge conformational space of peptides, a complete conformational sampling of this space for an oligopeptide is often not feasible.

In an elegant way, McDowell *et al.*[26] circumvent both problems by introducing an ensemble molecular dynamics method. The idea is to use information from an

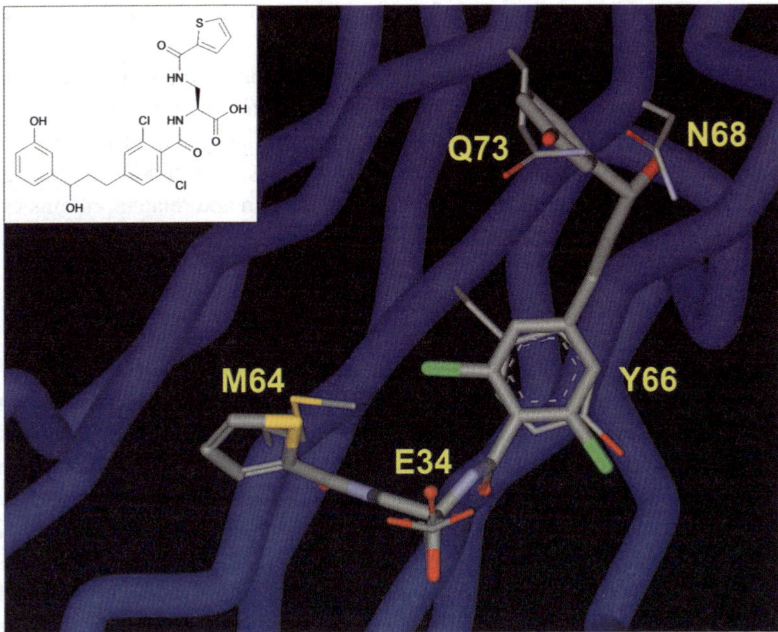

Figure 12 *Superimposition of an LFA-1 antagonist on the discontinuous epitope of ICAM-1 (pdb-code 1iam, protein backbone shown as blue tube), which binds to LFA-1. Residues of the protein indicated in yellow contribute to LFA-1 binding and are mimicked perfectly by the small molecule*

ensemble of diverse flexible peptides interacting with a common site. The primary goal is then to identify only consensus conformations that represent a common spatial arrangement of shared binding features. Technically, this goal is achieved by a dynamics simulation of a collection of molecules, which are tethered together at their corresponding binding features, while ignoring non-bonded interactions between the molecules. This procedure only generates conformations from the entire conformational space where the constrained binding features within all peptides of an ensemble occupy a similar location in space. As a result, the conformational space of each peptide is considerably reduced by limiting it to the part, which overlaps with the corresponding conformational space of all peptides in the ensemble. The resulting structures do not represent low energy structures for a single peptide but for the entire ensemble, since the behaviour of one peptide constrains that of each other. The appropriate choice of tethering features requires a rigorous structure–activity analysis.

This approach was used to calculate the bioactive conformation of the Arg-Gly-Asp recognition sequence that inhibits the glycoprotein II_bIII_a–fibrinogen interaction, which is important for platelet aggregation.[26] Based on this structural model, the compounds synthesised sharing a benzodiazepinedione scaffold showed efficient inhibition of platelet aggregation and oral bioavailability.[27] The ensemble dynamics approach also helped generate a leukocyte functional antigen-1 (LFA-1) antagonist based on a discontinuous epitope of intercellular adhesion molecule-1 (ICAM-1) and cyclic peptides (Figure 12).[28]

Structural information and computational methods have considerable potential in supporting the transformation of bioactive peptides into small molecules. Of major importance is combining all available information on the system of interest, including peptide SAR data, information on the consequences of mutations in the target on SAR, structural information from both the target protein and related proteins as well as the particular ligand and all other known ligands.

6   Conclusion

The design of "drug-like" small molecules from peptides is a good option for a number of different peptide–target interactions. This approach is especially well suited for target classes such as proteases and peptide or protein binding GPCRs, which are difficult to address using small molecules. However, although stepwise chemical transformation processes were successful in several cases, no general transformation strategies have emerged. In recent years, computational methods supported by information about the molecular structure of both targets and ligands have opened up new opportunities to overcome the common pitfalls of the transformation process. Even with these cutting-edge technologies, any transformation still remains a challenge. Several case studies have illustrated that the probability of success increases in parallel to the amount of knowledge about the system in question, such as SAR data, structural data, protein flexibility, binding kinetics and physiological context. The decreasing number of approved drugs produced by the pharmaceutical industry, accompanied by increasing expenses for research and development, demands alternative approaches to increase productivity. Peptide-based drug discovery may be one method to solve as yet unresolved problems.

References

1. A. Loffet, *J. Peptide Sci.*, 2002, **8**, 1.
2. H.M. Geysen, R.H. Meloen and S.J. Barteling, *Proc. Natl. Acad. Sci. USA*, 1984, **81**, 3998.
3. J.K. Scott and G.P. Smith, *Science*, 1990, **249**, 386.
4. L.C. Mattheakis, R.R. Bhatt and W.J. Dower, *Proc. Natl. Acad. Sci. USA*, 1994, **91**, 9022.
5. R.W. Roberts and J.W. Szostak, *Proc. Natl. Acad. Sci. USA*, 1997, **94**, 12297.
6. H. Wenschuh, R. Volkmer-Engert, M. Schmidt, M. Schulz, J. Schneider-Mergener and U. Reineke, *Biopolymers (Peptide Science)*, 2000, **55**, 188.
7. A. Furka, in *Combinatorial peptide and nonpeptide libraries*, G. Jung (ed), VCH Verlagsgesellschaft, Weinheim, 1996, 111.
8. K.S. Lam, S.E. Salmon, E.M. Hersh, V.J. Hruby, W.M. Kazmierski and R.J. Knapp, *Nature*, 1991, **354**, 82.

9. C. Pinilla, J. Appel, C. Dooley, S. Blondelle, J. Eichler, B. Dörner, J. Ostresh and R.A. Houghten, in *Combinatorial peptide and nonpeptide libraries*, G. Jung (ed), VCH Verlagsgesellschaft, Weinheim, 1996, 139.

10. R.A. Houghten, C. Pinilla, S.E. Blondelle, J.R. Appel, C.T. Dooley and J.H. Cuervo, *Nature*, 1991, **354**, 84.

11. V.J. Hruby, *Nat. Rev. Drug Discov.*, 2002, **1**, 847.

12. R.M.J. Liskamp, *Rec. Trav. Chim. Pays-Bas*, 1994, **113**, 1.

13. P.N. Lewis, F.A. Momany and H.A. Scheraga, *Biochim. Biophys. Acta*, 1973, **303**, 211.

14. V.J. Hruby and P.M. Balse, *Curr. Med. Chem.*, 2000, **7**, 945.

15. M. Feigel, *J. Am. Chem. Soc.*,1986, **108**, 181.

16. W.C. Ripka, G.V. De Lucca, A.C. Bach II, R.S. Pottorf and J.M. Blaney, *Tetrahedron*, 1993, **49**, 3593.

17. M. Kahn, *Synlett*, 1993, 821.

18. W.F. Huffman, J.F. Callahan, D.S. Eggelston, K.A. Newlander, D.T. Takata, E.E. Codd, R.F. Walker, P.W. Schiller, C. Lemieux, W.S. Wire and T.F. Burks, in *Peptides: Chemistry and Biology, Proceedings of the 10th American Peptide Symposium,* G.R. Marshall (ed); ESCOM, New York, 1988, 105.

19. K. Brickmann, Z. Yuan, I. Sethson, P. Somfai and J. Kihlberg, *Chem.-Eur. J.*, 1999, **5**, 2241.

20. G. Lauri and P.A. Bartlett, *J. Comp. Aided Mol. Design*, 1994, **8**, 51.

21. Y. Takano, M. Koizumi, R. Takarada, M.T. Kamimura, R. Czerminski and T. Koike, *J. Mol. Graphics Modeling*, 2003, **22**, 105.

22. D.R. Artis, C. Brotherton-Pleiss, J.H.B. Pease, C.J. Lin, S.W. Ferla, S.R. Newman, S. Bhakta, H. Ostrelich and K. Jarnagin, *Bioorg. Med. Chem. Lett.*, 2000, **10**, 2421.

23. J. Singh, H. van Vlijmen, Y. Liao, W.C Lee, M. Cornebise, M. Harris, I.H. Shu, A. Gill, J.H. Cuervo, W.M. Abraham and S.P. Adams, *J. Med. Chem.*, 2002, **45**, 2988.

24. S. Flohr, M. Kurz, E. Kostenis, A. Brkovich, A. Fournier and T. Klabunde, *J. Med. Chem.*, 2002, **45**, 1799.

25. E. Perola and P.S. Charifson, *J. Med. Chem.*, 2004, **47**, 2499.

26. R.S. McDowell, T.R. Gadek, P.L. Barker, D.J. Burdick, K.S. Chan, C.L. Quan, N. Skelton, M. Struble, E.D. Thorsett, M. Tischler, J.Y.K. Tom, T.R. Webb and J.P. Burnier, *J. Am. Chem. Soc.*, 1994, **116**, 5069.

27. R.S. McDowell, B.K. Blackburn, T.R. Gadek, L.R. McGee, T. Rawson, M.E. Reynolds, K.D. Robarge, T.C. Somers, E.D. Thorsett, M. Tischler, R.R. Webb and M.C. Venuti, *J. Am. Chem. Soc.*, 1994, **116**, 5077.

28. T.R. Gadek, D.J. Burdick, R.S. McDowell, M.S. Stanley, J.C. Marsters Jr., K.J. Paris, D.A. Oare, M.E. Reynolds, C. Ladner, K.A. Zioncheck, W.P. Lee, P. Gribling, M.S. Dennis, N.J. Skelton, D.B. Tumas, K.R. Clark, S.M. Keating, M.H. Beresini, J.W. Tilley, L.G. Presta and S.C. Bodary, *Science*, 2002, **295**, 1086.

Section 4

Operational Developments in Screening and High Throughput Assays

CHAPTER 9

High-Density Plates, Microarrays, Microfluidics

CHRISTOF FATTINGER AND GREGOR DERNICK

F. Hoffmann-La Roche Ltd, Grenzacherstrasse 124, Bldg 65-503, CH 4070 Basel, Switzerland

The creation of knowledge in the life sciences through automated high-throughput experiments using extremely small amounts of sample material has many applications in fundamental and applied research. The development of high-density well plates and microarrays for high-throughput experiments in pharmaceutical research is propelled by two guiding principles: "miniaturization" and "parallel processing". Miniaturization of the sample volume saves precious library compounds, biological material, and assay reagents. The arrangement of the sample compartments in a planar array enables process steps (such as making of the sample compartments, liquid handling, incubation, and readout of the analytical result) to be applied simultaneously to multiple samples in parallel. The conceptual and operational advantages of parallel processing of objects on a planar device are also well known in the semiconductor industry. Only through miniaturization and parallel processing has it been possible to find a solution for rapid and cost-effective manufacture of the millions of transistors that make up the electronic circuitries on silicon chips.[1,2]

In today's drug-discovery process, a wide variety of high-throughput experiments are carried out to explore the biological activity of small, drug-like molecules by high-throughput screening (HTS) of diverse compound collections, and to profile in secondary multidimensional assays the subsets of active compounds that have been identified. Particularly in cases when there was only a little known about the structure–activity relation (SAR) between the biological target of interest and the compounds interacting with the target, HTS has proven to be a valuable tool for the discovery of novel, target-specific lead structures.* Most of these high-throughput experiments are based on disposable well plates or microarrays. In recent years, a

---

* In this context a "lead structure" is a compound (or a subset of compounds) for which a well-defined SAR has been established.

wide variety of high-density well plates and microarrays have been developed for specific tasks in the drug-discovery process.

This article gives an overview on the functional design and the microfluidic features of well plates and microarrays, and explains why some high-throughput investigations in the pharmaceutical industry are carried out in well plates while others are carried out on planar arrays or microfluidic chips. This article also provides the general guidelines for assay miniaturization and discusses the requirements and limitations of assays that use high-density well plates, microarrays, and microfluidic chips. This article describes technical advances of high-density well plates and microarrays used in the research labs of F. Hoffmann-La Roche Ltd. It also provides the underlying thought processes and benchmarks for the implementation of planar device technology in pharmaceutical and biotech industries. It is the authors' intention to foster the talent and ability of young scientists, who are interested in (or already working in) pharmaceutical or biotech industries so that they can contribute to innovations and future developments in this fascinating, evolving discipline of the engineering sciences.

1 Functional High-Density Well Plates for High-Throughput Assays

A high-density well plate is a disposable, high-tech device integrating miniature sample compartments in a planar rectilinear array of cavities, the so-called "wells". The footprint of the well plate, the number and layout of wells on the plate, and the relative position of the wells has been standardized by The Society for Biomolecular Screening (SBS).[3] The number of wells in high-density plates is $16 \times 24 = 384$ or $32 \times 48 = 1536$.[†] The 384-well plate has a well-to-well pitch of exactly 4.5 mm. The well-to-well pitch of the 1536-well plate is exactly 2.25 mm. Disposability of the well plate is important because cross contamination between subsequent experiments needs to be avoided. The economic requirements of high-quality, disposable labware necessitate cost-effective high-volume production of the well plates and at the same time impose high demands on the production processes such as injection molding, interfacial bonding, and packaging.

It is important to note that the processing of samples in high-density plates has to cope with inherent surface-tension and capillary effects. For example, it is the surface tension of the liquid sample that affects the retention of the liquid in the high-density well plates, not gravity! Surface tension prevents liquid from flowing out of a 1536-well plate even when the plate is turned upside down. The menisci in low-volume 384- and 1536-well plates show a significant bending. Evaporation of a liquid droplet at the bottom of a well is significantly slower than evaporation of the same droplet volume on a planar surface. The relative evaporation dV/V_0 of a liquid sample in an open well during a given time is inversely proportional to the filling height in the well; it does not depend on the cross section of the well, at least to the first order.

[†] The 96-well plate, having a well-to-well pitch of 9 mm, is not considered as a high-density plate in the context of this chapter.

1.1 Sample Plates for Low-Volume High-Throughput Screening

To ensure the integrity and stability of compounds used in HTS assays and to maximize the number of HTS screens, which can be performed with microgram quantities of compounds, all research sites at Roche use "prealiquoted compound samples" that are delivered to the HTS labs in sealed, ready-to-use "sample plates". The sample plates contain aliquots of frozen millimolar compound solutions in water-free[‡] DMSO, 1 µL per well, corresponding to a small liquid droplet of approximately 1 mm diameter. The "sample plate" is a standardized, low-cost 384-well plate made from polypropylene with a tapered or parabolic well bottom forming a cavity for the small aliquot. This sample distribution concept for the HTS compound library is based on the following rationale:

- To achieve optimum compound stability, aliquoted compound solutions for HTS are stored in a *solid* solvent matrix (*frozen water-free* DMSO). Thereby, repetitive freeze-thaw cycles of the compound solutions in the aliquoting and sample distribution processes can be avoided. The sample plates for a screening campaign are transferred sealed and frozen to the HTS lab.
- The compound aliquots used for *multiple* HTS campaigns are prepared up-front in *one* replication cycle under dry nitrogen. Aliquoting of the screening library is a tedious but automation-friendly procedure, which is carried out most efficiently by high-throughput low-volume 384-well parallel liquid handling in a central plate-replication facility linked to the compound inventory of the research institution.
- For *each* full library screen, the HTS lab orders *one* copy of single-use sample plates from the compound inventory. The plates containing the compound aliquots are thawed and the sealing foils covering the wells are pierced or pealed off before the plates are processed on the HTS system.
- *Dilution* of the prealiquoted compound samples with assay buffer is performed *just in time* before the screen. A 384-well parallel pipettor on the HTS system adds assay buffer to the compound aliquots in the wells, mixes the diluted compound solutions in the wells, and transfers the amount of compound solution needed for the assay from the sample plate to one or multiple assay plates, which are processed in *one* screening run on the HTS system. This procedure minimizes degradation of unstable compounds caused by storage of compound collections in aqueous solution over days or even weeks. In addition, it allows very flexible dilution schemes in the assay plates without using intermediate dilution plates.

The aliquoted compound samples for one full library screen of 1 million compounds fit into the wells of 2841, 384-well plates (352 compounds per plate), or into

---

[‡] DMSO has a freezing point of 18 °C. The eutectic of water and DMSO (30% H_2O) has a freezing point of about −70 °C. To ensure freezing of the millimolar compound solutions in DMSO at −20 °C, the solvent DMSO has to be nearly water-free, *i.e.* it should contain no more than a few percent water. Details on the DMSO/water phase diagram can be found in Ref. 4.

the wells of 711 1536-well plates (1408 compounds per plate).[§] A high-throughput plate-replication process for preparation of microliter and submicroliter aliquots of compound solutions in DMSO is described in Section 2.2.

1.2 High-Density Assay Plates for HTS and Multidimensional Compound Profiling

A broad variety of assays with different detection technologies have been developed for HTS and for multidimensional profiling of the active compounds identified by HTS. Heterogeneous assays are multistep assays that can involve multiple liquid additions, incubations, washings, liquid transfer from a first assay plate to a second assay plate, filtration, and reading of the signal. Homogeneous assays, also called "mix-and-measure assays", are one-well assays without wash-, filtration-, or transfer steps. All constituents of the assay are added step by step or, in an ideal situation, already as mixtures to the assay plate and the signal is finally read in a plate reader. Both cell-free and cell-based assays can be carried out in a heterogeneous and homogenous format. Homogenous assays are preferred for HTS.

Generally speaking, an assay is a biophysical or biochemical method that is used to translate a biological effect into a measurable, physico-chemical signal. Assay methods that are compatible with high-density plates include "optical assays" detecting optical density, fluorescence emission, and luminescence signals, as well as "radioactive assays" with scintillation readout. All these assay methods use light to measure the assay signals in the miniature wells. A brief overview on "optical assays" that have been implemented in high-density plates is given below. A comprehensive review on HTS assay methods used for lead generation in drug discovery can be found in Ref. 5.

A precondition for precise readout of high-density plates is a highly transparent and well-defined optical interface between the cavity containing the sample and the optical reader. The bent liquid meniscus in miniaturized wells does not fulfill this requirement, hence most high-density assay plates are constructed as the so-called "view plates" in the form of a transparent well bottom of high optical quality, see Figure 1. The view plate consists of a black or white polymer body fabricated by injection molding and a thin transparent polymer film or glass sheet, which is bonded to the body of the plate. View plates are read out through the transparent well bottom, which eliminates meniscus effects at the optical interface. The smallest sample volume processed in today's high-density view plates is a few microliters per well. The widely used "low-volume 384-well plates" have an assay volume of 5–40 μL per well, and 1536-well plates have an assay volume of 2–8 μL per well.

Colorimetric assays use the optical density (OD) of the solution in the well or the change of the OD as a function of time as the readout parameter. These assays are typically applied to quantify the activity of enzymes by measuring the color change in the well originating from the differences in absorbance spectra between educt and

---

[§] Roche uses $2 \times 16 = 32$ wells in each 384-well assay plate for standards and controls; these wells are not filled with compounds. In the 1536-well assay plates $4 \times 32 = 128$ wells are reserved for standards and controls.

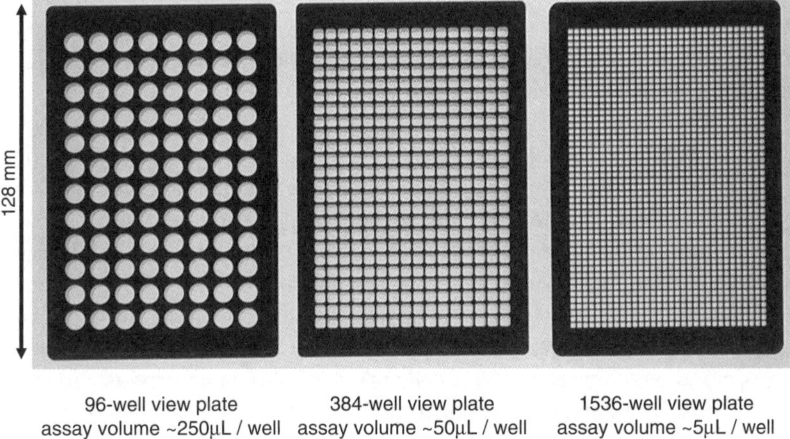

Figure 1 *Top view photograph of three view plates. The well-to-well pitch is 9 mm in the 96-well plate, 4.5 mm in the 384-well plate, and 2.25 mm in the 1536-well plate. The view plates are fabricated by injection molding and interfacial bonding of the transparent well bottom*

product of the enzymatic reaction. Disadvantages of colorimetric assays are: (i) colorimetric readout requires micromolar concentration of optical labels, which limits assay sensitivity and broad applicability of the technique, (ii) interference from colored compounds in the HTS libraries, and (iii) the colorimetric signal depends on the optical path length in the well (filling height of the well), which changes as a result of liquid additions. With the development of plate readers that allow kinetic absorbance measurements with high throughput, it is possible to overcome or to reduce these drawbacks. Today, absorbance-based assays can be successfully carried out even in 1536-well plates using advanced assay readers for high-density plates. Such a reader is the plate::vision™ multimode reader developed by Carl Zeiss AG (Jena, Germany) that is now sold by Evotec AG (Hamburg, Germany). The instrument comprises a modular, 96-channel parallel readout based on 96 mini-objectives that are arranged in a minilens-array. Different minilens-arrays can be selected according to the special needs of the readout mode. The absorbance minilens array provides an optimized detection scheme for OD measurements and makes it possible to carry out ultra-high-throughput, colorimetric readout in 1536-well plates. The fluorescence minilens array has a 96-channel readout scheme based on very high numerical aperture (NA = 0.5) epi-fluorescence optics, which allows an ultra-sensitive, quasi-confocal detection in well plates and eliminates both background emission from the plastic labware and fluorescence crosstalk between neighboring wells.

Fluorescence assays have become the most popular and powerful HTS methods because of their high sensitivity and broad applicability. Fluorescence is the process of cyclic absorption and emission of photons by certain dye molecules, called fluorophores. Fluorophores are generally rigid, planar molecules with extensively conjugated π-electron systems and molecular weights between 200 and 1000 Da.

A fluorophore is excited by absorption of a photon leading to an excited state; subsequent decay of the excited state emits a photon of longer wavelength. Fluorophores have been widely used to label small molecules, peptides, proteins, and nucleic acids, and to study molecular interactions; a review on fluorescence labeling is given in Refs. 6, 7. Today's fluorescence assays in high-density plates are based on various schemes of fluorescence detection, including fluorescence intensity (FI); fluorescence polarization (FP) or fluorescence anisotropy (FA); fluorescence resonance energy transfer (FRET); lifetime-based measurements (time-resolved fluorescence (TRF) and fluorescence lifetime measurements (FLT)); fluorescence correlation spectroscopy (FCS); and fluorescence intensity distribution analysis (FIDA).[5]

Fluorescence assays are 100 to 1000 times more sensitive than colorimetric assays and are ideally suited for assay miniaturization. The high sensitivity of fluorescence assays is based on the regenerative, cyclic nature of photon absorption and emission and on the high quantum efficiency of contemporary photon detectors. A single fluorophore can be excited many times and thereby generate thousands of detectable emission photons before "photobleaching" occurs. (Photobleaching is the irreversible reaction with photochemically generated, reactive oxygen species, which destroys the fluorophore and prevents further excitation–emission cycles.) Fluorescence detection offers a fundamental advantage for assay miniaturization, since sensitivity of fluorescence detection is determined neither by the optical path length in the well nor by the total number of dye molecules in the optical path. The sensitivity of fluorescence detection is determined by the concentration of the dye molecules in the assay and by the intensity of the fluorescence background. The fluorescence background originates from several sources: fluorescent molecules in the compound library, and weakly fluorescent but highly abundant molecules in the assay reagents and in the plastic material – trace impurities and the polymer itself – of the well plate. The sensitivity of fluorescence detection is also limited by weak scattering of excitation light in the reader or in the sample. Such scattering bleeds through the detection filters and increases the background signal. TRF is the most powerful detection technique for suppression of both fluorescence background and excitation light scattering.

View plates are also well suited for automated image analysis of fluorescently labeled cells in secondary and primary screening assays. This screening technique is denoted as image-based screening or "high-content screening" (HCS). Image-based screening utilizes an inverted, epi-fluorescence microscope for automated acquisition of fluorescence images from cells in cell culture plates. Sets of images serve as raw data from which various assay readouts are derived. Typical image parameters analyzed in HCS are the intracellular distribution or redistribution of fluorescently labeled proteins within the cell or on the cell membrane.[8]

1.3 Technical, Biological, and Economical Limits for Assay Miniaturization in High-Density Plates

The overall performance of miniaturized assays is predominantly influenced by the accuracy of the involved liquid-handling processes such as aliquoting of compound solutions and consecutive liquid additions to the analyzed sample. It is the precision of the low-volume pipetting and dispensing steps in the assay that limits assay miniaturization in high-density plates, not the optical readout of the plates. The precision of microliter and nanoliter liquid handling in high-density plates using state-of-the-art low-volume pipetting and dispensing technologies is discussed in Section 2. Fluorescence readout is applicable to sample compartments, which are orders of magnitude smaller than the wells of a high-density plate, *cf.* also Section 3 of this chapter.

Liquid handling in high-density plates tends to generate randomly appearing air bubbles in some wells of the plate. This effect is more pronounced in small wells and when the solutions contain proteins or detergents. The disturbing bubbles can be easily removed by a short (1 min) centrifugation of the plate. Therefore, high-density assay plates are often centrifuged before optical readout of the analytical result. A centrifuge should be integrated into an HTS system that processes high-density plates. Centrifugation also generates exceedingly uniform menisci over the entire plate and thereby yields assay data with higher precision, compared to readout without centrifugation. This difference can be attributed to the inevitable partial reflection of the light used for optical readout at the menisci.

The adsorption of the target protein to the surfaces of the disposables used in the assay needs to be analyzed and optimized in most assay development campaigns. In some assays, well plates with the so-called "nonbinding surface coatings" are used to reduce nonspecific binding of the target protein to the material of the well plate. The smaller the sample volume, the larger is the surface-to-volume ratio of the sample in the well. A very large "surface-to-volume ratio" might cause significant depletion of the investigated protein in the sample due to adsorption of the protein on the walls of the well.

An important consideration regarding the economics of assay miniaturization is the packing density of the sample compartments on the labware used for the assay. The higher the density of sample compartments (wells) on the planar device, the more samples can be analyzed per unit surface area and the lower is the cost per assay result associated with the production of the disposable labware. A 384-well plate holds 4.94 wells per cm^2; a 1536-well plate holds 19.75 wells per cm^2. The packing density in a rectilinear array of wells is fairly close to the best packing achievable in a two-dimensional array of cavities, which is the honeycomb structure. The high packing density of the sample compartments on high-density plates facilitates a cost-effective production of the disposable labware.

Cost reduction has been (and still is) a driving force for assay miniaturization in HTS. Well plates with higher densities than 1536 were proposed in the late 1990s but have not found their way to routine use in HTS. There are a few labs and biotech companies that use specially designed high-density plates like (9 $\times$ 384=) 3456- and 9600-well plates. Assay miniaturization beyond the "low-volume 384" and the "1536-well format" does not yield significant cost savings, because the expense of the disposable plates is already a significant cost component at the present degree of assay miniaturization, compared to the cost share for compounds, target proteins, and assay reagents used in most HTS campaigns. Including the additional cost for the acquisition of the needed high-precision liquid-handling instrumentation, a further miniaturization of microtiter plate-based assays beyond the 1536-well plate is not rational.

Miniaturization of the assay volume below the few microliter scale, which has been reached in high-density plates, can be achieved using disposable microfluidic devices and microarrays. The polymer-based microfluidic compact disk platform described in Ref. 9 integrates several microfluidic functions such as flow sequencing, cascade micromixing, and capillary metering into a disposable planar device. Precise liquid handling and metering of submicroliter volumes can be achieved by balancing the capillary force and the centrifugal force in the spinning microfluidic disk. This technology eliminates the need for costly instrumentation for high-precision sample metering, but imposes high demands on the cost effectiveness of the processes used for production of the disposable disk. This balance of these two considerations is important, especially when the number of samples that can be processed on the disk per unit surface area is smaller than that on high-density well plates.

However, it has to be kept in mind that the main motivations for the application of assays based on microfluidics and microarrays are (i) their ability to enable experiments with biological samples that are only available in very small quantities (few microliters), and (ii) the reduction of the number of liquid-handling steps compared to the same assay in a well plate. These aspects are further discussed in Section 3.

1.4 384-Microtube Plate for High-Throughput Retrieval of Compound Subsets

A further example of a functional high-density plate that serves a primary need in drug-discovery research is the 384-microtube plate. 384 microtubes are assembled in a planar array on an injection-molded frame having the shape of a microtiter plate with 384 rectangular, through holes. Each tube is held in a fixed and well-defined position in its hole on the plate. The functional design of the plate is depicted in Figure 2. It makes it possible to handle miniaturized compound sample aliquots (i) with high-throughput random sample access, (ii) at low temperature (frozen solutions), (iii) without intervening freeze-thaw cycles, and (iv) without exposure of samples to air or moisture. The compound-handling technology based on the 384-microtube plate has been developed for high-throughput retrieval of compound subsets from large chemical libraries.

Using a 384-well parallel pipettor, the 384 microtubes in the plate are filled with aliquots of compound solutions in water-free DMSO, a few microliters per microtube. The aliquoting process runs under dry nitrogen. After pipetting, all microtubes in the plate are sealed with sealing foil. The connecting foil between the tubes is punched off, leaving the 384 individually accessible tubes in the plate. The sealed 384-microtube plates are stored at -20 °C in a humidity-controlled cold room. When a set of particular compounds is ordered, a robot working in the cold room pushes the sealed microtube containing the ordered compound sample from a source plate (denoted as "Storage Plate") into a target plate (denoted as "Delivery Plate"). In this way, all the ordered compounds are assembled onto the Delivery Plate, which is handed over to the lab.

The storage and retrieval of aliquots in sealed microtubes at -20 °C ensures long-term stability of the compounds that are dissolved in the frozen DMSO matrix. This sample-handling concept also minimizes the risk of compound crystallization or

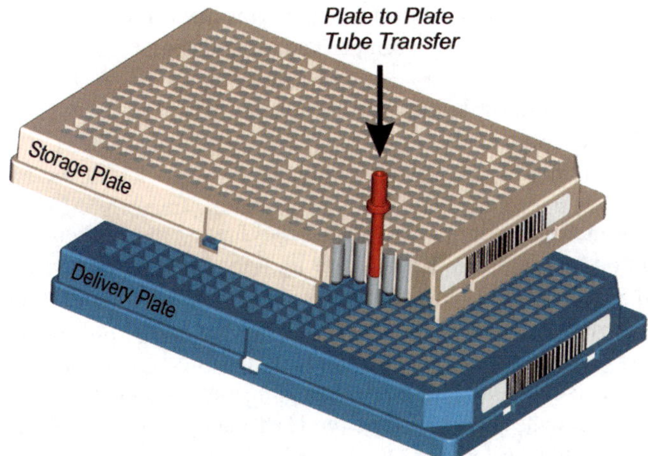

Figure 2 *The 384-microtube plate developed for high-throughput retrieval of compound subsets from large compound collections. At Roche, the 384-microtube plate is used for high-throughput cherry picking of compound samples from the HTS library for hit validation and secondary screening*

compound precipitation, which might result from repetitive freeze-thaw cycles. In recent years, the exploitation of single-use low-volume aliquots in sealed microtubes has evolved to an industry standard for high-throughput cherry picking of compounds from large screening libraries. The 384-microtube-plate technology was developed by collaboration between the three companies Remp AG (Oberdiessbach, Switzerland), Weidmann AG (Rapperswil, Switzerland), and F. Hoffmann-La Roche; it is now commercially available from Tecan AG (Männedorf, Switzerland).

1.5 Sample Management for HTS and Multidimensional Compound Profiling

Modern high-throughput synthesis methods deliver compounds in low-milligram quantities. Depending on the degree of assay miniaturization, this amount of material is sufficient for 1000 to 10,000 HTS projects. The size of the compound collections of pharmaceutical companies ranges from several hundred thousands to more than one million compounds. The vast number of samples processed in HTS are prepared and distributed in a staggered process, involving several consecutive sample transfer and sample-aliquoting steps. The chemical diversity and the quality of the compound samples in the HTS library of a pharmaceutical company are continuously extended and optimized. Newly synthesized compounds enter the HTS library. Compounds that show unfavorable properties or even insufficient purity in compound profiling and secondary assays are removed from the HTS library. Some compounds entering the HTS library are only available as DMSO solutions and not as powder samples.

The storage of the compound collection, the assembly of the HTS library, and the retrieval of the compound subsets required for hit validation, secondary screening, and multidimensional compound profiling at F. Hoffmann-La Roche Ltd is administered by three kinds of automated sample-handling systems:

- Solid Sample Store (installed at each research site),
- Liquid Master Store, and
- the so-called Smart Compound Depository.

The sample-management concept behind the three systems, their functions, and inter-connections are outlined in the following; see also Tables 1 and 2.

The Solid Sample Store is an automated sample bank facilitating storage and random access retrieval of glass vials containing powder samples. The preparation of the HTS library starts with the tedious process to prepare new solutions from the solid-compound collection. The Solid Sample Store assembles and provides the powder samples for the preparation of the HTS master solutions of defined molar concentration in pure, water-free DMSO. It is also used for the supply of powder samples to medicinal chemistry and biology labs.

The Liquid Master Store is an automated sample bank for managing the dynamic collection of HTS master solutions. The HTS master solutions are the seed samples of the aliquoting tree that generates the vast amount of samples used in HTS. The equal volume samples of master solutions in the Liquid Master Store constitute the "kernel" and the "smallest common denominator" of the aliquoting process for preparation of the small-volume compound aliquots needed for HTS. The Liquid Master Store periodically assembles a new edition of master solutions for preparation of a new set of compound aliquots for multiple HTS projects. The Liquid Master Store is vital for a continuous optimization of the HTS compound collection over many years.

The Liquid Master Store at Roche uses a 96-tube version of the microtube plate described in Section 1.4; a description of the 96-tube plate is found on the web site www.remp.com. The individually accessible tubes in 96-tube plates in the Liquid Master Store hold a few definite sample volumes, *e.g.* 26 µL and 140 µL per tube. Storage and retrieval of HTS master solutions with fixed sample volumes facilitate the sample logistics considerably. Upon retrieval from the Liquid Master Store, the master solutions in the 96-tube plates are thawed and transferred to 384-well plates before they enter the downstream aliquoting process. This sample-handling concept avoids repetitive freeze-thaw cycles of HTS master solutions and thereby improves the stability and integrity of the compound samples used in HTS.

The "aliquoting tree" for the production of the vast number of aliquots needed for multiple HTS projects is schematically depicted in Table 2. One edition of master solutions with equal sample volume is used as a source to produce multiple copies of prealiquoted compound samples for HTS. The copies of prealiquoted compound samples are prepared in the sample plate described in Section 1.1. The sample plates are stored at $-20\,^{\circ}\mathrm{C}$ in refrigerators; one copy of the library is allotted for one HTS screen. The optimum number of copies prepared in one aliquoting campaign depends on the number of HTS projects carried out between subsequent aliquoting campaigns. The prealiquoted compound samples for Roche's HTS projects are produced in one to two high-throughput aliquoting campaigns in every year.

The master solutions in the liquid master store are also used to produce multiple copies of single-use aliquots in 384-microtube plates, which are used for two processes: (i) loading of the Smart Compound Depository with new compound aliquots, and (ii) replenishment of frequently ordered compounds in this depository. The continuous replenishment of frequently ordered compounds through the Liquid Master Store

Table 1 *Management of large compound collections by three interconnected sample-handling systems. The three systems are used for automated storage and retrieval of powder samples, liquid samples, and liquid aliquots, respectively*

Solid Sample Store	Liquid Master Store	Smart Compound Depository
Samples stored in system		
Powder samples	Liquid samples stored as frozen compound solution	Liquid aliquots stored as frozen compound solution
Sample amount varies from sample to sample and over time	A few fixed sample volumes (e.g. 26μL and 140μL)	Fixed aliquot volume (e.g. 5μL aliquots)
Sample vessels entering and leaving the system		
4 mL glass vials with screw caps	Sealed tubes in 96-tube plates	Sealed microtubes in 384-microtube plates
Automated sample handling functions in system		
Storage and retrieval of glass vials	Storage and retrieval of 96-tube plates. In system plate-to-plate tube transfer	Storage and retrieval of 384-microtube plates. In system plate-to-plate microtube transfer
Typical storage capacity of system (depending on the size of the compound collection)		
1 million compounds	15 million aliquots	30 million aliquots
Random access cherry picking capacity (depending on number of robots in system)		
2 thousand vials per day	3 - 12 thousand tubes from different plates per day	3 - 12 thousand microtubes from different plates per day
Storage conditions in system (relative humidity and storage temperature)		
30% rel. humidity at +10°C	30% rel. humidity at −20°C	30% rel. humidity at −20°C
Main system functions		
Assembly of powder samples for preparation of HTS master solutions	Optimization of HTS library and assembly of focused sub-libraries	Cherry picking of compound samples for hit validation and secondary screening
Delivery of powder samples to medicinal chemistry and biology labs	Assembly of master solutions for preparation of compound aliquots for HTS projects	High-throughput retrieval of compound subsets from large compound collections
Long-term storage of powder sample collection	Replenishment of frequently ordered compounds in the Smart Compound Depository	

Table 2 *Sample aliquoting tree for the supply of the vast amount of samples needed for high-throughput screening. The Liquid Master Store assembles editions of master solutions for preparation of the compound aliquots needed for multiple HTS projects. The Smart Compound Depository delivers the compound subsets from the HTS library needed for hit validation and secondary screening*

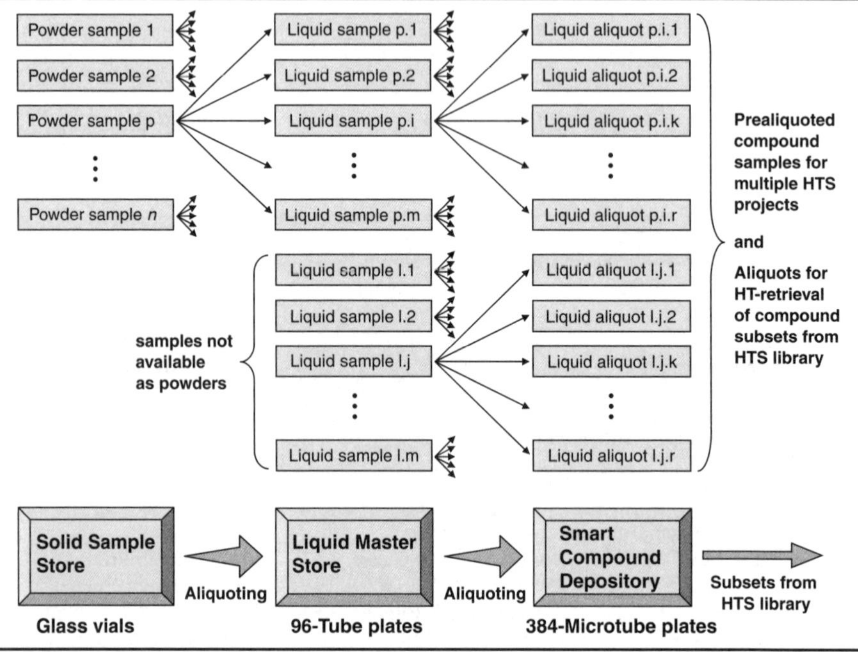

ensures very high availability of compounds. Typically, 99–99.9% of all compounds in the HTS library are available in random access.* The Smart Compound Depository fulfills two main tasks: (i) high-throughput cherry-picking of HTS hits from the HTS library for hit validation and secondary screening, and (ii) high-throughput retrieval of compound subsets from the compound collection for focused studies. The system handles all requests for random-access retrieval of individual HTS samples within Roche; it is based on the 384-microtube plate described in Section 1.4. The system is loaded with about 5 to 20 aliquots of each HTS compound in 384-microtube plates, typically 5 µL per microtube. Depending on the number of robots installed in the system, typically 3–12 thousand aliquots in microtubes can be retrieved from different plates in 24 h.

The plate-to-plate tube transfer at -20 °C in the Liquid Master Store and in the Smart Compound Depository are similar robotic operations that can be combined in one automated sample bank. The combination of 96-tube and 384-microtube plate-to-plate tube transfer in one system might be the preferred solution for sample management of small- and medium-sized HTS libraries.

---

* As soon as the tube count per compound drops below a defined threshold (e.g. 2 tubes) for a minimum of 384 compounds in the Smart Compound Depository, 384 of these compounds are ordered from the Liquid Master Store in order to prepare new aliquots for replenishment of samples.

2 Parallel Liquid Handling of Low-Volume Samples

A precondition for miniaturization of the assay volume to a few microliter is the availability of reliable and versatile liquid-handling tools for precise metering and delivery of microliter and nanoliter aliquots of compound solutions, assay reagents, and biological materials. Low-volume liquid handlers in the life sciences utilize a variety of basic microfluidic methods for precise dosage of the aliquot volume being transferred, including:

(1) Delivery of an aliquot through a pipette by positive displacement using a syringe or a micro pump that delivers precisely defined volume increments.
(2) Time-controlled actuation of a valve or a microfluidic pump in a liquid path connected to a pressurized reservoir.
(3) Ejection of an aliquot from an open capillary using a pressure pulse generated by a piezo actuator.
(4) Filling of a microcavity by capillary forces with subsequent draining by an air pressured pulse, centrifugation or a gentle touch of the microcavity onto the target surface.
(5) Ejection of a droplet from a meniscus by a focused ultrasound pulse.
(6) Evaporation of the liquid from the defined surface area of a capillary pump.

With the exception of the last-mentioned method, all these methods are used for liquid handling in high-density plates. A comprehensive overview on liquid-handling systems used in drug-discovery research is beyond the scope of this article. A pretty good overview on this subject can be gained by visiting the annual conference and exhibition of The Society for Biomolecular Sciences (SBS), *cf.* www.sbsonline.org.

A microfluidic well plate that makes use of method 4 is described in Section 2.3 and method 6 is explained in Section 3.1 of this article. Method 5, *i.e.* the acoustic ejection of a droplet from a meniscus, is described in Ref. 10. The interested reader finds a comprehensive tutorial on the physico-chemistry of fluids in microstructured systems and the basic principles of microfluidic operations and technologies in Ref. 11.

2.1 Pipetting and Dispensing in High-Density Plates

In "ultra-high-throughput screening" (uHTS) with more than 100,000 assays per day and an assay volume that is in the ideal case below 10 µL, liquid handling tends to be the rate limiting step and the pacemaker of the screening process. Parallel processing of (at least) 384 wells in 384- or 1536-well plates is vital to achieve ultra-high throughput. Most assays in high-density plates employ microliter or nanoliter liquid-handling operations to transfer compound aliquots from sample plates to assay plates. Good mixing of the assay components is crucial to the quality and reproducibility of the assays. Some assays require instantaneous mixing of the added components in the assay plate and simultaneous readout of a transient assay signal.[12] Cross contamination between consecutively processed plates needs to be avoided. Changing pipet tips, however, is limited for economical reasons. Therefore, careful and extensive parallel washing of the 384 tips is typically carried out after each compound addition.

Today's HTS and uHTS systems support two basic types of liquid-handling functions: pipetting and dispensing. This section describes pipetting and dispensing techniques used in HTS and uHTS. It also discusses their precision and provides quality criteria and benchmarks for state-of-the-art liquid handling in high-density plates.

In HTS, "pipetting" denotes a parallel transfer of liquid samples from a source plate to a target plate or from a reservoir trough to a target plate. Pipetting involves a bi-directional flow of liquid in the pipetting tips; it combines two alternating liquid-handling actions: (i) aspiration and (ii) dispensing. The first action is the aspiration of a certain volume from a well plate or a reservoir trough into the pipetting tips. The second action is dispensing of the aspirated volume into a target plate, or multidispensing of aliquots from the aspirated volume into multiple target plates. Parallel pipettors are equipped with a "tip wash station" for parallel washing of the tips between subsequent pipetting steps, or with a "tip changer" for changing sets of 384 disposable tips. Pipetting is well suited for high-throughput *parallel handling of different samples*. 384-well parallel pipettors are widely used in HTS and uHTS. Most 384-well parallel pipettors make use of the volumetric dosing principle based on synchronous displacement of 384 pistons in 384 syringes. A small air gap between piston and sample (or between system liquid and sample) prevents contamination of the piston (or dilution of the sample with system liquid). A minimized volume of the air gap is a key feature for high-precision pipetting. The compressibility of the trapped air between piston and liquid demands a rather gentle movement of the piston to ensure high pipetting accuracy. Fast movement of the piston would diminish the coherence between piston and fluid movement.

Pipettors make use of three methods to overcome the surface tension, which tends to retain the expelled aliquot volume at the tip orifice:

(1) A dip of the pipetting tip into the liquid in the well. The "liquid dip" ensures a complete release of the expelled volume from the tip and transfers it to the sample in the well.

(2) A gentle touch (or a submillimeter proximity approach) of the pipetting tip onto the bottom of an empty well. The "tip touch" releases the expelled droplet from the tip and transfers it to the bottom of the well.

(3) Contact-free dispensing using fast ejection of the aliquot through a narrow orifice at the very end of the tip.

Contact-free dispensing of submicroliter volumes does not allow for an air gap in the tip. The flexibility of the air gap diminishes the deceleration of the liquid column in front of the air gap and makes contact-free dispensing of nanoliter aliquots impossible.

In HTS, the "dispensing" function stands for a repetitive unidirectional transport of equal volume aliquots from one (or a few) reservoirs to a series of wells in a plate. Dispensers enable the "contact-free transfer" of assay reagents and biological materials without immersion of the dispensing nozzle into the liquid in the target well. Contact-free dispensing is based on a fast ejection of a liquid jet through a nozzle and the sudden stop of the liquid column movement in the nozzle. The rupture of the trailing edge of the ejected liquid jet at the nozzle orifice determines the dispensed volume. If the deceleration of the liquid column in the nozzle is not fast enough, a small droplet remains hanging at the rim of the orifice. This effect dramatically increases the volume variability between the dispensing shots. Dispensers for HTS and uHTS use typically 8 or 16 parallel nozzles for dispensing into 384- and 1536-well plates. Because the sample in the target well does not contaminate the nozzle, contact-free dispensing does not require washing or changing of the dispensing nozzle between subsequent dispensing steps into multiple wells.

The accuracy of liquid-handling operations in HTS is analyzed by three key parameters derived from test experiments in which equal volume aliquots are pipetted or dispensed over the plate. The *trueness over the plate* is a measure of the deviation of the average of the actual volume dispensed over a plate from the intended volume. It is measured gravimetrically and is defined as

$$\text{Trueness (\%Bias)} = \frac{V_a - V_0}{V_0} \times 100 \tag{1}$$

where V_a is the average of the actual volume dispensed and V_0 the intended volume.

The *precision over the plate* is a measure of the well-to-well variability of the dispensed volume over the plate. It is measured by a precise optical readout of the OD or the FI of a diluted solution of a dye (*cf.* Section 1.2). It is reported as a percentage coefficient of variation (%CV) defined as:

$$\text{Precision (\%CV)} = \frac{\sigma}{\mu} \times 100 \tag{2}$$

where σ is the standard deviation and μ the mean value of the signal measured in OD units or relative fluorescence units (RFU).

The *deviation from mean* (DFM) *in well n* is the relative deviation of the volume in one well from the mean volume over the plate. This value is obtained from precision OD or fluorescence measurements and is used for the detection of outliers and systematic errors. The DFM is defined as

$$\text{DFM in well } n \text{ (\%)} = \frac{V_n - V_{mean}}{V_{mean}} \times 100 \tag{3}$$

where V_n is the dispensed volume in well n and V_{mean} is the mean (average) volume dispensed over the plate. Outliers and systematic errors can be readily analyzed using a false color (or gray scale) map of the plate, which shows the DFM values of each well as color-coded (or gray-scale coded) patches; such maps are depicted in Figure 4. The histograms of the DFM values in Figure 4 reveal their distribution around the mean (average) aliquot volume dispensed over the plate. The minimum and maximum DFM values indicate the range between the smallest and the largest volume dispensed over the plate.

State-of-the-art quality standards for automated low-volume liquid handlers in HTS and uHTS are

$$\text{Trueness (\%Bias)} < \pm 10$$

$$\text{Precision (\%CV)} < 3 \tag{4}$$

$$\text{min/max DFM (\%)} < \pm 10$$

The stated values are current minimum requirement specifications for low-volume liquid-handling operations in high-density plates in Roche's screening labs. The trueness is a less critical specification for liquid handling in HTS and uHTS because the assay signal is calibrated for every assay plate using the data obtained from the standard and control wells on the plate (See footnote §). In test measurements of liquid-handling instruments carried out in Roche's screening labs, the following values have been measured for contact-free dispensing:[¶]

	Aliquots			
	5 µL	500 nL	200 nL	
Precision (%CV)	0.8	1.5	2.5	(5)
min/max DFM (%)	±3	±6	±10	

The precision values obtained for the 5 µL aliquots indicate an extremely small well-to-well variability for the optical readout and an excellent optical quality of the transparent well bottom of high-density view plates. The somewhat higher well-to-well variability obtained for readout of assay volumes in the submicroliter volume range is not determined by the readout process or by the well-to-well variability of the plate, it is mainly determined by the precision of liquid handling.

To assess the quality of HTS assays, Zhang *et al.*[13] introduced the z' factor

$$z' = 1 - \frac{3\sigma_S + 3\sigma_B}{\mu_S - \mu_B} \tag{6}$$

where σ and μ are the standard deviation and the mean of the assay response, respectively. The index "S" denotes the signal or positive control, *i.e.* the maximum assay response, and "B" is the blank or negative control, *i.e.* the minimum assay response. The z' factor compares the variations ($3\sigma_S + 3\sigma_B$) of the assay data with the signal window ($\mu_S - \mu_B$). The maximum value of the z' factor is 1. In this case, the standard deviations are negligible compared to the signal window. An assay with a $z' > 0.8$ is considered as very good and a value of $z' > 0.6$ is good. For $z' < 0.5$, the assay is not robust enough to be used for primary screening, *i.e.* the test of compounds at one

---

[¶] The stated accuracy values have been achieved with contact-free dispensing using eight dosing channels of the CyBi™-NanoJet dispenser from CyBio AG, Jena, Germany. The dispensed volume is metered by independent time-controlled actuation of eight solenoid valves, one valve for each dosing channel.

concentration will not lead to statistically relevant results. During the development and validation of a new assay, the z' factor is optimized in context of several parameters including DMSO tolerance, stability of assay reagents, and assay volume. The variability of the liquid-handling steps for the addition of assay reagents and biological material has a direct impact on the z' factor of the assay. Typical z' values obtained from the analysis of runs in Roche's HTS and uHTS laboratories range from 0.65 up to 0.95. It should be noted that the z' factor does not qualify the variability of the compound-aliquoting process.

2.2 High-Throughput Aliquoting of the HTS Library

The replication of microliter or nanoliter aliquots of compound solutions in DMSO in high-density plates with low well-to-well variability of the aliquot volume is a demanding liquid-handling task. The challenge is set forth by several prerequisites and requirements:

- Because the final DMSO concentration in the assay volume has to be kept low, compound aliquots for HTS and uHTS have a very small volume and are highly concentrated.
- DMSO is highly hygroscopic, but has to be kept nearly water-free to enable freezing of the compound solution at $-20\,°C$.
- To minimize the uptake of water during the aliquoting process, one replication cycle should be used to prepare multiple aliquots in high-density plates for multiple HTS and uHTS projects.
- To achieve high throughput in the plate-replication process for the preparation of aliquots in 384- or 1536-well plates, parallel processing of (at least) 384 wells is crucial.

To cope with these demands, Roche has set up a high-throughput aliquoting process for preparation of up to 80 replicas of its HTS library in high-density plates in one replication cycle, *cf.* Table 2. The number of aliquots prepared in one replication cycle is adjusted to the anticipated number of HTS projects in a time span of 6–12 months. The plate-replication system produces replicas of the HTS library in 384-well plates from one edition of master solutions in 384-well source plates. The preparation of replicas in 1536-well plates on this system is foreseen at a later stage. The system comprises a plate buffer holding several hundred plates, two gantry robots for the plate transport, and one (or four) 384-tip parallel pipettor(s).[‖] The pipetting system has a conveyor belt mechanism, which allows plate transport in an inert-gas tunnel floated with dry nitrogen and a plate-sealing unit for the replicated plates. An intermediate storage area is used for sorting out the series of replicas prepared in one replication cycle into stacks of replicated plates, so that each stack constitutes one copy of the HTS library.

[‖] One 384-well pipettor is sufficient for the preparation of aliquots in 384-well plates. For preparation of aliquots in 1536-well plates, four 384-well pipettors are used for transfer of the aliquots from four 384-well source plates to 1536-well target plates.

To produce the large number of replicas, a modified 384-well parallel pipettor (TeMO 384 from Tecan, Switzerland) was chosen as pipetting device.[14] The replicas are prepared by multidispensing of the master solutions aspirated from the source plate. Multidispensing of replicas (*i.e.* dispensing of multiple aliquots out of one aspiration) is a fast process; it has a duty cycle (plate-replication frequency) of about 30 s. The TeMO pipettor features a parallel flow-through wash of its 384 steel tips with DMSO system liquid. Washing of the pipetting tips by flushing DMSO from the back through the tips ensures a negligibly low carryover between subsequent replication cycles without the need for changing disposable tips. After dispensing the last aliquot in the replication cycle, the 384 Teflon-coated steel tips are washed from the inside and the outside with DMSO. The steel tips of the pipettor have an inner diameter of 1.2 mm and an orifice diameter of 200 μm.

The structure of the liquid column between the head of the piston and the tip orifice is shown schematically in Figure 3. The "partition volume" of pure DMSO is used for a "prewash step" prior to the wash step with DMSO system liquid. It is carried out by discarding the residual compound solution and the partition volume into waste after completion of each replication cycle. The 2-μL air gap (trailing air gap

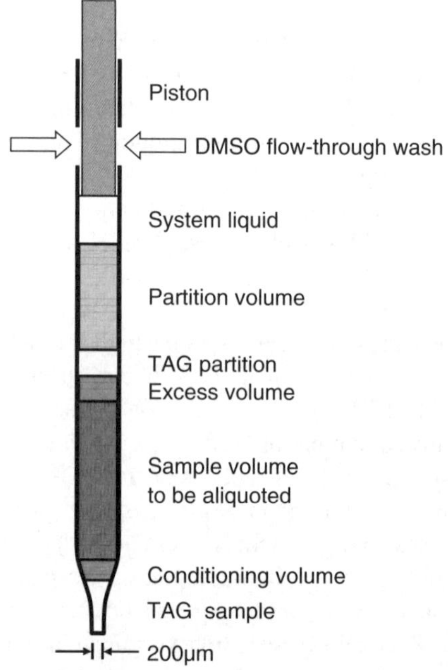

Figure 3 *Pipetting tip used for 384-channel parallel multidispensing of low-volume aliquots out of one aspiration. The structure of the liquid column in the tip is shown schematically. TAG stands for "Trailing Air Gap". After dispensing of the last aliquot in the plate-replication cycle, each tip is washed from the inside and the outside with DMSO. The piston movement is used to accurately meter the small aliquots being dispensed. The piston also acts as a valve regulating the flow of system liquid through the tip for the DMSO flow-through wash. Washing of the outside of the tip is not shown*

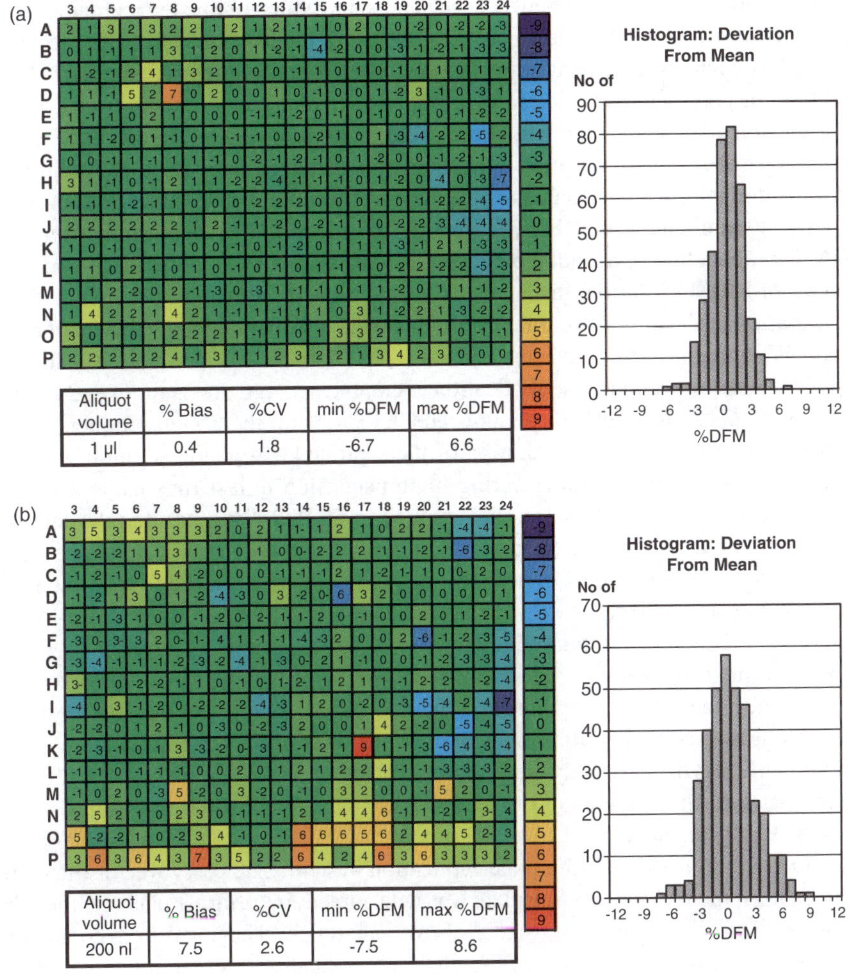

Figure 4 *The 384-channel parallel multidispensing of (a) 1 µL and (b) 200 nL aliquots out of one aspiration into empty wells. The false color maps on the left show the DFM values of the dispensed aliquots over the plate. The color code for the DFM values is shown in a look up table (vertical bars on the right-hand side of the two plate maps). The histograms show the distribution of the DFM values around the mean (average) aliquot volume. The false color map, the DFM histogram, and the stated values for %Bias, %CV, and min/max %DFM in (a) and (b) refer to one representative plate from the 24 replica plates produced in a replication cycle for (a) 1 µL and (b) 200 nL aliquots, respectively. Roche uses the first 2 columns (2 × 16 = 32 wells) in each 384-well plate for standards and controls; since columns 1 and 2 are not filled with compounds, they are not included in the false color maps*

partition) that is aspirated between the partition volume and the sample volume prevents mixing of the compound sample with pure DMSO in the partition volume. The small "conditioning volume" in front of the sample volume is dispensed into the waste before initiation of the replication cycle.

Parallel incremental movements of the pistons push the volume to be aliquoted out of the small orifice at the end of the tips, whereby the aliquot volume is precisely metered by each displacement of the pistons. The transfer of the small aliquots (1 µL and less) of compound solution into empty target plates has to prevail over the surface tension that tends to retain the dispensed droplet at the tip orifice. Precise release of the dispensed aliquots from the tips is achieved by using a self-adjusting tip touch method, which ensures an even contact between tip and well bottom in all 384 wells simultaneously.[14] Figure 4 shows the measured trueness, precision, and DFM for replication of a source plate by multidispensing of 24 aliquots of 1 µL and 200 nL of DMSO solution, respectively. Using the self-adjusting tip-touch method, the variation of the aliquot volume over the plate and from plate to plate is extremely small. We measured the precision over the plate (well-to-well variability) for the replica plates produced in one replication cycle, the average precision over 24 replicas was 1.7 %CV for the 1 µL aliquots and 3.1 %CV for the 200 nL aliquots.

We investigated a potential dilution of the aspirated compound solution with the DMSO in the partition volume during multidispensing in test runs using a source plate filled with a 500 µM fluorescein solution in DMSO. We measured the concentration of the indicator dye in the aliquots of the first and last target plate in one replication cycle. The fluorescein concentration difference between the first and the last target plate was found to be very small, indicating that there is no perturbing dilution of the compound solution with the DMSO in the partition volume.

We also investigated the tip-wash efficiency using two types of carryover test experiments: the first experiment uses source plates filled with pure DMSO and intervening source plates filled with fluorescein solution in DMSO. The measured fluorescein carryover from the source plate of the previous replication cycle to the target plates of the subsequent replication cycle was found to be less than 0.004% ($<4 \times 10^{-5}$). In the second test experiment, six different compound solutions in DMSO were processed on the plate-replication system. The carryover of the compounds from the previous source plate was determined in the target plates of the subsequent replication cycle using LC/MS/MS analytics. The measured compound carryover from the source plate of the previous replication cycle to the target plates of the subsequent replication cycle was always smaller than 0.008% ($<8 \times 10^{-5}$).[†]

2.3 A Microfluidic Well Plate for High-Throughput Solid/Liquid Separations

Proteomics is a modern parallel approach for investigating protein expression in experimental systems. In pharmaceutical research, this methodology is used to accelerate the discovery of novel drug targets and biomarkers.[15] Among the several possible techniques for the study of proteomes, the combination of two-dimensional polyacrylamide electrophoresis with mass spectrometric identification of the separated proteins has gained wide popularity. This technique involves the analysis of vast numbers of small gel spots that are excised from two-dimensional protein separation

---

[†] For the carryover measurements, the so-called "sticky compounds" were selected from the compound library. "Sticky compounds" tend to precipitate on the surface of pipetting tips; they have been identified by analysis of carryover effects in HTS data.

gels using automated spot-pickers. After appropriate sample preparation steps involving the digestion of the proteins in the gel spots and the extraction and transfer of the resulting peptides to a special sample support for mass spectrometry (MS), proteins can be readily analyzed using a technique called peptide mass finger printing.[16]

Our laboratory has developed a highly integrated microfluidic well plate for high-throughput sample preparation of the MALDI (matrix-assisted laser desorption/ionization) targets used for mass spectrometric analysis of the protein fragments. The plate enables high peptide extraction efficiency through efficient separation of the liquid phase from the immersed gel spots and comprehensive collection of the low-volume extracts for MS analysis. Figure 5a shows the top view of the densely packed

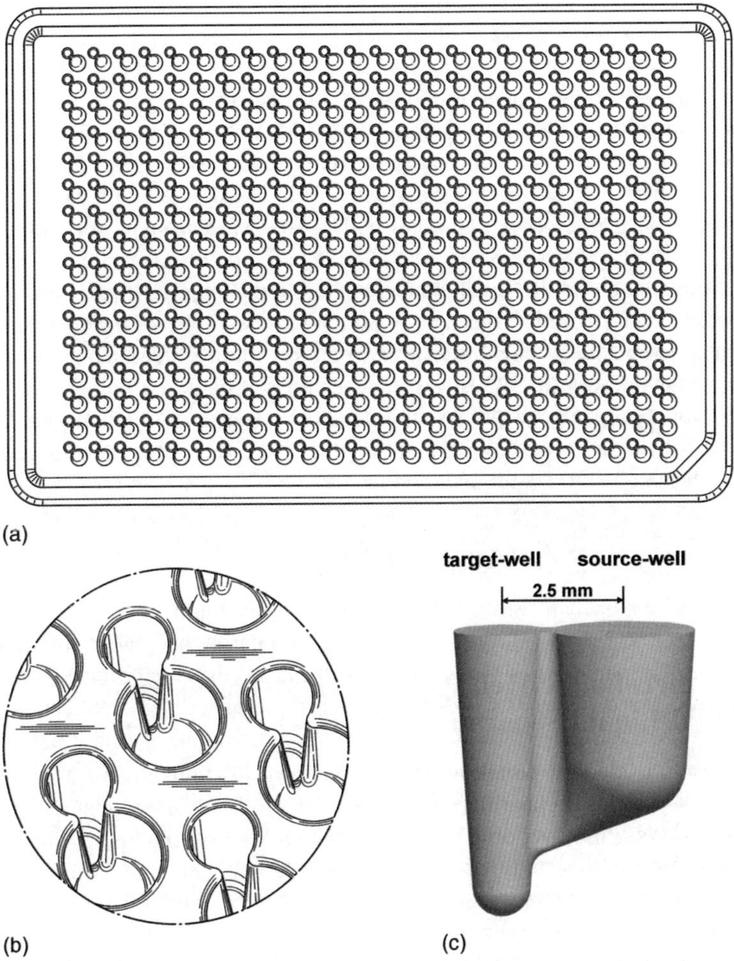

Figure 5 *(a) Top view of the microfluidic well plate for high-throughput solid/liquid separations. (b) Three-dimensional shape of the cavities. (c) Side view of one cavity. The narrow channel between source- and target-well allows passage of liquid, but does not allow passage of the solid-sample component (gel spot)*

array of 384 identical cavities on the microfluidic well plate. The three-dimensional shape of the cavities is shown in the perspective close-up of the array (Figure 5b) and in the side view of one cavity (Figure 5c). Each cavity comprises two sample chambers denoted as "source-well" and "target-well". The source-well receives the gel spot to be processed and the second – somewhat smaller – target-well is used for collecting washing buffer and the peptide extract. The bottom of the target-well is about 2 mm below the bottom of the source-well. A narrow channel fluidically connects the two wells with each other.

The 384 source-wells in the microfluidic well plate receive 384 gel spots that have been excised from a two-dimensional polyacrylamide gel. Liquid aliquots dispensed onto the gel spots in the plate reside in the source-wells; they are retained there by capillary forces. When a centrifugal force is applied to the plate, the narrow channels allow passage of liquid from the source-wells to the target-wells, but do not allow passage of the gel spots. Efficient separation of the liquid from the gel spots is achieved by a short centrifugation of the plate. Centrifugation transfers the entire liquid surrounding the gel spots to the target-wells, leaving the gel spots in the source-wells. A parallel pipettor aspirates 384 samples of 1.5 µL containing the peptides from the target-wells in the plate and transfers them onto the MALDI target. A so-called AnchorChip™ is used as MALDI target;[17] it has the footprint of a microtiter plate and carries 384 anchor spots for MALDI-MS analysis on its planar surface.

The rationale behind the development of the microfluidic well plate is to provide a disposable device for high-throughput parallel processing of the excised gel spots in a planar array of 384 cavities that can be fabricated economically by injection molding. The microfluidic 384-well plate enables reliable preparation of 50,000 MALDI target samples per day on an automated liquid-handling and plate-processing system. The microfluidic well plate might also be useful for solid-phase extractions in other high-throughput applications.

3   Microarray Assays on Chips

As stated in Section 1 of this chapter, the diversity of liquid samples is handled by confining each sample within the "walls" of a compartment. Planar arrays of such sample compartments are realized in the well plates discussed above. The liquid-handling process then needs to address each sample of a diverse collection individually, which imposes a limit for further miniaturization. Therefore, successful miniaturization concepts will lead to a reduction of the number of liquid-handling steps compared to well-based approaches. Miniaturization is particularly enabling in situations where precious biological fluids that are only available in limited quantities need to be tested against a set of samples.

This approach is realized in microarrays. The key is to "immobilize" the diversity of samples onto a solid substrate in discrete locations such that the walls that separate the samples become obsolete. An entire array of samples is thus parallel-processed with a single liquid-handling step without the need to miniaturize the liquid handling for the assay itself. However, some of the technologies to produce such arrays of immobilized samples do require precise liquid handling for volumes below 1 nL (1 nL = 10^{-9} L).

Arrays of immobilized samples are already fairly common in the life sciences. The large variety of different sample types includes nucleic acids, proteins, peptides, antibodies, cell lysates, entire cells, patches of tissue, or chemical compounds. The latter are discussed in a separate chapter of this book. A detailed review of the other types would be beyond the scope of this article but can be found elsewhere.[18]

There are general considerations for the choice of a microarray substrate. One is that the samples need to be well attached to the substrate without compromising the ability of the sample to interact with the molecules in the surrounding liquid during the assay. Another is a low autofluorescence of the substrates, since microarray experiments rely primarily on fluorescence detection. Last, the surface between the deposited samples needs to have a low susceptibility for nonspecific binding of signal-generating molecules. These three aspects require that the surface be tailored to the respective type of biomolecules of an assay. Unfortunately, there is no single substrate or surface that works optimally for all kinds of biomolecules.

Commercially available devices for nanoliter and subnanoliter deposition of liquids are based on two different operating principles. In pin printers (contact printing), a metallic needle (pin) is immersed into the sample and the attached liquid is deposited on the microarray surface by touching it. The throughput of such systems is increased by using up to 48 pins in parallel. The transferred volume, however, is influenced by many parameters: (i) viscosity and surface tension of the sample itself, which is determined by the biomolecules and the buffer in which they are dissolved, (ii) the relative interfacial forces between pin and liquid as well as between liquid and substrate, (iii) the speed of approaching and retracting the pin to the surface, and (iv) variations in pin geometry due to their manufacturing process. Moreover, the entire volume-metering process is surface tension dependent. And, whether the metal of the pins or the pressure between pin and substrate (5–10 kPa) alters the properties of the printed biomolecules, particularly of proteins, is still under debate.

In the nanoliter and subnanoliter range, the transferred volume can be controlled much better with noncontact printers, where a drop is shot on the substrate by a piezo-driven device. Here, the actuation energy from the piezo is coupled into the liquid, which leads to a droplet formation out of an orifice. The volume is defined by a dynamic process, which makes it less susceptible to surface-tension effects. Although viscosity and surface tension of the sample still influence the process, their contribution to variations in volume is thus lower than in pin printers, particularly because the microarray substrate itself is not involved in the volume metering. These devices are available with up to 16 nozzle-actuator assemblies arranged in parallel. Adjustment of the actuation parameters for all nozzles becomes increasingly complex with higher numbers of assemblies. The highest degree of parallelization is realized in a device that uses a microfabricated printhead with 24 or 96 nozzles and only one piezo actuator to operate all nozzles in parallel.[19] Within the past few years, the setup and operation of such devices became much easier and their performance more robust. Nevertheless, it is advisable to have them run by dedicated personnel in a core facility.

Nucleic acid arrays are used in the research labs at F. Hoffmann-La Roche in two varieties. One is represented by the commercially available high-density oligonucleotide arrays in which the samples are synthesized *in situ* and covalently bound to the substrate (Affymetrix™). The other is the so-called "custom spotted arrays" in which the droplets that contain different molecules are deposited on the substrate. If the nucleic acid sequence is known, but not available on a commercial chip, oligonucleotides are synthesized and deposited. In cases for which the sequence is not known, microarrays of cDNA fragments or plasmids are used.

For research on proteins, this article will only cover antibody-based technologies. The limitation in biological sample volume and the inability to amplify unknown proteins in a similar fashion as nucleic acids determines the type of array. Another limitation is the cross-reactivity of proteins and antibodies that do not allow multiplexed probing of a complex biological mixture with more than 30 to 40 antibodies with sufficient sensitivity.

In sandwich immunoassays, a pair of antibodies binds to different epitopes of an analyte and one of the antibodies (the so-called "capture antibody") is immobilized on the substrate. The second antibody is labeled with a probe (*e.g.* a fluorescent dye), it allows detection of the analyte molecules that bind to the capture antibody. We use this approach when the antigen is a well-characterized protein. Multiplexing is possible by immobilizing the capture antibodies in an array or on individually coded beads for medium- to low-volume amounts of biological liquids. The submicroliter version of this approach, in which capture antibodies are immobilized in stripes and the sample is guided to the capture sites by means of microfluidic networks, is described in Section 3.1 of this chapter.

In contrast to sandwich immunoassays, in which a capture *antibody* is immobilized on a substrate ("forward phase arrays"), in the so-called "reverse phase protein arrays" the complex mixture of *analytes* is immobilized on a substrate. The biological sample could be a cell lysate or from another source. Each spot then comprises the entire proteome of a biological system at a certain state or after a certain treatment. Each array contains a multitude of biological states. Several copies of such arrays are produced and each one is probed with a single antibody. Thus, each assay shows, in one experiment, the expression levels of a particular protein in all the different biological states represented on the array. Since the entire proteome is contained in each spot, essentially any protein can be analyzed in the context of different experimental foci.

3.1 Microchannel Assay: A New Generation of Miniaturized Multiplexed Bioassays

If proteins need to be detected with picomolar sensitivity out of a few microliters of a biological sample, even a microarray approach may not be sufficiently miniaturized. The mass transport by diffusion across the surface hinders the reaction because most of the analyte molecules in the biological sample are not brought in contact with the capture molecules within a reasonable amount of time. However, if mass transport is enhanced by flowing the entire sample along a miniaturized capture site, extremely high sensitivities can be achieved with minute amounts of sample.[20,21] Such an approach depends on the benefits of fluorescence detection, which can spatially resolve individual spots down to less than 1 μm separation, as long as they are well separated from each other and the background fluorescence between them is low enough. Even the detection of individual molecules is possible in laboratory setups. In other words, there is no practical limit for the miniaturization of the features for fluorescence detection. The detection limit of commercially available fluorescence readers these days is about 0.5 fluorophores per μm^2.

We reduced this concept to practice by developing a microfluidic system in collaboration with the Research Laboratory of IBM in Rüschlikon, Switzerland. The chip component of this system is depicted in Figure 6. On a silicon substrate, a number of capillary systems are etched. Each one consists of a fill port that allows

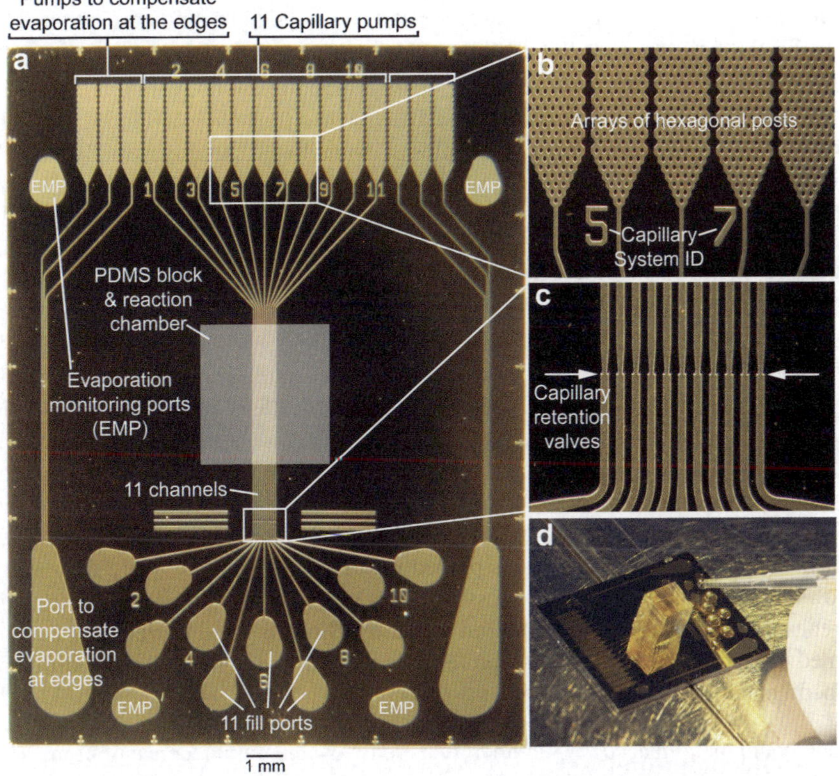

Figure 6 *Chip component of the microchannel assay system. (a) Overview of the entire chip, (b) detail of capillary pumps, which provide autonomous filling of each channel by capillary action, (c) each capillary retention valve prevents drainage of a channel's reaction zone after the fill port has emptied, and (d) a chip with an attached PDMS cube is filled by a standard pipette. By placing the chip on 2 aluminum blocks, whose temperature is individually controlled by 2 Peltier elements, differential evaporation allows extremely low flow rates*

the device to be interfaced to manual or automated pipetting, an appended microchannel, and a capillary pump. Filling of the capillary systems occurs autonomously by capillary forces. The liquid flow through the system is controlled by evaporation from the capillary pump. The device is placed on two Peltier elements such that the temperatures of fill ports and capillary pumps can be set independently. Evaporation from the fill ports is prevented by setting their temperature only slightly above the dew point. This also prevents condensation, which would dilute the small amounts of biological sample. The rate of evaporation is then dependent on the temperature difference between the filling ports and the capillary pump. Extremely low flow rates of <500 nL h^{-1} can be achieved with $\Delta T \approx 5$ K. A narrow constriction in the microchannel, called a capillary retention valve, prevents drainage of the channel once the liquid from the fill port has traversed the system.

The microchannels are 30 μm wide and 20 μm deep, open to the top, and sealed in the reaction zone by a cube of poly(dimethyl)siloxane (PDMS), a silicone elastomer. PDMS has a low autofluorescence and shows good protein adsorption. The PDMS cube thus serves as the substrate on which the capture molecules are immobilized. Lines of n different capture molecules (in this case capture antibodies) can be patterned with the device in the first step. The cube is then removed from the chip and placed onto a new chip turned by 90°. Now, m different biological samples with a typi-cal volume of 300 to 600 nL can be applied to the individual capillary systems. Upon flowing through the system and underneath the cube during 15 to 30 min, each sample will be exposed to a 30 μm wide stripe of the previously patterned capture molecules, resulting in a 30×30 μm^2 small interaction feature. The cube is then removed from the system and the captured analytes are detected in bulk by immersing the cube in a solution of the respective secondary antibodies. In this way, $n \times m$ molecular interactions can be tested with only $n + m$ pipetting steps. Such an interaction pattern is depicted in Figure 7, where each square of 30×30 μm^2 represents the interaction of capture molecules with different biological samples.

Besides saving liquid-handling steps, which we had identified as a prerequisite for successful miniaturization, the sensitivity we sought could be achieved by the effective mass transport of analyte molecules in the sample volume to a small capture site. In principle, the biological sample could be divided into a large number of incremental volumes of $20 \times 30 \times 30$ μm$^3 = 18$ pL. At the given flow rate of 1200 nL h^{-1} or a flow velocity of 0.7 mm s^{-1}, each incremental cube is exposed to the 30×30 μm^2 large capture area for about 40 ms. This time is sufficient to deplete the analyte molecules effectively from the sample solution. Every 40 ms, one volume element could be replaced by another that contains the initial concentration. The enhancement of sensitivity is so substantial that steady-state FI detection could be accomplished with dyes of the Cy- or Alexa-family, which would not yield a reasonable signal in macroscopic plate-based sandwich immunoassays.

The first assay we developed detected the cytokine tumor necrosis factor α (TNFα). We achieved a detection limit of 20 pg mL (1.1 pM) out of a sample of 600 nL.[22] Subsequent application to additional cytokines enables analysis of a cytokine panel as inflammatory markers for pharmaceutical research.

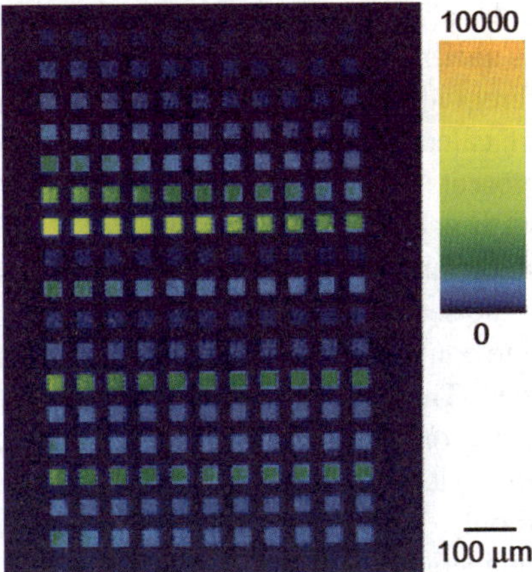

Figure 7 *Fluorescence image of a sandwich immunoassay that was generated with the microchannel assay system. Capture antibodies were patterned as 11 vertical lines. Subsequently, 17 different biological samples were patterned as horizontal lines. At the intersections, an interaction pattern of 11 × 17 = 187 features is thus generated with 11 + 17 = 28 liquid-handling steps*

4 Prospects for Multiparameter Assays

The advent of all the "-omics" disciplines in the life sciences is an indication that the simultaneous observation of multiple parameters of a biological system is not only feasible but also important. Advanced computational methods and computing power also allow the identification of a plurality of parameters whose changes are significant in a group but not necessarily as individual events.[23] The underlying essence of the "-omics" paradigm is the desire to develop a holistic picture of the complex and interdependent processes in biological systems. The wish to understand biological phenomena in their entirety rather than as individual unrelated events is the motivation to observe large numbers of parameters under the influence of many different factors. In the pharmaceutical sciences these factors are the large numbers of molecular entities.

The successful business concept of the semiconductor industry is miniaturization of functional features and their integration on a planar (or multilayered) device. The smaller the features, the more can be produced in parallel in a single processing step. In a way, this is simply economy of scale. Miniaturization and parallelization are also key paradigms for the life sciences, which will depend on the generation and interpretation of huge amounts of data in order to understand the complex relationships of biological phenomena. However, the strategy of "smaller-is-better-is-cheaper" by itself seems to be too simple to lead to this goal, as discussed in this article. Smart approaches will break down the different process steps for assays into optimized subprocesses. Each subprocess benefits from *different* strategies for miniaturization and parallelization, such that each subprocess and the whole process remain fairly simple while generating a vast diversity by intelligent nesting of the different processes.

Besides the microfluidic assay that was described in Section 3.1 of this chapter, we would like to discuss another example out of many others, where we feel that this strategy of nesting different approaches for miniaturization and parallelization in a clever way will lead or has led to substantial improvements.

This technology represents a novel method of fabrication and use of oligonucleotide ("oligo") arrays, it is described in the Refs. 24, 25. First, oligos are synthesized in a parallel fashion and attached in a bulk process to beads with 3 µm diameter. In addition to the sequence that serves to probe a gene sequence of a biological sample, each oligo contains an "address" sequence that allows each bead that contains it to be identified. Then, equal amounts of beads coated with different oligos are mixed and localized on a planar substrate. The substrate consists of an array of microfabricated cavities (denoted as "wells") with a diameter of 3 µm, which can accommodate one bead only. When it is immersed in the oligo-bead mixture, individual beads settle spontaneously into the wells to produce a randomly assembled array of thousands of beads on the substrate surface. Surface-binding interactions between the beads and the micrometer-sized wells hold each bead in place. The fixed localization of the beads on the planar substrate enables parallel identification of the address sequence attached to each bead with an optical imaging system. Using the address sequence of the oligos in a combinatorial way, the identity of each bead at each position in the array is determined by hybridization with "decoding" probe molecules. Owing to the combinatorial strategy of the decoding process, the number of decoding hybridizations is much smaller than the number of different bead types in an array. The array is then given to the user together with an "address list" that indicates which oligo is immobilized at which position.

In one type of experiment, the bead array is assembled at the end of a fiberoptic bundle. Ninety-six of these bundles are arranged in an 8 × 12 pattern with a 9 mm spacing in an "array matrix", such that they can all be immersed at once into the wells of a 96-well plate. This experiment allows 96 arrays to be hybridized and washed in parallel with simple handling steps. By probing the fluorescent signal from each immobilized bead in each bundle, a large amount of information, rather than a single data point, is gathered from every well in this single experiment.

The nested key steps of this technology are:

- Parallel production of many beads of one type. Although the beads are the miniaturized entities of the system, their bulk processing does not require miniaturization.
- Self-assembly of beads into cavities. All cavities are the same and their generation is a parallel, well-established microfabrication process. No miniaturized placement of individual beads to particular positions is necessary.
- Generation of an address list to identify each bead in the randomly assembled array in parallel with only a few hybridizations and combinatorial decoding of the addresses.
- Arrangement of arrays in an 8 × 12 matrix. This format allows parallel processing with standard lab equipment without miniaturization (96-well plates are not considered miniaturized).

The entire process works without the need for miniaturized liquid-handling or miniaturized spatial placement of samples to predefined positions. The miniaturization processes are the production of the beads and the production of the cavities. Both are done offline with standard microtechnology processes. Nevertheless, the entire system offers a feature density of more than 35,000 mm^{-2}. This density would allow several copies of probe sets for the entire human genome to be placed on a substrate of 25×75 mm^2, in another variation of this technology.

We imagine that, in a similar way, successful future technologies for the life sciences will allow enormous diversity to be managed in a miniaturized and parallelized way by intelligent combinatorial approaches without the need for miniaturized handling of individual entities during the diversity generating process itself.

Acknowledgment

The authors wish to thank Eva-Maria Gutknecht, Martin Brunner, Thilo Enderle, Joergen Nielsen, Martin Graf, Hanno Langen, Peter Berndt, Ulrich Certa, Patrick Iaiza, Remo Hochstrasser, and Klaus Mueller for stimulating and clarifying discussions on the various topics presented in the article and for proof reading of the manuscript.

References

1. I.M. Ross, The foundation of the silicon age, *Phys. Today*, 1997 (December issue), 34.
2. T.R. Reid, *The Chip: How Two Americans Invented the Microchip and Launched a Revolution*, Simon & Schuster, New York, 1984.
3. American National Standards Institute and Society for Biomolecular Screening, *ANSI/SBS Standard for Microplate – Well Positions*, April 2004; see http: //www.sbsonline.org/msdc/pdf/ANSI_SBS_1-2004.pdf.
4. D.H. Rasmussen and A.P. MacKenzie, *Nature*, 1968, **220**, 1315.
5. Ch.M. Apfel and Th. Enderle, Assays for high throughput screening in drug discovery, in *Combinatorial Chemistry – From Theory to Application*, W. Bannwarth and B. Hinzen (eds), Wiley-VCH, Weinheim, 2005.
6. J.R. Lakowicz, *Principles of Fluorescence Spectroscopy*, 2nd edn, Kluwer Academic/Plenum Press, New York, 1999.
7. R.P. Haugland, *Handbook of Fluorescent Probes and Research Products*, 8th edn, Molecular Probes, Seattle, 2001.
8. V.C. Abraham, D.L. Taylor and J.R. Haskins, High content screening applied to large-scale cell biology, *Trends Biotechnol.*, 2004, **22**(1), 15–22.
9. M.J. Madou, L.J. Lee, S. Daunert, S. Lai and C.-H. Shih, Design and fabrication of CD-like microfluidic platforms for diagnostics: microfluidic functions, *Biomed. Microdevices*, 2001, **3**(3), 245–254.
10. R. Ellson, M. Mutz, B. Browning, L. Lee, M.F. Miller and R. Papen, Transfer of low nanoliter volumes between microplates using focused acoustics, *JALA*, 2003, **8**, 29–34.

11. N.-T. Nguyen and S. Wereley, *Fundamentals and Applications of Microfluidics*, Artech House, Boston, MA, 2006.

12. K.S. Schröder and B.D. Neagle, *J. Biomolecular Screening*, 1996, **1**, 75–80.

13. J.-H. Zhang, T.D.Y. Chung and K.R. Oldenburg, *J. Biomolecular Screening*, 1999, **4**, 67–73.

14. Tecan Trading AG, Tecan Application Note Nr. 394 071, *Fast Multiple Plate Replication Using a 384-Channel Parallel Dispensing*, September 2005, retrieved from http://www.tecan.com/te-mo.

15. J.W. Zolg and H. Langen, How industry is approaching the search of new diagnostic markers and biomarkers, *Mol. Cell Proteomics*, 2004, **3**(4), 345–354.

16. P. Berndt, U. Hobohm and H. Langen, Reliable automatic protein identification from matrix-assisted laser desorption/ionization mass spectrometric peptide fingerprints, *Electrophoresis*, 1999, **20**(18), 3521–3526.

17. M. Schuerenberg, C. Luebbert, H. Eickhoff, M. Kalkum, H. Lehrach and E. Nordhoff, Prestructured MALDI-MS sample supports, *Anal. Chem.*, 2000, **72**(15), 3436–3442.

18. Microarrays Supplement of Drug Discovery Today, 2004, **9** (24, Suppl.).

19. O. Gutmann, R. Kuehlewein, S. Reinbold, R. Niekrawietz, C.P. Steinert, B. de Heij, R. Zengerle and M. Daub, *Lab Chip*, 2005, **5**, 675–681.

20. M. Zimmermann, E. Delamarche, M. Wolf and P. Hunziker, *Biomed. Microdevices*, 2005, **7**(2), 99–110.

21. R.P. Ekins, *Clin. Chem.*, 1998, **44**, 2015–2030.

22. S. Cesaro-Tadic, G. Dernick, D. Juncker, G. Buurman, H. Kropshofer, B. Michel, Ch. Fattinger and E. Delamarche, *Lab Chip*, 2004, **4**, 563–569.

23. G. Liao, J. Wang, J. Guo, J. Allard, J. Cheng, A. Ng, S. Shafer, A. Puech, J.D. McPherson, D. Foernzler, G. Peltz and J. Usuka, *Science*, 2004, **306**, **5696**, 690–695.

24. J.A. Ferguson, F.J. Steemers and D.R. Walt, *Anal. Chem.*, 2000, **72**(22), 5618–6624.

25. K.L. Gunderson, S. Kruglyak, M.S. Graige, F. Garcia, B.G. Kermani, C. Zhao, D. Che, T. Dickinson, E. Wickham, J. Bierle, D. Doucet, M. Milewski, R. Yang, C. Siegmund, J. Haas, L. Zhou, A. Oliphant, J.-B. Fan, S. Barnard and M.S. Chee, *Genome Res.*, 2004, **14**, 870–877.

CHAPTER 10

Fluorescence Technologies for the Investigation of Chemical Libraries

ERIC TRINQUET AND GÉRARD MATHIS

CISBio International, BP 84175, 30204 Bagnols sur Cèze cedex, France

1   Introduction

Fluorescent techniques have been developing dramatically this last decade and are now in position to represent more than 60% of the detection technologies used in High-Throughput Screening (HTS).[20] The reason for this hegemony originates from two intrinsic properties of fluorescence.

Sensitivity. the fluorescence intensity I_F is expressed by $I_F = I_0\,\varepsilon\phi\,Cl$ where I_0 stands for the excitation light source intensity, C for the fluorophore concentration, l for the light path. ε and ϕ are respectively the molar absorptivity and quantum yield and are characteristics of the fluorophore.[33]

It is obvious that, at low concentration, increasing the excitation power will lead to an increase of the measured signal and therefore, if background can be controlled, to a theoretically unlimited sensitivity.

A second consequence of this relation is that unlike radioactivity or luminescence, the fluorescence is proportional to the quantity of tracer in the observed volume.

High content of the fluorescent signal. The extraordinarily rich content of fluorescence output, characterized by energy (absorption and emission wavelength), lifetime, quantum yield, and anisotropy. Together with the large panel of possible interactions within the photo physical pathways, like electron or energy transfer and variation of rotational or translational correlation time, these characteristics have enabled the construction of homogeneous assays that do not require separation steps, for example, the very popular polarization and fluorescence resonance energy transfer (FRET) assays.

The possibility of having mix and measure formats that do not require separation steps as well as the extreme sensitivity and relative independence of the measurement on the volume of the sample, have made fluorescence techniques the core detection technologies for miniaturized ultra-HTS (uHTS).

This unique set of properties has convinced instrument makers to introduce versatile, multi-technology microplate readers adapted for HTS.

However, this ideal description should not overlook the difficulties encountered when using homogeneous fluorescent technologies in the real world.

In fact, it is worthwhile to note that all these homogeneous methods were developed in the 1970s and early 1980s with a targeted application in automated diagnostic immunoassay systems.[21,44] Despite the extraordinarily creative work of E. Ullman and co-workers at SYVA Company, none of the technologies they developed could be launched on this market, primarily as a result of the extremely variable optical and quenching behaviour of human serum. Among possible interferences, one can make distinctions between:[10]

- Optical interference generated by the variations of the optical properties of the media at the excitation or emission wavelength (inner filter effect).
- Interference with the intra- and intermolecular photophysical processes.
- Interference by the fluorescence of proteins, compounds, or nonspecific adsorption of the tracer on proteins.

Because of the very stringent reliability requirements for diagnostic applications, the only homogeneous techniques that have survived in this field are the fluorescence polarization (FP) methods originally developed by Abbott[21] and the homogeneous time resolved fluorescence (HTRF) and TRACE® approaches developed later by CIS bio international.[25,26]

The aim of this chapter is not to review in great details all the fluorescent technologies used in HTS because they have already been largely described in dedicated articles or books.

We will instead describe rapidly the principles of each one, making reference to more detailed descriptions while focusing more on the robustness and possible hurdles and drawbacks which may be encountered when these technologies are used to screen chemical libraries in the real world of HTS.

2 Dissociation-Enhanced Lanthanide Fluoroimmunoassay

The dissociation-enhanced lanthanide fluoroimmunoassay (DELFIA) technology uses lanthanide chelates as fluorescent probes. These chelates, which exhibit a long-lived fluorescence lifetime, are based on the noncovalent association of a chelate and a lanthanide ion, usually europium or terbium.[19] The chelate protects the lanthanide ion from potential quenching by its environment and acts as an antenna by collecting the energy from the excitation source before transferring it to the lanthanide ion. The long fluorescence lifetime of the complex is an intrinsic property of the lanthanide ion.

Such probes, which can be excited in the UV using a nitrogen laser or a Xenon flash lamp, emit their fluorescence between 500 and 700 nm, depending on the specific lanthanide ion used. Their fluorescence lifetime is usually longer than 100 µs. Their use as fluorescent tracers allows the establishment of a time resolved fluorescent detection using pulsed excitation sources (Figure 1). This detection mode make it possible to reject most of the fluorescent background given by biological media, by instrumentation components, or by the chemical compounds to be tested in HTS.

The luminescent lanthanide chelates designed for DELFIA are not stable in biological media. To circumvent this limitation, a heterogeneous assay based on a two-step approach was designed. In the first step, biomolecule conjugates bearing nonluminescent chelates (derivatives of EDTA or diethylenetriaminepentaacetate (DTPA)) were used to carry the lanthanide ions during the biological reaction. In the second step, the lanthanide ions were dissociated from the nonfluorescent chelates by adding a large excess of a different chelating agent that generates a strongly luminescent complex.[36]

Due to its heterogeneous nature (separation of the biological media prior to the measurement) and the use of a time-resolved detection, DELFIA is quite insensitive to fluorescence background, so very low detection limits can be achieved.[48] In this technology, the rare earth ion is in fact the tracer and can be present at very low concentration. Therefore, with DELFIA, the main risks for interference come from possible contamination by other rare earth ions, which can be released by the vessels (glass) or present in dust. DELFIA has been used to probe a large diversity of molecular interactions; however, the washing steps needed to run a DELFIA assay strongly limit its implementation in HTS.[10] Therefore, homogeneous technologies using fluorescence like enzyme fragment complementation (EFC), FP, or HTRF are progressively replacing it for a large number of applications.

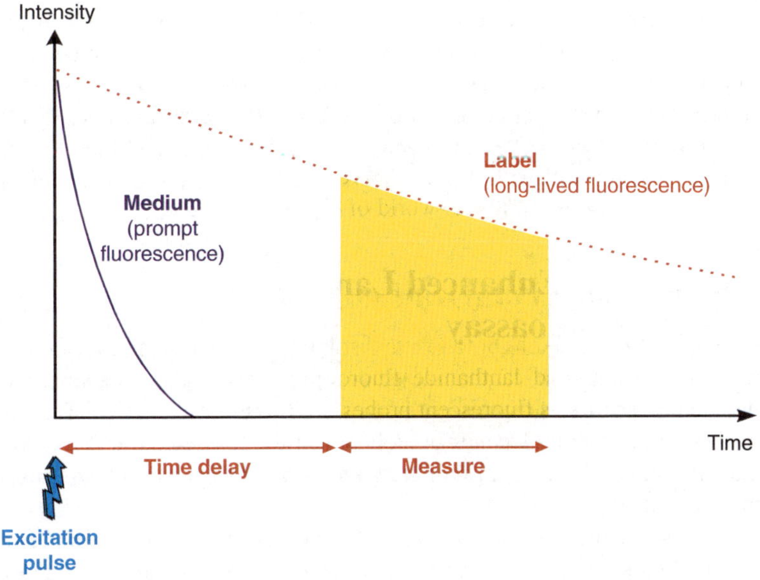

Figure 1 *Time-resolved detection of a fluorescent signal. The fluorescent signal from the long-lived tracer (red dashed line) is integrated after a fixed delay time to exclude the fluorescence background (black line) from the measurement window (yellow)*

3   Enzyme Fragment Complementation

This technology uses complementation of β-galactosidase fragments to generate a fluorescent or a chemiluminescent signal. When a small α fragment peptide of 4 kDa called ProLabel interacts with an ω-deletion mutant of the enzyme (called Enzyme Acceptor), an active tetrameric form of the β-galactosidase results.[11] The active enzyme can generate a signal by hydrolysing fluorescent substrates like 4-methy-lumbelliferyl β-D-galactopyranoside (4-MUG) or Resofurin, or other luminescent substrates. Since the interaction between the selected enzyme fragments occurs with a nanomolar affinity, detection limits in the same concentration range can be achieved without separation steps in homogeneous assays.

The ProLabel peptide can be chemically conjugated or recombinantly fused to various biomolecules. To probe molecular interactions, EFC assays are based on a competition between the free and the ProLabel peptide-conjugated form of the bio-molecule involved in the interaction under study. Bound to its interaction partner, the ProLabel-labelled biomolecule is not able to complement with the Enzyme Acceptor. Therefore the signal generated by the β-galactosidase is proportional to the concentration of the free biomolecule in the assay.[11]

EFC assays can be easily automated and miniaturized due to their homogeneous format and can therefore be used to screen large libraries of chemical compounds against a large diversity of biological targets, such as kinases or G Protein Coupled Receptors.[15]

However, depending on the substrate used in the enzymatic reaction, EFC is sensitive to compound interferences. Fluorescent compounds may increase the signal background of the assay, particularly if a green-fluorescent substrate like 4-MUG is used to generate the signal. The use of a red-shifted fluorescent substrate like Resofurin reduces the number of interferences from auto-fluorescent compounds.[49] Since the EFC signal readout is only based on a fluorescence or luminescence intensity, coloured compounds, or compounds which interfere with substrates by affecting their photo-physical properties can decrease the EFC signal and therefore may be interpreted as false positives in HTS.[16] Moreover, as for any enzyme, galactosidase inhibitors may be present in libraries and impair the quality of the results.

4   Fluorescence Polarization

In FP, the use of polarized light to excite a solution of fluorescent molecules results in preferential excitation of the molecules that have their transition moments (dipoles) parallel to the direction of the polarized light of excitation. The polarization of their emitted light depends upon how fast the fluorescent probes rotate during the lifetime of its excitation state:[9] the faster the rotation, the smaller the polarized signal (P).

The fluorescent probes usually used in FP possess fluorescent lifetimes shorter than 10 ns. For such probes, significant changes in their polarized signal can be detected in a range of molecular mass between 5 and 40 kDa.[33]

A large number of homogeneous assays have been developed using FP by inducing a dramatic change in the molecular mass of the moiety carrying the fluorescent probe. As an example, an increase of the polarized signal of a small fluorescently labelled peptide freely diffusing in solution could be detected after its binding to a receptor.[4]

A relatively simple instrument set-up can be used to detect such changes in the polarized signal (Figure 2). Fluorescence detected using the two polarization filters successively placed in the fluorescence emission pathway, one parallel ($I_{parallel}$) and one orthogonal ($I_{orthogonal}$) to the polarized excitation light, enables the polarized signal (P) to be calculated according the following formula:

$$P = (I_{parallel} - I_{orthogonal}) / (I_{parallel} + I_{orthogonal})$$

The ratiometric nature of the polarized signal makes FP independent of the fluorescence intensity and more tolerant to fluorescence quenching and light scattering than the fluorescence intensity-based measurements.[10,33]

However, fluorescence background from chemical compounds or biological media can affect the accuracy of FP. As such fluorescence background mainly arises in the wavelength range between 400 and 550 nm, the use of red dyes emitting above 550 nm can be an effective way to partially overcome this problem.[43]

The hydrophobic nature of some small fluorescent ligands can also induce their non-specific binding to various proteins like bovine serum albumin (BSA). This binding artificially increases the rotational correlation time and, as a consequence, their polarized signal and therefore may affect the assay result.

Despite a limited sensitivity in the low-range nanomolar, the large availability of FP-compatible instruments, the simplicity and low cost of the assay make FP quite popular in HTS. However, the short fluorescence lifetimes of the conventional fluorescent labels (below 10 ns) limit the use of FP to biological interactions involving molecules with a molecular weight below 15 kDa. To overcome this limitation, fluorophores possessing longer lifetime, like ruthenium (II) complexes with lifetimes around 300 ns, have been used to probe the interaction between larger proteins.[39]

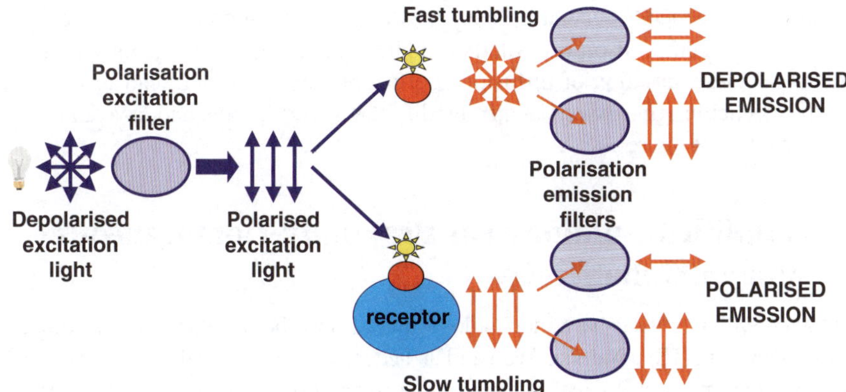

Figure 2 *Schematic diagram of the instrument set-up used to detect a polarized signal. Arrows indicate the orientation of the excitation light and the orientation of the fluorescence emitted by the probe, which is dependent on its tumbling speed*

However, these fluorescent labels are not very bright, which very much limits their use in HTS.[42]

5   Fluorescence Correlation Spectroscopy

Fluorescence correlation spectroscopy (FCS) is based on the measurement of intensity fluctuations. These fluctuations can be detected when the volume of solution under investigation is so small that the fluorescent signal measured is no longer randomized over many fluorescent molecules but arises when individual molecules pass through the detection volume, the signal is therefore related to their diffusion coefficient. Typically, the detection volume is around one femtoliter (confocal volume) and requires a confocal microscopic instrument set-up to be investigated. Fluorescent intensity fluctuations are analysed using autocorrelation techniques. The resulting autocorrelation function gives information regarding the concentration of fluorescent molecules, their diffusion time through the confocal volume, and the average brightness per molecule.[30]

Variations in the diffusion time of the fluorescent molecules can therefore be detected and used to probe molecular interactions in FCS. Such variations are induced by a large increase in the molecular mass of the fluorescence molecules ($>$ 10 fold) upon their binding to a specific partner (*e.g.* a large protein like a receptor).[33]

Due to the use of a confocal volume, FCS is particularly suited for miniaturization in HTS and relatively insensitive to auto fluorescent test compounds.[30] Moreover, in compound testing, the small path length of the confocal volume greatly limits any filter effects on fluorescence intensity.[33] As in FP, the requirement for large differences in mass in the assay design is a limitation to the applicability of FCS. However, it can be overcome by methods like Fluorescence Intensity Distribution Analysis (FIDA) or two-colour cross correlation derived from the original FCS concept.[17,47]

The high cost of a FCS instrument limits its use in HTS, as does the relatively long reading time (several seconds) required at the sub-nanomolar concentrations of fluorescent labels needed to accurately calculate the autocorrelation function.[33] Again, as for FP, non-specific adsorption of the labelled ligand onto endogenous proteins, inducing a large variation of diffusion coefficient, can occur and lead to impaired results. Quenching processes can also modify the tracer lifetime and have an impact on the quality of the results.

6   Amplified Luminescent Proximity Homogeneous Assay (AlphaScreen™)

AlphaScreen™ is a bead-based technology that allows the development of homogeneous assays in HTS.[35] A donor bead and an acceptor bead are used in AlphaScreen™, each one bearing a biological partner of the interaction under investigation. When the biological interaction brings the beads in close proximity, the 680 nm laser excitation of the donor beads induces a cascade of chemical reactions leading to the generation of a fluorescent signal from the acceptor bead. Upon laser excitation, the photosensitizer (phthalocyanine) in the donor bead excites ambient oxygen to the singlet state; the singlet oxygen then diffuses across to react with thioxene derivatives in the acceptor bead. Chemiluminescence is therefore generated by thioxene derivatives and activates fluorophores contained in the same acceptor bead through an energy transfer process. The fluorophores subsequently emit fluorescence in the 520–620 nm wavelength range.

Since the lifetime of singlet oxygen in a non-quenching media is around 4 µs, it can diffuse over a distance of about 200 nm. AlphaScreen™ can therefore probe interactions over distances similar to the dimensions of large biological molecules, allowing the investigation of a large number of biological targets. Because of the high concentration of photosensitizer in the donor bead, up to 60,000 singlet oxygen molecules can be generated per second resulting in a significant signal amplification after the donor bead excitation. Therefore, very low detection limits can be achieved.

Because of its homogeneous format, AlphaScreen™ can be easily miniaturized, making it suitable for screening chemical compound libraries.[7,46] However, special cautions have to be taken in handling the donor beads because of the sensitivity of the photosensitizer to ambient light.[10]

An excitation wavelength higher than the emission wavelength ensures a low assay fluorescent background by avoiding any auto fluorescence from biological media or compounds. However, AlphaScreen™ may be sensitive to other types of interferences. Antioxidants or other quenchers of reactive oxygen species like metal ions can strongly affect the emitted signal.[1,22,32] Moreover, since the AlphaScreen™ detection is only based on a fluorescence-intensity measurement, coloured compounds absorbing in the 500–600 nm wavelength range artificially decrease the AlphaScreen™ signal and therefore may be detected as false positives in HTS.[10]

7   Fluorescence Resonance Energy Transfer

FRET is a photophysical process based on a non-radiative energy transfer between a fluorescent donor and a suitable energy acceptor.[38] To occur, FRET requires an energetic compatibility between the two components of the FRET pair. Such compatibility exists if the absorption spectrum of the acceptor overlaps the emission spectrum of the donor. Since the efficiency of the FRET process is dependent on the inverse sixth power of the distance between the two molecules, donor and acceptor have to be in close proximity to allow a significant FRET.[38] For commonly used FRET pairs, a 5-nm R_0 value (corresponding to a 50% FRET efficiency) can be attained, giving an operative distance in the range of 1–8 nm.[28] Beyond this range, the FRET efficiency decreases very rapidly. FRET could therefore probe interactions over distances similar to the dimensions of biological molecules. For this analysis, both molecules involved in the interaction have to be labelled by one component of the FRET pair.

FRET allows the design of homogeneous assays in which different methods can be used to discriminate the specific signal from other signals emitted by the free donor and/or the free acceptor (Figure 3). Most commonly used FRET readouts are based on detecting variations in fluorescence intensity: a decrease in fluorescence emission from the donor,[37] increase in fluorescence emission from the acceptor (if

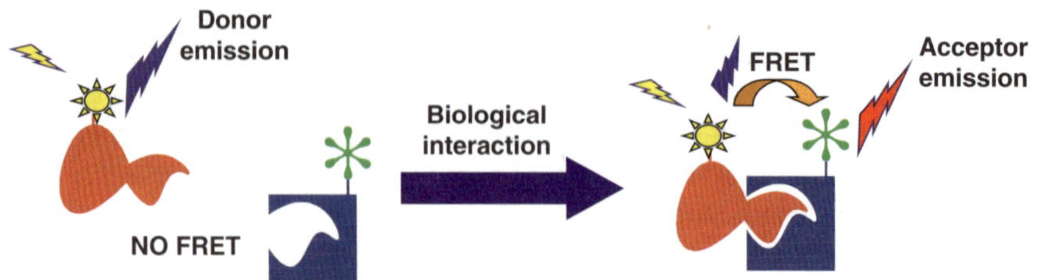

Figure 3 *Design of a homogeneous FRET assay. The spatial proximity between the donor and the acceptor induced by the binding of the biomolecules allows the FRET to occur. Upon FRET, the donor emission is decreased while the sensitized acceptor can emit fluorescence (if a fluorescent probe is used as acceptor)*

the chosen acceptor is a fluorescent probe),[8] or a change in their ratio (*e.g.* acceptor fluorescence emission/donor fluorescence emission).[29] Since the FRET process also changes the donor fluorescence lifetime (the higher the FRET efficiency, the lower the donor fluorescence lifetime), monitoring this parameter can also be used as a FRET detection method.[5] This analysis can be achieved using specific instruments based on modulation-phase techniques or on photon-counting techniques.

To screen chemical libraries, readouts based on a ratio of fluorescence intensities as well as on donor fluorescence lifetime variations can be of a great advantage to correct the FRET signal from the optical variations induced by some of the highly coloured compounds that are often found in libraries.[10] This correction can limit false-positive results in HTS; however, in practise, these analyses are difficult to implement.

The utility of ratiometric detection is strongly limited by the large overlaps in the excitation spectra and/or between the fluorescence emission spectra of the fluorescent probes commonly used as donor or acceptor in FRET. Because of these overlaps, the acceptor can be excited directly at the donor excitation wavelength and the donor can emit fluorescence at the acceptor fluorescence detection wavelength. This lack of spectral selectivity results in a significant background signal and therefore significantly reduces the signal/background ratio of the FRET assays using FRET pairs such as fluorescein/rhodamine or cyanine for *in vitro* applications, or cyan fluorescent protein (CFP)/yellow fluorescent protein (YFP) for intracellular applications.[13]

The use of the donor fluorescence lifetime as an FRET readout is currently limited by several parameters:

- The FRET process is not the only factor that can modify the donor fluorescence lifetime. The local environment of the donor or its quenching by any compound of the library under investigation can also change its lifetime significantly and therefore disturb the FRET detection.
- An accurate determination of the fluorescence lifetime of the donor molecules involved in an FRET process can be difficult to achieve, particularly when only a small percentage of the total number of donor molecules are involved in the FRET process.
- Until recently, none of the plate readers commonly used in HTS was able to determine the lifetime of a fluorescent probe.[16]

Therefore, due to the practical limitations described above, the use of a non-fluorescent acceptor (usually called a quencher) is still the most popular way to use FRET in HTS applications. Most of the time, FRET is used to probe a protease activity using a peptide substrate labelled at each end of its sequence with the fluorescent donor and the quencher. In that case, the FRET readout is an increase in the donor fluorescence emission upon peptide cleavage by the protease.[14] However, as discussed above, such FRET readouts can lead to a significant number of false positive results in HTS.

8   Bioluminescence Resonance Energy Transfer

Like FRET, bioluminescence resonance energy transfer (BRET) is based on non-radiative energy transfer between a donor and an acceptor. However, in BRET, the donor is a luminescent molecule, excited by the enzyme Renilla Luciferase (Rluc), and not a fluorescent molecule. The BRET acceptor can be a fluorescent protein like green fluorescent protein (GFP) or YFP. If the enzyme and the fluorescent protein are in close proximity (d < 10 nm), an energy transfer will occur between the Rluc substrate (coelanterazine) and the fluorescent protein, leading to emission from the later.[18]

Since Rluc and GFP recombinantly fused proteins can be expressed in living cells, BRET is an interesting tool for monitoring molecular interactions in cell-based assays. BRET has been particularly used for the study of GPCRs by probing receptor oligomerization or activation.[40]

The BRET signal is determined ratiometrically by dividing the acceptor emission intensity by the donor emission intensity. Moreover, the use of an enzyme to excite the luminescent donor (coelanterazine) avoids the use of any excitation source, thereby eliminating interferences from auto fluorescent compounds and inner filter effects at the excitation wavelength of the donor in HTS. However, like in FRET assays using ratiometric detection, the signal/background ratio of BRET assays is significantly affected by the overlap between the Rluc substrate and the GFP emission spectra. An improved BRET version, named BRET^2, has been developed to overcome this problem.[18] A coelanterazine analogue, called Deep Blue C, is used as Rluc substrate, resulting in a significant shift in the Rluc emission from 480 to 400 nm. A specially designed green fluorescent protein (GFP^2) mutant emitting at 510 nm is used as BRET^2 acceptor. Therefore, the spectral resolution of BRET^2 is significantly higher than that of BRET (110 nm instead of 50 nm), allowing its use in HTS to screen for GPCR targets.[45] However, the use of Deep Blue C as Rluc substrate gives a much lower quantum yield than coelanterazine, and as a result a highly sensitive instrument must be used for BRET^2.

9   Homogeneous Time Resolved Fluorescence

HTRF is a homogeneous technology which combines an FRET process with time-resolved fluorescence detection to probe biomolecular interactions.[27,44] This combination is made possible through the use of a long lifetime fluorescent FRET donor, a lanthanide cryptate. Cryptates are formed by the inclusion of a rare earth ion (*e.g.*

europium) into the 3-dimensional cavity of a ligand called cryptand.[3] The cryptand plays the same role as the chelate in the luminescent lanthanide chelates used in DELFIA; it protects the lanthanide ion from potential quenching by the environment, and it acts as an antenna by collecting the energy from the excitation source before transferring it to the lanthanide ion.[2] However, due to its caged structure, the cryptate kinetic stability is dramatically higher than the one of the lanthanide chelates. Rare earth chelates would be dissociated in acidic media or in the presence of divalent ions like Mn^{2+}, while rare earth cryptates can be used under drastic chemical conditions, like reverse phase chromatography in the presence of trifluoroacetic acid, and are not affected by the presence of divalent ions in the media.[6]

Europium cryptates are excited in the UV wavelength range either by a xenon flash lamp or by a nitrogen laser. Their fluorescence occurs in a wavelength range between 550 and 710 nm with typical narrow emission lines. Since the electronic transitions of the europium ion are forbidden by quantum mechanical rules, the cryptate fluorescence lifetime is exceptionally long, in the range of 100–1000 µs.

Europium cryptates can be associated with various acceptors, originally with a cross-linked allophycocyanin called XL665[25] or with small fluorescent near infrared dyes that have been selected so that their photophysical properties fit those of the europium cryptate for an efficient FRET process:

- High molar absorptivity at the cryptate emission wavelength, which allows a high transfer efficiency (R_0 value above 7 nm).
- Fluorescence emission in a spectral range (665 nm) where the cryptate signal is insignificant (Figure 4).
- A high fluorescence quantum yield, which allows an optimal FRET signal.

In FRET, the lifetime of the acceptor's emission contains a contribution equal to the donor's lifetime in the presence of energy transfer.[25,31,34] Therefore, the use of a

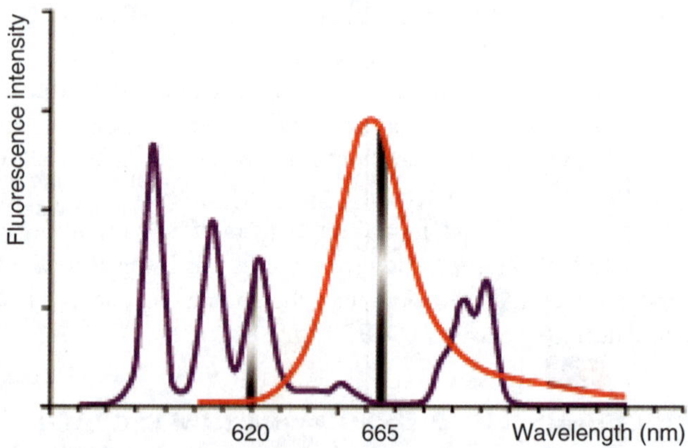

Figure 4 *Spectral selectivity of the HTRF signal. Characteristics of the emission spectra both of the europium cryptate (purple line) and the acceptor (red line) are responsible for the spectral selectivity between the acceptor fluorescence emission at 665 nm and the donor fluorescence emission at 620 nm*

long lifetime donor like the europium cryptate leads to a long-lived emission of the acceptor, which allows time-resolved fluorescence measurements. In the time domain, at the acceptor emission wavelength, a clear distinction can be made between the long-lived signal from the acceptor involved in the FRET process with the europium cryptate and the short-lived signal emitted by the freely diffusing acceptor. This temporal selectivity allows the isolation of a FRET signal free from the short-lived fluorescent background[25] (Figure 5).

The exceptionally large spectral selectivity obtained with the different FRET pairs used in HTRF allows the establishment of an efficient ratiometric signal detection. The measurement of the cryptate emission signal (usually done at 620 nm) reflects any absorption by the assay medium at the excitation wavelength. Since the acceptor signal at 665 nm is also inversely proportional to the media absorbance, it can be 'ratioed' to the cryptate signal to give a measurement, which is then independent of the media optical properties at the excitation wavelength. The ratio of acceptor signal to europium cryptate signal thus depends solely on the specific biological interactions under study.[23–25]

The high FRET efficiency obtained in HTRF assays enables a large variety of biological events to be probed. Protein/protein interactions, enzymatic activities (*e.g.* kinases or proteases), or a large number of biomarkers have been detected with a very low detection limit.[6] Like the other homogeneous fluorescent technologies like FP or FCS, HTRF can be easily automated and miniaturized down to a 1536# wells plate format for the HTS of large libraries of chemical compounds.[12]

Moreover, the HTRF signal readout minimizes the instances of compound interference in HTS: the ratiometric signal measurement overcomes the influence of highly coloured compounds while the time-resolved fluorescence detection discards

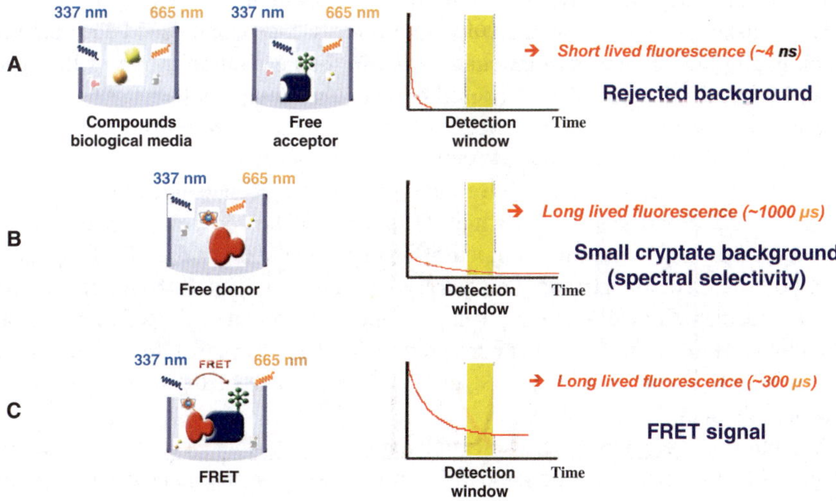

Figure 5 *Temporal selectivity of the HTRF signal. At the acceptor emission wavelength (665 nm), the FRET signal from the sensitized acceptor (C) and the small signal from the europium cryptate (B) can be easily distinguished from the free acceptor signal and from the background caused by chemical compounds or biological media (A)*

most of the fluorescence background originating from biological media and/or chemical compound autofluorescence.[10]

With HTRF, the only possible interferences that are not easily corrected are due to the inner filter effect at the acceptor emission wavelength. However, only very few compounds in the libraries absorb highly in the near infrared region. Possible cryptate quenching is even corrected to a certain extent by the signal ratio, while only acceptor quenching is not taken into account, similar to the situation with the other technologies.

10 Conclusion

As noted in the introduction, even if the fluorescence technologies are extremely sensitive, versatile, and well adapted to automation for HTS, one should not forget the intrinsic limitations of these technologies, which can be caused by three types of effects.

- Quenching generated by external interactions with the intramolecular excitation process (electron transfer, FRET, and bleaching).
- Optical interferences generated by coloured products absorbing at the excitation or emission wavelengths (inner filter effects), or light diffusion by small particles or imperfectly dissolved compounds.
- Fluorescence of the library compounds or the biological proteins.

It is important to realize that no studies have been published describing the fluorescent and absorption properties of library compounds in general, and analysis of libraries from this perspective are not routinely carried out.

Unpublished results[41] show that very high fluorescence can indeed be reached by a non-negligible number of compounds, even in the near infrared part of the spectrum. Light diffusion can also be caused by insoluble compounds.

What type of impact such compounds have on the assay response depends very much on the assay format and the specific technology used.

For example, in an FRET assay for the research of a receptor ligand inhibitor, the presence of a library compound which fluoresces at the acceptor wavelength will lead to a false negative result (compensation of the decrease of the FRET signal by the fluorescent compound); on the contrary, the same compound would give a false positive result in a cytokine assay using a sandwich format with two monoclonals (compound fluorescence will be interpreted as an FRET increase). In both cases, the use of HTRF would have corrected for the fluorescence of the compound in the acceptor channel.

It is our experience that the results obtained by using fluorescent techniques that do not use ratiometric measurement and/or time resolution should be carried out and validated with great care in order to avoid false positive or negative results.

Careful simulation of the influence of such drawbacks should be performed and proper references and thresholds set before implementing automatic interpretation of the data.

References

1. M.P. Alia and J. Matysik, *Amino Acids*, 2001, **21**(2), 195–200.
2. B. Alpha, R. Ballardini, V. Balzani, J.M. Lehn, S. Perathoner and N. Sabbatini, *Photochem. Photobiol.*,1990, **52**, 299.
3. B. Alpha, J.M. Lehn and G. Mathis, *Ang. Chem. Int. Ed. Engl.*, 1987, **26**, 266.
4. M. Allen, J. Reeves and G. Mellor, *J. Biomol. Screen.*, 2000, **5**(2), 63–69.
5. P. Bastiaens and P. Parker, *Patent WO 00/43780*, 2000.
6. H. Bazin, E. Trinquet and G. Mathis, *J. Biotechnol.*, 2002, **82**(3), 233–250 (Review).
7. R. Bossé, C. Illy, J. Elands and D. Chelsky, *Drug Discov. Today*, 2000, **1**, 42–47.
8. S. Chan, D. Arndt-Jovin and T. Jovin, *J. Histochem. Cytochem.*, 1979, **27**, 56.
9. W.J. Checovich, R.E. Bolger and T. Burke, *Nature*, 1995, **375**, 254–256.
10. J. Comley, *Drug Discov. World*, 2003, **4**(3), 91–97.
11. R.M. Eglen, *Assay and Drug Dev. Technol.*, 2002, **1**, 97–104.
12. M. Ferrer, P. Zuck, G. Kolodin, S. Mao, R. Peltier, C. Bailey, S. Gardell, B. Strulovici and J. Inglese, *J. Anal. Biochem.*, 2003, **317**, 94.
13. F. Gaits and K. Hahn, *Sci. STKE*, 2003, **2003**, pe3.
14. J. George, M.L. Teear and C.G. Norey, *J. Biomol. Screen.*, 2003, **8**(1), 72–80.
15. R. Golla and R. Seethala, *J. Biomol. Screen.*, 2002, **7**(6), 515–525.
16. P. Gribbon and A. Sewing, *DDT*, 2003, **8**(22), 1035–1043.
17. U. Haupts, M. Rüdiger, S. Ashman, S. Turconi, R. Bingham, C. Wharton, J. Hutchinson, C. Carey, K.J. Moore and A.J. Pope, *J. Biomol. Screen.*, 2003, **8**, 19–33.
18. A. Heding, *Expert Rev. Mol. Diagn.*, 2004, **4**(3), 403–411.
19. I. Hemmila, S. Dakubu, V.M. Mukkala, H. Siitari and T. Lövgren, *Anal. Biochem.*, 1984, **137**, 335–343.
20. Hightech Business Decisions, *High Throughput Screening*. 2003 survey.
21. M.E. Jolley, *J. Anal. Toxicol.*, 1981, **5**, 236–240.
22. T. Kon, T. Tanigawa, K. Hayamizu, M. Shen, T. Tsuji, Y. Naito and T. Yoshikawa, *Redox Rep.*, 2004, **9**(6), 325–330.
23. M. Mabile, G. Mathis, E.J.P. Jolu, D. Pouyat and C. Dumont, *Method for Measuring the Luminescence in a Luminescent Assay*. 1991 WO 9213264. US 5527684.
24. M. Mabile, G. Mathis, E.J.P. Jolu, D. Pouyat and C. Dumont, *Apparatus for Measuring Luminescence in a Luminescent Assay*. 2002 US 6 352 672.
25. G. Mathis, Clin. Chem., 1993, **39**(9), 1953–1959.
26. G. Mathis, F. Socquet, M. Viguier and B. Darbouret, *Anticancer Res.*, 1997, **17** 3011–3014.
27. G. Mathis, *Clin. Chem.*, 1995, **41**(9), 1391.
28. J.J. Meng, M. Rojas, W. Bacon, J.T. Stickney and W. Ip, *Method. Mol. Biol.*, 289, **34**, 341–357.
29. A. Miyawaki, J. Llopis, R. Heim, J. Mc Cafferty, J. Adams, M. Ikura and R. Tsien, *Nature*, 1997, **388**, 882.
30. K.J. Moore, S. Turconi, S. Ashman, M. Ruediger, U. Haupts, V. Emerick and A.J. Pope, *J. Biomol. Screen.*, 1999, **4**(6), 335–353.

31. L.E. Morrison, *Anal. Biochem.*, 1988, **174**, 101–120.
32. Packard Website, AlphaScreen™ Technology – *Application Note AC-012, Analysis of Potential Compound Interference of AlphaScreen™ Signal*, http://www.perkinelmer.co.jp/tech/tech_ls/protocol_collection/asc-012.pdf, 2001.
33. A.J. Pope, U.M. Haupts and K.J. Moore, *DDT*, 1999, **4**, 350–362.
34. P.W. Schiller, in *Biochemical Fluorescence Concepts*, Vol 1, F.C. Chem and H. Edeldoch (eds), Marcel Dekker, New York, 1976,285–303.
35. R. Seethala and F. Prabhavathi, *Handbook of Drug Screening*, Marcel Dekker, New York, 2001, 106.
36. E. Soini and T. Lovgren, *Crit. Rev. Anal. Chem.*, 1987, **18**, 105.
37. D. Sokol, X. Zhang, P. Lu and A. Gewirtz, *PNAS*, 1998, **95**, 11538.
38. L. Stryer, *Annu. Rev. Biochem.*, 1978, **47**, 819.
39. E. Terpetschnig, H. Szmacinski and J.R. Lakowicz, *Anal. Biochem.*, 1995, **227**, 140–147.
40. S. Terrillon, T. Durroux, B. Mouillac, A. Breit, M. Ayoub, M. Taulan, R. Jockers, C. Barberis and M. Bouvier, *Mol. Endocrinol.*, 2003, **17**(4), 677.
41. E. Trinquet and U. Böhmer, to be published.
42. S. Turconi, R.P. Bingham, U. Haupts and A.J. Pope, *DDT*, 2001, **6**, 12, S27–S39.
43. T.C. Turek-Etienne, E.C. Small, S.C. Soh, T.A. Xin, P.V. Gaitonde, E.B. Barrabee, R.F. Hart and R.W. Bryant, *J. Biolom. Screen.*, 2003, **8**(2), 176–184.
44. E.F. Ulmann, *J. Chem. Ed. Chem. Visc. Edu.*, 76, **6**, 781–788.
45. M. Vrecl, R. Jorgensen, K. Pogacnick and A. Heding, *J. Biomol. Screen.*, 2004, **9**(4), 322.
46. A. Von Leoprechting, R. Kumpf, S. Menzel, D. Reulle, R. Griebel, M.J. Valler and F.H. Buttner, *J. Biomol. Screen.*, 2004, **9**(8), 719–725.
47. T. Winckler, U. Kettling, A. Koltermann and M. Eigen, *PNAS*, 1999, **96**, 1375–378.
48. F.B. Wu and C. Zhang, *Anal. Biochem.*, 2002, **311**, 57–67.
49. X. Zhao, S. Panah, H. Pham, J. Araujo and R. Eglen, www.discoverx.com.

The Use of Genetically Engineered Cell-Based Assays in in-vitro Drug Discovery

RENATE SCHNITZER AND WOLFGANG SOMMERGRUBER

Drug Discovery, Boehringer Ingelheim Austria, Dr. Boehringergassse 5—11, A-1121 Vienna, Austria

1 Introduction

Cell-based assays are of increasing importance in drug discovery. The completion of the sequence of the human genome has delivered a large number of potential novel targets, and cell-based assays provide a good way to elucidate their functions in relevant pathways in the cellular environment. With the increase of compound collections to hundreds of thousands of structures that can be screened for new drugs in a high-throughput mode, the identification of the most promising lead candidates as early as possible is crucial. This analysis includes a more detailed validation of hits with regard to toxicity, efficacy, and selectivity on the cellular level. Novel technological advances during the last few years allow many of these questions to be addressed with good sensitivity and reasonably high throughput in the cellular environment. These cell-based assays not only cover reporter gene assays to investigate signaling events and receptor activities, but also cover assays that quantify protein-protein interactions and track the relocalization of proteins to different intracellular compartments, although the latter can only be quantified in imaging-based assay systems. Until recently, cell-based assays have been primarily used in target validation, toxicity, and secondary testing; however, the scope of the assay and miniaturization technologies that are now available also allow a broader use for high-throughput screening (HTS). Although only 35% of HTS assays were cell-based in 1999, an increase to 50% is expected for 2005 according to a recent survey.[1] A large proportion of these assays are for (GPCRs) G protein-coupled receptors, the major class of drug targets, for which a variety of cellular functional assays has become available. The use of genetically engineered cell lines for such systems is usually more promising, since limitations encountered with untransfected cell lines, which express the target of interest naturally often cause problems.[2] This article will review recent developments and important approaches in the generation and use of genetically engineered cells for drug discovery. It will not cover the broad, progressing field of cellular fluorescent imaging, which makes it possible to monitor the spatial organization of molecular interactions and compartmental translocation events by the use of genetically encoded green fluorescent protein (GFP) fusion proteins, and which also offers great potential for *in vivo* studies.[3]

2 Genetic Engineering for Cell-Based Assays

2.1 Expression Systems

The cell lines used in drug discovery for target and lead identification and optimization are usually immortalized cell lines, which, depending on the assay system to be developed, are transfected with different reporter genes, target fusion proteins, and additional accessory proteins, *e.g.* a specific type of G protein. Since untransfected cell lines often do not express sufficient amounts of the target to obtain high-quality assays, transfection with the target of interest is also required in these cases. Another approach is to transfect with transcription factors that activate the expression of the endogenous genes of interest.[4] In contrast to transient gene expression, which is mainly used for preliminary testing of expression vectors, the establishment of stable cell lines that constitutively express the gene of interest depends on the stable integration of a plasmid into the chromosome. However, it is also possible to generate a plasmid that harbors vectors extrachromosomally, such as expression plasmids carrying the Epstein-Barr virus nuclear antigen (EBNA-1) and the origin of replication (oriP). Examples of this approach are vectors pREP, pCEP4, or pEBVhis (Invitrogen), which can be maintained episomally in primate and canine, but not in rodent, cell lines.[5] Furthermore, an episomal replicating vector has been described that lacks expression of viral proteins, thus avoiding cell transformation. The vector contains the SV40 origin of replication and the scaffold/matrix-attached region (S/MAR) from the human interferon-β gene. S/MARs are DNA sequences associated with chromosomal origins of bidirectional replication. The vector was shown to replicate at very low copy numbers (<20) in Chinese hamster ovary (CHO) cells and was stably maintained without selection for more than 100 generations.[6] A comprehensive listing and description of currently available vectors for gene expression in mammalian cells is found in the article by Makrides.[7]

An interesting and more recent expression system for the development of cell-based assays is based on BacMams.[8] These recombinant baculoviruses containing mammalian cell-active expression cassettes seem to be an efficient strategy to speed up assay development. These viruses are produced in insect cells and transiently express but do not replicate in transduced mammalian cells. The expression level can be well controlled by titrating the amount of virus. Highly reproducible transient expression levels, which are a prerequisite to use transient transfection for HTS, might thus be an attractive alternative for some approaches.

2.2 Choice of Cell Line and Promoter

The choice of the host cell line has a significant impact on gene expression levels. Although, many promoters are transcriptionally active in a wide range of cell types, a defined cell line often has to be used for the generation of a cellular reporter system because of its specific transduction cascade. On activation, the resulting signal leads to transcriptional activation of only those downstream target/reporter genes containing the corresponding transcription factor binding motifs in their promoter regions. Therefore, the specific responsiveness of cell lines must be carefully considered prior to the construction and use of expression vectors.[9,10] For example, the widely used cytomegalovirus (CMV) promoter exhibits low transcriptional activity in hepatocytes.[11] Strong constitutive and rather unspecific promoters are the adenovirus major late promoter, the human CMV immediate early promoter, and the SV40 and Rous sarcoma virus promoter. For many cell-based screening assays, a strict cell type specific system is required. Therefore, only cell lines that are derived from cancers and are dependent on the particular growth-signaling cascade can be used.[12] Additionally, transcriptional activity of weak promoters can be enhanced without loss of tissue specificity. Nettelbeck and colleagues developed a strategy for enhancing the transcriptional activity of weak cell type-specific promoters.[13] In this approach, a cell type-specific promoter was used to drive the simultaneous expression of the gene of interest and an artificial transcriptional activator to stimulate transcription through binding sites in the promoter ("positive feedback loop"). A fusion transcription factor composed of the Herpes simplex virus VP16 transcriptional activation domain and the DNA-binding domain of LexA was designed, and it was shown that the transcriptional activity of different promoters could be increased ~10- to 100-fold. For many HTS screens, an inducible promoter system is needed either because the specific inducing signaling cascade is itself the molecular target (*e.g.* vascular endothelial growth factor receptor (VEGFR) signaling) or because the production of a gene of interest has to be correlated with a phenotypical read-out in a high-content screen (*e.g.* surface antigens and their impact on migration of tumor cells). Therefore, inducible promoters should exhibit a minimal level of basal transcriptional activity and should be capable of substantial induction via natural (*e.g.* addition of growth factor agonists) or artificial non-toxic inducers (*e.g.* tetracycline or doxycycline) in a simple and cost-effective manner. The tetracycline-induced expression system is one of the most widely applied inducible systems today.[14] Similarly, the functionality of the bacterial lac operator-repressor system in mammalian cells has been exploited for the inducible expression of heterologous genes in mammalian cells facilitating a high specificity in target gene regulation and the level of induction of gene expression.[15,16] The disadvantage of this system is the cytotoxicity of the artificial inducer IPTG (isopropyl-β-D-thiogalactopyranoside). Besides promoter and enhancer regions, additional important factors influence the level of expression, such as the presence of introns, the polyadenylation signals, transcriptional terminator sequences, translational control elements, such as the 5′- and 3′-non-coding region, the Kozak sequence for optimal translation initiation, translational termination codons, species-specific codon usage in general, and mRNA stability. Sometimes it may also be valuable to amplify the gene of interest to generate a more efficient reporter signal. Amplifiable expression systems using CHO cells have been used extensively. There are many genes that confer drug resistance upon amplification.[17] The most widely used amplification systems are based on the dihydrofolate reductase and glutamine synthetase genes. By growing cells in the presence of increasing amounts of selection drugs, it is possible to amplify significantly the copy number of the cotransfected gene of interest. The wide variation in the level of expression and amplification of the transfected gene depends on the chromosomal site of plasmid integration.[18] For example, in CHO cells, gene amplification frequency in one transformation experiment varied 100-fold; genes inserted near a centromere resulted in chromosome instability and rearrangements.[19]

2.3 Chromosomal Integration Site

Generally, a single expression vector utilizing a construct with both the reporter gene and the selectable gene marker on the same plasmid, hence the same chromosomal integration site, does not appear to be advantageous over the use of separate vectors, which result in different integration sites. However, in our experience, various cell-based reporter systems exhibit higher and more stable expression when we utilize a two-vector-based system for selection based on permanent expression (unpublished observation). To circumvent a possible silencing positional effect and the clonal variability of expression levels, site-specific DNA recombination systems have been used for directed DNA-transfer techniques.[20] Based on theses studies, Invitrogen has developed Flip-In™ host cell lines bearing a single Flp recombinase target site (FRT) located at a transcriptionally active genomic locus. Isogenic expression cell lines are then generated by co-transfection with an Flp-In™ expression vector and the Flp recombinase expression vector, pOG44. The transiently expressed recombinase integrates the expression vector at the genomic FRT site.

3 Reporter-Based Assays

Reporter-based assays measure the activity of a gene, known as the reporter gene, which is synthesized in response to a specific promoter and associated response elements. Fusion of a reporter gene with an inducible promoter, such as that responding to specific signal transduction cascades, specific chemicals (as inducers), and various types of stresses, provides a means by which the expression levels of a gene native to the promoter can be measured. Due to their relative simplicity, reporter-based assays are widely used in academia and industry to investigate gene transcription and regulation, translation, interactions of pathogen with host cells, signaling pathways, RNAi silencing effects, and to screen for new drugs. Since the activity of GPCRs, which represent a major class of drug targets, initiates signaling events in the cell, reporter-based assay approaches can also be used to measure receptor activity. This approach not only allows the identification of agonists and antagonists, but also makes it possible to distinguish partial agonists,[21] inverse agonists and allosteric inhibitors,[22] which cannot be identified in classical radioactive receptor-binding assays. This ability, in addition to the avoidance of radioactive detection reagents, are marked advantages of the functional cellular approach. Reporter-based assays necessarily require genetic engineering of the cells with a reporter gene construct. Depending on the target of interest and their endogenous expression levels in the cell, transfection with additional genes may be necessary, *e.g.* a specific GTP-ase interacting with a distinct GPCR in the signaling cascade. Usually, the reporter gene construct consists of a gene encoding a reporter molecule that can easily be measured and quantified. The reporter gene is under the control of a promoter sequence linked to the transcriptional control region of interest. Specificity and sensitivity of the system are controlled by the choice of the promoter and the transcriptional control region, which determine the basal expression level and the degree of transcriptional response. Other factors influencing the sensitivity of the system are the stability of the reporter molecule in the cell and the dynamic range of the reaction. A short half-life of the reporter molecule is important for minimal baseline accumulation of the signal and hence a high dynamic range. Enzymes as reporters have the advantage of signal amplification, which improves sensitivity and dynamic range. The ideal reporter enzyme is not endogenously produced in the cell, thus eliminating background expression. These criteria are discussed for the most important reporter genes: secreted alkaline phosphatase, β-galactosidase, GFP, luciferase, and β-lactamase, out of a larger number of available genes that have been reviewed previously (*e.g.* β-glucuronidase, human growth hormone).[23]

3.1 Chloramphenicol Acetyl Transferase, Secreted Placental Alkaline Phosphatase, β-Galactosidase

The longest-established reporter gene system is the bacterial enzyme chloramphenicol acetyl transferase (CAT).[24] CAT can detoxify chloramphenicol (an inhibitor of prokaryotic protein synthesis) by catalyzing the transfer of acetyl moieties from acetyl CoA to the 3-hydroxyl position of chloramphenicol. The enzyme is stable, and there is no endogenous expression in mammalian cells. However, it has a rather high detection limit of 5×10^7 molecules and a low dynamic range. Furthermore, it is a radioactive, labor-intensive assay, which makes it difficult to automate for HTS. Due to these disadvantages, its present use in drug discovery is rather limited. The activity of SEAP (secreted placental alkaline phosphatase),[25] which is a mutated form of the human placental enzyme, can be monitored easily by colorimetric, fluorescent, and chemiluminescent detection systems. The chemiluminescence assay is based on the use of 1,2-dioxetane substrates. Its sensitivity in several orders of magnitude is greater than that of the other detection methods and comparable to that of the luciferase reporter system. SEAP is secreted into the medium, which allows repeated measurements to be taken from one sample. The cells remain intact and viable for further analyses.[26] The gene for β-galactosidase, commonly referred to as lacZ, forms a tetramer and is responsible for the cleavage of lactose into galactose and glucose. It is also slightly promiscuous, resulting in enzymatic cleavage of other substrates. Different substrates for detection have been developed[27,28] including those for fluorometric analysis[29] or laser-induced fluorescence detection of β-gal protein levels. A luminescence-based assay even lowers the detection limit of β-gal to ~2 fg and is about three times more sensitive than the above-mentioned fluorescence assays.[30] The limitations of this reporter are the lysis of cells required prior to the assay and the addition of expensive chemicals.

3.2   Green Fluorescent Protein

Another possible reporter gene is GFP. GFP from the jellyfish, *Aequoria victoria*, transforms the blue chemiluminescence of aequorin into green fluorescent light with an emission maximum at 508 nm.[31] GFP has been studied extensively by mutational analysis leading to different variants of this protein that show increases in thermostability, intensity, and shifts within the fluorescence spectra.[32] The universal application of GFP to highly different organisms and the fact that it requires no substrate for its activity makes it the reporter of choice in eukaryotes, especially for the analysis of developmental processes and as a biosensor for environmental biomonitoring.[33] However, as GFP usually requires a long time for the fluorophore to form, more time is needed to evaluate differences between the expression levels of samples. On the other hand, the robustness of this protein also limits its denaturation. Due to the lack of enzymatic signal amplification, the signal-to-background ratio of GFP is usually lower than that of enzymatic reporter genes and the assays are less sensitive, which clearly restricts the application of GFP.

3.3   Luciferase

A number of advantages to the luciferase gene from *Photinus pyralis*[34] make it still the most commonly used reporter gene. In a two-step reaction, the enzyme catalyses the oxidation of luciferin, which results in an emission of a short flash of light at 550–570 nm. Fast decay of the signal, within seconds, initially required injection of ATP and luciferin into the well immediately before reading; however, optimized reaction buffers have since been developed that lead to a prolonged half-life of the signal to a so-called "glow reaction," which is stable for more than 1 h. Appropriate buffers are commercially available. The assay protocol is very simple and only requires the addition of the reaction buffer, which contains ATP and luciferin and lyses the cells. Luciferase is not endogenously expressed and has a short half-life of 3–4 h.[35] Hence, there is no intrinsic background in cells, and the short half-life makes it possible to track the reduction of expression efficiently. The assay is very sensitive, with a detection limit in the range of 10^3 molecules and a large dynamic range over 7 orders of magnitude. It is thus amenable to miniaturized assay formats, *e.g.* 384-well and even 1536-well plates. Renilla luciferase from the sea pansy *Renilla reniformis* is also a monomeric enzyme that catalyses the oxidation of coelenterazine to give a blue light of 480 nm. Coelenterazine can easily cross eukaryotic cell membranes;[36] hence, the activity of Renilla luciferase can be monitored in live cells, thus eliminating the need for cell lysis of cells and enabling multiple measurements to be made. However, this system has a lower signal intensity than beetle luciferase and, moreover, coelenterazine shows autoluminescence, which results in a background signal. Very recently, the novel Renilla luciferase substrate EnduRen™ has been introduced by Promega. This compound is an inactive form of coelenterazine, which is metabolized to the active substrate by intracellular esterases. It is therefore significantly more stable, which should allow prolonged real-time measurements and, furthermore, autoluminescence is significantly reduced. Despite this improvement, Renilla luciferase still has a lower sensitivity than the firefly luciferase, but it is very useful in a dual reporter assay together with beetle luciferase, where the activity of both the enzymes is measured in two consecutive steps. Dual bicistronic reporter gene assays with simultaneous measurement of the activities of both the reporter genes can also be generated by using click beetle luciferase as the second reporter gene. Several luciferases from the click beetle *Pyrophorus plagiophtalamus*[37,38] with luminescence over a wavelength range from 546 to 593 nm are available for this purpose. Dual bicistronic reporter assays allow the activity of two distinct reporter genes driven by different promoter elements to be monitored in one experiment. This type of assay system can be used to measure the activity of two different signaling pathways or receptors at the same time, and it can also be applied for a ratiometric analysis of the activity of two differentially regulated reporter genes and thus normalize the activity of one reporter gene to the second one. By normalization to the number of cells per well, this approach can control for the experimental variability resulting from *e.g.* edge effects or for viability, which may be affected by toxicity of the compounds being tested. This approach is also a valuable tool in molecular biology to investigate ribosomal frameshifting,[39] replication, or translational regulation. Dual reporter assays were used to characterize the regulation of hepatitis C virus translation[40] and to develop a HTS assay for inhibitors of the West Nile virus (MNV) replication.[41]

3.4 β-Lactamase

Bacterial β-lactamase is another enzyme with characteristics desirable for use as reporter gene[42] and for which colorimetric substrates have been known for a long time. The β-lactamase cell-based assay system couples molecular and biological events to a fluorescence resonance energy transfer (FRET)-based detection method in single live cells (GeneBLAzer™ technology, Invitrogen). Since there is no mammalian counterpart, its expression is background-free. Its half-life is similar to that of beetle luciferase. The development of a membrane-permeable, self-quenched, fluorescence resonance energy-transfer substrate a few years ago allows visualization of enzyme activity in living cells[43] and also delivers a ratiometric readout. The substrate consists of a fluorescein moiety as the acceptor fluorophore attached to a coumarin moiety, which is the fluorescence donor. Excitation of the donor at 409 nm leads to emission of green light (529 nm) from the uncleaved substrate. β-lactamase cleavage releases fluorescein and leads to the emission of blue light (447 nm). The ratio of the emission at 447 nm to that at 520 nm is a concentration-independent measure of the extent of the reaction. An esterified derivative of the substrate is non-fluorescent and only metabolizes to the fluorescent version when cleaved by cytoplasmic esterases in the cell. The high sensitivity of the assay, which requires only 100 reporter molecules per cell, allows flourescene-activated cell sorter (FACS) analysis of cell suspensions and clonal selection of transfected cells, thus significantly shortening the time to genetically engineer the reporter cell line. The ratiometric readout normalizes the signal to cell number, which is especially useful for compound screening. As described before, it can correct for edge effects, which can occur during longer incubation times of plated cells and for cells of different viability. These characteristics together with the very high sensitivity of the differential assay make it amenable to remarkable miniaturization beyond the

384–3456-well plate format.[44] The properties described before make the β-lactamase a very attractive reporter gene, especially for screening purposes, since edge effects and toxic effects of test compounds can significantly impact the test results. However, utilizing the β-lac reporter system, we observed that the substrate CCF2-AM exhibits a general instability in aqueous solution increasing the background fluorescence and resulting in a non-β-lac-induced conversion of the substrate in the cell, which leads to an increase of the blue fluorescence even in the absence of β-lac. However, this effect depends very much on the cell type used for the assay.

3.5   Examples of Applications

In cases where the activation of one specific signaling cascade is under investigation, a synthetic promoter containing the respective response element may be the best approach. Multiple copies of the response element are usually coupled to enhance the regulatory effect. A whole variety of promoter elements that are responsive to the activation of different second messenger pathways are available. One such response element is TRE (tissue plasminogen activator (TPA) responsive element), which is activated by the signal transduction pathways mediated by PKC.[45] GPCRs coupling to $G\alpha_i$ and $G\alpha_S$ can be investigated by CREs (using cAMP responsive elements),[46,47] which bind CRE-binding (CREB) protein phosphorylated by PKA.[35] Those coupling to $G\alpha_q$ signal through calcium, which activates the NFAT response element.[48] Genetic engineering of a luciferase reporter cell line under the control of multiple copies of the hypoxia-responsive element (HRE) allow one to investigate the totally different mechanism of regulation of the response to hypoxia and to search for appropriate inhibitors.[49,50] This mechanism is involved in the regulation of angiogenesis and anaerobic metabolism. Alternatively, natural promoter sequences can be used. The c-fos[51] promoter contains several different responsive elements, which are regulated by various signaling pathways and hence can reflect the activity of different classes of receptors like GPCRs (histamine receptor, gastrin-releasing peptide receptor, unpublished results) and growth factor receptors (EGF, HGF, unpublished results). We have also successfully used the ICAM-1 receptor, which confers response to several signaling pathways,[52,53] for the construction of reporter gene assays for different GPCRs, *e.g.* NK2 and 5-HT$_2$.[54] Reporter gene assays can also be a tool in the elucidation of virus host interactions and the identification of novel drugs for intervention. Reporter gene constructs carrying viral IRES (internal ribosome-entry site) sequences, such as that derived from the hepatitis C virus (HCV), allow the regulation of virus translation to be examined in host cells and provide the base for an assay to screen for virus-specific translation inhibitors.[40] Engineering reporter genes into virus replicons of MNV[41] or HCV[55] generate powerful screening assays to identify specific inhibitors of virus replication. The technology is also applied to identify modulators of orphan GPCRs.[56] An artificial multiple regulatory element containing the major response elements, through which different G-proteins can signal, allows a reporter assay to be set up without knowledge of the coupling mechanism of the orphan GPCR.[57] The examples given above demonstrate the scope of applications of reporter gene assays in drug discovery, covering the whole process from gene to function, target identification and validation to lead identification, which includes primary HTS. Given the large number of compounds screened per target, which typically lies in the range of 600,000 to a million, the prerequisites for HTS suitability are easy automation in a high-throughput mode and miniaturization to reduce reagent consumption and cell-culturing efforts. Since reporter gene assays, like those based on luciferase, require only one reagent addition step prior to measurement, they are readily amenable to automation. Miniaturization to 384-well plates and even 1536-well plates and beyond have been possible,[57,58] but assay noise and variability due to edge effects, uneven distribution of cells and sensitivity of expression levels toward cell density and culture conditions are specific hurdles to be surmounted with cellular assays. Highly standardized cell-culturing procedures and suitable equipment on the automated screening system are required to cope with this challenge. Since the β-lactamase reporter gene delivers a ratiometric measurement that normalizes the result to the number of viable cells, this reporter gene seems to be specifically suited for HTS. One recent approach to obtain a unique stock of cells of uniform quality for HTS is to combine the BacMam technology with frozen stocks of cells for plating. This approach makes it possible to prepare cells in bulk and store them for up to 6 months before the start of HTS. Cell supply and HTS are thus uncoupled and assay quality is improved as a result (oral presentations at SBS conference 2004, Orlando by D. Finnigan, GlaxoSmithKline, and C. Cowan, Norak Biosciences). However, the examples referred mostly to cell lines measured by the fluorescent imaging plate reader (FLIPR) platform (Molecular Devices). The strategy is unlikely to be applicable to all kinds of cell lines, but may be an attractive approach in a number of cases.

One important issue with respect to the development of cell-based assays is the fact that cells display a multiplicity of targets in addition to the target of interest. This potential for additional interactions has to be considered specifically when using cellular assays in HTS to identify new hits. As the majority of compounds in a compound collection cannot be expected to be selective for the target of interest, many of them will hit additional targets within the monitored signaling cascade. These off-target interactions will affect the test results to different degrees and will usually lead to an increased number of false positive hits. It is therefore necessary to profile identified hits in another independent assay format to verify whether the compound has the desired mode of action. Reporter gene assays can be designed to monitor a gain or a loss of signal. Compounds interacting with molecules within the cell, which induce apoptosis, necrosis, or lead to reduced growth will reduce the measured signal as well. They will therefore show up as hits in assays with a loss of signal readout. As a result, significantly higher hit rates are found for assays with a readout involving reduction in the signal, as compared to assays that monitor a gain of signal (see Table 1). The effect becomes stronger with increasing incubation time and may already be obvious as early as 1–2 h of incubation. Generally speaking, assays with fast readouts, in the range of seconds or minutes, *e.g.* Ca^{2+} release, suffer less from this problem. In the β-lactamase reporter assay, interference from cytotoxic substances should also be reduced, since the results are normalized to the number of viable cells. There are several other strategies to reduce the number of false positive hits.

One approach is to test the compound in a control cell line in parallel. Ideally, this involves the same cell line in which the reporter gene is induced differently. Another option is to use a different cell line of the same origin as that of the target. It should express the same reporter gene driven by the same signaling cascade but involve another receptor, in case of GPCRs, or be driven by a promoter not regulated by the signaling cascade of interest. If two different cell lines are used, it is important that

Table 1 *Primary hit rate of different luciferase reporter assays (los: loss of signal; gos: gain of signal) at the given hit criterion*

Assay	Type	Incubation time (hours)	Compounds tested	Hit criterion	Hit rate (%)
Translational regulation	los	24	422100	<50%	17
Signaling	los	10	480959	<50%	11.6
Growth factor receptor	los	5	254767	<50%	11
GPCR	los	5	460047	<50%	7.4
GPCR	gos	12	565259	>200%	0.04

both show similar sensitivity. Another strategy is to run a cytotoxicity assay in parallel on the target cell line and simply eliminate cytotoxic hits.

4 Assays to Measure Intracellular Calcium

Two methods to measure changes in intracellular calcium concentration in a high throughput mode in microtiter plates have also been developed.[59–61] Their advantage in comparison to the reporter-based assays is that the concentration changes can be monitored very rapidly, within seconds or minutes after activation of the signaling cascade, and the measured interaction point is not as far downstream as a reporter gene response. Therefore, toxic effects of compounds and unspecific interactions at the different steps of the signaling cascade are reduced. These technologies offer an attractive alternative in cases where the signaling cascade results in a change of intracellular calcium concentration. The activity of GPCRs coupling to G_q through the phospholipase C-signaling cascade (*e.g.* 5-HT2 receptor subfamily, bombesin receptor), which leads to changes of the intracellular calcium concentration, can therefore be monitored with these approaches.[62] Usually the receptor of interest is overexpressed in the test cell line. Cotransfection together with promiscuous and chimeric G-proteins like $G_{\alpha 16}$ or G_q with the 5'C-terminal amino acids from G_S or G_i allows these approaches to be applied to Gi-coupled receptors[59,60] and has thus significantly broadened the scope of receptors that can be investigated by this technology. One approach is based on cotransfection with apoaequorin. This protein forms a complex with coelenterazine, which is added as cofactor to the reaction. Binding of calcium leads to an oxidation reaction generating coelenteramide, CO_2, and luminescence at 469 nm.[63] The other technology uses the addition of fluorescent calcium indicators such as FLUO-3 and FLUO-4,[64] which change fluorescence intensity dependent upon changes in calcium concentration, or a novel "no-wash calcium assay kit (Molecular Devices)." The convenience of the second approach is that it does not require the cotransfection of apoaequorin. This strategy, together with the development of a powerful FLIPR(Molecular Devices), equipped with an argon laser, a cooled CCD camera, fast simultaneous liquid transfer, plate handling capabilities, and data evaluation software, enables homogeneous kinetic cellular fluorescent assays with time resolution of seconds in formats down to 1536-well plates in high throughput. This technology has thus become the first choice for high-throughput applications in the field of intracellular calcium measurement. It will be very useful to identify and characterize endogenous ligands and to screen for novel inhibitory and activating compounds for pharmaceutical development.

5 Assays to Monitor Protein–Protein Interactions

During the last few years, several technologies have been developed for the study of protein–protein interactions in living cells in real time. The assays are homogeneous and can be run in microtiter plates in higher throughput. The identification and elucidation of the function of molecular interactions helps to elucidate cellular function. Furthermore, the technology offers the potential for identifying novel inhibitors and activators through HTS. These approaches thus provide a novel approach with considerable potential in the development of novel pharmaceuticals.

5.1 Bioluminescence Resonance Energy Transfer and Fluorescence Resonance Energy Transfer

The general principle is based on the transfection of the cells with genes of the interacting proteins fused to two different fusion partners. The latter are molecules between which resonance energy transfer or enzyme complementation occurs if both are in close proximity. One of these technologies is FRET, for which a number of color variants of GFP fused to the proteins of interest have been used. Another very important technology for intracellular analysis is BRET (bioluminescence resonance energy transfer), which occurs naturally in marine organisms. This technology was first reported in 1999 as a tool to study the interaction of circadian clock proteins *in vivo* in bacteria.[65] Resonance energy transfer takes place between donor and acceptor molecules if the emission spectrum of the donor and the absorption spectrum of the acceptor overlap sufficiently. This effect is inversely related to the distance with the power of six and thus only works efficiently at distances in the range of 1–10 nm,[66] a distance which can be taken as evidence of molecular binding. In contrast to FRET, where the donor molecule is fluorescent and must be excited with monochromatic light, BRET uses a bioluminescent donor molecule, which does not require an external light source. The most frequently used donor–acceptor pairs are Renilla luciferase and different color variants of GFP from *Aequoria victoria* that have been developed during the last few years. Renilla luciferase generates light with a wavelength of 480 nm as a result of the oxidation of coelenterazine to coelenteramide, as described before. A larger Stokes shift can be obtained by using a modified version of coelenterazine, coelenterazine DeepBlueC™, (emission peak 395 nm), which gives better sensitivity in the assay. The signal is measured as the ratio of emission at 510 nm to that at 395 nm. The disadvantage, however, is a reduced quantum efficiency of DeepBlueC™. The recently introduced, novel Renilla luciferase substrate EnduRen™, mentioned above, reduces autoluminescence and allows longer real-time measurements due to the higher stability of the substrate. FRET has been applied in the investigation of conformational changes within GPCRs upon agonist binding by incorporating two different color variants of GFP, cyan- and yellow-flourescent proteins (CFP and YFP, resp.), into different parts of the receptor.[67] It is a better tool than

BRET for monitoring interactions in single living cells, due to the better spatial resolution as a consequence of the higher light emission. This technology requires the application of fluorescence spectroscopy on the microscopic level, which will not be discussed in more detail here. Despite the lower light emission of luciferase as compared to fluorescent proteins, BRET has several advantages over FRET. First of all, it does not require an excitation light source, which also eliminates the problem of photobleaching of the donor molecule. Furthermore, the illumination may induce auto-fluorescence of intracellular components, which can result in a significant background signal. BRET can be measured in a fluorometer and does not need imaging technology for detection. This application has been applied successfully *in vitro* and *in vivo* to identify and investigate a variety of molecular interactions, *e.g.* hetero and homodimerization of GPCRs, receptor tyrosine kinases and nuclear receptors,[68,69,70] the interaction of transcriptionally active nuclear matrix proteins[71] and the interaction of receptors with β-arrestin,[69] which only binds to the activated form of the receptor. BRET plays an increasingly important role in the GPCR field, especially in probing receptor oligomerization. A recent overview of the applications of BRET and FRET in the area of GPCR complex formation is given by Pfleger and Eidne.[72] Agonist-mediated β-arrestin translocation by GPCR occurs with most GPCRs independent of the class of G-proteins to which they couple. The interaction of GPCRs with β-arrestin is therefore a general method to study the activity of GPCRs, irrespective of their activated signaling pathway. Based on this approach, a universally applicable screening assay has been developed.[73] Apart from being independent of the signaling pathway, this approach has several additional advantages over the traditional reporter-based assays. First of all, it is a ratiometric measurement and thus not dependent on the number of cells per well, which significantly reduces variability. Because of this normalization, and because interaction with the targets of interest is monitored directly, the number of false positive hits is reduced in comparison with reporter gene assays, which measure the signal of a signaling cascade with many more steps. However, the establishment of an appropriate high-quality HTS assay needs significant optimization work. Since the resonance energy transfer efficiency is dependent on the proper orientation of donor and acceptor molecule, an unfavorable geometry can lead to a very poor signal. Therefore, different combinations of N- and C-terminal fusion constructs usually have to be tested to identify those that deliver the best result. Furthermore, the DNA amounts of donor and acceptor fusion constructs have to be titrated carefully to determine the ratio that leads to the optimal energy transfer signal. One general concern is the requirement that the fusion proteins be overexpressed, which can result in unnatural interactions with other molecules within the cell and may thus lead to non-physiological events and a high-signal background. Since donor and acceptor molecules are rather large, they could also change the conformation and binding behavior of the proteins of interest. Therefore, proper controls are needed to ensure that the monitored signal has the desired specificity.

5.2 Enzyme Complementation

Another technology for the study of protein–protein interactions in cells is based on intracistronic enzyme complementation. β-galactosidase is the most frequently used enzyme for this approach. The pair of target proteins is fused to two different inactive deletion mutants of β-gal, an N-terminal deletion delta alpha and a C-terminal deletion delta omega. These two mutants have a very low affinity toward each other and can only form a stable complex if the fused target proteins interact. In the complex, both mutant forms complement each other and form an active enzyme. Since very sensitive fluorescent and chemiluminescent substrates are available for β-gal, the enzyme activity can easily be measured in standard laboratory readers and the assay is amenable to automation. β-galactosidase enzyme complementation has been used to investigate EGF receptor dimerization[74] and the interaction of EGF receptor with the downstream signaling protein Grb2. It has also been demonstrated that the assay designed to monitor EGF receptor dimerization can be adapted successfully to HTS condition.[75] As can be expected and has been explained before for BRET, in this case as well the number of false positive hits is significantly lower than in corresponding reporter gene assays, which is a clear benefit if the technology is used to screen for new drugs. One advantage of enzyme complementation as compared to BRET or FRET is that enzymatic signal amplification reduces the need for overexpressing the chimeric proteins to obtain a sensitive assay system, thus reducing the probability of artefactual protein interactions. One disadvantage of this approach in comparison to BRET or FRET is that it cannot be used to investigate interaction in living cells, since no cell-permeable substrate for β-gal is available. Although the assay principle seems very well suited for the analysis of protein–protein interactions, there have not been many new publications on the successful use of this technology. A lot of different fusion constructs have to be tested and it does not always seem easy to generate an appropriate cell line. Recently, a β-gal complementation assay has been developed to investigate HIV envelope glycoprotein-mediated fusion.[76] In a similar way, we used this system for an HTS search for inhibitors of viral proteinase 2A of human rhinovirus serotype 2 (HRV2). The assay is based on the fact that many virally encoded proteinases cleave themselves out of a polyprotein, with cleavage usually occurring at their own N terminus. Briefly, the intramolecular *cis*-activity of this proteinase type was used to develop an *in vivo* screening system in *E. coli* based on the alpha-complementation of the lacZ gene fragment of M13mp18. When a fusion protein of the alpha fragment of beta-galactosidase and an active 2A proteinase of HRV2 was expressed, alpha complementation was not affected, as the 2A proteinase cleaved itself off the alpha fragment. However, fusion of an inactive 2A prevented alpha complementation, as the 2A polypeptide remained fused to the alpha fragment. This versatile system thus allowed the screening of substances inhibiting the proteolytic activity.[77]

6   Conclusions and Outlook

As described above, a large variety of different cell-based assay technologies are now available for target validation, lead identification, and optimization, allowing many biochemical processes to be analyzed in the cellular environment. These approaches can thus be of more physiological relevance than traditional biochemical assays if properly designed and characterized. One potential issue is the fact that overexpression of proteins can significantly change the cellular physiology. Furthermore, normal cellular background, the level of the analyzed signaling cascade, and cross talk to other signaling cascades have an important impact on the measured signal. Different cellular functional assays for the same target may therefore deliver different results, *e.g.* agonist potencies. Therefore, whether the transfected cells show a meaningful pharmacological behavior must be analyzed carefully. The more the signaling steps between the target of interest and the measured signal, the greater the possibility that non-specific interaction sites can contribute to interference with tested compounds. Cell-based assays therefore usually require additional assay systems to specify exactly the point of interaction of an interfering compound in the second step. If these issues are taken into account, cell-based assays offer a large potential for the advancement of the drug-discovery process. Imaging technologies, which have not been discussed here and which are still on the way to showing their final potential open additional perspectives, *e.g.* relating enzymatic activities to morphological processes, examining the variability of responses within a population of cells and investigating spatial distribution changes within cells.

References

1. S.J. Fox, *High Throughput Screening 2003: Improving Strategies, Technologies, and Productivity*, Executive Summary Report, High Tech Business Decisions, Moraga, California, 2003.
2. P. Johnston, *Methods Mol. Biol.*, 2002, **190**,107.
3. A.K. Hadjantonakis, M.E. Dickinson, S.E. Fraser and V.E. Papaioannou, *Nat. Rev. Genet.*, 2003, **4**, 613.
4. P. Liu, M.F. Morton, A. Reik, R. de la Rosa, M.C. Mendel, X.-Y. Li, C.C. Case, C.O. Pabo, V. Moreno, A. Kempf, J. Pyati and N.P. Shankley, *J. Biomol. Screen.*, 2004, **9**, 44.
5. J.L. Yates, N. Warren and B. Sugden, *Nature*, 1985, **313**, 812.
6. C. Piechaczek, C. Fetzer, A. Baiker, J. Bode and H.J. Lipps, *Nucleic Acids Res.*, 1999, **27**, 426.
7. S.C. Makrides, *Protein Expres. Purif.*, 1999, **17**,183.
8. J.L. Pfohl, J.F. Worley, J.P. Condreay, G. An, C.J. Apolito and T.A. Kost, *Receptors Channels*, 2002, **8**, 99.
9. R.J. Kaufmann, *Methods Enzymol.*, 1990, **185**, 487.
10. M. Krieger, *Methods Enzymol.*, 1990, **185**, 512.
11. S.M. Najjar and R.E. Lewis, *Gene*, 1999, **230**, 41.
12. J. Taipale, J.K. Chen, M.K. Cooper, B. Wang, R.K. Mann, L. Milenkovic, M.P. Scott and P.A. Beachy, *Nature*, 2000, **406**,1005.
13. D.M. Nettelbeck, V. Jerome and R. Muller, *Gene Ther.*, 1998, **5**, 1656.
14. M. Gossen and H. Bujard, *Proc. Natt. Acad. Sci. USA*, 1992, **89**, 5547.
15. M.C. Hu and N. Davidson, *Cell*, 1987, **48**, 555.
16. A. Simons, D. Tils, B. von Wilcken-Bergmann and B. Muller-Hill, *Proc. Natt. Acad. Sci. USA*, 1984, **81**, 1624.
17. R.J. Kaufmann, *Methods Enzymol.*, 1990, **185**, 537.
18. J.R. Lambert and S.K. Nordeen, *J. Biol. Chem.*, 1998, **273**, 32708.
19. G.M. Wahl, B. Robert de Saint Vincent and M.L. DeRose, *Nature*, 1984, **307**, 516.

20. B. Sauer, *Curr. Opin. Biotechnol.*, 1994, **5**, 521.

21. D.M. Kemp, S.E. George, P.J. Bungay and L.H. Naylor, *Eur. J. Pharmacol.*, 1999, **373**, 215.

22. S. Litschig, F. Gasparini and D. Rueegg, *Mol. Pharmacol.*, 1999, **55**, 453.

23. C.M. Suto and D.M. Ignar, *J. Biomol. Screen.*, 1997, **2**, 7.

24. C.M. Gorman, L.F. Moffat and B.H. Howard, *Mol. Cell. Biol.*, 1982, **9**, 1044.

25. J. Berger, J. Hauber, R. Hauber, R. Geiger and B.R. Cullen, *Gene*, 1988, **66**, 1.

26. R.E. Jones, D. Defeo-Jones, E.M. McAvoy, G.A. Vuocolo, R.J. Wegrzyn, K.M. Haskell and A. Oliff, *Oncogene*, 1991, **6**, 745.

27. V.K. Jain and I.T. Magrath, *Anal. Biochem.*, 1991, **199**, 119.

28. L.H. Naylor, *Biochem. Pharmacol.*, 1999, **58**, 749.

29. H. Labrousse, J.L. Guesdon, J. Ragimbeau and S. Avrameas, *J. Immun. Methods*, 1982, **48**, 133.

30. I. Bronstein, C.S. Martin, J.J. Fortin, C.E. Olesen and J.C. Voyta, *Clin. Chem.*, 1996, **42**, 1542.

31. A.B. Cubitt, R. Heim, S.R. Adams, A.E. Boyd, L.A. Gross and R.Y. Tsien, *Trends Biochem. Sci.*, 1995, **20**, 448.

32. R.Y. Tsien, 1998, *Annu. Rev. Biochem.*, 1998, **67**, 509.

33. M.B. Gu, R.J. Mitchell and B.C. Kim, *Adv. Biochem. Eng. Biotechnol.*, 2004, **87**, 269.

34. J.R. De Wet, K.V. Wood, M. DeLuca, D.R. Helinski and S. Subramani, *Mol. Cell. Biol.*, 1987, **7**, 725.

35. A. Himmler, C. Stratowa and A.P. Czernilofsky, *J. Recept. Res.*, 1993, **13**, 79.

36. O. Shimomura, *Biochem. J.*, 1997, **326**, 297.

37. K.V. Wood, Y.A. Lam, H.H. Seliger and W.D. McElroy, *Science*, 1989, **244**, 700.

38. U. Stolz, S. Velez, K.V. Wood, M. Wood and J.L. Feder, *Proc. Natt. Acad. Sci. USA*, 2003, **100**, 14955.

39. G. Grentzmann, J.A. Ingram, p.J. Kelly, R.F. Gesteland and J.F. Atkins, *RNA*, 1998, **4**, 479.

40. Y. He, W. Yan, C. Coito, Y. Li, M.J. Gale and M.G. Katze, *J. Gen. Virol.*, 2003, **84**, 535.

41. M.K. Lo, M. Tilgner and P.Y. Shi, *J. Virol.*, 2003, **77**, 12901.

42. J.T. Moore, S.T. Davis and K.D. Inderjit, *Anal. Biochem.*, 1997, **247**, 203.

43. G. Zlokarnik, P.A. Negulescu, T.E. Knapp, L. Mere, N. Burres, L. Feng, M. Whitney, K. Roemer and R.Y. Tsien, *Science*, 1998, **279**, 84.

44. O. Kornienko, R. Lacson, P. Kunapuli, J. Schneeweis, I. Hoffman, T. Smith, M. Alberts, J. Inglese and E. Strulovici, *J. Biomol. Screen.*, 2004, **9**, 186.

45. P. Sista, S. Edmiston, J.W. Darges, S. Robinson and D.J. Burns, *Mol. Cell. Biochem.*, 1994, **141**, 129.

46. M.R. Montminy, G.A. Gonzalez and K. Karen, *Trends Neuroscie.*, 1990, **5**, 184.

47. S.E. George, P.J. Bungay and L.H. Naylor, *J. Neurochem.*, 1997, **69**, 1278.

48. V. Boss, D.J. Talpade and T.J. Murphy, *J. Biol. Chem.*, 1996, **271**, 10429.

49. M. Fukasawa, T. Tsuchiya, E. Takayama, N. Shinomiya, K. Uyeda, R. Sakakibara and S. Seki, *J. Biochem.*, 2004, **136**, 273.

50. A. Rapisarda, B. Uranchimeg, D.A. Scudiero, M. Selby, E.A. Sausville, R.H. Shoemaker and G. Melillo, *Cancer Res.*, 2002, **62**, 4316.

51. R. Janknecht, *Immunobiology*, 1995, **193**, 137.

52. G. Voraberger, R. Schäfer and C. Stratowa, *J. Immunol.*, 1991, **147**, 2777.

53. C. Stratowa and M. Audette, *J. Immunol.*, 1995, **193**, 293.

54. U. Weyer, R. Schäfer, A. Himmler, S.K. Mayer, E. Bürger, A.P. Czernilofsky and C. Stratowa, *Receptors Channels*, 1993, **1**, 193.

55. P. Zuck, E.M. Murray, E. Stec, J.A. Grobler, A.J. Simon, B. Strulovici, J. Inglese, O.A. Flores and M. Ferrer, *Anal. Biochem.*, 2004, **334**, 344.

56. A. Cacace, M. Banks, T. Spicer, F. Civoli and J. Watson, *DDT*, 2003, **8**, 785.

57. J. Chin, A.D. Adamsn, A. Bouffard, A. green, R.G. Lacson, T. Smith, P.A. Fischer, J.G. Menke, C.P. Sparrow and L.J. Mitnaul, *Assay*, 2003, **1**, 777.

58. A.M. Maffia, I. Kariv and K.R. Oldenburg, *J. Biomol. Screen.*, 1999, **128**, 721.

59. J. Stables, A. Green, F. Marshall, N. Fraser, E. Knight and M. Sautel, *Anal. Biochem.*, 1997, **252**, 115.

60. P. Coward, S.D.H. Chan, H.G. Wada, G.M. Humphries and B.R. Conklin, *Anal. Biochem.*, 1999, **270**, 242.

61. E.L. Poul, S. Hisada, Y. Mizuguchi, V.J. Dupriez, E. Burgeon and M. Detheux, *J. Biomol. Screen.*, 2002, **7**, 57.

62. J.C. Jerman, S.J. Brough, T. Gager, M. Wood, M.C. Coldwell, D. Smart and D.N. Middlemiss, *Eur. J. Pharmacol.*, 2001, **414**, 23.

63. M. Brini, R. Marsault, C. Bastianutto, J. Alvarez, T. Pozzan and R. Rizzuto, *J. Biol. Chem.*, 1995, **270**, 9896.

64. K.R. Gee, K.A. Brown, W.N. Chen, J. Bishop-Stewart, D. Gray and I. Johnson, *Cell Calcium*, 2000, **27**, 97.

65. Y. Xu, D.W. Piston and C.H. Johnson, *Proc. Natl. Acad. Sci. USA*, 1999, **96**, 151.

66. P. Wu and L. Brand, *Anal. Biochem.*, 1994, **218**, 54.

67. J.P. Vilardaga, M. Bunemann, C. Krasel, M. Castro and M.J. Lohse, *Nat. Biotechnol.*, 2003, **21**, 807.

68. E. Michelini, A. Mirasoli, M. Karp, M. Virta and A. Roda, *Anal. Chem.*, 2004, **76**, 7069.

69. S. Angers, A. Salahpour, E. Joly, S. Hilairet, D. Chelsky, M. Dennis and M. Bouvier, *Proc. Natl. Acad. Sci. USA*, 2000, **97**, 3684.

70. N. Boute, K. Pernet and T. Issad, *Mol. Pharmacol.*, 2001, **60**, 640.

71. D. Germain-Desprez, M. Bazinet, M. Bouvier and M. Aubry, *J. Biol. Chem.*, 2003, **278**, 22367.

72. K.D. Pfleger and K.A. Eidne, *Biochem. J.*, 2004, **385**, 625.

73. M. Vrecl, R. Jorgensen, A. Pogacnik and A. Hedning, *J. Biomol. Screen.*, 2004, **9**, 322.

74. B.T. Blakely, F.M.V. Rossi, B. Tillotson, M. Palmer, A. Estelles and H.M. Blau, *Nature Biotech.*, 2000, **18**, 218.

75. D.L. Graham, N. Bevan, P.N. Lowe, M. Palmer and S. Rees, *J. Biomol. Screen.*, 2001, **6**, 401.

76. A.U. Holland, C. Munk, G.R. Lucero, L.D. Nguyen and N.R. Landau, *Virology*, 2004, **319**, 343.

77. H.D. Liebig, T. Skern, M. Luderer, W. Sommergruber, D. Blaas and E. Kuechler, *Proc. Natl. Acad. Sci. USA*, 1991, **88**, 5979.

NMR-Based Screening: A Powerful Tool in Fragment-Based Drug Discovery

JOCHEN KLAGES[a], MURRAY COLES[b] AND HORST KESSLER[a]

[a]Department Chemie, TU München, Lichtenbergstr. 4, 85747 Garching, Germany
[b]Department of Protein Evolution, Max-Planck-Institute for Developmental Biology, Spemannstrasse 35, 72076 Tübingen, Germany

1 Introduction

NMR has become a powerful tool in the pharmaceutical industry for a variety of applications. Instrumental improvements in recent years have contributed significantly to this development. Digital recording, cryogenic probes, autosamplers, and higher magnetic fields shorten the time for data acquisition and improve the spectral quality. In addition, new experiments and pulse sequences make a vast amount of information available for the drug discovery process. As a consequence, new screening technologies involving NMR have become indispensable for modern drug discovery. These developments have been summarized in several recent reviews.[1–12]

The drug development process is usually divided into the following familiar phases. Initially, a target is identified that may be involved in initiating or blocking a certain physiological effect. Most drug targets are proteins, but RNA and DNA also can serve as targets for drug discovery. When a suitable target is identified, it is screened for compounds that may cause the desired biological effect. Assays involving binding to the target, either directly or in competition with labelled ligands, are often used in place of functional (cellular) tests. The molecules that are identified as hits from binding screens are validated for the desired biological effect, *e.g.* inhibition of enzymatic activity. The remaining leads are then optimized by structural refinement to increase the affinity and selectivity for the receptor and for bioavailability and other *in vivo* pharmacological and physiological properties, which may eventually result in a *drug candidate*.

NMR can assist the above process at different stages, namely, in hit finding, hit validation, and lead optimization.[13] These three areas of application will be discussed in detail in the following sections. The drug discovery process is summarized in Figure 1, in which shaded boxes indicate where incorporation of NMR techniques is possible.

NMR offers some unique features that make it an attractive alternative for applications in drug research. The primary, intrinsic advantage, however, is the ability to

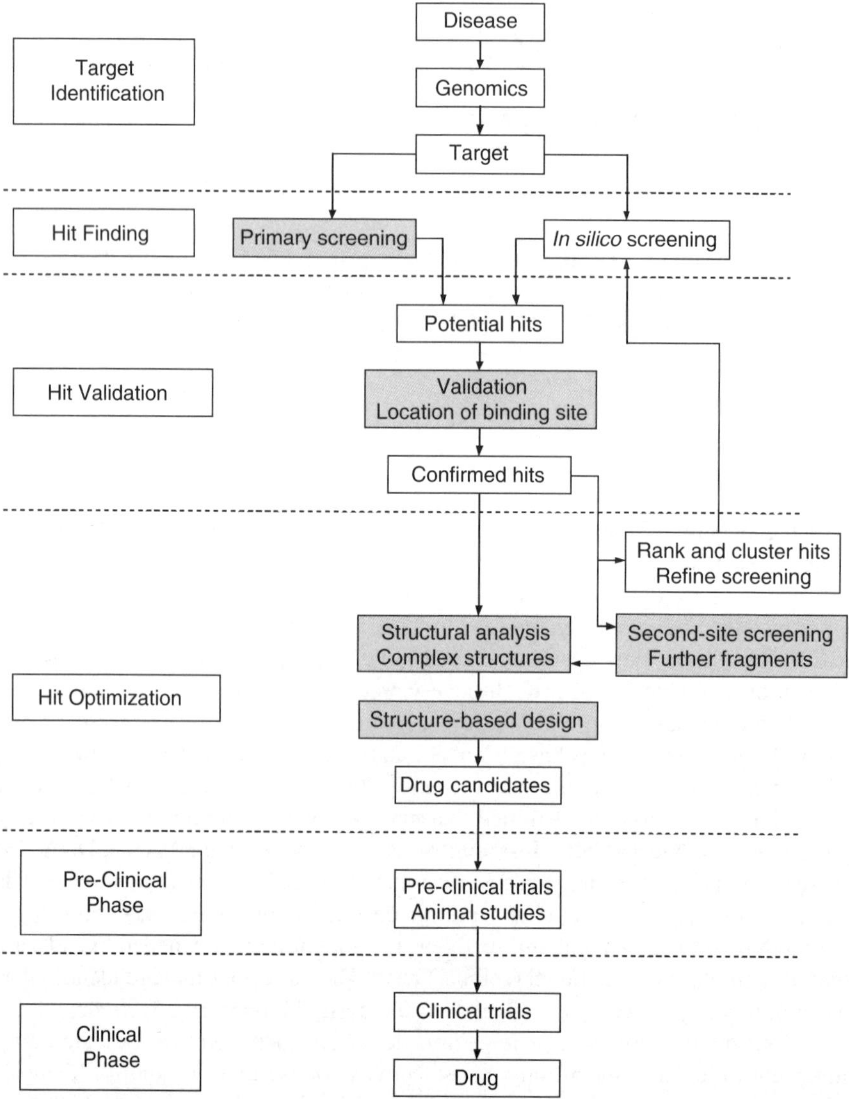

Figure 1 *A flowchart outlining the drug discovery process; steps allowing the application of NMR techniques are shaded*

detect weak intermolecular interactions, *e.g.* between a ligand and a target, with unmatched sensitivity. This ability makes NMR ideal for fragment-based screening.[1, 14–18] In a fragment-based approach, comparably small and simple molecules are screened for binding to a target. These compounds often reveal only a weak affinity. However, when several fragments for different binding sites of the target are identified, they can be linked to form higher affinity ligands. Although the linkage of fragments is an additional difficulty, this approach accesses a larger structural space because each fragment can be optimized separately.

In contrast, structural space is significantly restricted in conventional high throughput screening (HTS), in which the identified hits often are hydrophobic and possess relatively high molecular masses already, and it is difficult to improve their activity without further increasing either their hydrophobicity or molecular weight, or both. Lipinski's "rule of five"[19] (Table 1) requires a certain window of lipophilicity, a maximum of five hydrogen bond donors and <10 hydrogen bond acceptors, and in addition a molecular weight below 500 Da. Clearly, the fragment-based approach carries a higher potential for designing efficient drug candidates within the limits set by the "rule of five". The advantage of the fragment-based approach over other techniques like HTS is illustrated in Figure 2.

Additionally, high throughput screens usually work as functional or competitive binding assays with labelled ligands that are often derived from the natural binding molecules. Establishing a functional assay can be tedious, whereas binding can be measured easily. An advantage of NMR is that it is applicable even when no functional assay or known binders are available; moreover, NMR can be used for validation of hits from functional HTS screening.

A further key advantage of NMR is its ability to provide additional structural information, which is welcome even at the earliest stages of the drug development process. Structural information supports rational lead design or the design of biased libraries based on known structure activity relationships (SARs).

The number of compounds that can be screened by NMR is limited to about 10^3–10^4 depending on the technology used – much lower than for HTS and *in silico* screening methods (10^5–10^6 compounds). Hence, it is not the number of compounds that makes NMR an attractive method, but rather the possibility to use fragment-based screening.

Table 1 *Properties of orally available drug-like compounds*

"Rule of five" criteria[19]
- Molecular weight $\leq$500 Da
- Log $P \leq 5$
- Hydrogen bond donors (OH and NH) $\leq$ 5
- Hydrogen bond acceptors (lone-pairs of hetero-atoms, like O and N) $\leq$10

Other criteria (Veber-rules[77])
- Number of heavy atoms 10–70
- Rotatable bonds 2–8
- Number of rings 1–6, aromatic $\leq$3
- Molar refractivity 40–130
- Size of hydrophobic area

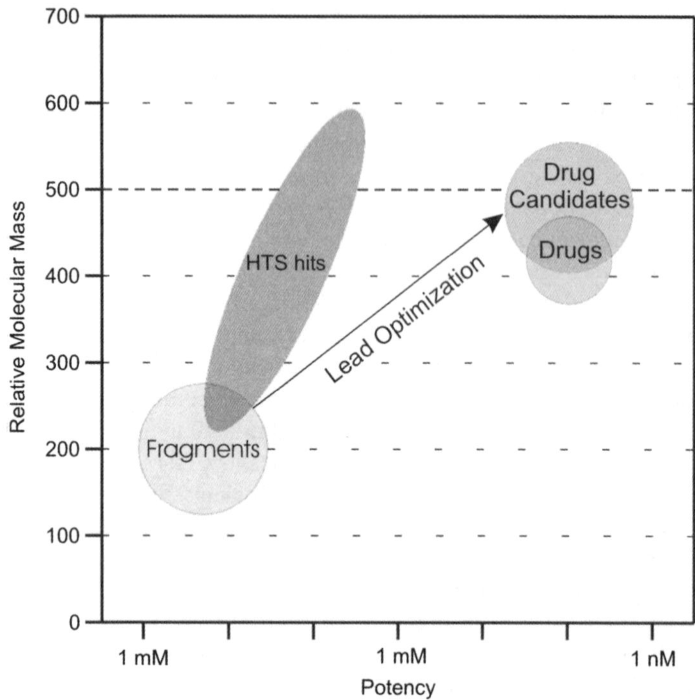

Figure 2 *Schematic representation of the advantage of the fragment-based approach over conventional HTS screening.[14] The range of affinities and molecular weights of average HTS hits and fragments are illustrated. The thickly dashed line indicates the size limit set by Lipinski's "rule of five". Whereas HTS hits tend to have high molecular masses, fragments show much lower molecular weights but only slightly lower affinities, which conveys higher lead-like character*

In combination with the structural information that can be extracted simultaneously, NMR experiments can open up new, unexplored regions of chemical diversity in the search for drugs.

2   NMR Screening: General Aspects

It is clear that to carry out ligand screening by NMR, the library compounds (ligands) must be distinguished from the target. One of the most important methods of achieving this is through their vastly different molecular masses. Several NMR parameters are dependent on the rotational and translational diffusion rates of the subject molecule, and therefore on molecular mass. Notably, the transverse (entropic) relaxation of the NMR signal is dependent on the rotational diffusion rate. Large, slowly tumbling target molecules relax much faster than rapidly tumbling small molecules. As the transverse relaxation rate is directly related to linewidth, small molecules can be distinguished by their narrow lines. Also, filters suppressing broad signals can be used to minimize signals from the target. Other relaxation-related parameters, such as the nuclear overhauser effect (NOE), are also related to rotational diffusion rates; small molecules show a positive enhancement and large molecules a negative enhancement. In contrast to rotational diffusion, translational diffusion can be measured directly by NMR. This technique allows rapidly diffusing small molecules to be distinguished from the slowly diffusing target by suppressing the signal from the former.

These differences in NMR properties between small and large molecules form the basis of many NMR screening experiments. When a small ligand binds to a large target, it adopts the properties of the target to an extent dependent on the residence time of the binding event (Figure 3). NMR screening experiments can therefore be designed to detect molecules with intermediate properties. It is clear that distinction on the basis of molecular mass would fail if the ligands were not considerably smaller than the target. It should also become clear in the following sections that the use of relaxation or translational diffusion filters to suppress ligand and/or target signals places considerable affinity limits on the techniques involved.

A second major method of distinguishing ligand and target signals does not rely on molecular mass, but on the much greater dispersion of chemical shifts usually observed for the target. This property can be exploited to create non-equilibrium magnetization specifically on the target, which can be transferred to binding ligands and subsequently detected. By far, the most robust method of distinguishing ligands and target, however, takes this a step further, using specific isotope labelling of one or the other component to create a unique chemical shift range. This strategy allows some of the simplest NMR screening techniques to be used, relying on the change in chemical environment induced by the binding ligand. This change in environment affects the chemical shifts of both the target and a binding ligand, predominantly at the binding site. Thus, chemical shift changes observed in a (^{15}N- or ^{13}C-labelled) target can localize and partially characterize the binding site. Distinction via isotope labelling avoids the use of relaxation- and/or diffusion-based filters to suppress the unwanted component, giving these techniques a much wider affinity range. These techniques are also ideally suited to competitive screening applications, which represent some of the most promising new developments in the field.

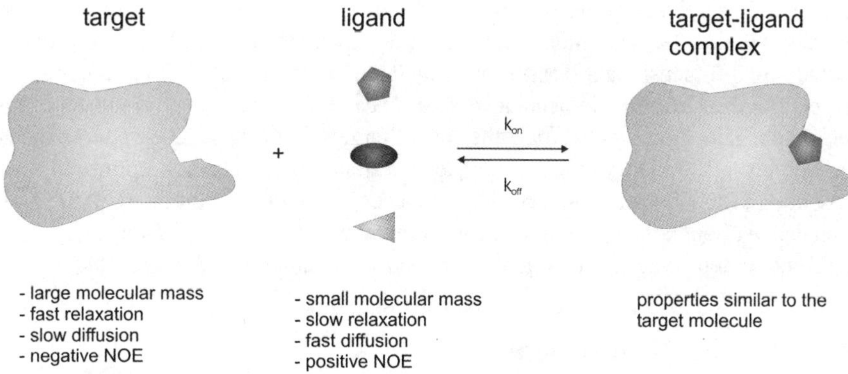

Figure 3 *Alteration of the physicochemical properties during the process of binding. On binding, the ligand adopts the properties of the large target molecule due to drastic increase in the effective molecular mass*

3 Ligand- *vs.* Target-Detected Methods

NMR screening can be divided into ligand- and target-detected methods. In the first class, changes induced in the ligand's NMR signals by binding to a large target are observed, whereas in the second class, the influences of ligands on the spectra of the target are detected. Both techniques have their intrinsic advantages for different applications, making them largely complementary.

Ligand-based methods predominantly make use of one-dimensional (1D) NMR spectra and therefore are comparably fast, allowing higher throughput in screening. The detection of the ligand signals offers the opportunity to screen mixtures of ligands without the need for deconvolution, as long as the signals of the ligands do not overlap. Moreover, the amount of target required for the screening process is smaller and there are fewer restrictions on the properties of the target. Commonly, the target is not isotopically labelled and its molecular mass is practically unlimited; in some techniques (*e.g.* saturation transfer difference (STD), see below), the target can even be immobilized in membranes. Finally, some crude information about the binding epitope and the binding mode can be extracted by ligand-based methods.[20]

Most ligand-detected methods are limited to low and medium affinities because of the need to suppress the signals of the target molecule. Ligands that bind too tightly are indistinguishable from the target and thus are suppressed as well, resulting in false negatives. In contrast, non-specific binding can result in the appearance of false positives.

Target-based methods can easily distinguish between non-specific and specific binding. In addition, effects caused by aggregation and pH-changes can be excluded. The detectable affinity range is much higher than in ligand-detected techniques, particularly at the high-affinity end. However, the most valuable advantage is the possibility of extracting detailed structural information about the ligand–target complex from the spectra. For target-detected experiments comparably large amounts of the protein target are needed, because the observation of its NMR signals requires concentrations in the range of 0.1–1 mM. These targets usually have to be labelled with NMR active isotopes (^{15}N and/or ^{13}C) making these techniques quite costly. In contrast to ligand-detected screening methods, it is not possible to screen mixtures of compounds without deconvolution. The standard implementation of these experiments includes two-dimensional (2D) spectra, which require longer acquisition times. As the targets usually have high molecular masses ($>$10 kDa), the signals are subject to fast relaxation. This limitation restricts the size of targets that can be observed to molecular weights $<$100 kDa, even if techniques like transverse relaxation-optimized spectroscopy (TROSY)[21] or cross-relaxation-induced polarization transfer (CRIPT)[22] are included. Although this limit may be overcome with specific methyl group labelling in the future, interesting targets might still remain intractable with these techniques.

3.1 Sample Requirements

Because the interaction between the ligand and the target usually takes place in aqueous solution, the ligands have to be soluble in water to an appropriate level; otherwise, aggregation effects lead to false positives during the screening process. For standard ligand-based techniques, a concentration of 100 μM is found to be satisfactory for a reasonable signal-to-noise ratio. The target, which is usually unlabelled, is sufficiently concentrated when it is within a range of 1–50 μM; for a 0.5 mL solution, this corresponds to 0.1–5 mg of protein with a molecular weight of 20 kDa. In target-detected experiments, however, labelling is essential to avoid signal overlap. Moreover, higher concentrations are necessary, raising the amount of target needed to 100–300 μM. Depending on the size and the relaxation properties of the target, different labelling schemes are necessary. For relatively small proteins, uniform ^{15}N-labelling is sufficient. For larger proteins, ^{13}C-labelling and/or deuteration may be required. Site- or amino acid-selective labelling[23] can extend the size of the proteins, which can be studied as overlap is reduced. Selective ^{13}C-labelling of the methyl groups of valine, leucine, and isoleucine[24–27] has proved to be especially advantageous because these residues are at the surface of hydrophobic areas, which are often involved in binding events. In addition, methyl group signals are usually narrower and therefore better resolved.

4 Incorporation of NMR into the Drug Discovery Process

In the following detailed discussion of the application of NMR techniques to drug discovery, a short description for hit finding, hit validation, and hit (lead) optimization is given. In addition, the corresponding requirements for the NMR experiments and the resulting implementations are described. Only typical experiments are given explicitly, but other techniques are summarized in Table 2.

Table 2 *Overview of common screening techniques and their application[10]*

Method	Application	Limits and requirements			Identification of		
		Target MW limit	*Affinity limit*	*Labelled target req.*	*Binding site on target*	*Binding epitope on ligand*	*Binding comp. in mixtures*
Diffusion filtering	Hit finding	Lower	U/L	No	No	No	Yes
Relaxation filtering	Hit finding	Lower	U/L	No	No	No	Yes
TrNOE	Hit finding	Lower	U/L	No	No	No	Yes
NOE pumping	Hit finding	Lower	U/L	No	No	No	Yes
Rev. NOE pumping	Hit finding	Lower	U/L	No	No	Yes	Yes
WaterLOGSY	Hit finding	Lower	U/L	No	No	Yes	Yes
STD	Hit finding	Lower	U/L	No	No	Yes	Yes
^{19}F-screening	Hit finding	None	None	^{19}F ligand	No	No	Yes
CSM	Hit validation	Upper	None	^{15}N or ^{13}C	Yes	No	No
Comp. screening	Hit optimization	None	None	No	Yes	No	No
Comp. ^{19}F-screening	Hit optimization	None	None	^{19}F ligand	Yes	No	No

The classification of drug discovery into different phases is to a certain extent artificial – the different phases overlap or can be applied simultaneously – but the general outline indicated here shows that more complex NMR techniques are needed when the drug discovery process is more advanced.

4.1   Hit Finding

In hit finding, a large number of compounds, the so-called library, is screened *vs.* a target to identify components that bind. Libraries can meet different types of requirements, and efforts are made to optimize them with respect to diversity, solubility, drug-like character (see Table 1), and the synthetic accessibility of the compounds they contain.[28,29] Substance libraries can contain small organic fragments, synthetic compounds of diverse chemical structures, or natural products.[30]

Experiments for hit finding by NMR spectroscopy should allow high throughput, since substance libraries can easily contain up to 10^4 compounds. Hence, time-optimized techniques are required that not only reduce experimental time, but also the time for sample preparation and data analysis. Consequently, lengthy 2D or three-dimensional (3D) techniques are unfavourable, as are complex experiments that require fine tuning for each sample. For an industrial application, the ability to identify binding components from large mixtures without deconvolution is nearly indispensable. Moreover, labelled samples and target-consuming techniques should be avoided for cost reduction. These requirements make 1D ligand-based experiments the most favourable alternative.

4.1.1   STD and WaterLOGSY

STD[31,32] and WaterLOGSY[33,34] both take advantage of the fact that the *intermolecular* NOE transfer is strongly negative. A non-equilibrium magnetization is created on the receptor–ligand –complex, and the subsequent alteration of the free ligand signal intensities is monitored. Two principle implementations are imaginable. Either all resonances (target, binding, and non-binding ligands) are perturbed simultaneously, followed by selective suppression of some signals, or the excitation itself is already selective. Selective suppression for instance can be implemented by a diffusion filter,[35] eliminating signals of small molecules or by a relaxation element,[36] filtering out target resonances.

STD is one of the most useful experiments for NMR screening. Here, non-equilibrium magnetization is achieved by selective excitation (on-resonance) of target resonances. The magnetization is transferred within the target via spin diffusion and eventually to a bound ligand (Figure 4). When this ligand dissociates from the target into solution, the magnetization change transferred in the bound state is retained in the free ligand. The difference from a reference spectrum taken without on-resonant irradiation thus yields a spectrum containing only those ligands that have been perturbed by binding to the target. The reference spectrum is obtained by off-resonance irradiation in a spectral window where no target signals appear (Figure 5).

STD spectroscopy offers several advantages. First, there is almost no upper limit for the size of the target, even membrane proteins or immobilized targets are applicable.[32]

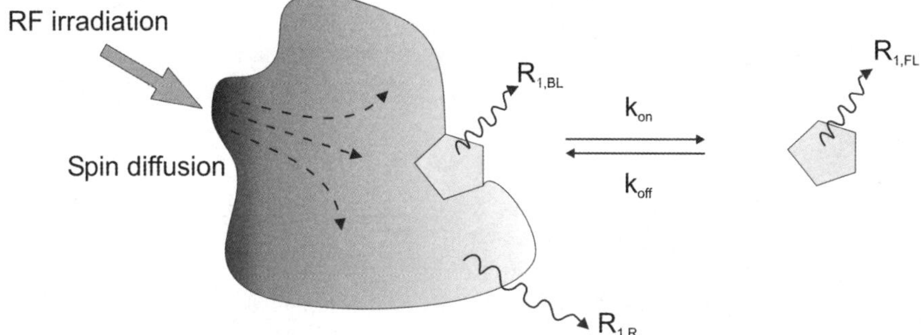

Figure 4 *Illustration of the effect of on-resonant irradiation of protein signals. Due to spin diffusion, the resulting non-equilibrium magnetization spreads out across the target molecule and is transferred to the ligand. The magnetization of the bound ligand decays rapidly with a rate of $R_{1,BL}$, which is of the same order as the rate for the target molecule $R_{1,R}$. On dissociation into solution, the relaxation properties of the ligand change and the acquired non-equilibrium magnetization now decays with a rate of $R_{1,FL}$ for the free ligand[12]*

On the contrary, since the correlation time increases with increasing molecular mass, spin diffusion becomes more and more effective, which in turn leads to stronger STD effects. For this technique, the amount of target and ligand needed to give reasonable results is relatively small. In larger ligands, the part of the molecule that is in direct contact with the target is most strongly affected. Hence, structural information of the complex from the ligand side is provided. Estimations of the binding constants can also be drawn.[20] STD spectra are usually recorded as 1D spectra, but they can be easily extended to 2D or 3D versions if necessary.[37] This option offers a diverse tool for a variety of different tasks in screening. STD fundamentally depends on an effective spin-diffusion mechanism and therefore on proton density. Hence, the observed STD effect might not be large enough to be detected, especially when small targets are used. As it is the free fraction of the ligand that is observed, one normally sees a dramatic amplification effect if the ligand concentration is higher than that of the target. However, similar to other ligand-based techniques, STD is not applicable to high-affinity binders.

If the target does not provide a large enough proton density, as is the case with nucleic acids,[12] the WaterLOGSY (water-ligand observed via gradient spectroscopy) technique might be preferred. The large bulk water magnetization is used for an effective transfer via the ligand–target complex to the free ligand in solution. By selectively inverting or saturating the water resonance, the magnetization is transferred to the target and then finally to the ligand. The exact details of the transfer are more complicated and will not be discussed in this review. The important point is the difference in the cross-relaxation properties of the water with non-binding ligands and binding ligands. In the first case, the relaxation includes small correlation times leading to positive cross-relaxation rates, in contrast to the second case in which positive cross-relaxation rates are observed. As a consequence, non-binding and

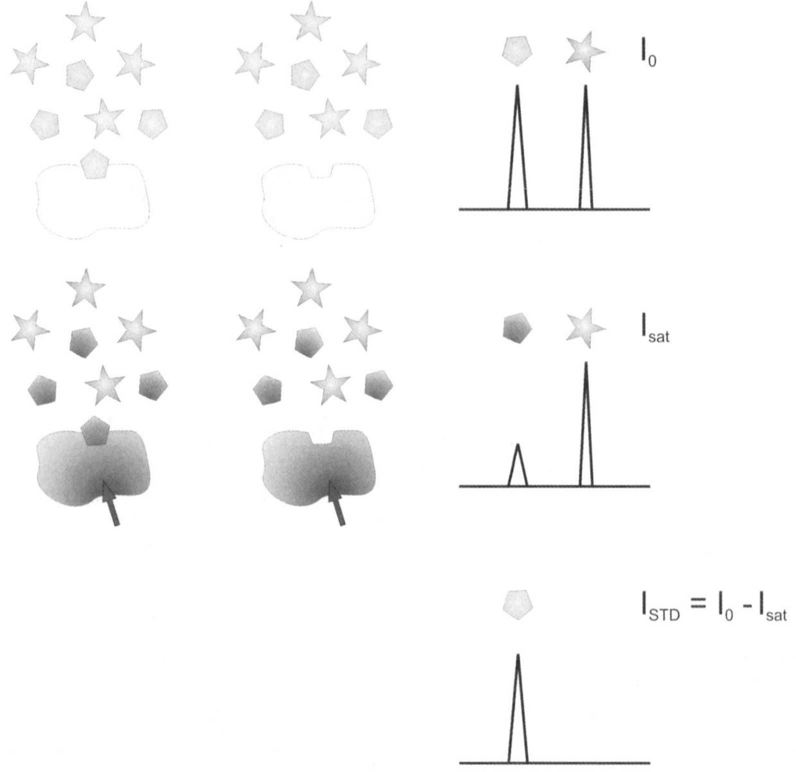

Figure 5 *Schematic representation of the procedure for the measurement of STD experiments.[12] Polygons and stars represent binding and non-binding components, respectively. During the off-resonant experiment (top panel), no magnetization is transferred from the binding ligand to the target molecule. If the irradiation changed to "on-resonant frequencies" (middle panel), magnetization transfer from the bound ligand to the target occurs, which lowers the signal intensity of the ligand. The difference between the two spectra only contains the signals of binding components*

binding ligands show different signs in the spectrum. This combination can lead to an erroneous interpretation if positive and negative peaks cancel each other out.

4.1.2 Libraries of ^{19}F-Containing Ligands

A comparably new technique is the screening of ligands that contain ^{19}F-labels, for example, in the form of fluorinated aromatic or trifluoromethyl groups.[38–40] The ^{19}F-nucleus has some unique features that make it an attractive probe for screening. It has a high gyromagnetic ratio ($\gamma_F \sim 0.94\ \gamma_H$) and occurs at 100% natural abundance making it a very sensitive NMR nucleus. It has broad chemical shift dispersion, allowing the use of large mixtures without signal overlap. The target, additives, and solvents generally do not contain fluorine, meaning only ligand signals are observed in the spectra. Upon binding, the chemical shift and linewidth of the ligand ^{19}F-signal is strongly affected (as a result of the large chemical shift anisotropy (CSA) of the ^{19}F-nucleus). Hence, the acquisition of 1D spectra with and without the target molecule is sufficient to screen for binding.

Of course, the need for ^{19}F-labels seems to be a profound drawback. However, about 10% of all drugs on the market already contain fluorine[10] mainly for modification of metabolic stability. Additionally, the small fluorine atom can be often replaced by hydrogen without loss of binding affinity.

4.2 Hit Validation

For all compounds found in the primary screen, the binding has to be validated to remove false positives, which is especially important for hits of functional assays. False positives might originate from changes in the pH, aggregation, or non-specific binding. By monitoring binding on a molecular level, interactions with the wrong binding site and chemical reactions with the drug target can also be excluded. The number of compounds investigated during the validation process is much lower than for the hit finding process. Therefore, the problems of deconvolution, concentration, and experimental time can be less rigorous, whereas a large window of affinities is still essential. Even more important is the lack of ambiguity of the experiment with respect to the binding of the ligands. To extract true binders from the hits, it is advisable to apply several experiments to confirm the mode of interaction.

4.2.1 Chemical Shift Mapping

A comparably simple approach uses perturbations of the target's chemical shifts to confirm binding. Multidimensional spectra are usually used, as signal overlap obscures the analysis in 1D spectra. In these spectra, the displacement of peaks due to the binding event is monitored. Obviously, the use of 2D- or 3D spectra results in an increased experimental time. As an additional feature of this technique, the dissociation constant K_D may be extracted from the experiment. This analysis is possible if the displacement is related to the concentration of the ligand via a titration. Of course, these experiments require a reference spectrum without the ligand.

Chemical shift perturbations are most commonly tracked in ^{13}C- and ^{15}N-HSQC spectra, although HNCO spectra have also been used.[15] ^{15}N-labelling is relatively cheap if an appropriate overexpression system has been established, while in contrast, uniformly ^{13}C-labelling is expensive. Nowadays, new labelling schemes enable the selective incorporation of ^{13}C-labels in the methyl groups of valine, leucine, and isoleucine.[27] Isotopic enrichment with ^{13}C at specific locations forms a convenient alternative to ^{15}N-labelling, especially considering that the apparent signal intensity of methyl groups is three times that of amide protons. The chemical shift dispersion of the methyl moieties is merely somewhat smaller than that of the amide protons. The amide protons visible in the ^{15}N-HSQC spectrum are hydrophilic, whereas the methyl groups of the above-mentioned residues are hydrophobic. Therefore, both techniques are partially complementary.

4.3 Hit Optimization

For hit optimization, the confirmed hits from the validation process are first ranked and clustered. Specific properties, *e.g.* Lipinski's "rule of five", have to be considered to develop a promising lead structure (Table 1).[13,41] This assessment includes the solubility, molecular weight, chemical accessibility, and affinities (binding constants). For a further improvement of the binders, structural information is indispensable.[15] The exact binding site as well as the precise binding mode has to be extracted from the experiments. For the design of new ligands, analogs of the lead structure are explored, often assisted by combinatorial chemistry. The resulting compounds are subjected to a screening comparable to the hit finding process, but with more thorough evaluation. As targets often have more than one binding site, these have to be identified and screened too. If ligands for adjacent binding sites can be linked, large increases in their affinities can be expected.

Depending on the stage of the hit optimization process, the experiments should provide structural information of different qualities. Whereas at the beginning, even crude SARs are sufficient, more detailed information is required at an advanced stage of drug development. Therefore, the techniques change from ligand- to target-based methods and more complex pulse sequences are applied. Target-based techniques have the additional advantage of offering the possibility of monitoring high-affinity ligands, as low-affinity binders are of minor interest at this stage. Of course, the target needs to be NMR accessible, which sets some limitations in size.

4.3.1 Chemical Shift Mapping

The above-mentioned implementations for chemical shift mapping (see hit validation) usually allow the structural characterization of the binding epitope as well, if a sequential assignment is available. A simple mapping of the displaced signals identifies the binding site. This method is of particular interest if two fragments are to be combined as proposed by the SAR-by-NMR technique.[1] In this approach, two fragments are fine tuned separately to give higher affinity ligands. Both molecules bind to the same target but at different binding sites. The subsequent linkage of the two results in a high-affinity ligand (Figure 6). SAR-by-NMR is a prominent example for structure- or fragment-based drug design. It is also possible to draw conclusions about the orientation of the ligand within the binding pocket from the experiments.[42,43] The ligand is docked into the binding site of the target by computational techniques. A theoretical spectrum is calculated and compared to the experimental spectrum. If several spectra are calculated for different conformations of the ligand, information about the binding mode can be generated.

Another interesting approach is the observation of binding within a living cell.[44–46] ^{15}N-HSQC spectra are recorded directly of the bacterial slurry, resulting in spectra of practical resolution. This result is of particular interest because many compounds have high affinity *in vitro* but show much weaker effects *in vivo*.

2D spectra sometimes are not sufficient to visualize the binding process because of remaining signal overlaps. If the binding site is already known, site-selective

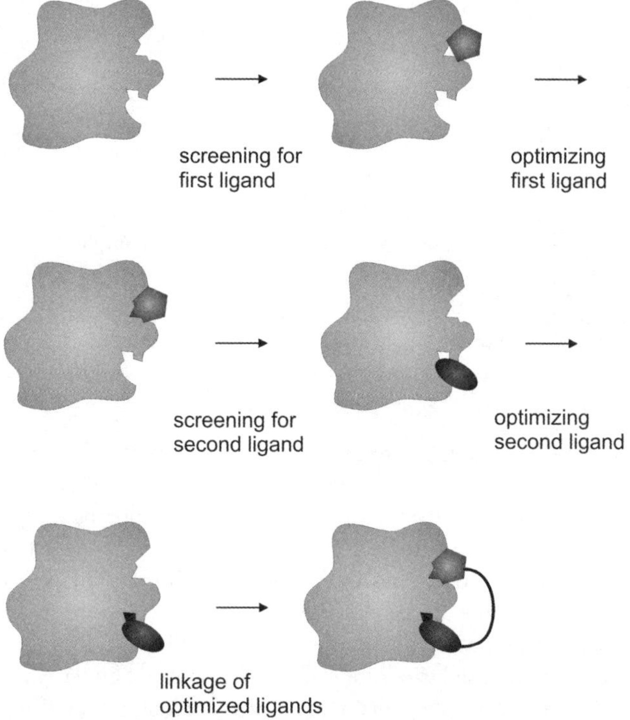

Figure 6 *Illustration of the SAR-by-NMR technique. High-affinity binders are identified on the basis of two medium-affinity binders of different but adjacent binding sites*

labelling might help to avoid this problem. Single types of amino acids are selectively labelled to simplify the spectra and focus it on the binding pocket.

The labelling schemes described are of an immense value for protein targets; however, there are still no comparable labelling techniques that make target-based experiments appropriate for nucleic acid targets.

4.3.2 Competition-Based Screening

The major drawback of ligand-based experiments is the restricted ability to screen higher affinity ligands. Because strong binders possess a long residence time in the binding pocket of the target, their exchange rate is very low. This slow exchange results in the loss of the information about the bound state as the non-equilibrium magnetization that is created on the ligand decays before it can be detected via the free form. Tighter binders can be investigated to a certain extent if the concentration of the ligand is lowered. Of course, this approach is limited by the sensitivity of the technique used. Thus, high-affinity binders usually cannot be detected by simple ligand-based techniques and lead to false negatives.

An alternative approach is competition-based screening.[47–49] A known medium-affinity binder (reporter ligand) is added to the solution of the target. If other ligands

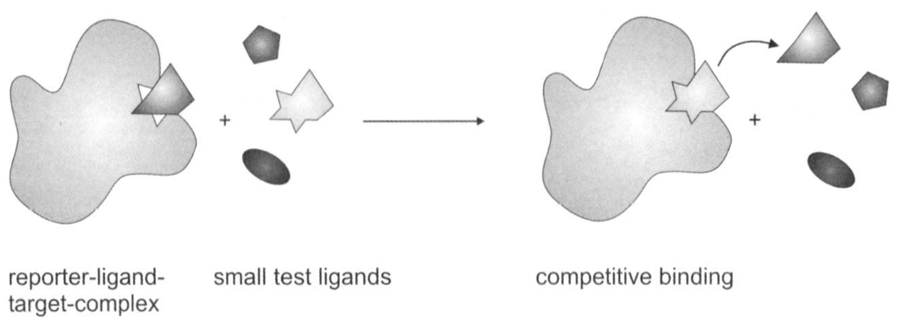

Figure 7 *Depiction of competition-based screening. Low- to medium-affinity binders are displaced from their binding site by higher affinity binders. In this experiment, only the signals of these low- to medium-affinity binders (reporter ligands) are monitored*

(screening ligands) are present in this solution as well, the reporter ligand will be displaced according to the affinity of the screening ligands (Figure 7). The prerequisite of fast exchange is only valid for the reporter ligand and therefore the range of affinities for competing ligands is not limited. For the detection of weaker binders, the concentration of the reporter ligand has to be decreased. During the experiment the focus is on the reporter ligand, since only signals of this species are detected. This limitation raises the problem of deconvoluting mixtures of ligands if binders are discovered. With competition-based experiments, only ligands that bind to the same site as the reporter ligand are detectable, also making this application relevant to the hit validation process. All competition experiments can be used to estimate the dissociation constant of the screened ligands if the K_D of the reporter ligand is known. An NMR titration has to be performed with the ligand of interest while the concentration of the target and the reporter ligand are kept fixed. The change of the intensity of the reporter ligands signal and subsequent fitting to corresponding equations[12] gives the required information.

In principle, all ligand-based techniques can be extended to a competition type of experiment. However, experiments using simple 1D spectra are especially worth considering, *e.g.* WaterLOGSY,[47] STD, and ^{19}F-screening,[40] which rank among the most powerful ligand-based screening techniques. Their advantage that they only require a small amount of target and their high sensitivity is adopted by the competition experiment. Competition-based fluorine screening has to be highlighted in this context as it combines the advantages of most techniques. Fluorine chemical shift anisotropy and exchange for screening (FAXS) is one implementation of this concept. The relaxation parameters of the reporter ligand change dramatically on the displacement by other ligands. Clean spectra are easily obtained without suppression of target or solvent signals making it a highly effective tool for screening, even in hit finding.

4.3.3 Paramagnetic Spin Labels

Paramagnetic spin labels offer the opportunity to screen for binding sites remote from a known binding site.[50] The introduction of these labels increases the T_2-relaxation rate of binding components and therefore results in an observable

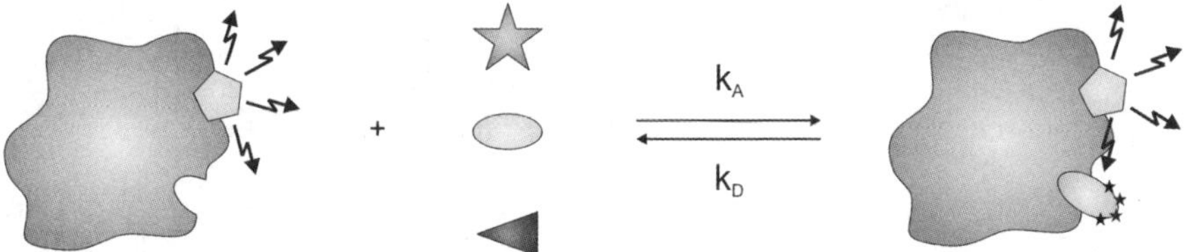

Figure 8 *Introduction of paramagnetic spin labels for second site screening. Paramagnetic spin labels drastically alter the relaxation properties of binding components. The large gyromagnetic ratio of unpaired electrons amplifies the relaxation via dipole–dipole interaction, which is represented by the arrows. Stars at the edges of the binding ligand illustrate the increased transverse relaxation*

broadening of their signals (Figure 8). The origin of this modified T_2-time is the increased relaxation via the electron–proton dipole–dipole interaction. Because of the vastly larger gyromagnetic ratio of the electron, the amount of target and ligand needed is reduced. Also, shorter contact times and comparably large distances (~20 Å) between the two binding sites are practicable. Obviously, the general handicap is the introduction of a spin label that does not alter the binding properties of the binding site. Therefore, a detailed knowledge of the 3D structure of the target is essential.

5 Representative Case Studies

In the following section, examples of NMR-based screening are presented. The first two examples include the screening of libraries containing 400 and 10,000 compounds, respectively. One involves fluorine techniques as a method for primary screening, while the other uses target-based methods for detection. In the second example, the fragment-based approach is described where NMR methods for hit finding, hit validation, and hit optimization are included. The last example deals with a new technique (saturation transfer double difference (STDD)) that might have a considerable impact on drug development in future. The examples reflect the techniques we regard as the most promising alternatives for the drug discovery process.

5.1 Fluorine Screening

Recently, we screened riboflavin synthase (RiSy) against a small library of ^{19}F-containing ligands. The library encompasses about 400 ligands containing either an aromatic fluorine or a trifluoromethyl moiety.

RiSy catalyzes the final step of riboflavin (vitamin B_2) biosynthesis. During the last step, two lumazine units (**1**, DMRL) are fused to form riboflavin (**2**) and 5-amino-6-ribitylamino-2,4-(1*H*,3*H*)-pyrimidinedione (**3**, ARP) (see Figure 9). ARP is recycled afterwards in the biosynthetic pathway via lumazine synthase. The disproportionation reaction that results in the formation of vitamin B_2 involves the transfer of a C_4-fragment between the two DMRL molecules. The xylene ring system of riboflavin is obtained by a head-to-tail assembly of the two substrate molecules.

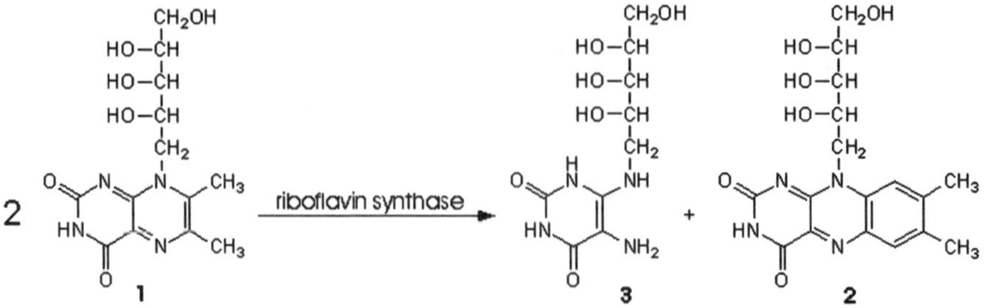

Figure 9 *Biosynthesis of riboflavine (**2**) by catalysis of RiSy. Two molecules of lumazine (**1**) form one molecule of riboflavine and one molecule of 5-amino-6-ribitylamino-2,4-(1H, 3H)-pyrimidinedione (**3**) by disproportionation*

RiSy consists of three identical subunits denoted α.[51,52] They form a complex of pseudo-D_3 symmetry with a total molecular mass of about 75 kDa. The α-subunit (23 kDa) shows an internal sequence similarity resulting in a 26% identity of the C- and N-terminus. The N-terminal domain can be expressed separately and forms a homo-dimer that binds the substrate with a similar affinity to the complete trimeric protein.[53–55]

This enzyme forms an attractive target for medical chemistry to create an anti-infective drug. Selective suppression of this enzyme will result in a deficiency of vitamin B_2. While mammals have the ability to absorb riboflavin from their nutrition, gram-negative bacteria and various yeasts lack an equivalent mechanism[56–59] and hence depend completely on the endogenous biosynthesis of this vitamin. This requirement facilitates the development of a specific drug, as the target is present in bacteria and fungi but not in the host.

Several inhibitors mimicking the scaffold of lumazine (**1**) have been identified,[60] but none of them consisted of new structural features. To overcome this drawback, a screening of fluorinated ligands was performed.

The ^{19}F-chemical shift was measured in advance for each member of the library. These reference experiments were carried out in D_6-DMSO solutions with trifluoroethanol (TFE) as an internal standard. The 400 ligands were divided into groups of about 10–30 compounds, resulting in 15 mixtures. The combinations were chosen such as to avoid reactions and overlap of ^{19}F-chemical shifts. For the composition of these mixtures, stock solutions in D_6-DMSO were set up and added to buffer solutions (50 mM KH_2PO_4, 10 mM EDTA, 10 mM Na_2SO_3, pH 7.1). To test for binding, a portion of the stock solution was added to a solution of RiSy (0.1 mM) in the same buffer. For both samples 1D ^{19}F-spectra were acquired at 14.09 T (564 MHz ^{19}F-frequency).

Compounds binding to RiSy experienced a significant line broadening (Figure 10); this effect was observed for 11 out of the 400 ligands (Table 3). The six best (approximated from the amount of broadening) were combined in a new mixture, which was added in excess to a small amount of enzyme. In the case of competitive binding it is assumed that the binder with the highest affinity displaces all the other ligands. From this experiment it was possible to identify the two strongest binders, compounds **4** and **5** (Figure 11).

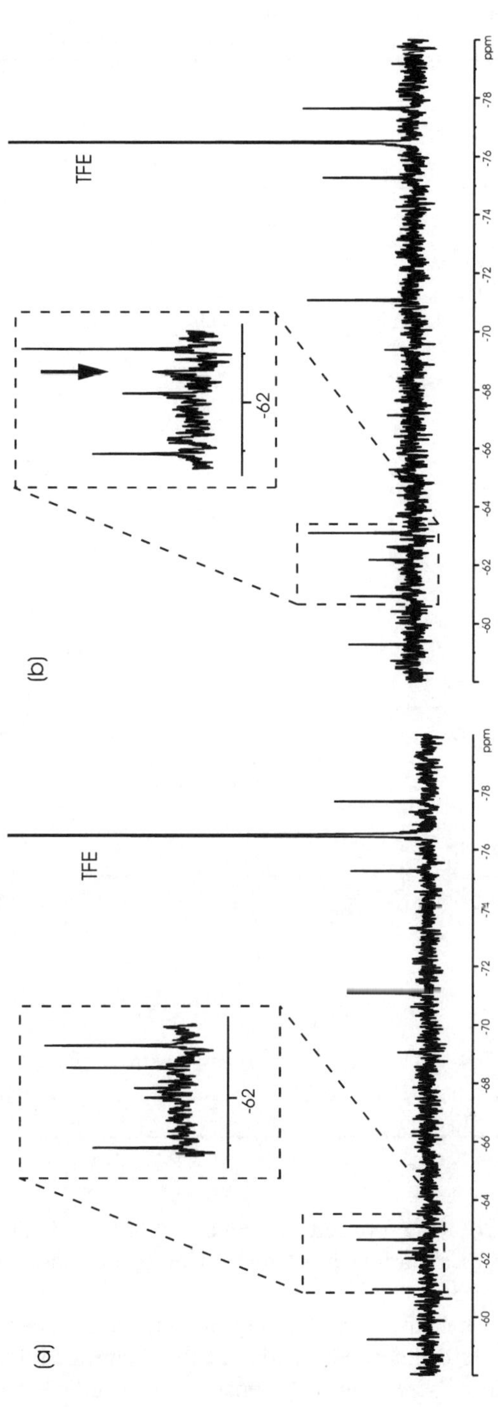

Figure 10 *Typical ^{19}F spectrum of a screening mixture with (a) and without (b) protein. In the expansion, the change of the chemical shift as well as the line broadening of the centre signal is visible*

Table 3 *Hits found by ^{19}F-screening*

No	Compound	No	Compound
4	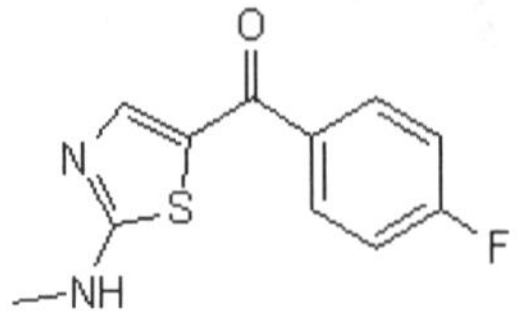	10	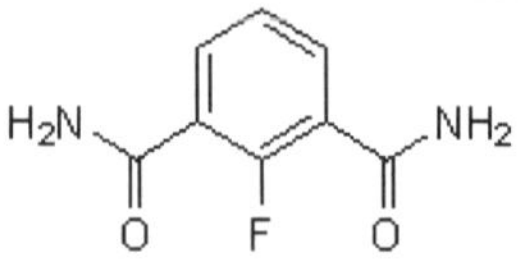
5	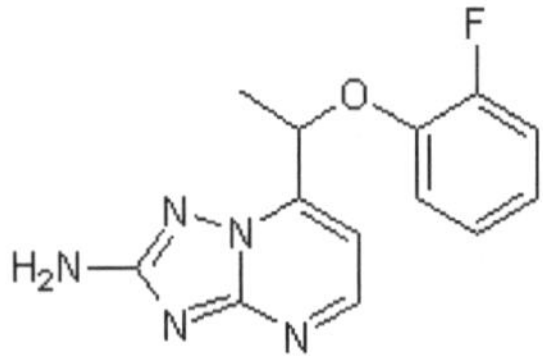	11	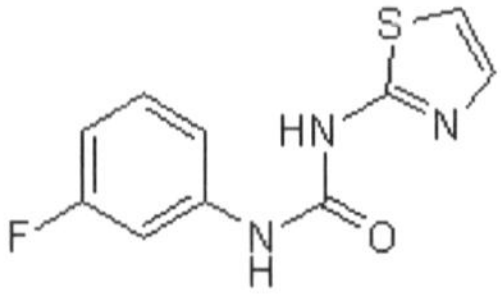
6	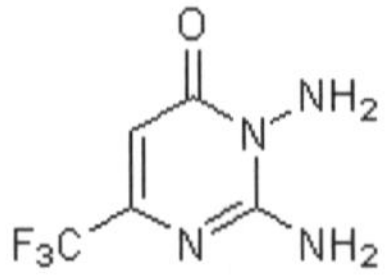	12	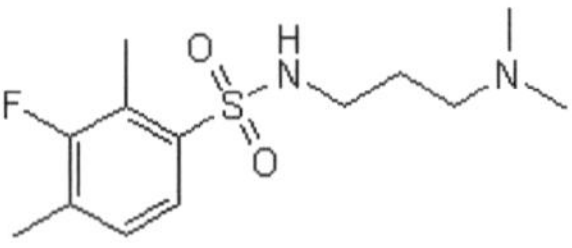
7	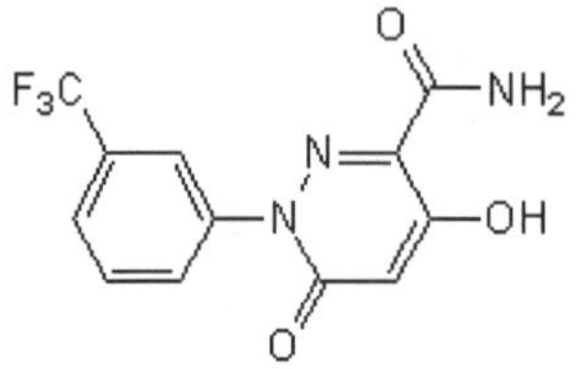	13	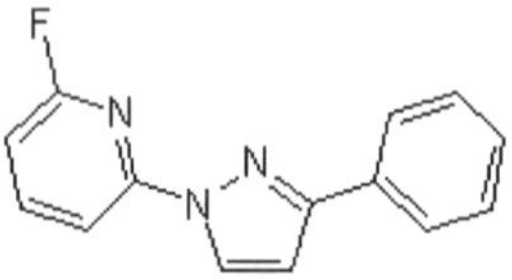
8	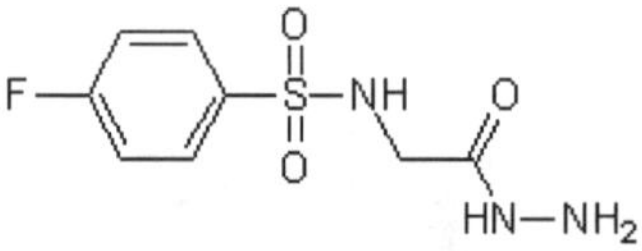	14	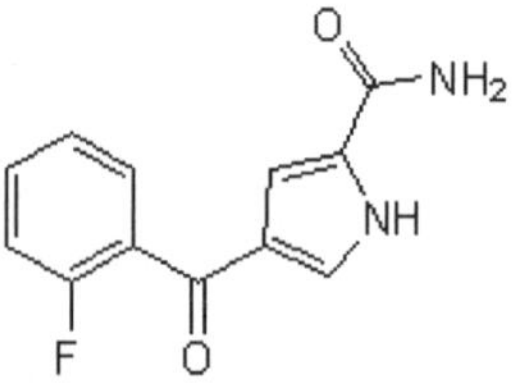
9	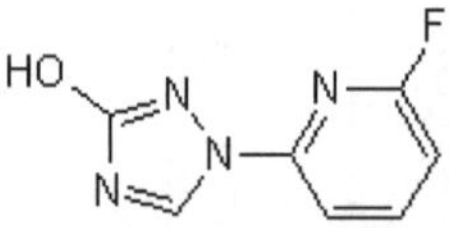		

An enzymatic assay of RiSy revealed a K_i value of about 20–30 μM for compound
4, which could serve as a starting point for further optimizations to generate an
appropriate lead structure.

This example shows the ease of performing fluorine-based screenings. Once the
^{19}F-library is characterized (knowledge about all ^{19}F-chemical shifts) and standard
mixtures can be used for screening, it becomes a rather efficient technique. The
experimental time to acquire a spectrum of one mixture took about 75 min in this

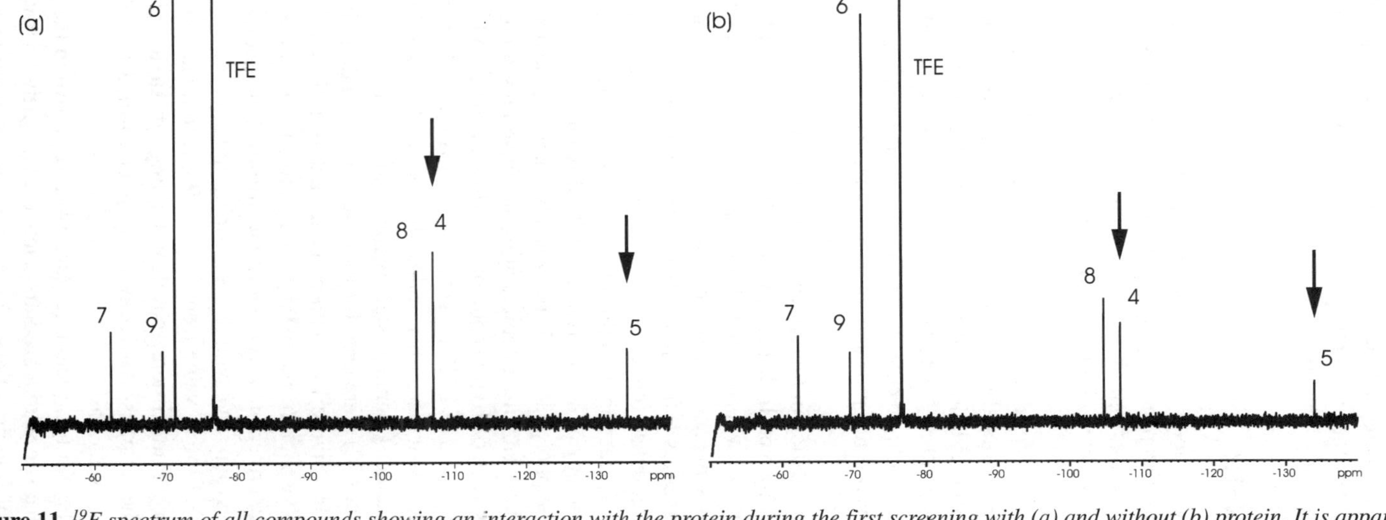

Figure 11 ^{19}F spectrum of all compounds showing an interaction with the protein during the first screening with (a) and without (b) protein. It is apparent that compounds **4** and **5** are the highest affinity binders among the ligands investigated

case, but with higher concentrations, experimental times can be reduced to 30 min or even less. This estimate includes the sample set up, data acquisition, and the change of the sample by the autosampler. Therefore, a throughput of 400–1000 compounds pcr day can easily be achieved if 20 components are assumed per mixture.

5.2 SAR-by-NMR

SAR-by-NMR[1] represents a very intuitive way of rational, fragment-based drug design. Two or more independently optimized ligands of different binding sites are combined to give a high-affinity binder. The example presented here deals with the development of a selective inhibitor for the protein tyrosine phosphatase 1B (PTP1B).[61]

PTP1B operates in the insulin-signalling pathway. When insulin binds to its receptor, there is a conformational change of intracellular region of the protein, which results in the *O*-phosphorylation of three tyrosine residues as the first step in the cascade of the insulin signalling.[62] PTP1B is reckoned to be responsible for the dephosphorylation of the insulin receptor.[63–67] Dephosphorylation results in the down regulation of the insulin receptor and therefore selective inhibition of PTP1B may enhance insulin activity.

The exogenous regulation of this enzyme is of general interest as disturbed insulin signalling may lead to type II diabetes. Currently, about 130 million people suffer from this disease and it is expected that this number will rise during the next years.[68] Nowadays, the treatment encompasses the application of insulin and hypoglycaemic drugs, but none of them is sufficient to assuage the disease completely.[69]

Many processes in eukaryotic cells are regulated by reversible phosphorylation. Hence, compounds interacting with this class of proteins are regarded as promising candidates for the therapy of many diseases. PTP1B belongs to one subclass of these proteins, the tyrosine phosphatases. Unfortunately, the catalytic domain is generally conserved among these enzymes, complicating selective inhibition.

First, a primary screen of 10,000 compounds was performed to find new scaffolds that bind to the active site of PTP1B. This screen involved a shortened form of the protein (292 residues) that was either uniformly ^{15}N-labelled or selectively labelled with δ-^{13}CH$_3$ isoleucine. Chemical shift perturbations in the ^{15}N-HSQC and ^{13}C-HSQC spectra (Figure 12) were used for the detection of binding ligands. From the screen, compound **15** (Figure 13) could be identified as a ligand. Comparison with the natural ligand phosphotyrosine (pTyr) revealed that both ligands have equivalent binding modes. For further improvement of the affinity, a derivative of the new ligand with a bulkier group was synthesized (naphthyloxamic acid, **16**). This ligand showed an increase of the K_i value from 293 to 39 μM in a *para*-nitrophenyl phosphate (*p*NPP) assay. Furthermore, covalent binding of **16** to PTP1B could be excluded by enzyme kinetics. An X-ray analysis confirmed that naphthyloxamic acid also binds in the same fashion as the natural ligand.

Apparent from the X-ray structure (Figure 14) is a groove in the protein surface connecting the active site with a second, inactive site. Further derivatization of the lead with the goal of finding a linker for the two sites resulted in compound **17**. A diamido chain at 4-position gave a 40-fold boost in affinity ($K_i = 1.1$ μM).

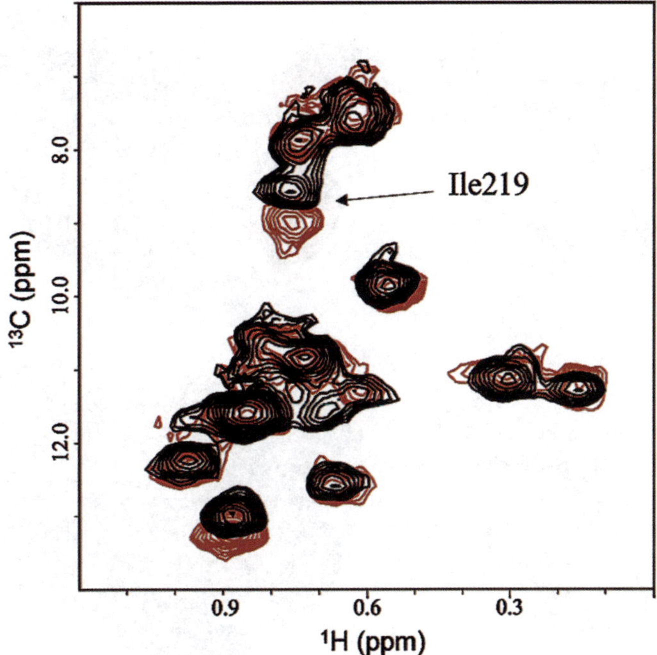

Figure 12 *^{13}C-HSQC of PTP1B with ligand (red) and without (black) ligand. The chemical shift change for the isoleucine residue 219 indicates binding of the ligand to the corresponding amino acid*

Figure 13 *Structure evolution of a high-affinity binder (18) for PTP1B via SAR-by-NMR screening*

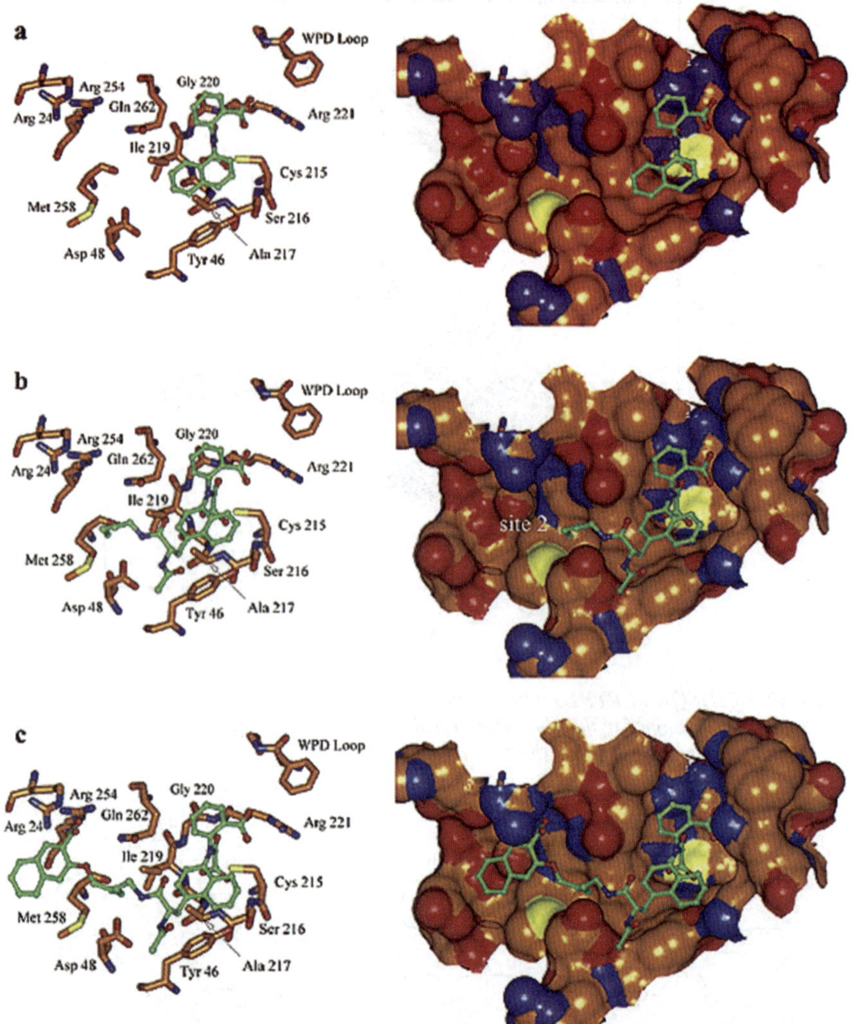

Figure 14 *X-ray structure of compounds **15–18** with PTP1B (a–c) revealing the same binding mode. Also visible is the groove in the protein surface that connects the catalytic and the non-catalytic binding sites*

The second, non-active site is also a binding site for pTyr. This non-catalytic site offers great potential to increase the selectivity of the new ligand for PTP1B. Another NMR screen, also involving 10,000 compounds, provided several hits, mostly small fused-ring aromatics. This time, the screening was performed with the truncated protein consisting of selectively ^{13}C-labelled methionine. Residue Met258 is present in the second binding site and is therefore an excellent probe for a binding event. The fusion of 3-hydroxy-2-naphtoic acid and **17** resulted in a nanomolar binder (**18**) with a 50-fold higher affinity ($K_i = 22$ nM).

Table 4 *Relative sensitivities of the hits to different phosphatases*

Phosphatase	Compound			
	15	*16*	*17*	*18*
PTP1B	1	1	1	1
TCPTP	1	1	1	2
LAR	>3	1	6	36
SHP-2	>3	>7	27	104
CD45	>1	>4	380	2700
Calicineurin		>7	>270	>13000

Finally, this inhibitor was cross-checked against five other phosphatases to determine the selectivity. Among those analyzed, TCPTP is closely related to PTP1B on the basis of sequence homology. Therefore, a difference in the affinity of the evaluated compounds for these two enzymes would be a remarkable achievement. The relative selectivities are summarized in Table 4, indicating that compound **18** is indeed able to "distinguish" between PTP1B and TCPTP.

This example underscores the efficiency of SAR-by-NMR and moreover illustrates the fragment-based approach.

5.3 Saturation Transfer Double Difference

STD spectroscopy[9,31] is a frequently used tool to test ligands for binding. The reason is the large variability and the high sensitivity of this experiment and because it can also be applied to the investigation of integral membrane proteins.[32] These proteins are especially difficult to study as they usually loose their biological activity if removed from their natural environment. Therefore, this class of proteins is reintegrated into liposomes for NMR analysis to simulate the physiological conditions. Nevertheless, most screening techniques are not able to monitor the interaction between these targets and a ligand.

Membrane-bound proteins are of highest interest for the pharmaceutical industry, as about 30% of proteins in the mammalian cells belong to this class. A subclass of these proteins are the G-protein coupled receptors (GPCRs),[70] which play an important role in signal transfer pathways. One example is the $\alpha_{IIb}\beta_3$ integrin, which occurs on the surface of platelets where it is the most prevalent component of the exposed glycoproteins. The integrin is composed of two non-covalently linked α and β subunits and is able to bind proteins and peptides consisting of an RGD motif.[71,72] Platelets are of crucial importance in the blood clotting process by forming platelet plugs. This process can be inhibited by RGD-containing peptides and non-peptidic peptidomimetics. A cyclic peptide that shows high activity (low nanomolar range) for the $\alpha_v\beta_3$ integrin is cyclo(-RGDfV-).[73,74] However, this peptide shows the desired lower affinity (micromolar range) for the $\alpha_{IIb}\beta_3$ integrin, which is required for the STD experiment to achieve a sufficiently high STD amplification by rapid exchange of bound and free ligand molecules.

For a proof of concept, Meinecke *et al.*[75] reinvestigated this interaction by STD spectroscopy on liposome, embedded $\alpha_{IIb}\beta_3$ integrin. Recently, a further example

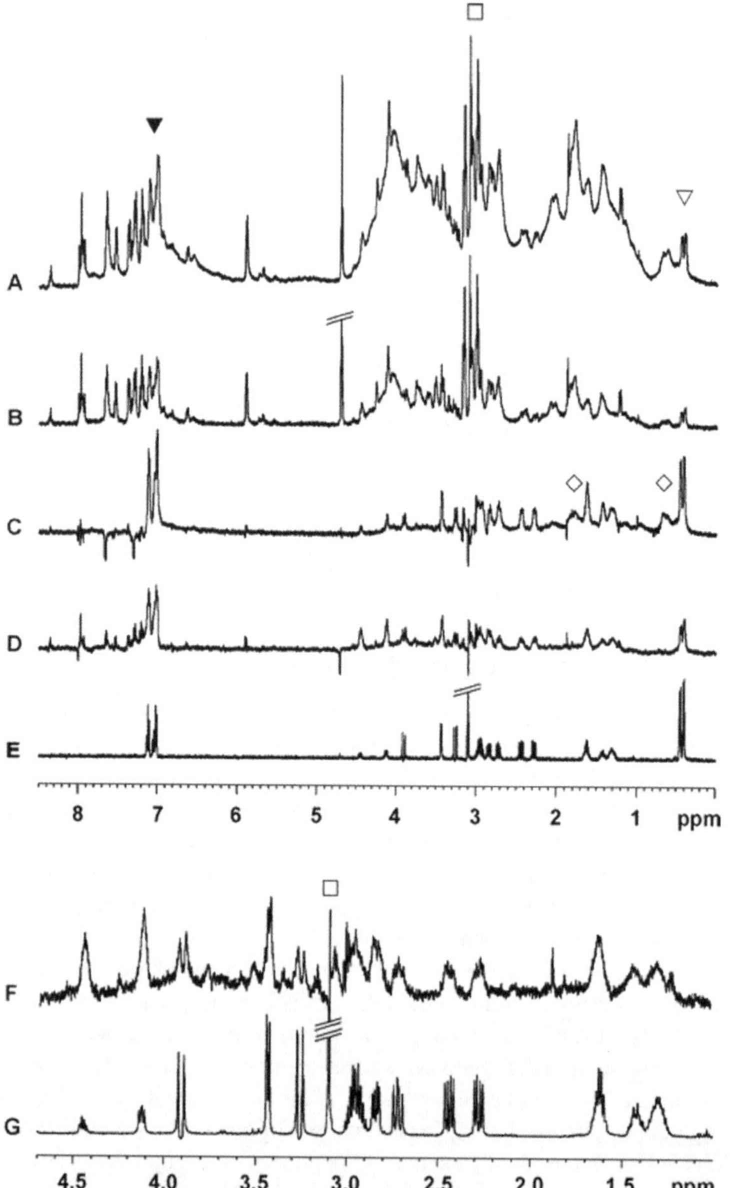

Figure 15 *Filter effect of the STDD technique. In panel A, the crowded ^1H STD NMR spectrum of a platelet suspension with cyclo(-RGDfv-) is shown, where ▼ and ▽ denote the only quantifiable signals. The unsatisfactory filter effect of a $T_{1\rho}$-element to suppress signals of the platelets resulted in spectrum B. Spectrum C (700 MHz) and D (500 MHz) show the filter effect of the saturation double difference, when compared to the spectrum of the ligand only (E). The platelet signals are marked with ◇. Spectrum F and G show expansions of spectra D and E to highlight the strong filter effect of the STDD method. Signals of TRIS included in deuterated TBS buffer are indicated by ▢*

that probably represents a key step towards in-cell screening was published by Claasen *et al.*[76] They investigated the binding of the above-mentioned system *in vivo*. Intact human platelets were mixed with cyclo(-RGDfV-) and the resulting solution was analyzed with a new technique called STDD spectroscopy (see below).

Suspensions of human platelets were prepared from concentrated blood donations. The work up yielded two samples, both containing the same composition of platelets in a D_2O-TBS (deuterated TRIS saline) buffer. The amount of platelets was estimated to $7–10^9$, representing an effective concentration of 100–600 pM if a number of $1–5–10^4$ receptors is assumed per platelet. One sample (A) was treated with the cyclic peptide at a concentration of 150 nM, while the other one (B) remained unaltered. The normal STD analysis of sample A containing a 250–1500-fold excess of the ligand did not result in an interpretable spectrum. Even a $T_{1\rho}$-filter element integrated into the pulse sequence to suppress resonances of large-size macromolecules did not lead to significant improvement. The problem was primarily the result of the fact that the suspension consisted of many small and large molecules showing a lot of binding events. The resulting STD spectrum obviously contained signals of these compounds. To filter these contributions out of the spectrum, a reference STD spectrum that included exactly these ligands was recorded and subtracted afterwards. For this reference, an STD spectrum of sample B was acquired. Subtraction of the two resulted in the spectrum shown in Figure 15D. This double difference method also has the advantage that residual signals of macromolecules are eliminated without the need for a relaxation filter element, thus increasing the signal-to-noise ratio. The quality of the resulting spectra is apparent by the comparison of the STDD spectrum and the simple 1D spectrum of the pure cyclic peptide (Figure 15E).

Compared to the STD spectrum of cyclo(-RGDfV-) in the presence of integrin $\alpha_{IIb}\beta_3$ integrated into liposomes, the STDD spectrum of the native integrin showed a fivefold higher STD effect. Moreover, a difference in the STD response can be observed, reflecting a slightly different binding mode. These findings emphasize the value of this experiment, particularly if all of the capabilities are considered.

Of course, this technique is not useful for a primary screen, as the ligands may interact with many targets and in addition to the desired one. However, if ligands are found with a high enough affinity and an accordingly high selectivity *in vitro*, this experiment may serve as a control to distinguish scaffolds of high activity *in vivo* from those with a low activity at an early stage of the drug discovery process.

6 Conclusion

NMR spectroscopy has evolved into an important method for screening ligand mixtures for binders to medicinal relevant protein or nucleic acid targets. A number of new technologies have been established for this purpose in the last decade; the most efficient ones are STD, waterLOGSY, and the screening of fluorine-containing libraries. Moreover, NMR spectroscopy can assist the hit validation process and provide structural information on the ligand–target complex.

To a large extent, NMR is, complementary to other techniques. The ability to detect weak binders under quasi-natural conditions is a particular advantage, making it ideally suited for the fragment-based approach for developing lead structures.

References

1. S.B. Shuker, P.J. Hajduk, R.P. Meadows and S.W. Fesik, *Science*, 1996, **274**, 1531.
2. P.A. Keifer, *Curr. Opin. Biotechnol.*, 1999, **10**, 34.
3. G. Roberts, *Drug Discov. Today*, 2000, **5**, 230.
4. A. Ross, G. Schlotterbeck, W. Klaus and H. Senn, *J. Biomol. NMR*, 2000, **16**, 139.
5. T. Diercks, M. Coles and H. Kessler, *Curr. Opin. Chem. Biol.*, 2001, **5**, 285.
6. M. Heller and H. Kessler, *Pure Appl. Chem.*, 2001, **73**, 1429.
7. B.J. Stockman and C. Dalvit, *Prog. Nucl. Magnetic Reson. Spectrosc.*, 2002, **41**, 187.
8. M. Pellecchia, D.S. Sem and K. Wüthrich, *Nat. Rev. Drug Discov.*, 2002, **1**, 211.
9. B. Meyer and T. Peters, *Ang. Chem.-Int. Ed.*, 2003, **42**, 864.
10. M. Coles, M. Heller and H. Kessler, *Drug Discov. Today*, 2003, **8**, 803.
11. W. Jahnke and H. Widmer, *Cell. Mol. Life Sci.*, 2004, **61**, 580.
12. J.W. Peng, J. Moore and N. Abdul-Manan, *Prog. Nucl. Magnetic Reson. Spectrosc.*, 2004, **44**, 225.
13. J.R. Huth and C.H. Sun, *Combinat. Chem. High Throughput Screening*, 2002, **5**, 631.
14. D.C. Rees, M. Congreve, C.W. Murray and R. Carr, *Nat. Rev. Drug Discov.*, 2004, **3**, 660.
15. M. van Dongen, J. Weigelt, J. Uppenberg, J. Schultz and M. Wikstrom, *Drug Discov. Today*, 2002, **7**, 471.
16. J. Fejzo, C.A. Lepre, J.W. Peng, G.W. Bemis, Ajay, M.A. Murcko and J.M. Moore, *Chem. Biol.*, 1999, **6**, 755.
17. J.W. Peng, C.A. Lepre, J. Fejzo, N. Abdul-Manan and J.M. Moore, *Nucl. Magnetic Reson. Biol. Macromol., Pt A*, 2001, **338**, 202.
18. P.J. Hajduk, A. Gomtsyan, S. Didomenico, M. Cowart, E.K. Bayburt, L. Solomon, J. Severin, R. Smith, K. Walter, T.F. Holzman, A. Stewart, S. McGaraughty, M.F. Jarvis, E.A. Kowaluk and S.W. Fesik, *J. Med. Chem.*, 2000, **43**, 4781.
19. C.A. Lipinski, F. Lombardo, B.W. Dominy and P.J. Feeney, *Adv. Drug Deliver. Rev.*, 1997, **23**, 3.
20. M. Mayer and B. Meyer, *J. Am. Chem. Soc.*, 2001, **123**, 6108.
21. K. Pervushin, R. Riek, G. Wider and K. Wuthrich, *Proc. Natl. Acad. Sci. USA*, 1997, **94**, 12366.
22. R. Riek, G. Wider, K. Pervushin and K. Wüthrich, *Proc. Natl. Acad. Sci. USA*, 1999, **96**, 4918.
23. J. Weigelt, M. Wilkstrom, J. Schultz and M.J.P. van Dongen, *Combinat. Chem. High Throughput Screening*, 2002, **5**, 623.
24. K.H. Gardner and L.E. Kay, *J. Am. Chem. Soc.*, 1997, **119**, 7599.
25. N.K. Goto, K.H. Gardner, G.A. Mueller, R.C. Willis and L.E. Kay, *J. Biomol. NMR*, 1999, **13**, 369.
26. V. Tugarinov and L.E. Kay, *J. Am. Chem. Soc.*, 2003, **125**, 13868.
27. P.J. Hajduk, D.J. Augeri, J. Mack, R. Mendoza, J.G. Yang, S.F. Betz and S.W. Fesik, *J. Am. Chem. Soc.*, 2000, **122**, 7898.
28. C.A. Lepre, *Drug Discov. Today*, 2001, **6**, 133.

29. C.A. Lepre, J. Peng, J. Fejzo, N. Abdul-Manan, J. Pocas, M. Jacobs, X.L. Xie and J.M. Moore, *Combinat. Chem. High Throughput Screening*, 2002, **5**, 583.

30. F.E. Koehn and G.T. Carter, *Nat. Rev. Drug Discov.*, 2005, **4**, 206.

31. M. Mayer and B. Meyer, *Ang. Chem.-Int. Ed.*, 1999, **38**, 1784.

32. J. Klein, R. Meinecke, M. Mayer and B. Meyer, *J. Am. Chem. Soc.*, 1999, **121**, 5336.

33. C. Dalvit, P. Pevarello, M. Tato, M. Veronesi, A. Vulpetti and M. Sundstrom, *J. Biomol. NMR*, 2000, **18**, 65.

34. C. Dalvit, G. Fogliatto, A. Stewart, M. Veronesi and B. Stockman, *J. Biomol. NMR*, 2001, **21**, 349.

35. A. Chen and M.J. Shapiro, *J. Am. Chem. Soc.*, 1998, **120**, 10258.

36. A.D. Chen and M.J. Shapiro, *J. Am. Chem. Soc.*, 2000, **122**, 414.

37. M. Vogtherr and T. Peters, *J. Am. Chem. Soc.*, 2000, **122**, 6093.

38. J.W. Peng, *J. Magnetic Reson.*, 2001, **153**, 32.

39. C. Dalvit, P.E. Fagerness, D.T.A. Hadden, R.W. Sarver and B.J. Stockman, *J. Am. Chem. Soc.*, 2003, **125**, 7696.

40. C. Dalvit, M. Flocco, M. Veronesi and B.J. Stockman, *Combinat. Chem. High Throughput Screening*, 2002, **5**, 605.

41. G.M. Rishton, *Drug Discov. Today*, 2003, **8**, 86.

42. B. Wang, K. Raha and K.M. Merz, *J. Am. Chem. Soc.*, 2004, **126**, 11430.

43. M.A. McCoy and D.F. Wyss, *J. Am. Chem. Soc.*, 2002, **124**, 11758.

44. Z. Serber, A.T. Keatinge-Clay, R. Ledwidge, A.E. Kelly, S.M. Miller and V. Dotsch, *J. Am. Chem. Soc.*, 2001, **123**, 2446.

45. Z. Serber and V. Dötsch, *Biochemistry*, 2001, **40**, 14317.

46. Z. Serber, R. Ledwidge, S.M. Miller and V. Dotsch, *J. Am. Chem. Soc.*, 2001, **123**, 8895.

47. C. Dalvit, M. Fasolini, M. Flocco, S. Knapp, P. Pevarello and M. Veronesi, *J. Med. Chem.*, 2002, **45**, 2610.

48. C. Dalvit, M. Flocco, S. Knapp, M. Mostardini, R. Perego, B.J. Stockman, M. Veronesi and M. Varasi, *J. Am. Chem. Soc.*, 2002, **124**, 7702.

49. W. Jahnke, P. Floersheim, C. Ostermeier, X.L. Zhang, R. Hemmig, K. Hurth and D.P. Uzunov, *Ang. Chem.-Int. Ed.*, 2002, **41**, 3420.

50. W. Jahnke, L.B. Perez, C.G. Paris, A. Strauss, G. Fendrich and C.M. Nalin, *J. Am. Chem. Soc.*, 2000, **122**, 7394.

51. A. Bacher, R. Baur, U. Eggers, H.D. Harders, M.K. Otto and H. Schnepple, *J. Biol. Chem.*, 1980, **255**, 632.

52. S. Gerhardt, A.K. Schott, N. Kairies, M. Cushman, B. Illarionov, W. Eisenreich, A. Bacher, R. Huber, S. Steinbacher and M. Fischer, *Structure*, 2002, **10**, 1371.

53. D.I. Liao, Z. Wawrzak, J.C. Calabrese, P.V. Viitanen and D.B. Jordan, *Structure*, 2001, **9**, 399.

54. W. Meining, G. Tibbelin, R. Ladenstein, S. Eberhardt, M. Fischer and A. Bacher, *J. Struct. Biol.*, 1998, **121**, 53.

55. V. Truffault, M. Coles, T. Diercks, K. Abelmann, S. Eberhardt, H. Lüttgen, A. Bacher and H. Kessler, *J. Mol. Biol.*, 2001, **309**, 949.

56. S.V. Bandrin, M.Y. Beburov, P.M. Rabinovich and A.I. Stepanov, *Genetika*, 1979, **15**, 2063.

57. J.C. Sircar, T. Capiris, S.J. Kesten and D.J. Herzig, *J. Med. Chem.*, 1981, **24**, 735.
58. O. Oltmanns and F. Lingens, *Z. Naturforsch. Part B-Chem. Biochem. Biophys. Biol. Verw. Gebiete*, 1967, **B22**, 751.
59. J.C. Roberts, H. Gao, A. Gopalsamy, A. Kongsjahju and R.J. Patch, *Tetrahedron Lett.*, 1997, **38**, 355.
60. M.K. Otto and A. Bacher, *Eur. J. Biochem.*, 1981, **115**, 511.
61. B.G. Szczepankiewicz, G. Liu, P.J. Hajduk, C. Abad-Zapatero, Z.H. Pei, Z.L. Xin, T.H. Lubben, J.M. Trevillyan, M.A. Stashko, S.J. Ballaron, H. Liang, F. Huang, C.W. Hutchins, S.W. Fesik and M.R. Jirousek, *J. Am. Chem. Soc.*, 2003, **125**, 4087.
62. M.P. Czech and S. Corvera, *J. Biol. Chem.*, 1999, **274**, 1865.
63. K.A. Kenner, E. Anyanwu, J.M. Olefsky and J. Kusari, *J. Biol. Chem.*, 1996, **271**, 19810.
64. H. Chen, S.J. Wertheimer, C.H. Lin, S.L. Katz, K.E. Amrein, P. Burn and M.J. Quon, *J. Biol. Chem.*, 1997, **272**, 8026.
65. F. Ahmad, P.M. Li, J. Meyerovitch and B.J. Goldstein, *J. Biol. Chem.*, 1995, **270**, 20503.
66. H. Chen, L.N. Cong, Y.H. Li, Z.J. Yao, L. Wu, Z.Y. Zhang, T.R. Burke and M.J. Quon, *Biochemistry*, 1999, **38**, 384.
67. S. Walchli, M.L. Curchod, R.P. Gobert, S. Arkinstall and R.H. van Huijsduijnen, *J. Biol. Chem.*, 2000, **275**, 9792.
68. P. Zimmet, K. Alberti and J. Shaw, *Nature*, 2001, **414**, 782.
69. J.M. Nuss and A.S. Wagman, *Annu. Rep. Med. Chem.*, 2000, **35**, 211.
70. A. Wise, K. Gearing and S. Rees, *Drug Discov. Today*, 2002, **7**, 235.
71. R.O. Hynes, *Cell*, 1987, **48**, 549.
72. E. Ruoslahti and M.D. Pierschbacher, *Science*, 1987, **238**, 491.
73. M. Pfaff, K. Tangemann, B. Muller, M. Gurrath, G. Muller, H. Kessler, R. Timpl and J. Engel, *J. Biol. Chem.*, 1994, **269**, 20233.
74. L. Marinelli, A. Lavecchia, K.E. Gottschalk, E. Novellino and H. Kessler, *J. Med. Chem.*, 2003, **46**, 4393.
75. R. Meinecke and B. Meyer, *J. Med. Chem.*, 2001, **44**, 3059.
76. B. Claasen, M. Axmann, R. Meinecke and B. Meyer, *J. Am. Chem.l Soc.*, 2005, **127**, 916.
77. D.F. Veber, S.R. Johnson, H.Y. Cheng, B.R. Smith, K.W. Ward and K.D. Kopple, *J. Med. Chem.*, 2002, **45**, 2615.

Screening Chemical Microarrays: Methods and Applications

PAPPANAICKEN R. KUMARESAN AND KIT S. LAM

Division of Hematology & Oncology, Department of Internal Medicine, UC Davis Cancer Center, University of California Davis, 4501 X Street, Sacramento, CA 95817, USA

1 Introduction

Although the development of planar peptide microarrays in 1990[1] precedes that of DNA microarrays,[2,3] it is the latter that have popularized microarray technology in the late 1990s. However, it was not until the early 2000s that peptide, small molecule, carbohydrate, and protein planar microarrays became an important tool for proteomics and diagnostics. Chemical microarrays generally involve the synthesis and immobilization of chemical compounds as microspots on a planar solid support in a spatially addressable manner. These microarrays can be prepared by (i) *in situ* synthesis, using a photolithographic light-directed synthesis method,[1] a maskless approach with micro-mirrors devices[4–7] or SPOT synthesis with standard solid-phase chemistry[8,9] or (ii) pre-synthesizing the chemical compounds, which are then spotted onto solid support with high-speed arrayer.[10,11] In this latter approach, compounds can be generated either by parallel synthesis (*e.g.* in 96 deep-well plates) or by "split-mix" synthesis using macrobeads,[12,13] the IRORI radiofrequency-tagged system[14,15] or bead aggregates.[16] Since chemical microarray consists of a large number of chemical compounds, it can also be considered as a combinatorial chemical library.[10] The "one-bead-one-compound" (OBOC) combinatorial library first reported by us in 1991 involves the use of a "split-mix" synthesis method to generate compound-beads, such that each bead displays only one chemical entity.[17,18] The OBOC combinatorial bead-library is essentially a high-density chemical microarray format that is spatially separable but not addressable. Each bead measures approximately 80–100 μm in diameter, the size of a typical high-density microarray spot. The compound-beads can be screened in suspension in screening buffer[17,18] or immobilized in a petri dish with soft agar for solution-phase releasable assays.[19,20]

This approach is highly efficient, as thousands to hundreds of thousands or even millions of compounds can be generated easily. In this review, we shall briefly discuss the preparation of these chemical microarrays and follow with a more detailed discussion on the various screening methods. Applications of these microarrays to various biological systems will also be described. The approaches that have been employed for preparing microarrays and the methods used to detect activity in different applications are summarized in Table 1. Schematic representations of the generation of planar chemical arrays and bead-arrays are shown in Figure 1. DNA and

Table 1 *Diverse options for microarray formats, detection methods and applications*

Compounds synthesized in situ, directly on solid support	*Microarray formats for compounds synthesized in situ*	*Detection methods used for screening*	*Examples of applications*
Photolithographic synthesis	Glass slide	Fluorescence	Protein binding
Micromirror device	Nitrocellulose membrane	Chemiluminescence	Receptor/ligand interactions
SPOT synthesis	Microtiter plate	Radioisotope	Cell signaling
		Colorimetric	Enzyme specificity
		Label-free optical	& inhibition
Compounds pre-synthesized spotted on solid support	*Microarrays of pre-synthesized compounds*	Mass spectrometry	Chemical detection
		Fiber optic bead	Medical
Parallel synthesis	Peptide/peptoid	Laser detection	applications
Split-mix synthesis	Carbohydrate	Electrochemical	Drug discovery
	Small molecule	Biosensor-based methods	
	One-bead-one-compound (OBOC)		

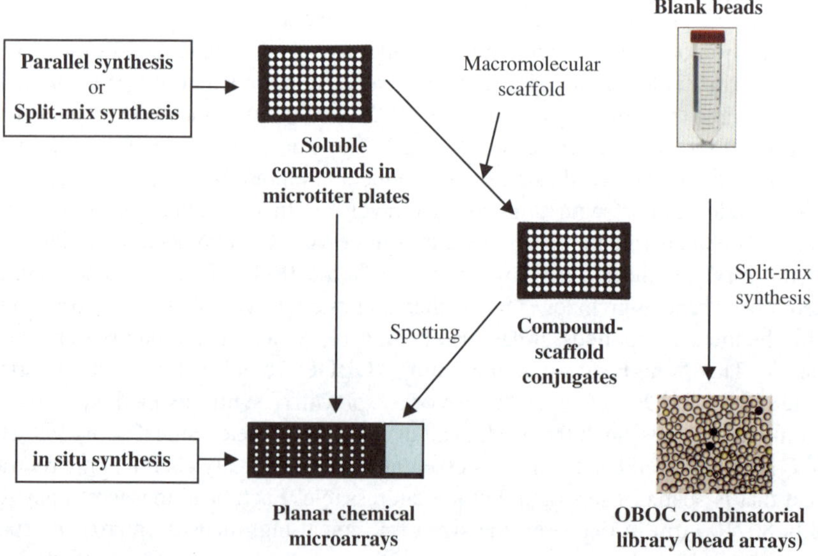

Figure 1 *Schematic representation of generation of planar chemical arrays and bead-arrays*

protein microarrays will not be discussed, although many of the screening methods described here can also be applied to protein microarrays.

1.1 *In situ* Synthesis of Peptide and Non-Peptide Microarrays

Because the coupling efficiency of solid-phase peptides synthesis, like that of oligonucleotide synthesis, is extremely good, high-density peptide microarrays can be synthesized *in situ* using light-directed deprotection approaches on glass chips.[4–6] These light-directed approaches are generally incompatible with most carbohydrate or small-molecule microarrays. Another *in situ* synthesis method is the SPOT array, in which low-density peptide microarrays are synthesized as spots on cellulose paper. Since light-directed deprotection is not required in the synthesis of a SPOT array, the chemistry is less restrictive and small molecule arrays can, in principle, be prepared in this way. Recent developments, such as the introduction of novel polymeric surfaces, new linker and cleavage strategies, as well as automated robot systems, have increased the versatility of the SPOT microarray.[9,21,22] This method has also been adapted to the parallel synthesis of other compounds, such as peptide nucleic acids,[22] peptidomimetics such as peptomers[23] or peptoids,[24] and small heterocyclic compounds.[25] Combining a synthetic method with high-throughput solid- and solution-phase screening assays, the SPOT technique has become a valuable tool in biomedical research. Selection of an appropriate planar surface represents an important prerequisite for the successful "SPOT-wise" assembly of peptides and subsequent screening. This technology has been reviewed recently.[21]

1.2 Spotting of Pre-Synthesized Small Molecules and Peptides

Over the last few years, small molecule planar microarrays have become an important tool for high-throughput experimentation in biological research.[10,26] Methods for immobilization of small molecules on a planar surface have been recently reviewed.[27] Except for the SPOT array, in which some heterocyclic libraries can be synthesized *in situ*, most other planar small-molecule microarrays require pre-synthesis of the small molecule compounds with an appropriate functional group for subsequent ligation. Immobilization of small-molecule compounds on the planar support can be accomplished by three different approaches. In one approach, the compounds can be covalently ligated directly to the surface non-specifically[28] or through a chemo-selective ligation strategy[29] via a hydrophilic linker. A second method is to first biotinylate the compounds and then spot them on a streptavidin or anti-biotin antibody-coated surface.[30] A third approach is to first chemoselectively ligate the compounds to a macromolecular scaffold, such as agarose or human serum albumin, followed by spotting of the compound-macromolecular scaffolding onto the solid support.[31]

1.3 Carbohydrate Microarrays

Carbohydrates are key components of cell-surface glycolipids and glycoproteins, which play an important role in cellular functions such as cell adhesion, fertilization, differentiation, development, and tumor-cell metastasis. By analogy to peptide and small molecule microarrays, carbohydrates can be immobilized onto solid surfaces via covalent linkage, non-specific adsorption, or specific but non-covalent binding. Covalent ligation is required for small carbohydrates, but large carbohydrates or carbohydrate-containing macromolecules can be arrayed onto solid supports via non-specific adsorption. Four types of sugar arrays have been reported: (i) polysaccharide and glycoconjugated microarrays,[32] (ii) monosaccharide chips,[33] (iii) natural and synthetic oligosaccharide arrays,[34] and (iv) synthetic oligosaccharides in microtiter plate format.[35] Immobilization of glycans on planar surfaces has been recently reviewed.[26,36]

1.4 One-Bead-One-Compound Combinatorial Library Bead-Arrays

OBOC combinatorial bead-libraries can be considered as chemical microarrays that are spatially separable but non-addressable. The identity of the chemical compound on the positive beads can be determined directly with an automatic sequencer if it is an N-terminally unprotected peptide,[37] by mass spectroscopy,[38–40] or through chemical encoding.[15,41,42] A synthetic scheme for the OBOC library is shown in Figure 2. Using the highly efficient "split-mix" synthesis method,[17,18,43–45] literally hundreds of thousands to millions of compounds can be prepared within a week. The recent

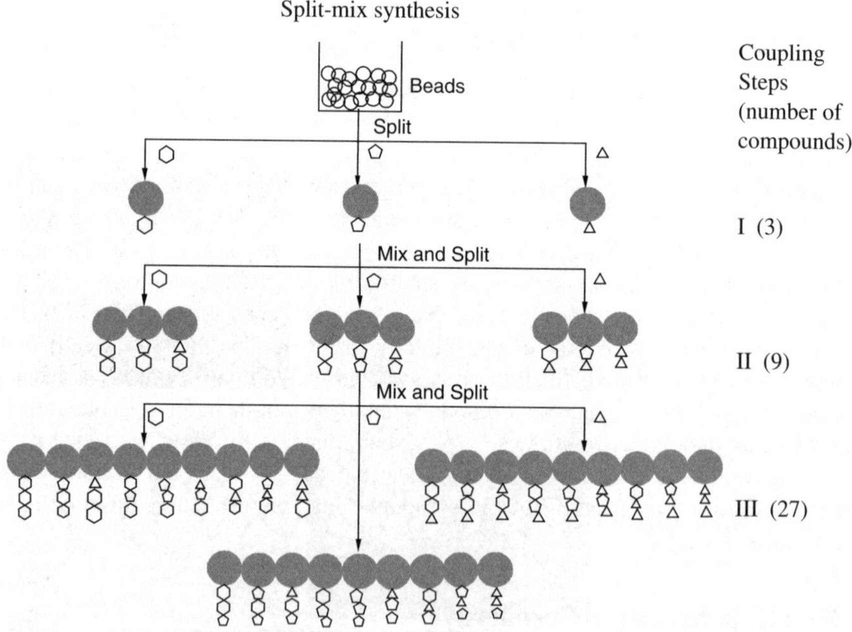

Figure 2 *Synthetic scheme of the "split-mix synthesis" method to generate an OBOC combinatorial library*

development of the bilayer beads[46] enables one to encode the OBOC library very efficiently.[46–49]

2 Screening of Chemical Microarrays

Many of the screening methods that have been developed for OBOC combinatorial peptide or small molecule libraries (bead-arrays)[17,50] have also been applied to planar chemical microarrays.[51] These screening methods (Figure 3) generally involve the binding of target proteins to immobilized ligands. Based on the detection technique, screening methods are broadly classified into two groups.

(i) Labeling method in which the analytes contain target proteins that are usually tagged with reporter groups such as alkaline phosphatase [Figure 3(A)], fluorescent dyes, radionuclides, or epitope tags such as polyhistidine, glutathione-S-transferase, or biotin, which in turn can be detected by secondary reagents containing fluorescent dyes or radionuclides. Functional assays for

Screening Methods	Low magnification	High magnification
A. Colorimetric assay: Biotinylated target protein followed by streptavidin-alkaline phosphatase are used to screen bead library; positive beads turned dark upon adding of BCIP substrate.	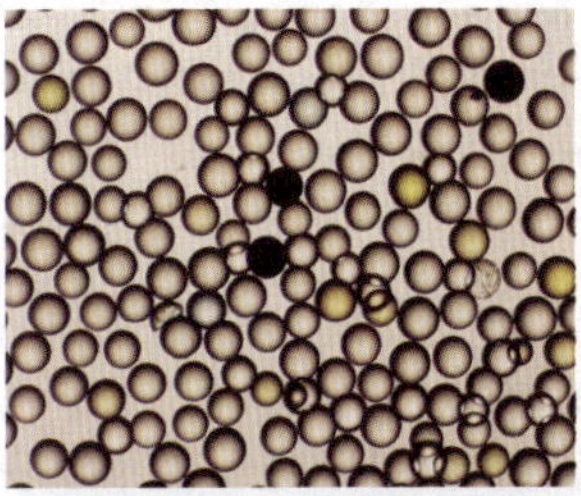	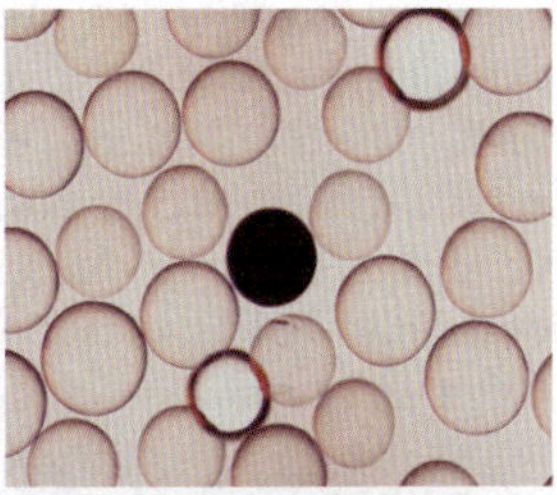
B. Fluorescent assay: Protease is used to screen bead library for protease substrate; positive beads in this fluorescent-quench assay fluoresce because the quencher is cleaved.	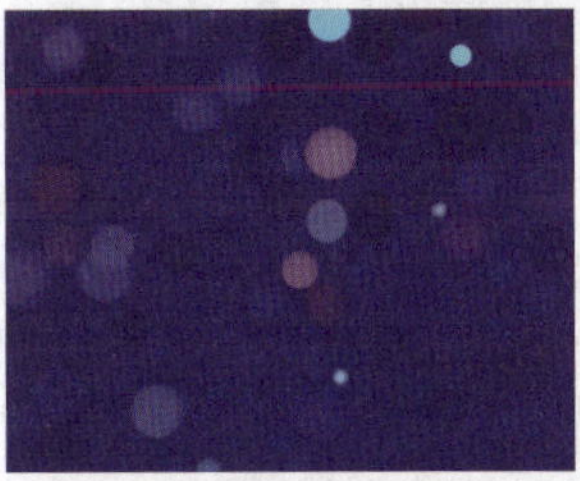	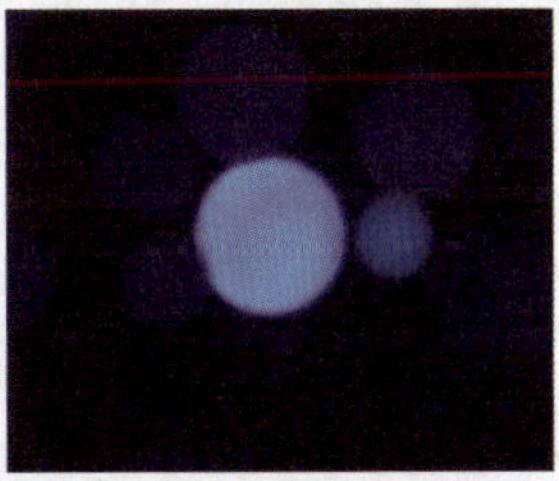
C. Cell binding assay: Live cells are used to screen bead library for cell surface binding ligands; positive beads are coated with a monolayer of cells.	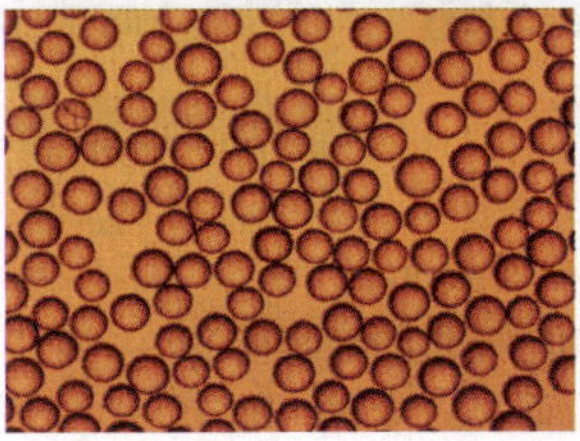	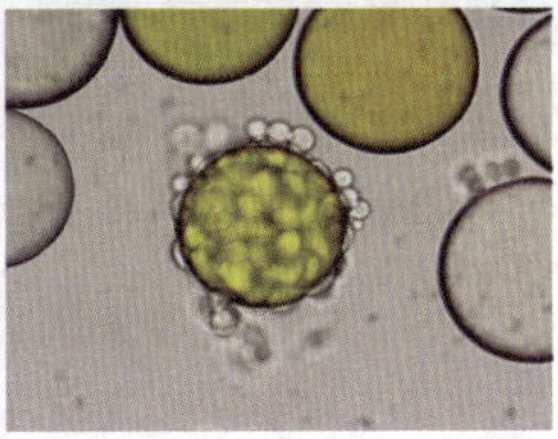

Figure 3 *Some of the on-bead screening methods for OBOC combinatorial libraries (bead-arrays)*

substrates of enzymes such as protein kinases and proteases have also been developed for microarrays. In these assays, modification of the immobilized compounds by the enzymes can be detected with a fluorescent quench assay [Figure 3(B)] or radiolabel assay.

(ii) Other label-free detection methods for binding target proteins to the immobilized chemical microarrays have also been developed. These include surface plasmon resonance (SPR),[52] grating couplers,[53,54] and reflectometry,[55] among others. Whole cell-binding assays on beads have been developed for the discovery of binding ligands for various cancers[30,51] [Figure 3(C)], and planar ligand-arrays can be used to profile-binding activity of ligands to various cell types. *In situ* releasable solution-phase cell-based assays have also been developed for both bead-arrays[56] and planar microarrays.[57]

2.1 Labeling Methods

2.1.1 *Fluorescence Method*

For binding assays with soluble proteins, fluorescence methods are often preferred because they are simple, safe, extremely sensitive, and compatible with the commercially available microarray scanners. The most widely used fluorescent techniques include laser-induced fluorescence, fluorescence resonance energy transfer (FRET), fluorescence polarization, homogenous time-resolved fluorescence, and fluorescence correlation spectroscopy.[58] In standard fluorescence methods, analytes can be directly labeled with a fluorescent probe or indirectly with biotin, followed by detection with fluorescent-labeled streptavidin or fluorescent-labeled anti-biotin antibodies. However, most of the chemical conjugation methods are not site-specific and there is always a concern that chemical labeling of a protein may negatively affect its ability to bind its ligand. Recently, using molecular biology techniques, proteins have been expressed either as a fluorophore-puromycin conjugate or a red/green fluorescent fusion protein and detected by a conventional fluorescence slide scanner.[58,59] A similar approach has been adopted to identify protein–carbohydrate interactions on carbohydrate chips.[36] These methods, while useful, require additional steps, work in some assay systems, but do not work in all proteins. Therefore, alternative methods to detect the bound proteins in their native form are sometimes preferred. One approach is to use fluorescently labeled antibodies to detect specific proteins (unlabeled) bound to the microarrays.

Recently, a small molecule fluorophore phosphosensor technology referred as Pro-Q Diamond dye has been developed to detect and quantitate phosphorylated amino acids within peptides and proteins in microarrays.[60] In addition to binding assays, fluorescence detection methods have also been developed for functional assays. For example, microarrays of quenched fluorescent substrates can be used to detect protease or esterase activities in the analytes. In this method, quenched fluorescent substrates are prepared by coupling the peptide substrate to coumarin, a fluorescent dye. These peptide substrates are then spotted onto the solid support and incubated with proteases such as caspase, matrix metalloprotease-2 (MMP-2), and trypsin. Peptides that are susceptible to proteolysis fluoresce while others do not[61,62] [Figure 3(B)]. These methods are very similar to the on-bead screening assay for OBOC combinatorial libraries developed by Meldal's group several years ago.[63]

2.1.2 Chemiluminescence Method

Chemiluminescent methods have been used for detection in chemical microarrays for their simplicity and sensitivity. They have widely been used in peptide SPOT arrays printed on cellulose membranes or nitrocellulose-coated glass slides. The microarrays are first exposed to either cell lysates or testing solution, followed by horseradish peroxidase-conjugated antibodies against the targeted antigen. After adding the luminescent substrate, the generated light is captured either by a charge-coupled device (CCD) camera or X-ray autoradiography film.[25,64]

2.1.3 Radiolabeling Methods

Radioisotope methods for detecting modifications of the immobilized compounds are the most sensitive methods and have been used successfully to probe the activities of specific post-translational modification of enzymes (*e.g.* proteases, protein kinases, esterases, glycosyl transferases, and acetylase) present in an analyte.[51,65] Zhu *et al.* analyzed the kinase–substrate specificity of almost all (119 of 122) yeast kinases using 17 different protein substrates. The substrates were first covalently immobilized on the surface of individual nanowells, and individual protein kinases in kinase buffer with [γ^{33}P]. ATP were incubated with the substrates. After washing, the nanowell chips were analyzed for[33] P-labeled substrates using a phospho-imager (Molecular Dynamics, Inc.).[66] We described similar radioisotope approaches for kinase assays using immobilized peptide substrates.[51] For efficient phosphorylation, the peptide substrates need to be tethered to the glass surface via a long hydrophilic linker.

2.1.4 Colorimetric Methods

Colorimetric methods have not been used extensively in planar microarrays due to their low sensitivity. Recently, a unique colorimetric method was described for the analysis of proteins bound to immobilized spots on the microarrays. In this method, after incubating the microarrays with the analytes (unlabeled or biotinylated), gold-conjugated antibody or streptavidin is added. Final detection is achieved by silver enhancement to produce a black image of microarray spots.[67] Rakow and Suslick have developed a colorimetric sensor array for detection of volatile chemicals at a concentration below two parts per million.[68] We have been using enzyme-linked colorimetric assays to screen OBOC combinatorial bead libraries for over a decade. Target proteins tagged directly or indirectly with alkaline phosphatase are mixed with the bead-arrays. Color development is achieved by adding bromochloroindoyl phosphate [Figure 3(A)].[69]

Chemically labeling a protein always carries the possibility of disrupting its ligand-binding activity or blocking the epitope of the protein, thereby preventing secondary antibody attachment. Therefore, an alternative method to detect the bound proteins in their native form is sometimes preferred.

2.2   Label-Free Optical and Mass Spectrometry Methods

Label-free optical techniques for detecting bound proteins on microarrays have been recently reviewed.[70–72] The advantage of these methods over labeling methods is that the native form of the analyte is preserved. These methods include SPR,[52] surface-enhanced laser desorption/ionization mass spectrometry (SELDI-MS),[73] atomic force microscopy[74] and fiber-optic methods.

2.2.1   Surface Plasmon Resonance

SPR has now matured as a versatile detection tool for analysis of the kinetics of protein–ligand interactions over a wide range of molecular weights, affinities, and binding rates.[75–79] SPR detection optics can be adapted to the requirements of array imaging. The array is illuminated homogeneously with monochromatic light tuned to a wavelength close to the resonance condition, and an image is taken by CCD camera. Upon binding of macromolecules, the affected array spots loose reflectance, and a differential image reveals darkened spots. Houseman and Mrksich[33] reported the use of SPR as a detection method to profile carbohydrate–lectin interactions on carbohydrate arrays. As an alternative detection method, Sapsford *et al.*[80] used a planar waveguide to develop an antibody array biosensor and studied the kinetics of antigen–antibody interactions in parallel.

2.2.2   Oblique-Incidence Optical Reflectivity Difference Microscopy

The label-free oblique-incidence optical reflectivity difference (OI-RD) detection method for analyzing microarrays described by Zhu and his colleagues is based on the change in reflectivity (defined as the ratio of the reflected to the incident electric field), specifically the disproportionate changes in reflectivity at oblique incidence for transverse electric mode (the s-polarized) and transverse magnetic mode (the *p*-polarized) arising from such an array. The difference in fractional reflectivity changes between the two polarization components that can be measured in microscope configuration.[81–83]

2.2.3   Surface-Enhanced Laser Desorption/Ionization Mass Spectrometry

In recent years, various MS technologies have evolved as the dominant tools for identification and characterization of bound proteins.[84,85] In the future, microarrays in combination with MS will provide new directions in biotechnology research. The microchip development for MS microarray and its application have been reviewed.[86] In this method, captured proteins bound to the metal surface (SELDI protein array) are vaporized with a laser beam and the proteins are then identified by MS.[87] However, the ability to detect minute mounts of bound proteins on high-density microarrays using this approach still remains to be developed.

2.2.4 Atomic Force Microscopy

AFM has been used to detect surface topological changes on the imprinted compounds that occur when the imprinted compounds bind the analytes.[88] However, AFM is very slow and the method is impractical unless the chip can be scanned with a large array of AFM probes in parallel.

2.2.5 Fiber-Optic Bead Methods

Walt *et al.* developed the fiber-optic microarray biosensor technology that has been commercialized and is now referred to as the Illumina bead-array. This microarray is composed of bundles of selectively etched glass fibers, which are dipped into the OBOC oligonucleotide library of microspheres (3 or 5 µm). For some reason, these microspheres are captured randomly and spontaneously at the end of each optical fiber.[89,90] The array is reusable, allows rapid responses, and has extremely low detection limits. However, because the microarray is spatially non-addressable, an encoding method is needed.[90–93] Recently, encoded fiber-optic microsphere arrays of carbohydrates have been reported.[94] SRU Biosystem has very recently developed a technology in which a narrow-band guided-mode resonance reflectance filter is incorporated with the matrix. When the light beam passes through, the reflected color in the resonance filter is modulated by the attachment/detachment of biochemical material to the surface.[95]

2.2.6 Laser-Detection Methods

A compact disc-based microarray system was developed by Kıdo *et al.*[96] and used for immunoassays. They used a piezoelectric inkjet applicator to deposit the proteins onto a polycarbonate disc. Recently, Clair *et al.* reported the attachment of small molecules to a polycarbonate compact disc (CD) surface via phosphodiester bonds. Molecular interactions between analytes and some of these molecules can be detected with a conventional CD player.[97]

2.2.7 Electrochemical Biosensor Method

The electrochemical biosensor method has shown high potential in microarray usage for its sensitivity, reproducibility, selectivity, and reversibility. Unlike other methods, the biosensor method depends on the electrochemical property of the coating receptors and binding analytes. It is prepared by electrochemical synthesis of polymers *in situ* on platinum electrodes.[98,99] This technology, however, will require further development before it can be applied to high-density microarrays. Potentially, it can be applied in the field to identify airborne pathogens in a hand-held device.

2.2.8 *Cell-Based Assays*

We have reported the printing of cell-binding peptide microarrays and their use to profile cell–surface-binding properties in a number of different cell types.[30,51] In these experiments, the polystyrene slide is first coated with neutravidin, and biotinylated peptides are then printed on the surface of the slide. A cell suspension containing a lymphoma cell line, normal lymphocytes, or cells derived from cancer biopsy specimens or a leukemia patient's peripheral blood is then applied to each of the replicate sets in peptide microarrays. After incubating for about 30 min, the slide is gently washed, and the bound cells are fixed with formaldehyde and stained with crystal violet. The slide is then scanned, and the relative amount of bound cells is measured. By probing the bound cells with appropriate fluorescent-labeled antibody, cell signaling initiated by binding immobilized ligand can also be detected.[51]

Another cell-based assay is a solution-phase releasable assay, in which the immobilized compounds are released from the chemical spot and then exert their biological effects on the living cells surrounding the chemical spot. This assay will be discussed in the section on drug-discovery arrays below.

3   Applications of Chemical Microarrays

Chemical microarray applications are extensive and have been applied successfully in basic research as well as in translational medicine. They have been used not only to identify small molecules that can bind biological molecules such as DNA, proteins, lipids, and carbohydrates but they have also been used to modulate protein function. Most importantly, chemical microarrays can perform many different assays in parallel and with minimal consumption of reagents and samples. As shown in Table 2, the OBOC combinatorial chemical bead-arrays are highly efficient as tens of thousands to millions of compound-beads can be easily constructed. Because the bead-arrays are prepared in suspension and are non-addressable, generating replicates of the same array is not feasible. Therefore, a diverse OBOC combinatorial library (array) is best used for lead identification, and focused OBOC arrays for lead optimization. In contrast, planar chemical microarrays are addressable and replicates can be readily prepared. However, the number of compounds one can generate on a planar array is generally much more limited. The light-directed *in situ* synthesis approaches,[1,5] even though they are more efficient, are limited mostly to peptide or peptoid arrays assembled with simple chemistry and high coupling efficiencies. As a result, planar chemical microarrays are generally more useful for lead optimization, profiling, and diagnostics.

Chemical microarrays are used extensively in both basic science and applied science applications.

(i) In basic science it has been used mainly to study the ligand-binding domain of receptor proteins and to characterize its signaling pathways.

Table 2 *Planar chemical microarrays vs. OBOC combinatorial library bead-arrays*

	Planar chemical microarrays	*OBOC bead-arrays*
Synthesis	Less efficient (except for light- directed synthesis methods, which are more efficient but restricted to peptides or peptoids)	Highly efficient using the split synthesis method
Compound arrangement	Spatially separable on a two-dimensional array	Spatially separable on beads (beads in suspension or immobilized on solid support or in soft agar)
Addressability	Addressable	Non-addressable
Number of compounds	Hundreds to a few thousands	Thousands to a few millions
Screening methods	Mostly on-slide screening, solution phase assay also possible	Mostly on-bead screening, solution phase assay also possible
Structure determination of compounds in microarrays	Not needed because the compounds are identified by addressable location	Direct sequencing, MS, or chemical encoding
Replicates	Replicates of addressable microarrays can be prepared easily	Not possible, as the library is non-addressable and in suspension
Applications	(1) Optimization of leads (2) Profiling (3) Diagnostics	(1) Diverse libraries for lead discovery (2) Focused libraries for lead optimization

(ii) In applied science its application is phenomenal. Especially in medical field it has been in usage from diagnostic to drug discovery.

Below is a brief description of various applications of chemical microarrays, with a focus on peptide/peptidomimetics, small molecule, and carbohydrate microarrays.

3.1 Basic Science Applications

3.1.1 Protein-Binding Arrays

In the original high-density peptide microarrays described by Fodor *et al.* a series of YGGFL peptide analogues were synthesized *in situ* for epitope mapping of an anti-β-endorphin monoclonal antibody.[1,100] Others used the SPOT synthesis approach to prepare low-density peptide arrays consisting of overlapping peptides for epitope mapping[101] and mimotope identification. For example, through screening a small library of 5520 randomly generated 15-mer peptides, Reineke *et al.* were able to identify epitopes for three distinct monoclonal antibodies, and a mimotope for one of the antibodies.[102,103] Pellois *et al.*[104] reported the use of high-density peptide microarrays for epitope mapping of an anti-p53 monoclonal antibody. Rodriquez *et al.* recently described the use of an oriented peptide array library to study the specificities of a number of protein–protein interactions, including antibodies, protein domains such as Src homology 2 domains, and protein kinases.[105]

Although there are many reports of peptide microarrays being used to probe interactions between peptides and cellular proteins, there have been only a limited number of reports on the use of small molecule microarrays to probe cellular proteins. Barnes-Seeman *et al.*[106] identified new calmodulin-binding small molecules by screening a 6336-member microarray of phenol-containing fused heterocyclic molecules. Using SPOT synthesis, Reuter *et al.*[107] generated an array of many overlapping dodecapeptides of endonuclease EcoRII, and screened it with a ^{32}P-labeled oligonucleotide. A motif of KXRXXK was identified in two separate DNA-binding regions of EcoRII.

3.1.2   Carbohydrate Microarrays for Cell Receptors

Carbohydrate microarrays have been used to identify molecules that interact with glycoproteins and glycolipids.[108–110] Galustian *et al.* described the immobilization of glycan-bovine serum albumin conjugates, glycan-lipid conjugates, and glycoproteins on nitrocellulose membranes to form glycan microarrays. These microarrays were then used to determine the binding profile of the 80 glycans to three C-type lectin receptors on antigen-presenting cells: murine SIGN-R1, SIGN-R3, and Langerin. Mannose- and fucose-related glycans were found to be the dominant specificities for SIGN-R1 and SIGN-R3 receptors.[111]

3.1.3   Cell-Signaling Arrays

Signal transduction in mammalian cells is mediated by complex networks of interacting proteins. Understanding these networks at a circuit level requires methods to quantitate the activities of multiple proteins in a rapid and accurate manner. Multiplex antibody microarrays have been used to study the receptor-tyrosine-kinase-signaling cascade in crude cell lysates.[112] Similarly, Diks *et al.* used a 192 peptide microarray to study the phosphorylation signaling in the lipopolysaccharide-stimulated peripheral blood mononuclear cells.[113] These peptide sequences were derived from the Phosphobase resource (phosphor.elm.eu.org). Schreiber and *et al.* have reported the use of high-density small-molecule arrays of 1,3-dioxanes to identify chemical ligands for the Ure2p transcriptional repressor in yeast. The identified ligand, Uretupamine, was capable of modulating Ure2p signaling function inside the yeast cells. They also identified small molecules that interact directly with a signaling protein (calmodulin) and a yeast transcriptional factor (Hap3p), and demonstrated that the haptamide A inhibited the Hap3p functions in a dose dependent manner.[115] We have recently reviewed the applications of OBOC-combinatorial libraries (bead-arrays) and chemical microarrays in signal-transduction research in our laboratory,[116] which include the identification of protein kinase substrates and inhibitors, and the identification of small molecules that interact with intracellular proteins.

3.1.4 Enzyme Substrate/Inhibitor Arrays

Houseman *et al.* prepared a peptide chip by Diels–Alder-mediated reaction of kinase-peptide substrates with a self-assembled monolayer of alkanethiolates on gold surface. Peptide phosphorylation was determined by incubating the peptide microarrays with c-src protein-tyrosine kinase followed by quantitation with a phosphoryl imager. In the presence of soluble inhibitors at a range of concentrations, dose-dependent inhibition of phosphorylation against a number of peptide substrates could be determined on a single chip.[117]

Carbohydrate arrays have been used successfully to identify the potential donor and acceptor pair in enzymatic reactions. Houseman's group developed monosaccharide arrays and demonstrated that the plant lectin concanavalin A binds specifically to mannose by SPR spectroscopy. They further identified N-acetylglucosamine as a good acceptor for galactosylation by β-1, 4-galactosyltransferase in the presence of the donor substrate UDP-galactose.[33,36] Winssinger *et al.* used spatially addressable small molecule arrays to study the activity-based profile of proteases in crude cell extracts. In their method, a small molecule was covalently tethered to a peptide nucleic acid (PNA) tag, whose sequence could be used to decode the chemical identity of the small molecule by hybridization to an oligonucleotide microarray. Using this method, they were able to identify a small molecule that binds to caspase-3.[118]

Small molecule or peptide substrates can also be used to profile protease and other enzyme activities. Salisbury *et al.* described a protease-substrate microarray in which the carboxyl end of the peptide substrates was conjugated to 7-amino-4-carbamoylmethyl coumarin, a fluorogenic compound. The conjugate was non-fluorescent when the electron-donating group on the coumarin was attached to the peptide. Upon proteolysis, the peptide was released and the microarray spot fluoresced.[61] Zhu *et al.*[62] demonstrated that small molecule microarrays could be used to detect enzyme activities of epoxide hydrolases and phosphatases.

3.1.5 Chemical-Detection Arrays

Microarray technology can be used not only to characterize biological compounds, but also to study the properties of chemical reactions. Park and Clark[119] described a sol–gel encapsulated enzyme array to screen for biocatalytic activity or enzyme inhibition. Rakow and Suslick[68] have developed a colorimetric sensor array for detection of volatile chemicals at concentrations below two parts per million. In this method, a library of vapor-sensing metalloporphyrin dyes were immobilized on solid support. Visual identification of color change was easily achieved while a ligand was bound to the metalloporphyrin dyes. A wide range of ligating vapor can be detected with this method, even for weakly ligating vapor, such as halocarbons and ketones. This type of sensing array is of practical importance for general-purpose vapor dosimeters and analyte-specific detectors. A chemical microarray technique has also been applied to monitor chemical reactions by determining the enantiomeric excess of thousands of samples.[120]

3.2   Medical Applications

DNA microarrays have been a very useful genomic tool for the identification of molecular markers for pathological conditions. Peptide and chemical microarrays have proven to be an emerging but very useful tool in proteomics and in the identification and optimization of drug leads and development of diagnostics.[51] In the near future, it is likely that many diagnostic tests will be performed in a microarray format suitable for personalized medicine.[121]

3.2.1   Diagnostic Arrays

In clinical medicine, diagnostic tests for determining serum antibody titers to a number of infectious agents, autoantigens, or other exogenous molecules have been used for many years. These tests are commonly performed one at a time in the clinical laboratory and require a large quantity of serum and reagents. All of these tests could potentially be miniaturized by immobilizing the antigens (proteins or peptides) in a microarray format. Antibodies, specific to cytokines or other biological molecules, can be immobilized on chips, which can then be used as diagnostic tools to evaluate a patient's serum, or serve as research tools in proteomics.[122,123] It is conceivable that biochips will be available for clinical diagnosis within a decade. Biochips could test hundreds to thousands of compounds using only a small amount of a blood sample. Recent studies also suggest that peptide or small molecule microarrays are useful tools in the discovery of biomarkers for various diseases, such as autoimmune diseases and cancers.[124,125] Microarrays containing hundreds of autoantigens, including peptides, proteins, and other biomolecules, have been developed to determine autoantibody profiles and IgE reactivity profiles in serum.[11,126]

3.2.2   Immunological Arrays

Glycoproteins play an active role in the innate and adaptive immune response and, in many cases specific glycoforms are involved in the immune process.[127] These glycans are displayed on macromolecules and the surface of cells, where the information they encode is deciphered by glycan-binding proteins in numerous processes, such as antigen-recognition machinery, bacterial and viral adhesion to host cells and evasion from the host immune system, and protein folding, stability, and trafficking.[128,129] Cells of the immune system use glycans on the surface of the cells they encounter to identify whether they belong to self or are foreign.[26] Wang *et al.*[130] and others have used carbohydrate-based microarrays to analyze the different types of anticarbohydrate antibodies in human and mammalian sera.[130] A novel anticellulose antibody was detected that binds specifically to beta-4-linked saccharides with a preference for glucopyranose over galactopyranose residues.[131] Interestingly, many of the carbohydrates that react with the sera are normally present in microbial pathogens, suggesting that the individuals may have acquired these antibodies during a microbial infection. The importance of carbohydrate arrays has been reviewed recently.[32,132] Carbohydrate arrays can also be used to profile other protein and lectin-binding activities, as demonstrated by Houseman and Mrksich.[33]

3.2.3   Cell-Binding Arrays

Chemical or carbohydrate microarrays can be used to profile surface receptors or to study the biological function of a living cell. We have used OBOC combinatorial library methods to identify peptide ligands that bind to intact cells.[133] The selected peptides or small molecules were then immobilized on solid support (plastic or glass slide) and used for cell-adhesion studies. Nimrichter *et al.* used a panel of 45 different glycans to study the cell adhesion properties of human CD4+ T cells. Through screening, he noticed that CD4+ T cells bind stronger and more specifically to Sialyl Lewis x (SLe^x) antigen (Neu5Acα3Galβ4(Fucα3)-GlcNAcβ than other glycans.[134] Chemical microarrays can be used as a diagnostic tool to profile cancer cells from patients, so that the physician can tailor an appropriate peptide cocktail for targeted therapy.[10,30,51]

3.2.4   Drug-Discovery Arrays

Screening chemical microarrays will provide potential drug leads for a variety of biological targets. Probing cancer cells or their lysates with diversified peptide or glycan microarrays can enable one to identify targets as well as drug leads for cancer drug discovery.[135,136] Kunimatsu *et al.* identified a peptide inhibitor for Angiotensin II (Ang II), by screening a peptide array derived from the Ang II receptor sequence. The peptide (VVIVIY) derived from the first transmembrane region exhibited the highest affinity to Ang II. The soluble form of VVIVIY peptide, which was found to have an 84% inhibitory effect on Ang II-induced aorta contraction, could be used as a potential drug for cardiovascular disease.[137] A non-covalent carbohydrate microarray was used to screen for possible inhibitors to fucosyltransferases, which are critical to the synthesis of inflammation mediators like SLe^x. Four inhibitors with nanomolar K_is were discovered, with three of the top five inhibitors exhibiting a common architecture.[138] As enzymatic modification by bacteria is the most common mechanism for their resistance to the aminoglycoside antibiotics, Disney *et al.* used a glycan array to determine the binding profile of antibiotics to resistance-causing proteins. They have screened a number of aminoglycoside mimetics against 2$'$-aminoglycoside acetyltransferase, an enzyme from antibiotic resistant *Mycobacterium tuberculosis*, and 6$'$-acetyltransferase from *Salmonella enterica*, and they were able to identify 6$'$-β-Ala-guanidinoribostamycin as a strong binder.[139]

Stockwell *et al.* developed a microarray-based system for screening small molecules in mammalian cells. This system is compatible with image-based screens and requires less than 100 cells per compound. Each compound is impregnated in a 200 µm-diameter disc composed of biodegradable poly-(D),(L)-lactide/glycolide copolymer. Cells are seeded on top of these discs, and compounds slowly diffuse out, affecting proximal cells. In contrast to microtiter-based screening, this system does not involve the use of wells or walls between each compound-treated group of cells. By this method they have identified a compound (macbecin II) that has reduced activity in cells with RNA interference-mediated decrease in the expression of tuberous sclerosis 2.[57] These reports suggest that screening chemical microarray strategy is an effective approach for identifying novel drug leads.

Over 10 years ago, our group[19] and Lerners group[140] described the immobilization of OBOC chemical bead-arrays in soft agar together with live cancer cells, followed by controlled release of immobilized compounds from each bead. Lerner's group used a melanin-production-reporter system to detect antagonists and agonists for G-protein coupled receptors. We used an MTT assay to detect anti-cancer agents. In both the cases, a halo was seen surrounding the positive beads. Similar approaches have been applied to the discovery of antibiotics, and certainly can be applied to many cell-based reporter systems that have been developed for a variety of signaling pathways over this last decade.

4   Conclusion

Peptide and chemical microarrays have emerged as powerful tools for the fields of proteomics, diagnostics, and drug discovery.[9,141] OBOC combinatorial bead-arrays and planar microarrays are highly complementary to each other. The former is particularly useful for lead discovery, whereas the latter is well-suited for profiling and diagnostics. Many immobilization chemistries for planar microarrays have been developed. Most of these methods involve the synthesis of large numbers of chemical compounds prior to their immobilization on solid support. Parallel synthesis of these compounds is labor-intensive, but can be facilitated by robotics. Alternatively, these compounds can be synthesized by the highly efficient "split-mix" synthesis, but a chemical or physical encoding method is needed for decoding such libraries. The next challenge is to develop efficient methods of immobilization so that these compounds can be printed on a solid support with high uniformity in concentration. We anticipate that the field of chemical microarrays will continue to grow, and biochips consisting of hundreds to thousands of immobilized biomolecules, such as peptides, proteins, carbohydrates, lipids, or small molecules, will be used for the development of high-throughput diagnostics. Although the number of publications on small molecule planar microarrays is still rather limited, this will change in the next few years, and we anticipate that such microarrays will play an important role in drug-target identification and drug optimization. The ultra-high-throughput OBOC combinatorial peptide and small-molecule library bead-array method is well established, particularly for on-bead screening. One major challenge in this chemical microarray approach is to develop robust *in situ* releasable chemistry, and reliable multi-parametric cell-based assays so that over 100,000 diverse small molecule compounds can be screened concurrently against multiple assays in one single 10 cm Petri dish, in a day or two. We believe this can be achieved within the next few years.

Acknowledgments

This work was supported by NIH R33CA-86364, NIH R33CA-99136, R01CA-098166, NSF CHE-0302122

References

1. S.P. Fodor, J.L. Read, M.C. Pirrung, L. Stryer, A.T. Lu and D. Solas, *Science*, 1991, **251**, 767.
2. M. Schena, D. Shalon, R. Heller, A. Chai, P.O. Brown and R.W. Davis, *Proc. Natl. Acad. Sci. USA*, 1996, **93**, 10614.
3. A.C. Pease, D. Solas, E.J. Sullivan, M.T. Cronin, C.P. Holmes and S.P. Fodor, *Proc. Natl. Acad. Sci. USA*, 1994, **91**, 5022.
4. S. Singh-Gasson, R.D. Green, Y. Yue, C. Nelson, F. Blattner, M.R. Sussman and F. Cerrina, *Nat. Biotechnol.*, 1999, **17**, 974.
5. K. Komolpis, O. Srivannavit and E. Gulari, *Biotechnol. Prog.*, 2002, **18**, 641.
6. X. Gao, P. Yu, E. LeProust, L. Sonigo, P. Pellois and H. Zhang, *J. Am. Chem. Soc.*, 1998, **120**, 12698.
7. X. Gao, X. Zhou and E. Gulari, *Proteomics*, 2003, **3**, 2135.
8. R. Frank, *Tetrahedron*, 1992, **48**, 9217.
9. R. Frank, *J. Immunol. Methods.*, 2002, **267**, 13.
10. K.S. Lam and M. Renil, *Curr. Opin. Chem. Biol.*, 2002, **6**, 353.
11. W.H. Robinson, C. DiGennaro, W. Hueber, B.B. Haab, M. Kamachi, E.J. Dean, S. Fournel, D. Fong, M.C. Genovese, H.E. de Vegvar, K. Skriner, D.L. Hirschberg, R.I. Morris, S. Muller, G.J. Pruijn, W.J. van Venrooij, J.S. Smolen, P.O. Brown, L. Steinman and P.J. Utz, *Nat. Med.*, 2002, **8**, 295.
12. H.E. Blackwell, L. Perez, R.A. Stavenger, J.A. Tallarico, E. Cope Eatough, M.A. Foley and S.L. Schreiber, *Chem. Biol.*, 2001, **8**, 1167.
13. P.A. Clemons, A.N. Koehler, B.K. Wagner, T.G. Sprigings, D.R. Spring, R.W. King, S.L. Schreiber and M.A. Foley, *Chem. Biol.*, 2001, **8**, 1183.
14. K.C. Nicolaou, X. Xiao, Z. Parandoosh, A. Senyei and M.P. Nova, *Angew. Chem. Int. Ed. Engl.*, 1995, **34**, 2289.
15. A.W. Czarnik, *Curr. Opin. Chem. Biol.*, 1997, **1**, 60.
16. J. Marik, Q. Xu, X. Wang, L. Peng and K.S. Lam, Peptides; Peptide Revolution: Genomics, Proteomics and Therapeutics *Proceedings of the Eighteenth American Peptide Symposium*, Boston, 2004, (p. 849).
17. K.S. Lam, M. Lebl and V. Krchnak, *Chem. Rev.*, 1997, **97**, 411.
18. K.S. Lam, S.E. Salmon, E.M. Hersh, V.J. Hruby, W.M. Kazmierski and R.J. Knapp, *Nature*, 1991, **354**, 82.
19. S.E. Salmon, R.H. Liu-Stevens, Y. Zhao, M. Lebl, V. Krchnak, K. Wertman, N. Sepetov and K.S. Lam, *Mol. Divers.*, 1996, **2**, 57.
20. C.K. Jayawickreme, H. Sauls, N. Bolio, J. Ruan, M. Moyer, W. Burkhart, B. Marron, T. Rimele and J. Shaffer, *J. Pharmacol. Toxicol. Methods*, 1999, **42**, 189.
21. H. Wenschuh, R. Volkmer-Engert, M. Schmidt, M. Schulz, J. Schneider-Mergener and U. Reineke, *Biopolymers*, 2000, **55**, 188.
22. S. Matysiak, F. Reuthner and J.D. Hoheisel, *Biotechniques*, 2001, **31**, 896.
23. T. Ast, N. Heine, L. Germeroth, J. Schneider-Mergener and H. Wenschuh, *Tetrahedron Lett.*, 1999, **40**, 4317.
24. N. Heine, T. Ast, J. Schneider-Mergener, U. Reineke, L. Germeroth and H. Wenschuh, *Tetrahedron*, 2003, **59**, 9919.

25. D. Scharn, H. Wenschuh, U. Reineke, J. Schneider-Mergener and L. Germeroth, *J. Comb. Chem.*, 2000, **2**, 361.

26. C. Ortiz Mellet and J.M. Garcia Fernandez, *Chembiochem*, 2002, **3**, 819.

27. M. Uttamchandani, D.P. Walsh, S.Q. Yao and Y.T. Chang, *Curr. Opin. Chem. Biol.*, 2005, **9**, 4.

28. A. Ulman, J.F. Kang, Y. Shnidman, S. Liao, R. Jordan, G.Y. Choi, J. Zaccaro, A.S. Myerson, M. Rafailovich, J. Sokolov and C. Fleischer, *J. Biotechnol.*, 2000, **74**, 175.

29. M.L. Lesaicherre, M. Uttamchandani, G.Y. Chen and S.Q. Yao, *Bioorg. Med. Chem. Lett.*, 2002, **12**, 2079.

30. O.H. Aina, T.C. Sroka, M.L. Chen and K.S. Lam, *Biopolymers*, 2002, **66**, 184.

31. Q. Xu, S. Miyamoto and K.S. Lam, *Mol. Divers.*, 2004, **8**, 301.

32. D. Wang, *Proteomics*, 2003, **3**, 2167.

33. B.T. Houseman and M. Mrksich, *Chem. Biol.*, 2002, **9**, 443.

34. S. Fukui, T. Feizi, C. Galustian, A.M. Lawson and W. Chai, *Nat. Biotechnol.*, 2002, **20**, 1011.

35. M.C. Bryan, O. Plettenburg, P. Sears, D. Rabuka, S. Wacowich-Sgarbi and C.H. Wong, *Chem. Biol.*, 2002, **9**, 713.

36. S. Park and I. Shin, *Angew. Chem. Int. Ed. Engl.*, 2002, **41**, 3180.

37. R.A. Laursen, *Eur. J. Biochem.*, 1971, **20**, 89.

38. B.T. Chait, R. Wang, R.C. Beavis and S.B. Kent, *Science*, 1993, **262**, 89.

39. D.T. McLachlin and B.T. Chait, *Curr. Opin. Chem. Biol.*, 2001, **5**, 591.

40. R.S. Youngquist, G.R. Fuentes, M.P. Lacey and T. Keough, *Rapid Commun. Mass Spectrom.*, 1994, **8**, 77.

41. R.L. Affleck, *Curr. Opi. Chem. Biol.*, 2001, **5**, 257.

42. C. Barnes and S. Balasubramanian, *Curr. Opin. Chem. Biol.*, 2000, **4**, 346.

43. A. Furka, F. Sebestyen, M. Asgedom and G. Dibo, *Int. J. Pept. Protein. Res.*, 1991, **37**, 487.

44. R.A. Houghten, C. Pinilla, S.E. Blondelle, J.R. Appel, C.T. Dooley and J.H. Cuervo, *Nature*, 1991, **354**, 84.

45. K.S. Lam and Z.G. Zhao, *Hematol. Oncol. clinical N A*, 1997, **11**, 1007.

46. R. Liu, J. Marik and K.S. Lam, *J. Am. Chem. Soc.*, 2002, **124,** 7678.

47. X. Wang, L. Peng, R. Liu, S.S. Gill and K.S. Lam, *J. Comb. Chem.*, 2005, **7**, 197.

48. X. Wang, L. Peng, R. Liu, B. Xu and K.S. Lam, *J. Pept. Res.*, 2005, **65**, 130.

49. A. Song, J. Zhang, C.B. Lebrilla and K.S. Lam, *J. Am. Chem. Soc.*, 2003, **125**, 6180.

50. C.L. Chen, P. Strop, M. Lebl and K.S. Lam, *Method. Enzymol.*, 1996, **267**, 211.

51. J.R. Falsey, M. Renil, S. Park, S. Li and K.S. Lam, *Bioconjug. Chem.*, 2001, **12**, 346.

52. C.E. Jordan and R.M Corn, *Anal. Chem.*, 1997, **69**, 1449.

53. F.F. Bier and F.W. Scheller, *Biosens. Bioelectron.*, 1996, **11**, 669.

54. F. Morhard, J. Pipper, R. Dahint and M. Grunze, *Sensor. Actuator.*, 2000, **B70**, 232.

55. A. Brecht and G. Gauglitz, *Biosens. Bioelectron.*, 1995, **10**, 923.

56. S.E. Salmon, K.S. Lam, M. Lebl, A. Kandola, P.S. Khattri, S. Wade, M. Patek, P. Kocis, V. Krchnak, D. Thorpe and S. Felder, *Proc. Natl. Acad. Sci. USA*, 1993, **90**, 11708.

57. S.N. Bailey, D.M. Sabatini and B.R. Stockwell, *Proc. Natl. Acad. Sci. USA*, 2004, **101**, 16144.

58. Y. Kawahashi, N. Doi, H. Takashima, C. Tsuda, Y. Oishi, R. Oyama, M. Yonezawa, E. Miyamoto-Sato and H. Yanagawa, *Proteomics*, 2003, **3**, 1236.

59. T. Kukar, S. Eckenrode, Y. Gu, W. Lian, M. Megginson, J.X. She and D. Wu, *Anal. Biochem.*, 2002, **306**, 50.

60. K. Martin, T.H. Steinberg, L.A. Cooley, K.R. Gee, J.M. Beechem and W.F. Patton, *Proteomics*, 2003, **3**, 1244.

61. C.M. Salisbury, D.J. Maly and J. Ellman, *J. Am. Chem. Soc.*, 2002, **124**, 14868.

62. Q. Zhu, M. Uttamchandani, D. Li, M.L. Lesaicherre and S.Q. Yao, *Org. Lett.*, 2003, **5**, 1257.

63. M. Meldal, I. Svendsen, K. Breddam and F.I. Auzanneau, *Proc. Natl. Acad. Sci. USA*, 1994, **91**, 3314.

64. A. Roda, P. Pasini, M. Guardigli, M. Baraldini, M. Musiani and M. Mirasoli, *Fresenius J. Anal. Chem.*, 2000, **366**, 752.

65. H. Zhu, M. Bilgin and M. Snyder, *Annu. Rev. Biochem.*, 2003, **72**, 783.

66. H. Zhu, J.F. Klemic, S. Chang, P. Bertone, A. Casamayor, K.G. Klemic, D. Smith, M. Gerstein, M.A. Reed and M. Snyder, *Nat. Genet.*, 2000, **26**, 283.

67. R.Q. Liang, C.Y. Tan and K.C. Ruan, *J. Immunol. Method.*, 2004, **285**, 157.

68. N.A. Rakow and K.S. Suslick, *Nature*, 2000, **406**, 710.

69. D.A. Knecht and R.L. Dimond, *Anal. Biochem.*, 1984, **136**, 180.

70. B. Lin, J. Qiu, J. Gerstenmeier, P. Li, H. Pien, J. Pepper and B. Cunningham, *Biosens. Bioelectron.*, 2002, **17**, 827.

71. M.A. Cooper, *Anal. Bioanal. Chem.*, 2003, **377**, 834.

72. D. Vetter, *J. Cell Biochem. Suppl.*, 2002, **39**, 79.

73. C.E. Forde and S.L. McCutchen-Maloney, *Mass Spectrom. Rev.*, 2002, **21**, 419.

74. L.T. Mazzola and S.P. Fodor, *Biophys. J.*, 1995, **68**, 1653.

75. P.R. Morrill, R.B. Millington and C.R. Lowe, *J. Chromatogr. B Analyt. Technol. Biomed. Life Sci.*, 2003, **793**, 229.

76. J.S. Shumaker-Parry, M.H. Zareie, R. Aebersold and C.T. Campbell, *Anal. Chem.*, 2004, **76**, 918.

77. G.J. Wegner, H.J. Lee, G. Marriott and R.M. Corn, *Anal. Chem.*, 2003, **75**, 4740.

78. G.J. Wegner, A.W. Wark, H.J. Lee, E. Codner, T. Saeki, S. Fang and R.M. Corn, *Anal. Chem.*, 2004, **76**, 5677.

79. M. Zizlsperger and W. Knoll, *Prog. Colloid Prog. Sci.*, 1998, **109**, 244.

80. K.E. Sapsford, Z. Liron, Y.S. Shubin and F.S. Ligler, *Anal. Chem.*, 2001, **73**, 5518.

81. J.P. Landry, X.D. Zhu and J.P. Gregg, *Opt. Lett.*, 2004, **29**, 581.

82. J.P. Landry, X.D. Zhu, J.P. Gregg and X.W. Guo, in Microarrays and combinatorial techniques: design, fabrication, and analysis II, *Proceedings of the Society of Photo-Optical Instrumentation Engineers,* Vol. 5358, D.V. Nicolau and R. Raghavachari (eds.), SPIE, Bellingham, WA, 2004, 121–128.

83. J.P. Landry, J.P. Gregg, X.W. Guo and X.D. Zhu, *Mater. Res. Soc. Symp. Proc.*, 2003, **773**, 1.

84. S.R. Weinberger, T.S. Morris and M. Pawlak, *Pharmacogenomics*, 2000, **1**, 395.

85. D. Figeys, L.D. McBroom and M.F. Moran, *Methods*, 2001, **24**, 230.

86. G. Marko-Varga and T.E. Fehniger, *J. Proteome. Res.*, 2004, **3**, 167.

87. E. Scrivener, R. Barry, A. Platt, R. Calvert, G. Masih, P. Hextall, M. Soloviev and J. Terrett, *Proteomics*, 2003, **3**, 122.

88. V.W. Jones, J.R. Kenseth, M.D. Porter, C.L. Mosher and E. Henderson, *Anal. Chem.*, 1998, **70**, 1233.

89. J.R. Epstein and D.R. Walt, *Chem. Soc. Rev.*, 2003, **32**, 203.

90. K.L. Michael, L.C. Taylor, S.L. Schultz and D.R. Walt, *Anal. Chem.*, 1998, **70**, 1242.

91. J.M. Yeakley, J.B. Fan, D. Doucet, L. Luo, E. Wickham, Z. Ye, M.S. Chee and X.D. Fu, *Nat. Biotechnol.*, 2002, **20**, 353.

92. K.L. Michael, L.C. Taylor and D.R. Walt, *Anal. Chem.*, 1999, **71**, 2766.

93. J.A. Ferguson, F.J. Steemers and D.R. Walt, *Anal. Chem.*, 2000, **72**, 5618.

94. E.W. Adams, J. Ueberfeld, D.M. Ratner, B.R. O'Keefe, D.R. Walt and P.H. Seeberger, *Angew. Chem. Int. Ed. Engl.*, 2003, **42**, 5317.

95. B.T. Cunningham, P. Li, S. Schulz, B. Lin, C. Baird, J. Gerstenmaier, C. Genick, F. Wang, E. Fine and L. Laing, *J. Biomol. Screen.*, 2004, **9**, 481.

96. H. Kido, A. Maquieira and B.D. Hammock, *Anal. Chim. Acta.*, 2000, **411**, 1.

97. J.J. La Clair and M.D. Burkart, *Org. Biomol. Chem.*, 2003, **1**, 3244.

98. V. Mirsky, *Macromol. Rapid Commn.*, 2004, **25**, 253.

99. S. Szunerits, J.M. Tam, L. Thouin, C. Amatore and D.R. Walt, *Anal. Chem.*, 2003, **75**, 4382.

100. J.W. Jacobs and S.P. Fodor, *Trends Biotechnol.*, 1994, **12**, 19.

101. Z.X. Wang and R.A. Laursen, *Pept. Res.*, 1992, **5**, 275.

102. U. Reineke, C. Ivascu, M. Schlief, C. Landgraf, S. Gericke, G. Zahn, H. Herzel, R. Volkmer-Engert and J. Schneider-Mergener, *J. Immunol. Methods*, 2002, **267**, 37.

103. U. Reineke, R. Volkmer-Engert and J. Schneider-Mergener, *Curr. Opin. Biotechnol.*, 2001, **12**, 59.

104. J.P. Pellois, X. Zhou, O. Srivannavit, T. Zhou, E. Gulari and X. Gao, *Nat. Biotechnol.*, 2002, **20**, 922.

105. M. Rodriguez, S.S. Li, J.W. Harper and Z. Songyang, *J. Biol. Chem.*, 2004, **279**, 8802.

106. D. Barnes-Seeman, S.B. Park, A.N. Koehler and S.L. Schreiber, *Angew. Chem. Int. Ed. Engl.*, 2003, **42**, 2376.

107. M. Reuter, J. Schneider-Mergener, D. Kupper, A. Meisel, P. Mackeldanz, D.H. Kruger and C. Schroeder, *J. Biol. Chem.*, 1999, **274**, 5213.

108. C. Galustian, R.A. Childs, M. Stoll, H. Ishida, M. Kiso and T. Feizi, *Immunology*, 2002, **105**, 350.

109. O. Blixt, S. Head, T. Mondala, C. Scanlan, M.E. Huflejt, R. Alvarez, M.C. Bryan, F. Fazio, D. Calarese, J. Stevens, N. Razi, D.J. Stevens, J.J. Skehel, I. van Die, D.R. Burton, I.A. Wilson, R. Cummings, N. Bovin, C.H. Wong and J.C. Paulson, *Proc. Natl. Acad. Sci. USA*, 2004, **101**, 17033.

110. B.S. Bochner, R.A. Alvarez, P. Mehta, N.V. Bovin, O. Blixt, J.R. White and R.L. Schnaar, *J. Biol. Chem.*, 2005, **280**, 4307.

111. C. Galustian, C.G. Park, W. Chai, M. Kiso, S.A. Bruening, Y.S. Kang, R.M. Steinman and T. Feizi, *Int. Immunol.*, 2004, **16**, 853.

112. U.B. Nielsen, M.H. Cardone, A.J. Sinskey, G. MacBeath and P.K. Sorger, *Proc. Natl. Acad. Sci. USA*, 2003, **100**, 9330.

113. S.H. Diks, K. Kok, T. O'Toole, D.W. Hommes, P. van Dijken, J. Joore and M.P. Peppelenbosch, *J. Biol. Chem.*, 2004, **279**, 49206.

114. F.G. Kuruvilla, A.F. Shamji, S.M. Sternson, P.J. Hergenrother and S.L. Schreiber, *Nature*, 2002, **416**, 653.

115. A.N. Koehler, A.F. Shamji and S.L. Schreiber, *J. Am. Chem. Soc.*, 2003, **125**, 8420.

116. K.S. Lam, R. Liu, S. Miyamoto, A.L. Lehman and J.M. Tuscano, *Acc. Chem. Res.*, 2003, **36**, 370.

117. B.T. Houseman, J.H. Huh, S.J. Kron and M. Mrksich, *Nat. Biotechnol.*, 2002, **20**, 270.

118. N. Winssinger, S. Ficarro, P.G. Schultz and J.L. Harris, *Proc. Natl. Acad. Sci. USA*, 2002, **99**, 11139.

119. C.B. Park and D.S. Clark, *Biotechnol. Bioeng.*, 2002, **78**, 229.

120. G.A. Korbel, G. Lalic and M.D. Shair, *J. Am. Chem. Soc.*, 2001, **123**, 361.

121. D.S. Wilson and S. Nock, *Angew. Chem. Int. Ed. Engl.*, 2003, **42**, 494.

122. C.A. Borrebaeck, *Immunol. Today*, 2000, **21**, 379.

123. A. Sreekumar and A.M. Chinnaiyan, *Curr. Opin. Mol. Ther.*, 2002, **4**, 587.

124. E.F. Petricoin, A.M. Ardekani, B.A. Hitt, P.J. Levine, V.A. Fusaro, S.M. Steinberg, G.B. Mills, C. Simone, D.A. Fishman, E.C. Kohn and L.A. Liotta, *Lancet*, 2002, **359**, 572.

125. S. Fournel and S. Muller, *Curr. Protein. Pept. Sci.*, 2003, **4**, 261.

126. R. Hiller, S. Laffer, C. Harwanegg, M. Huber, W.M. Schmidt, A. Twardosz, B. Barletta, W.M. Becker, K. Blaser, H. Breiteneder, M. Chapman, R. Crameri, M. Duchene, F. Ferreira, H. Fiebig, K. Hoffmann-Sommergruber, T.P. King, T. Kleber-Janke, V.P. Kurup, S.B. Lehrer, J. Lidholm, U. Muller, C. Pini, G. Reese, O. Scheiner, A. Scheynius, H.D. Shen, S. Spitzauer, R. Suck, I. Swoboda, W. Thomas, R. Tinghino, M. Van Hage-Hamsten, T. Virtanen, D. Kraft, M.W. Muller and R. Valenta, *Faseb. J.*, 2002, **16**, 414.

127. P.M. Rudd, M.R. Wormald and R.A. Dwek, *Trends Biotechnol.*, 2004, **22**, 524.

128. P.R. Crocker and T. Feizi, *Curr. Opin. Struct. Biol.*, 1996, **6**, 679.

129. A. Helenius and M. Aebi, *Annu. Rev. Biochem.*, 2004, **73**, 1019.

130. D. Wang, S. Liu, B.J. Trummer, C. Deng and A. Wang, *Nat. Biotechnol.*, 2002, **20**, 275.

131. M. Schwarz, L. Spector, A. Gargir, A. Shtevi, M. Gortler, R.T. Altstock, A.A. Dukler and N. Dotan, *Glycobiology*, 2003, **13**, 749.

132. T. Feizi, F. Fazio, W. Chai and C.H. Wong, *Curr. Opin. Struct. Biol.*, 2003, **13**, 637.

133. M.E. Pennington, K.S. Lam and A.E. Cress, *Mol. Divers.*, 1996, **2**, 19.

134. L. Nimrichter, A. Gargir, M. Gortler, R.T. Altstock, A. Shtevi, O. Weisshaus, E. Fire, N. Dotan and R.L. Schnaar, *Glycobiology*, 2004, **14**, 197.

135. J. Schlessinger, *Nat. Biotechnol.*, 2002, **20**, 232.
136. H.M. Osborn, P.G. Evans, N. Gemmell and S.D. Osborne, *J. Pharm. Pharmacol.*, 2004, **56**, 691.
137. R. Kato, M. Kunimatsu, S. Fujimoto, T. Kobayashi and H. Honda, *Biochem. Biophys. Res. Commun.*, 2004, **315**, 22.
138. M.C. Bryan, L.V. Lee and C.H. Wong, *Bioorg. Med. Chem. Lett.*, 2004, **14**, 3185.
139. M.D. Disney and P.H. Seeberger, *Chem. Biol.*, 2004, **11**, 1701.
140. C.K. Jayawickreme, G.F. Graminski, J.M. Quillan and M.R. Lerner, *Proc. Natl. Acad. Sci. USA*, 1994, **91**, 1614.
141. D.S. Wilson and S. Nock, *Curr. Opin. Chem. Biol.*, 2002, **6**, 81.

Section 5

Conceptual Advances in Lead Evaluation: Screen Early and Often

Screen/Counter-Screen: Early Assessment of Selectivity

MARTYN N. BANKS[a], LITAO ZHANG[b] AND JOHN G. HOUSTON[a]

[a]Applied Biotechnology, Bristol Myers Squibb Co., Pharmaceutical Research Institute, Wallingford, CT 06492-7660, USA
[b]Lead Evaluation, Applied Biotechnology, Bristol Myers Squibb Co., Pharmaceutical Research Institute, Pennington, NJ 08534, USA

1 Introduction

Identifying and successfully optimizing a drug candidate is a multistep, complex process that relies on a seamless mix of high quality, chemical and biological knowledge, combined with a strong technology-based R&D process. During the past decade, pharmaceutical companies have been looking for ways of improving this drug discovery process to make it more efficient, productive and successful.[1]

The clearest examples in recent years of areas where these process improvements have had a positive impact are in high throughput screening (HTS) and compound management. Over the last 10 years, investment in automated systems, novel approaches to assay design and informatics integration tools have turned a previously inefficient, manually intensive process into a high capacity, highly efficient drug discovery asset. The groups involved took advantage of the potential economies of scale to increase productivity and improve quality while driving down, or at least, maintaining costs.[2,3] At Bristol-Myers Squibb (BMS), a rigorous analysis of the discovery process, tracking the life of a compound from its initial synthesis through various biological screens and tests, allowed the discovery team to identify gaps and inefficiencies in the overall process of lead discovery and optimization. This analysis resulted in a more streamlined, efficient plan for early drug discovery.

The first major step in enabling this plan was building an integrated early discovery process that spanned gene discovery to lead discovery. This process would be underpinned by a robust technology platform of automated compound management and screening systems. Around this integrated process and infrastructure was a concomitant growth in highly trained, multiskilled staff with a blend of functional expertise and technology savvy. The aim was to integrate scientific and technological excellence into a highly efficient drug discovery process by blending diverse but critical skills such as disease area expertise, engineering, HTS automation, logistics and production management (Figure 1).

The overall goal of this redesign was to create a more efficient discovery process to deliver diverse 'chemotypes' to project groups as leads to enable them to start up medicinal chemistry programs. Creation of centralized core functions and leveraging state-of-the-art technology would also allow staff to provide significant productivity gains, while increasing the quality of the data being passed onto project teams.

The next step in this 'root and branch' analysis was taking this 'leveraging technology' approach from the initial discovery of compounds through to their optimization. Using the lessons learned in previous years, of changing and enhancing the HTS and compound management environments, directed us to deploy technologies and processes that were 'tried and tested' into the lead optimization process. The premise behind this objective was that generating more and more data on a particular compound or chemotype series earlier in the discovery process would allow chemists and biologists to make more informed decisions around which compound to make next. Providing a full array of safety and selectivity profiles on every project compound as quickly after synthesis or discovery as possible would allow scientists to rank and compare compounds against a multiparameter matrix, not just potency.

To have maximum impact, these multiparameter profiles would need to be generated in a rapid, high quality, high throughput manner. This would allow project team

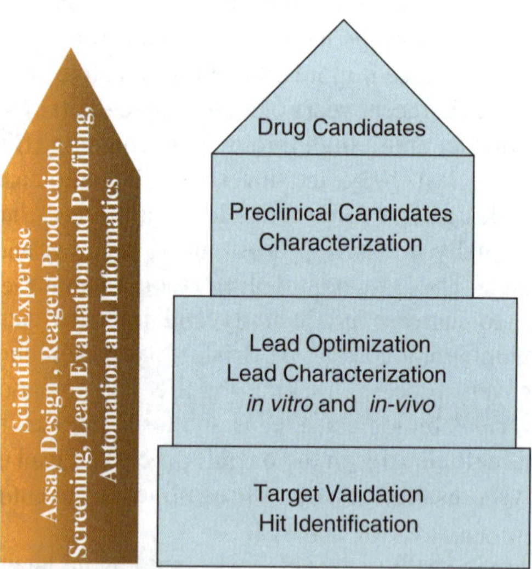

Figure 1 *A highly efficient process by integrating critical skills including scientific expertise, engineering, automation, logistics, and production management into drug discovery throughout the life of the discovery program*

leaders to run true multiparameter structure activity relationship (MSAR) and structure liability relationship (SLR) programs. HTS and liability profiling aligned with rapid structure activity relationship (SAR) support would allow a significant increase in the data available to scientists at the earliest stage of drug discovery.

Supplying this type of data early in the discovery process would allow a program to select and progress initial interesting hits and leads, not based on potency alone, but also by general and specific selectivity criteria, as well as by potential liabilities, *e.g.*, cytochrome P450 profiles. The application of HTS technology and processes would enable lead optimization programs to track this full package of information throughout the life of the program, through to the successful selection of a pre-clinical drug candidate, while maximizing the efficiency of decision-making processes.

Applying HTS assay design and reagent development approaches to the most common target families that would constitute the selectivity and liability panels was a key component of this new approach to lead optimization. Evaluating leads in a more comprehensive fashion this early in the drug discovery process would require the use of HTS paradigms for compound testing, assay scheduling and data integration.

This chapter will briefly describe the supporting technologies and processes used in lead evaluation and lead profiling. In addition, the rationale behind particular assay designs used to construct selectivity panels for common target families will be discussed in detail.

2   Approaches Used for Selection of Drug Candidates

Although the drug discovery model changes from one company to another, the goal of drug discovery is the same – to find the best way to turn an organic compound into a highly valued drug candidate in a timely fashion. However, this process is an extremely complex endeavor with numerous hurdles that have to be overcome. In the past, the early evaluation of compounds was mainly directed toward driving the potency down rather than removing off-target selectivity, metabolic liabilities, *etc.* Today, it is important to increase the knowledge applied to the design of compounds through each synthesis cycle by providing early, complete and parallel SAR data. The collection of critical information about lead compounds is vital for the rapid progression of compounds with the necessary properties required for clinical efficacy and safety, into the drug discovery pipeline. Evaluating leads emanating from the lead discovery process and tracking liabilities throughout the life of the project are now essential components of lead optimization.

2.1   Lead Evaluation and Liability Profiling

The lead evaluation and profiling process at BMS has taken advantage of high throughput technologies in assay design, miniaturization and automation to design a rapid, integrated approach to SAR and SLR.

The aim of lead evaluation and profiling is to produce high quality, multiparameter data on every compound emanating from HTS or within a discovery program. The key components of this process are described in Figure 2, but basically include centralized capabilities for testing compounds in broad *in vitro* specificity panels and liability assay panels, as well as processes for continued tracking of multiparameter SAR and SLR throughout the life of a program. The concept behind this approach is that early selectivity and liability profiling of compounds will allow better direction for medicinal chemists. These richly annotated compounds could also become a valuable source of future program leads. Table 1 summarizes some common *in vitro* screening approaches that are commonly used for profiling of drug candidates.

2.1.1   ADME Liability Profiling

During the lead optimization process, the major focus of medicinal chemistry is to design a progressible compound that is potent and selective. Additionally, the chemist needs to optimize the properties of the compounds so that they are viable drug candidates using multiparameter SAR and SLR data. Typically, early *in vitro* SLR data includes, inhibition of cytochrome P450 isoenzymes, drug metabolism, the potential for cytochrome P450s induction, activity against certain ion channels and

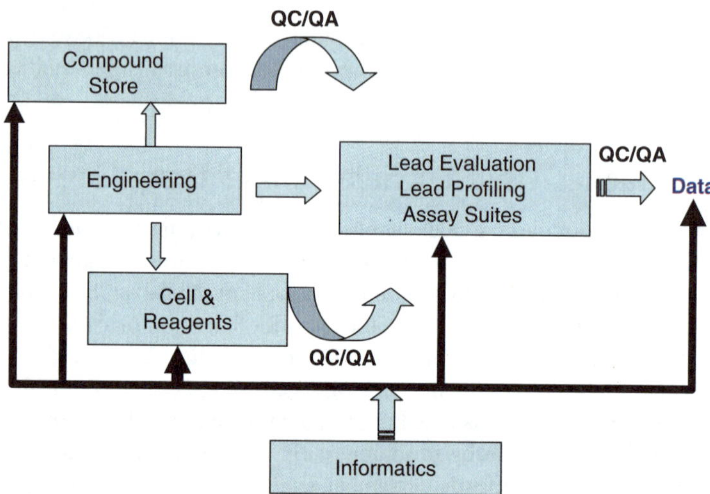

Figure 2 *The key components of lead evaluation and lead profiling process: this integrated process allows rapidly delivering high quality, multiparameter data on every compound produced within a discovery program*

Table 1 *In vitro tests*

In vitro tests	Assay type
	Cell-free assays
	Cell assays
	Protein binding
In vitro target specificity	Drug-drug interaction
	Metabolic stability
	Cell-based toxicity
	Membrane permeability

receptors known to produce toxic events in human, cytotoxicity and the absorption potential.

Unlike program-specific lead optimization assays, data from these SLR assays transcends multiple drug discovery programs. Generally speaking, there are two modes in which one wishes to use SLR assays: first, to scan a number of general chemotypes to identify potential issues, and second, having identified a liability, to track the SLR through cycles of medicinal chemistry. With more advanced compounds, the liability assays need to help define potential adverse events in clinical trials. For the determination of multiparameter SAR relationships, it is important to process a liability assay in parallel with the target activity assay. By using a core facility, the compound-routing mechanism for lead optimization can readily incorporate liability assays, enabling the data to be derived from exactly the same compound at the same time, thus increasing the fidelity of the data. Most of the *in vitro* liability assays use similar readouts to conventional target activity assays, hence the same infrastructure can serve both functions. However, there are exceptions, *e.g.*, in situations where LC-MS/MS is employed to quantify the amount of compound, as in metabolic stability assays and cell-based assays that predict intestinal absorption. In metabolic stability assays, a compound is incubated with microsomes and the rate of metabolism is calculated from compound concentration determinations at a range of time points. It is common to profile the metabolism by different species reflecting the *in vivo* model that will be used to assess compound efficacy.

The type of assay may vary depending on the exact data that is required. Early in a drug discovery program, it is useful to understand the SLR associated with a particular liability, *e.g.*, inhibition of a cytochrome P450 may be observed, suggesting that the compound may cause drug–drug interactions. Under these circumstances, using recombinant cytochrome P450s may be more suitable as a source of enzyme because the assay is 'biochemically cleaner'. As a compound progresses down the drug discovery pipeline, one needs to understand the probability that the observed cytochrome P450 inhibition will have a potential adverse drug–drug interaction in humans. Under these circumstance, human liver microsomes, which are 'biochemically complex' would be a preferred source of enzyme. It is critical with all of these types of liability assays to clearly understand the context of the data; these are *in vitro* assays that only suggest a potential *in vivo* issue.

2.1.2 The Lead Evaluation Process: Technologies and Methods

The lead evaluation process at BMS is run through centralized groups that have core strengths in high throughput assay design and screening. They are able to test compounds in broad, selectivity panels as well as program-specific selectivity assays at a capacity and speed that cannot be attained in individual project teams. The advantages of an integrated lead evaluation process with automated platforms and parallel processing of compounds include: rapid turn around of data (days, instead of weeks), standardization of assay formats and high-quality data packages. By parallel processing, compounds can be readily tested in assays that might be related to, but not the same as, those performed within, a therapeutic team. For example, a project may be focused on a specific member of a common therapeutic target class, *e.g.*, a kinase, but it is useful to have data from a whole range of kinases or proteins that contain nucleotide-binding domains to help progress important compounds. The core teams can also focus on bioassay technology scouting, evaluation and procurement to facilitate the continuous need to improve productivity, efficiency and the quality of the data. Costs can be contained because bulk-buying abilities aid negotiation and expensive equipment can be used continuously, giving a better return on investment.

The efficient movement and management of project compounds is one of the most important parts of the entire lead evaluation process. Compounds from medicinal chemistry need to be tracked through the entire process and associated with the data. The compounds may need to be tested in lead evaluation panels as well as other secondary and liability assays and so an efficient routing mechanism needs to exist. For quality control, one of our approaches is to include automated liquid chromatography and/or MS analysis on all compounds tested in lead optimization. This policy confirms the identity of the actual compound tested in the bioassay and its purity. Additionally, we use automated nephalometry to measure particle counts in the assay buffer, which serves as a surrogate for aqueous compatibility and helps to put the bioassay data in context. This step is important if selectivity or liability assays are being used with relatively high compound concentrations to show adequate differential activity from the primary target assay.

Having established an effective process for getting the compound into the assay, the next important input is the supply of reagents. Most biochemical assay reagents, *i.e.* proteins, peptides, membranes *etc.*, can be prepared and stored ready for assay at a future date; this approach removes the need to provide 'just-in-time' reagents. Often one can consider outsourcing this supply to external vendors. However, for cellular assays, there is often a need to supply cells directly to the assay. This dependency on the 'just-in-time' supply of cells for the assay is always challenging and can be resource intensive. We utilize two automated platforms to ensure the supply of cells for assays, shown in Figures 3 and 4.

The Cellmate™ (The Technology partnership, Cambridge, UK) together with a flask loader mechanism is ideal for a relatively large number of flasks for a single cell line.[4] A lot of the routine media changes and harvesting can be performed using this platform. The SelecT™ (The Technology partnership, Cambridge, UK) is a more sophisticated platform that maintains upward of 50 different cell lines that require different culturing protocols. The robot delivers viable cell counts and dispenses cells directly into microtitre plates ready for assay. Ideally, one would like to minimize the need for 'just-in-time' cell supply and one solution has been cryopreservation of cells with resurrection and recovery protocols that are used just before the assay.[5] In our hands, this is a more efficient and robust process that works well for transient transfection cellular assays.

Over the last decade, there has been a significant increase in laboratory automation and miniaturization, primarily driven by the needs of HTS laboratories.[2] However, the real challenge for lead evaluation is somewhat different from HTS. In HTS, one bioassay tests many hundreds of thousands of compounds in a short span of time. In lead evaluation, smaller numbers of compounds – typically 10–100 – are put through an array of assays that remain relatively constant over a longer time

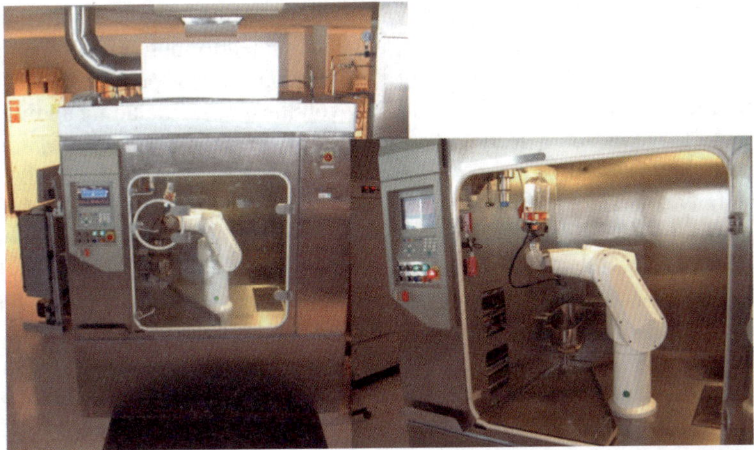

Figure 3 *Photograph of the Cellmate™ robot built by The Technology Partnership, Cambridge, UK. This robot performs cell seeding in either large T flasks or roller bottles, exchanges media, rinses cell sheets, purges gases, harvests cells by trypsinization or scraping, and supports transfections*

Figure 4 *A photograph of the SelecT™ robot built by The Technology Partnership, Cambridge, UK. This robot maintains between 1 and 50 cell lines, performs passaging, cell counting and viability measurements, and direct plating into microtitre plates for bioassay*

frame. Nevertheless, important drivers for miniaturization are common between lead evaluation and HTS, *e.g.*, cost savings per compound tested. Reagents are a significant cost and by reducing the 100-μL 96-well assay format to a 25-μL 384-format, there is an obvious saving. Reduced amounts of compounds, whether individually synthesized or from a direct SAR library, will be consumed in a lower volume assay, hence material from one synthesis run can be submitted to more assays, increasing the information content and improving the decision-making process. There are a number of important considerations with low volume assays, including adding compounds and reagents and detecting the product of the reaction. The compound presentation to the low volume assay can be challenging, *e.g.*, in a 25-μL cellular assay that can only tolerate a final concentration of 0.1% dimethylsulphoxide (DMSO), only 25 nL of a 100% DMSO solution can be added. One way out is to use intermediate dilution: compound is diluted into an aqueous solution, mixed and then transferred to the bioassay. However, this process has significant disadvantages; extra plastic consumables are used, and more importantly the solubility of the compound in the intermediate aqueous stage can often cause precipitation and thereby reduce the apparent IC50. To enable low volume liquid transfer, a variety of equipment options, based on solenoid or piezo technologies, are now available [6,7] for accurately and precisely dispensing low volumes (5–250 nL). These dispensing tips still need to be cleaned between each compound addition, but can rapidly generate a concentration response curve because the number of droplets fired into the well can be varied. Elimination of the tips, to which some compounds can absorb, is an ideal solution that has been achieved using acoustic wave dispensing.[8] In our experience, this system is a highly precise, rapid method of transferring compounds from a compound source plate to a bioassay plate. The instrumentation also has a number of other useful attributes; the volume of compound in the compound source plate need not be constant, and the plate wells can be randomly accessed. An important aspect of acoustic wave technology is the ability to accurately determine the water content within the DMSO solutions. DMSO is hydroscopic and will rapidly absorb water from the air; hence monitoring the water content is an important quality control measure.

Bulk inexpensive reagents can be accurately added to a 384-well microtitre plate using pump-based instrumentation like the Multidrop™. More expensive reagents are delivered using tip-based automated pipetting heads from an environmentally controlled reservoir to the assay plate; for larger compound collections, 96- or 384-pipette heads can be incorporated to deliver the reagents. The detection technologies will be described later, but, in summary, multimodal plate readers (fluorescence, fluorescence polarization, absorbance, *etc.*) or charge-coupled device (CCD) chip-based imagers[9] provide the functionality required in a modern lead optimization laboratory. By putting all these elements together with a plate input store, an incubator and scheduling software, one can envision a robot doing the job of moving the plates around (see Figure 5).

For the higher demand assays, this approach maybe a good solution because multiple assays can be scheduled and interleaved with each other. Alternatively, a series of flexible workstations and operators that move the plates to and from liquid handlers, incubators and readers may be a preferred option. In a modern lead optimization laboratory, both platforms are routinely used.

Having described some of the technologies and processes being used to support lead evaluation, a critical element for the overall effectiveness of such an operation is the range of bioassays used for profiling and selectivity testing.

Figure 5 *The lead evaluation robot above is loading a 384-well microtitre plate into a liquid-handling device. The robot move up and down a track, move plates between reagent addition stations, incubators and readers*

2.2 Specificity of Drug Candidates and the Construction of *In Vitro* Specificity Panels

In the following sections, we focus on the *in vitro* target screening strategies that are used for monitoring specificity of drug candidates. The first part will concentrate on the target specificity assays for receptors, this section is followed by the bioassay technologies used for kinases, and then finally we illustrate the *in vitro* methods for ion channel targets.

2.2.1 Receptors

The concept of a receptor as a therapeutic target is central to drug discovery. Receptors sense extracellular chemical signals and transduce a signal cascade within the cell that can be stimulated or inhibited by specific pharmacological agents. Understanding pharmacological properties of receptors has led to an explosion in the number and type of chemicals available as medicines.[10] A general property of all receptors, including G protein-coupled receptors (GPCRs) and nuclear receptors (NRs), is the ability to interact with their endogenous ligands to alter cellular responsiveness without changing the chemical nature of the ligand. Among them, GPCRs rank higher than any other class of receptors in terms of drug discovery potential based on historical precedence. Today, more than 30% of the top 50 products marketed are targeting GPCRs.[11]

2.2.1.1 G Protein-Coupled Receptors.   GPCRs constitute the largest superfamily of the receptors and mediate the majority of transmembrane signal transduction in living cells. About 600 GPCR genes have been identified from genomic sequencing, and

200 GPCRs have known ligands.[12] These receptors respond to a wide range of relatively small and structurally diverse chemicals such as biogenic amines, peptides and hormones. GPCRs have been divided into three classes based on the similarity of sequences and the ligands that activate them.[13,14] Class 1 is the largest subgroup and includes rhodopsin-like receptors. Class 2 includes secretin-like receptors with a relatively long amino terminus. Class 3 contains the metabotropic-glutamate-receptor-like receptors with a long amino terminus and carboxyl tail. The ligands that activate these three class GPCRs are summarized in Table 2. The receptors for which no ligands are known are referred to as 'orphan' receptors. The identification of orphan receptors has led to new opportunities for novel drug discovery targets.[15–17]

We often face two challenges for achieving GPCR ligand specificity. The first challenge is when molecules developed against one GPCR cross-react with other members of the same family. For example, there are 17 known chemokine receptors and specificity for a selected chemokine target is a common hurdle that needs to be overcome for the design of highly selective antagonists (unpublished internal observation). The second challenge is that these molecules may also show cross-reactivity with members of different GPCR families. For example, in our experience, chemokine receptor antagonists show cross-reactivity with the monoamine GPCR receptors including dopamine, serotonin and adrenergic receptors. This cross-activity with the monoamine GPCRs can limit the potential therapeutic use of these antagonists. The overall goal is to remove these undesired activities while maintaining potency of the antagonists. Therefore, once a GPCR drug target is chosen, screening for target-specific ligands needs to include both target and target-related family members. In addition, emphasis on the detection of liability issues that may be associated with, *e.g.*, the monoamine GPCR receptors, early in the discovery process leads to increased probability of success for drug candidates progressed to first-in-human studies. To select the right screening strategies for GPCR drug discovery, it is vital to understand the pharmacological activities of a given GPCR's ligand. The mechanism of GPCR and ligand interaction can be illustrated by the two-state receptor model[18] (Figure 6). In the two-state model, a GPCR can exist in

Table 2 *Known ligands for GPCRs*

GPCRs	*Biological ligands*
Class 1	Biogenic amines
	Chemokines
	Prostanoids
	Neuropeptides
	Secretin
Class 2	Parathyroid hormone
	Glucagon
	Adrenomedullin
	Calcitonin
Class 3	Calcitonin gene-related peptide
	Metabotropic glutamate
	Ca^{2+}-sensing reagents
	γ-aminobutyric acid type B

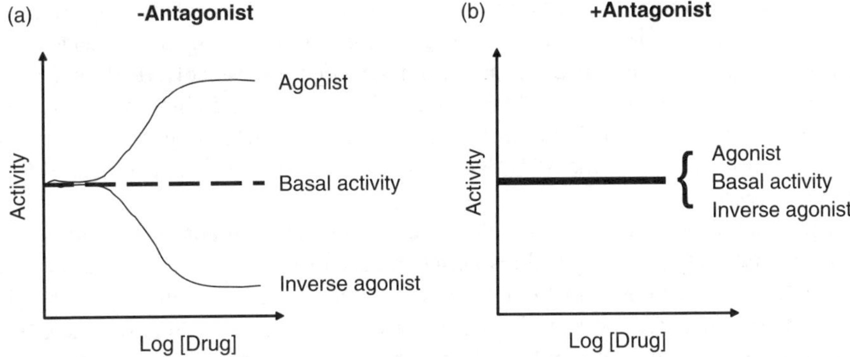

Figure 6 *Classification of GPCR ligands: GPCR ligands can be divided into three classes: agonists (panel a), which increase the proportion of active receptor states to cause a biological response; inverse agonists (panel a), which decrease the proportion of active receptor states to reduce constitute (basal) activity; and antagonists (panel b), which inhibit the action of other ligands*

the inactive state (R) and the active state (R*). A ligand can bind selectively to either R or R*. The affinity of a ligand to R or R*, therefore can influence the signal transduction through the GPCR. Three types of ligands can be classified (Figure 7) based on the ligand receptor interaction: agonists, which increase the proportion of active receptor states (R < R*) to cause a biological response; inverse agonists, which decrease the proportion of active receptor states (R > R*), and thereby reduce basal receptor activity; and antagonists, which inhibit the action of other ligands interacting with the GPCR. An antagonist has an equal affinity to R and R*, so that it does not disturb the equilibrium and therefore does not alter basal response. An inverse agonist binds selectively to R, resulting in a decreased biological response. This model has been further developed by Leff *et al.*[19] into a three-state model that predicts that agonists acting at the same receptor may show different affinity and efficacy depending upon the assay system used to measure the response and the signal pathways to which the receptor couples.[20,21]

Another important aspect for selecting a GPCR screening platform is an awareness of the GPCR signal transduction pathway, which is modulated by a three-component system including receptors, G proteins and downstream effectors. The cell surface receptor is first coupled to heterotrimeric G protein, which consists of GTP-hydrolyzing Gα subunit and Gβγ dimer. Binding of an agonist to the receptor leads to a conformation change in the receptor, resulting in the exchange of GTP for GDP by Gα subunit and the release of Gβγ subunit. GTP-bound Gα or Gβγ then interacts with downstream effectors, resulting in the generation of second messengers. These effectors include adenylate cyclase, phosphodiesterases, guanylate cyclase, phospholipase A2, phospholipase C, phosphatidylinositol-3-kinases and ion channels. The secondary messengers include cyclic AMP, cyclic GMP, diacylglycerol, inositol-1,4,5-trisphosphate, arachidonic acid and intracellular calcium.[22,23]

One technology that has had a major impact on GPCR drug discovery during the past decade has been the use of fluorescence labels as a tool to monitor GPCR activity. The common detection techniques based on these labels include fluorescence polarization, fluorescence intensity distribution analysis, fluorescence resonance energy transfer (FRET) and homogeneous time-resolved fluorescence (HTRF). These techniques, in addition to radiolabelled ligand binding assays, have led to the study of the molecular mechanism underlying GPCR–ligand interactions. Table 3 summarizes the methods that are commonly used to characterize GPCRs activity *in vitro*.

The technologies listed in this table detect whether compounds under study either block or stimulate physiological responses in cells or membranes.[24,25] Each method has its own advantages and limitations that should be clearly understood by the assayist. The major advantages for these assays are that they are relatively straightforward and facilitate the rapid screening of hundreds to thousands of compounds with putative agonist and antagonist activity against GPCRs. These assay techniques have also been enormously useful in defining basic properties of GPCR systems. The cell-based technologies allow GPCRs to be expressed in large quantities in multiple cellular backgrounds, enhancing pharmacological analyses.

However, there are limitations to be considered when selecting a screening method for lead evaluation of GPCR targets. For example, the binding assays are all based on the displacement of labeled ligands and cannot differentiate between agonists, antagonists or inverse agonists. Cell-based techniques used for GPCR assays can be somewhat artificial due to the overexpression of the receptor required to give

$$\begin{array}{ccccc}
\mathbf{L} & + & \mathbf{R} & \rightleftharpoons & \mathbf{LR} \\
 & & \updownarrow & & \updownarrow \\
\mathbf{L} & + & \mathbf{R^*} & \rightleftharpoons & \mathbf{LR^*} \longrightarrow \text{Biological Response}
\end{array}$$

Figure 7 *The two-state receptor model: L = ligand, R = inactive GPCR, and R* = active GPCR*

Table 3 *In vitro bioassay techniques used for characterizing GPCRs*

Characterization	Bio-assay techniques	Sensitivity	Throughput
Receptor/ligand affinity	Radio-label filtration	High	Low
	SPA	High or medium	High
	FP	Medium or low	High
G protein activation	Non-hydrolysable analog [35S]-GTPγS filtration binding	Medium	Low
	Time-resolved Eu-GTP filtration binding	Medium	High
	FP cAMP	High or medium	High
	αScreen cAMP	Medium	High
	HTRF cAMP	High	High
	SPA cAMP	Medium	High
Downstream effector response	Fluorescence discoveRx	High	High
	IP3 αscreen	High	High
	Calcium mobilization	High	High
	Reporter assay	High	High
	Aequorin flash luminescence	High	High

an adequate signal. Therefore, the data obtained using them should be interpreted with caution. The receptor expression levels may have a huge impact on the efficacies of partial agonists and inverse agonists.[26] GPCR signaling *in vivo* is influenced by tissue location and signaling complexes can be cell-type specific.[27] Finally, a GPCR can interact with different proteins when overexpressed in a cell system compared with its interaction partners in native tissue. The most obvious shortcoming of all these techniques, even for the best cell-based assays, is that they can neglect liability and toxicity concerns that only become evident in *in vivo* experiments.

2.2.1.2 Nuclear Receptors.   NRs constitute a large superfamily of intracellular ligand-dependent transcription factors. There are about 50 distinct members in the human genome.[28] NRs can be divided into three classes. Cytoplasmic steroid receptors including androgen receptor (AR), mineralcorticoid receptor (MR), estrogen receptor (ER), glucocorticoid receptor (GR) and progesterone receptor (PR) are active as homodimers and undergo nuclear translocation upon ligand binding. The retinoid X receptor (RXR), heterodimeric receptors containing thyroid hormone receptor (TR), vitamin D receptor (VDR) and retinoic acid receptor (RAR) bind the transcription elements as heterodimers with RXR and are retained in the nucleus regardless of the presence of ligands.[29,30] NRs share six common functional and structural domains including a variable A/B amino-terminal region (50–500 amino acids): a highly conserved C domain (~70–80 amino acids) responsible for specific binding of the receptor to its target DNA sequence and weak dimerization; a flexible linker D region containing the nuclear localization signal, and a flexible hinge between the C and E regions; the moderately conserved carboxy-terminal E domain (~250 amino acids), and the ligand binding domain (LBD) responsible for hormone binding and dimerization.[31]

In simple terms, the mode of action of NRs consists of three steps. The first step is repression, where NRs recruit a corepressor complex. The second step is the derepression that is triggered when this complex binds its normal ligand and that results in chromatin decondensation, which is believed to be necessary but not sufficient for activation of the target gene. The third step is transcription, when a second coactivator complex is assembled which is able to establish contact with the basal transcription machinery, resulting in transcription activation of the target gene. This mechanism is not ubiquitous, since some NRs may act as activators without a ligand, whereas others are unable to interact with target gene promoter in the absence of ligands.[31,32]

Ligands for NRs can be classified into naturally occurring ligands and synthetic ligands. The natural ligands (such as 17-β-estradiol (ER), testosterone (AR), progesterone (PR), cortisol (GR), and aldosterone (MR), all-trans retinoic acid (RAR), L-3,5,3'-L-triiodothyronine (T3) (TR), 9-*cis*-retinoic acid (RXR) and 1,25-dihydroxy vitamin D3) are all small and fairly rigid molecules with a high degree of hydrophobicity. This overall hydrophobic nature facilitates diffusion across cell membranes. The importance of the NRs as medicinal targets has resulted in the generation of an enormous number of synthetic compounds that have advanced understanding of the NR ligand requirements.[31] Agonist and antagonists of NRs are terms often used to describe ligands that either activate or repress transcription in transient transfection reporter gene assays; however, these are inadequate for the description of NR ligands.

The problem is that many ligands function as agonists in certain tissues and antagonists in others.[31,33,34] In addition to ligands that act as pure agonists and antagonists, NRs can also bind partial (or mixed) agonists (or antagonists). These compounds typically bind with a lower affinity than pure agonists and induce unique conformational changes in the receptors, resulting in specific biological responses. The ability to selectively modulate the receptors has led to the search for selective ER modulators, selective androgen receptor modulators (SARMs), selective liver X receptor (LXR) modulators (seLRMs), selective peroxisome proliferator-activated receptor (PPAR) modulators, among others. Together, these make NRs attractive drug targets.[35,36] Figure 8 illustrates the interactions involved as NRs function as transcription factors, and Table 4 summarizes the common bioassay techniques used to characterize ligand specificities for NRs.

2.2.2   Protein Kinases

Protein kinases regulate significant aspects of cell life, including metabolism, cell cycle progression, cytoskeletal rearrangement and cell movement, apoptosis and

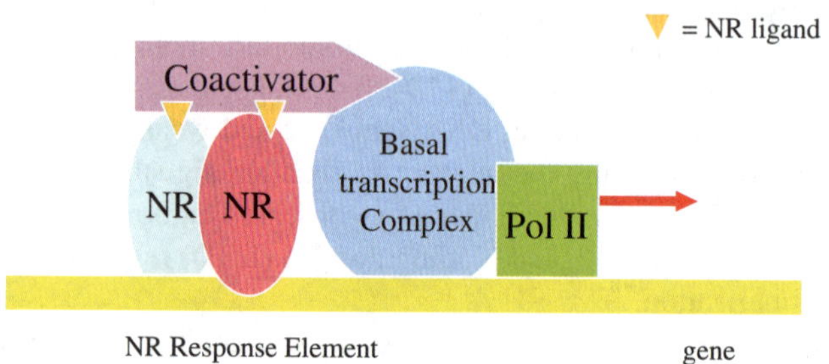

Figure 8 *Schematic diagram of the nuclear hormone receptor transcription mechanism*

Table 4 *The common bioassay techniques used to characterize ligand specificities for NRs*

Purpose	Assay type	Assay techniques	Sensitivity	Throughput
Primary		Filtration	High	Low
	Cell-free binding assay	SPA	High or medium	High
		FP	High or medium	High
Secondary	Cell-free coactivator recruit assay	FRET	High or medium	High
		Luciferase	High	Medium
	Reporter transcription assay	Secreted alkaline phosphatase	High or medium	Low
Tertiary	Gene induction through mRNA detection	Taqman	High	Low
		bDNA	Medium	Medium

differentiation processes through reversible phosphorylation of proteins. As a result, a number of diseases, including cancer, diabetes and inflammation, are associated with an abnormal phosphorylation event mediated by protein kinases.[37–39] As of today, 518 protein kinases are thought to be represented in the human genome,[40] hence these enzymes share the potential to become the second most important group of drug targets after GPCRs.

Almost all protein kinases contain a catalytic domain (250–300 amino acid) with a binding pocket for ATP.[41] The catalytic domains present in various protein kinases share homologous structural features. They are responsible for (1) binding and orientation of the protein (or peptide) substrate, (2) binding of the ATP (or GTP) phosphate donor as a complex with Mg^{2+} and (3) transfer of the γ-phosphate from ATP (or GTP) to the acceptor hydroxyl residue (serine, threonine or tyrosine) of the protein substrate. This common catalytic mechanism across the many families of kinases raises the question as to whether any selectivity and specificity can ever be achieved by protein kinase inhibitors competitively targeting the ATP binding site. Considering the fact that there are as many as 518 protein kinases, it is not surprising that selectivity is one of the major obstacles for kinase drug discovery. The success of Gleevec™ for the treatment of chronic myelogenous leukemia is clear validation of a protein kinase target. Gleevec™ is a relatively specific, ATP-competitive inhibitor of the Abelson tyrosine kinase (ABL); however, it also inhibits other kinases, including cKIT and platelet-derived growth factor (PDGF) receptor tyrosine kinases with similar potency.[42,43] Several protein kinase inhibitors that have entered human clinical trials are also not very specific for a single kinase.[37–39,44–46] These examples suggest that it may be impossible to design completely specific ATP site-based inhibitors, and that perhaps it is more practical to look for an inhibitor with a preferred kinase inhibition profile, rather than a completely specific one. It is clearly important to evaluate lead compounds against a broad panel of protein kinases to aid the future development of protein kinase inhibitors.

Since the protein kinases are bisubstrate enzymes, the binding of ATP affects the binding of protein substrates (or peptides) and *vice versa*. Therefore, it is important to define the appropriate kinetic parameters and understand the limitations of the assay technologies that are chosen for kinase lead evaluation. Discovery of kinase inhibitors has been focused on three different types: ATP-competitive inhibitors, non-ATP or allosteric inhibitors and inhibitors that bind the inactivate kinase form. The kinase assay technologies (see Table 5) can be divided into radioactive and non-radioactive approaches and can be homogenous or heterogeneous. The traditional 'gold' standard methods include filtration and HPLC, which are based on the separation of substrate and product; both are defined as heterogeneous assays because they include a separation step. These methods provide powerful tools to characterize the mode of action of protein kinase inhibitors, due to separation of phosphorylated products from unphosphorylated substrates with little or no interference from the compound under evaluation. However, they are laborious techniques and not ideally suited for screening large numbers of compounds. The non-separation, homogeneous, radiometric scintillation proximity assay[47] and FlashPlate™ assay[48] have been used for HTS of kinase inhibitors. In addition, methods that utilize fluorescence measurement with much higher capacity have become available for assaying kinase

Table 5 *Assay technologies for protein kinases*

Quantitation	Methods	Assay Technology	Assay Principle	Throughput
Phosphorylated substrates or peptides	Radiometric	Filtration	Phosphorylated substrates are captured on the filtration plates	Low
	Radiometric	SPA	Phosphorylated, biotinylated substrates are captured by strepavidin (SA)-coated SPA beads	High
	Radiometric	Flashplate	Phosphorylated, biotinylated substrates are captured on the SA-coated FlashPlate	High
	Time-resolved fluorescence	Lance kinase assay	Phosphorylated, biotinylated substrates are captured through Lance Eu-W1024 labeled P-tyr-100 antibody (Eu-Ab)and SA-conjugated APC (APC-SA)	High
	Time-resolved fluorescence	HTRF	Phosphorylated substrates form a complex with a labeled antibody	High
	Fluorescence Polarization	FP	Specific antibody recognizes fluorescence labelled peptides or substrates	High
	Luminescence	α Screen kinase assay	SA donor beads and antibody acceptor beads are brought to proximity through phosphorylated, biotinylated substrates	High
	Fluorescence Polarization	IMAP	Phosphorylated substrates are detected at high salt concentration through specific binding of metal coordination complexes	High
	Time-resolved fluorescence	Z'-LYTE	A site-specific protease recognizes and cleaves non-phosphorylated FRET-peptides. Cleavage disrupts FRET signals, whereas uncleaved, phosphorylated FRET-peptides maintain FRET	High
	Fluorescence	Off-chip microfluidic assay	The electrophoretic seperation and detection of substrate and product take place in the channels of the chip	Medium
ATP consumption	Fluorescence	Kinase Glo-luminescent	This method quantifies the ATP remaining in solution after the kinase reaction is terminated	High

inhibitors. These include dissociation-enhance lanthanide fluorescence immunoassay (DELFIA®),[49] time-resolved fluorescence resonance energy transfer (LANCE™),[49] fluorescence polarization (FP),[50] homogeneous time-resolved fluorescence (HTRF™),[51] immobilized metal ion affinity-based fluorescence polarization (IMAP™),[52] Z-LYTE™[53] and off-chip microfluidic assays.[54] Kinase-Glo™ technology[55] is based on luminescent readout for quantitation of ATP consumption during protein kinases reaction. Among these technologies, FP can be used for detecting allosteric inhibitors and measuring inhibitors that bind inactivated kinases.[56,57]

2.2.3 Ion Channels

Ion channels play a critical role in maintaining cellular physiology by providing a route for the transport of ions into and out of as well as within the cell. At the simplest level, ion channels can be subdivided into two major classes, the ligand-gated ion channels and the voltage-sensitive ion channels. Ligand-gated channels include nicotinoid receptors, ATP-gated channels and cationic channels activated by excitatory amino acids.[58] The voltage-sensitive ion channels include sodium, calcium and potassium channels, as well as the transient receptor potential channels.[59] These transmembrane multisubunit proteins show remarkable selectivity for the ion that they transport. The magnitude of the current is proportional to the channel conductance, the density of channels and the time the channel stays open. For voltage-sensitive channels, the time that the channel remains open is dependent on the membrane potential.

Diseases associated with aberrant ion channel function, either loss or gain of function, have fueled the desire to find drugs to modulate these biochemical lesions.[60] Additionally, in a broader drug discovery context, there is a need to avoid compounds that inadvertently interact with critical ion channels and cause adverse events. A well-publicized cardiotoxic event is the prolongation of the QT interval in the ECG, for which the hERG potassium channel appears to play a critical role.[61,62] In a lead evaluation function, it is important to enable ion channel assays as direct support of drug discovery programs as well as to monitor the potential for adverse events. Consequently, there is a real need to build an ion channel assay capability. Current thinking indicates that patch-clamp electrophysiology is the highest quality method for evaluating ion channel modulation. This method, although precise and accurate, has a major limiting factor: throughput; a single operator can typically generate 20–50 data points in a single day. Recently, new instrumentation has become commercially available that automates electrophysiological measurements, *e.g.*: IonWorks from Molecular Devices and the PatchXpress from Axon/Molecular Devices.[63] In our hands, the PatchXpress can increase operator throughput by 5–15-fold. This throughput is still not sufficient to routinely support a lead evaluation team, hence a variety of surrogate methods are used to measure ion channel activity.

At the simplest level, competitive binding assays can be used with radiolabelled ligands using filter binding or scintillation proximity assays. These types of assays are relatively high throughout, miniaturizable and easy to automate. The major disadvantage is that if the compound under study binds to a distal site relative to that of the radioligand, the data would suggest that the compound is not binding to the ion channel. Additionally, binding will not distinguish an agonist from channel-blocker activity. To overcome the disadvantages of radioligand binding assays, a variety of ion flux assays are commonly used.

The ion flux assay methods for measuring ion channel activity fall into two broad categories: direct ion measurement or indirect measurement through the use of fluorescence sensors. With direct ion measurement, radioactive tracers can be used, *e.g.*,^{45}Ca^{2+}, ^{22}Na^{+}, ^{86}Rb^{+} (a surrogate for K^{+}) and ^{36}Cl^{-}. These methods are cumbersome because cells need to be loaded and washed and only the steady state function of the ion channel can be measured. For K^{+} channels, commercially available atomic absorption spectrometers can quantify Rb^{+} flux, *e.g.*, for the hERG channel,[64] avoiding the need to use radiochemicals. Two basic fluorescence sensor designs tend to be commonly used for ligand- and voltage-gated channel high throughput assays: direct ion sensors and membrane potential dyes. Fluorescent dyes, *e.g.*, Fluo-4, can be loaded within cells and give a fluorescence intensity that is proportional to the free intracellular calcium concentration. The change in fluorescence intensity on influx of Ca^{2+} can be readily measured in a CCD-based reader, *e.g.*, FLIPR^{TM} (Molecular Devices, Sunnyvale, CA). With potassium channels, dyes have been reported[65] that monitor the flux of Tl+ as a surrogate for K^{+} surrogate =, again using instruments like the FLIPR^{TM} or FDSS^{TM} from Hamamatsu (Hamamatsu city, Japan).[66]

In contrast, the membrane potential dyes sense the polarization state of the cell. In the resting state, negatively charged oxonol dyes partition into the outer layer of the hyperpolarized cell membrane. After the opening of an ion channel, the inner membrane becomes positively charged and the oxonol dye repartitions to the inner surface. This movement can be measured in two ways: using FRET[67] or through the application of a fluorescence-quenching dye in the external media.[66]

Although these fluorescent methods aid throughput, they may not always display the exact pharmacological activity seen in electrophysiology. In our experience with a thallium flux assay for hERG channel activity, a poor correlation with electrophysiological measurements is found for approximately 15% of the compounds evaluated. It is recommended that 'spot checking' of compounds that are positive in a flux assay, by using electrophysiology, be adopted for routine practice.

3 Summary

The application of efficient processes and technologies that were originally developed for HTS are now showing great utility in lead optimization. By using a bioassay suite that delineates MSARs and SLRs early in the drug discovery process, there is greater potential for informed decision-making that could lead to greater success in developing drug candidates.

References

1. B. Booth and R. Zemmel, *Nat. Rev*, 2004, **3**, 451.
2. J.G. Houston and M.N. Banks, High throughput screening for lead discovery, in *Burger's Medicinal Chemistry and Drug Discovery*, 6th edn, vol. 2, D.J. Abraham (ed), Wiley, New York, 2003, 37–69.

3. P. McGee, *Drug Discov. Dev.*, 2005, **8**, 26.

4. C. Bernard, D. Conners, L. Barber, S. Jayachandra, A. Bullen and A. Cacace, *J. Assoc. Lab. Automation*, 2004, **9**, 209.

5. A. Weinberg, L. Zhang, D. Brown, A. Erice, B. Polsky, M. Hirsch, S. Owens and K. Lamb, *Clin. Diagnos. Lab. Immunol.*, 2000, **7**, 714.

6. J. Comley, *Drug Discov. World*, 2002, **3**(3), 33.

7. J. Comley, *Drug Discov. World*, 2004, **5**(3), 43.

8. S. Keating, *Drug Discov. Dev.*, 2005, 46–48.

9. P. Ramm, *Drug Discov. Today*, 1999, **4**, 401.

10. J. Drews, *Science,* 2000, **287**, 1960F.

11. F. Scussa, *Med. Adv. News*, 2002, **21**, 1.

12. J. Vender *et.al.*, *Science*, 2001, **291**, 1304.

13. S. Food, *Curr. Opin. Pharmacol.*, 2002, **2**, 561.

14. Gaulton and T.K. Attwood, *Curr. Opin. Pharmacol.*, 2003, **3**, 114.

15. S. Shaaban and B. Benton, *Curr. Opin. Drug Discov. Dev.*, 2001, **4**, 535.

16. S. Katugampola and A. Davenport, *Trends Pharmacol. Sci.*, 2003, **24**, 30.

17. E.S. Huang, *Drug Discov. Today*, 2005, **10**, 69.

18. U. Geth and B.K. Kobilka, *J. Biol. Chem.*, 1998, **273**, 17979.

19. P. Leff, C. Scaramellini, C. Law and K. McKechnie, *Trends Pharmacol. Sci.*, 1997, **18**, 355.

20. C.B. Brink, B.H. Harvey, J. Bodenstein, D.P. Venter and D.W. Oliver, *J. Clin. Pharm.*, 2004, **57**, 373.

21. T. Kenakin, *FASEB J.*, 2001, **15**, 598.

22. G. Gilman, *Annu. Rev. Biochem.*, 1987, **56**, 615.

23. M. Rodbell, *Nature*, 1980, **284**, 17.

24. P.A. Johnson, *Drug Discov. Today*, 2002, **7**, 353.

25. I.A. Hemmila and P. Hurskainen, *Drug Discov. Today*, 2002, **18**, 150.

26. R. Seifert and K. Wenzei-Seifert, *Naunyn-Schmiedeberg's Arch. Pharmacol.*, 2002, **366**, 381.

27. S.F. Steinberg and L.L. Brunton, *Annu. Rev. Pharmacol. Toxicol.*, 2001, **41**, 751.

28. M. Robinson-Rechavi, A.S. Capentier, M. Duffraisse and V. Laudet, *Trends Genet.*, 2001, **17**, 554.

29. A. Aranda and A. Pascual, *Physiol. Rev.*, 2001, **81**, 1269.

30. J. Mangelsdorf and R.M. Evans, *Cell*, 1995, **83**, 841.

31. V.R. Weatherman, R.J. Fletterick and T.S. Scanlan, *Annu. Rev. Biochem.*, 1999, **68**, 559.

32. G.A. Fancis, E. Fayard, F. Picard and J. Auwerx, *Annu. Rev. Physiol.*, 2003, **65**, 261.

33. J.I. MacGregor and V.C. Jordan, *Pharmacol. Rev.*, 1998, **50**, 151.

34. J.A. Gustafsson, *Curr. Opin. Chem. Biol.*, 1998, **2**, 508.

35. M.S. Frances, *Expert Opin. Ther. Targets*, 2003, **7**, 679.

36. F. Valledor and M. Ricote, *Biochem. Pharmacol.*, 2004, **67**, 201.

37. J. Dancey and E.A. Sausville, *Nat. Rev. Drug Discov.*, 2003, **2**, 296.

38. M.E.M. Noble, J. Endicott and L.N. Johnson, *Science*, 2004, **19**, 1800.

39. P. Cohen, Protein kinases – the major drug target of the twenty-first century? *Nat. Rev. Drug Discov.*, 2002, **1**, 309.

40. G. Manning, D.B. Whyte, R. Martinez, T. Hunter and S. Sudarsanam, *Science*, 2002, **298**, 1912.

41. S.K. Hanks and T. Hunter, *FASEB J.*, 1995, **9**, 576.

42. M.P. Wymann and L. Pirola, *Biochim. Biophys. Acta*, 1998, **8**, 127.

43. B.C.L. Cioffi, N. Law, D. Stover, S. Ohno-Jones, B.J. Druker and N.B. Lydon, *J. Pharmacol. Exp. Ther.*, 2000, **295**, 139.

44. M.A. Fabian, W.H. Biggs, D.K. Treiber, C.E. Atteridge, M.D. Azimioara, M.G. Benedetti, T.A. Carter, P. Ciceri, P.T. Edeen, M. Floyd, J.M. Ford, M. Galvin, J.L. Gerlach, R.M. Grotzfeld, S. Herrgard, D.E. Insko, M.A. Insko, A.G. Lai, J.M. Lelias, S.A. Mehta, Z.V. Milanov, A.M. Velasco, L.M. Wodicka, H.K. Patel, P.P. Zarrinkar and D.J. Lockhart, *Nat. Biotechnol.*, 2005, **3**, 329.

45. S.P. Davies, H. Reddy, M. Caivano and P. Cohen, *Biochem. J.*, 2000, **351**, 95.

46. P.M. Ficher, *Curr. Med. Chem.*, 2004, **11**, 1563.

47. Y.W. Park, R.T. Cummings, L. Wu, S. Zhang, P.M. Cameron, A. Woods, D.M. Zaller, A.I. Marcy and J.D. Hermes, *Anal. Biochem.*, 1999, **269**, 94.

48. C. Sun, Y. Newbatt, L. Douglas, P. Workman, W. Aherne and S. Linardopoulos, *J. Biomol. Screening*, 2004, **9**(5), 391.

49. J.R. Beasley, P.M. McCoy, T.L. Walker and D.A. Dunn, *Assay Drug Dev. Technol.*, 2004, **2**, 141.

50. C. Eggeling, L. Brand, D. Ullman and S. Jäger, *Drug Discov. Today*, 2003, **8**, 632.

51. M.A. Sills, D. Weiss, Q. Pham, R. Schweitzer, X. Wu and J.J. Wu, *J. Biomol. Screening*, 2002, **7**, 191.

52. G.J. Zaman, A. Garritsen, T. de Boer and C.A.V. Boeckel, *Combinat. Chem. High Throughput Screening*, 2003, **6**, 313.

53. H. Jin, J.J. Emanuele Jr., R. Fairman, J.G. Robertson, M.E. Hail, H.-T. Ho, P.J. Falk and J.J. Villafranca, *Biochemistry*, 1996, **35**, 1423.

54. J. Dunne, H. Reardon, V. Trinh, E. Li and J. Farinas, *Assay Drug Dev. Tech.*, 2004, **2**, 121.

55. M. Koresawa and T. Okabe, *Assay Drug Dev. Tech.*, 2004, **2**, 153.

56. D.C. Dalgarno, C.A. Metcalf III, W.C. Shakespeare and T.K. Sawyer, *Curr. Opin. Drug Discov. Dev.*, 2000, **3**, 549.

57. M. Huse and J. Kuriyan, *Cell*, 2002, **109**, 275.

58. N. Le Novere and J.-P. Changeux, *Nucl. Acid Res.*, 2001, **29**, 294.

59. W.A. Catterall, K.G. Chandy, D.E. Clapham, G.A. Gutman, F. Hofmann, A.J. Harmar, D.R. Abernethy and M. Spedding, *Pharmacol. Rev.*, 2003, **55**, 573.

60. C.A. Huber and T.J. Jentsch, *Hum. Mol. Genet.*, 2002, **11**, 2435.

61. R. Netzer, A. Ebneth, U. Bischoff and O. Pongs, *Drug Discov. Today*, 2001, **6**, 78.

62. C.E. Clancy, J. Kurokawa, M. Tateyama, X.H.T. Wehrens and R.S. Kass, 2003, **43**, 441.

63. C. Wood, C. Williams and G.J. Waldron, *Drug Discov. Today*, 2004, **9**, 434.

64. G.C. Terstappen, *Anal. Biochem.*,1999, **272**, 149.
65. C.D. Weaver, D. Harden, S.I. Dworetzky, B. Robertson and R. Knox, *J. Biomol. Screening*, 2004, **9**, 671.
66. W. Zheng, R.H. Spencer and L. Kiss, *Assay Drug Dev. Tech.*, 2004, **2**, 543.
67. J.E. Gonzalez, K. Oades, Y. Leychkis, A. Harootunian and P.A. Negulescu, *Drug Discov. Today*, 1999, **4**, 431.

Concepts for In Vitro Profiling: Drug Activity, Selectivity and Liability

MICHAEL B. BOLGER[a,*], ROBERT FRACZKIEWICZ[a], MICHAEL ENTZEROTH[b] AND BOYD STEERE[a]

[a]Simulations Plus, Inc., 1220 W Avenue J, Lancaster, CA 93534, USA
[b]S*BIO Pte Ltd, 1 Science Park Road, #05-09 The Capricorn, Singapore Science Park 2, Singapore 117528

1 Introduction

Concepts and approaches for drug discovery and development have undergone major changes over the last 50 years. In the 1950s and 1960s, basic pharmacology and the observation of behavioral phenomena in rodents were the standard means to discover new drugs and evaluate their action in animals and humans. Although the concept of discrete molecular targets responsible for drug action, chemoreceptors as defined by Ehrlich, was known for almost 70 years,[1] it was not until the 1980s, when molecular biology was able to supply tools for target cloning that a paradigm shift in drug discovery began to change the way new drugs were identified. Since then, research has become more and more focused on the molecular level. These years have witnessed strong technologically based efforts to shorten the ever-increasing time it takes to translate an initial target idea into a development candidate in the clinic. High Throughput and Virtual Screening, Combinatorial Chemistry and Molecular Modeling, just to mention a few, were all designed to increase the output of pharmaceutical R&D. As a result, pharmaceutical companies have been screening a large number of chemical entities, sometimes even millions, creating hundreds of hits for every target.

However, it is recognized that, despite all these efforts, the approval of new chemical entities (NCEs) has stagnated and reached an historical low in 2002 with 17 new molecule entities approved by the US Food and Drug Administration (US FDA);[2] the approval rates only increased marginally the following year as 21 new entities were approved for the market. A major concern was raised when dropout rates were analyzed.[3,4] It became obvious that lead selection, a critical step in the drug-development process, was far from optimal and that not all candidates who moved into development fulfilled the necessary requirements.

In order to cope with these shortcomings and to increase the quality of drug leads in preclinical R&D, modern drug discovery has come up with a number of new strategies for optimizing the selection criteria by which hits can be identified that have a higher probability for subsequent advancement as chemical leads for medicinal chemistry programs. To understand the rationale, it is first necessary to take a closer look at the reasons for failure in the development of certain drug candidates and the respective frequency of those events. Prentis,[4] in 1988, published the first detailed study based on drugs sent into development over a decade in the United Kingdom; a similar study was laid out in 1997 by Kennedy.[3]

Although published approximately 10 years apart, both studies revealed the same pattern of liabilities that impair the success of the development of NCEs indicating a lack of awareness or failure to remedy. Approximately, 40% of the drugs advanced into development but failed due to insufficient pharmacokinetic properties. Since pharmacokinetics received the most attention among the various parameters, these studies indicated that there was a definite need for more predictive data and models in this area. The information on the pharmacokinetic and physicochemical properties of the lead compounds was incomplete in the very early stages of development. This realization caught researchers' attention, since failure at this stage is still not yet associated with major costs and filtering and eventually dropping a candidate is still reasonable from the economic point of view. The subsequent paradigm shift changed the preclinical research activities from a sequential, efficacy-driven sequence to a more parallel, efficacy and profiling-oriented process.[5] Pharmaceutical companies focused attention on the early evaluation of drug properties in addition to increasing the potency and selectivity of the compounds. This parallel process provides early recognition of potential liabilities of drugs through profiling. Potency, selectivity and compound properties are the key drivers in the hit-to-lead and candidate selection processes, which translate the early screening hit with limited structure-activity-relation data into the final clinical candidate. This strategy contrasts with the earlier sequential approach, in which the only drivers were potency and selectivity. Potential liabilities were only evaluated when the clinical candidate was about to be selected or had already been moved forward into development. Indeed, it has been shown that this approach has led to increased dropout rates in pharmaceutical development.

The application of early *in vitro* studies using human sources of tissue and *in silico* methods as surrogates for *in vivo* measurements has been shown to be an effective alternative, with overall cost savings realized as well. A number of physicochemical parameters determine the fate of chemicals when administered to animals and humans. A detailed study on the physicochemical parameters of marketed drugs was published in 2000 by C.A. Lipinski.[6] These findings became known later as Lipinski's Rule-of-five. Today, Lipinski's Rule-of-five was intended as a way to assess which compounds are most drug-like and thus the best potential candidates. The study gave evidence and guidelines to medicinal chemists as to which parameters influence absorption, distribution and finally drug action. The optimal molecular weight for drugs is considered to be below 500 and the lipophilicity coefficient or partition coefficient (log P) should remain below 5. It is best if the number of proton acceptors is does not exceed 10, and donor functions in the molecule should be limited to 5. The number of rotatable bonds was not considered in Lipinski's original study; later, however, it was added as an additional criterion for drug-like properties.[7] Almost all of the drug-like compounds have no more than 10 rotatable bonds. These general guidelines are now widely considered when compounds are selected for drug screening and hits are being evaluated for moving forward to become medicinal chemistry leads. Unfortunately, when the number of hits in the Rule-of-five is compared to a large database of fraction absorbed (Fa), the value of the rule breaks down due to the significant number of poorly absorbed compounds that do not show any hits in the Rule-of-five. An alternative set of rules dubbed the J-Alert has been shown to be much more accurate in compound screening for absorption.[8]

It is worth mentioning that during the last 15 years, modern technologies in combinatorial chemistry and high-throughput screening have come up with hits that have much higher molecular weights, increased lipophilicity and more hydrogen-bond donor and acceptors than the traditional drugs already on the market[9] and thereby have added additional hurdles to any medicinal chemistry project. The focus on large numbers of compounds, rather than their quality as leads, combined with information-based screening and lead selection, was guiding pharmaceutical R&D in the 1990s in the wrong direction,[10,11] even with the recognition that no company will ever be able to synthesize or screen a significant part of the theoretically possible low-molecular weight compounds, roughly estimated to range from 10^{40} to 10^{100} chemical entities.[6,12] The remaining questions are how densely should a compound set cover chemical space, and how should the selection criteria be applied. Larger pharmaceutical companies such as Pfizer[13] have limited their screening file to 2–3 million and have applied stringent filters on the resulting hits and leads. Most companies now apply the following major approaches to drug characterization and profiling in order to increase early awareness and to select the optimal medicinal chemistry starting point of a program:

- determination of physicochemical properties
- evaluation of *in vitro* metabolism
- investigation of compound selectivity
- establishment of the drug-toxicology profile.

Since implementation of these approaches for early evaluation and understanding of drug properties, there has been a slight reduction in the failure rates associated with pharmacokinetic properties.[2] The extent of effort put into compound profiling varies among institutions since it is very much subject to the company philosophy and the budget available. The selection of the profiling panel applied to a compound set is governed by the size of the compound collection as well as the available human and monetary resources. However, there are some standard tests that have emerged and are applied to almost all lead series.

2   Physicochemical Parameters

2.1   Partition Coefficient

Biological nomenclature and the definition of biopharmaceutical properties are fundamental to the understanding of surrogates for *in vivo* properties. Partitioning or distribution of a molecule between an organic solvent and water is often considered to be a surrogate for the most fundamental biopharmaceutical property, membrane penetration/permeability. Distribution (log D) is always a measured property, while partitioning (log P) is often calculated as the distribution coefficient of the unionized molecular species.[14] The importance of lipophilicity for property-based design has been reviewed in a number of papers highlighting correlations with CNS permeability, absorption and pharmacokinetic parameters.[15–18] The lipophilicity of compounds is most often estimated by calculations of partition coefficients such as C log P or S log P. These values are the logarithm of the octanol/water-partition coefficient (including implicit hydrogens), which can be obtained by an atomic contribution model that calculates log P from the given structure. C log P[19] is based on molecule fragments, which unfortunately cannot be parameterized in a larger number of entities in compound libraries. S log P[20] is an improved linear estimator model based on and accounting for all atoms in the molecule, including explicit hydrogen atoms. The calculated values help the chemist to optimize compound properties, to increase solubility and to improve absorption.

It is important to distinguish between membrane penetration and membrane permeability. Penetration represents movement of a molecule from outside of a membrane into the membrane bilayer but not through the membrane bilayer. Penetration into a membrane is proportional to partition coefficient in an octanol/water biphasic solvent experiment.[21] Permeability is a more complicated process that involves transfer of the molecule from one side of the bilayer to the opposite side. Most biological membranes that are important for permeability of drug molecules during oral absorption or in the processes of elimination are found at the apical or basolateral side of an epithelial cell. Epithelial cells form asymmetric barriers in an organism, where one side is exposed to the external environment and is called the apical side and another side is exposed to the internal systemic environment and is called the basolateral side.[22] For reference, endothelial cells (from which blood vessels are constituted) form asymmetric barriers within an organism, where one side is exposed to the internal systemic circulation and the other side is exposed to the interstitial fluid that surrounds the cells of a tissue. By contrast, epithelial cells form a barrier between interstitial fluid and the external environment. For example, the lumen of the intestine, which is lined with epithelial cells, is exposed to the external environment of an organism by virtue of continuous contact between the oral cavity and the anus. In the same regard, the lumen of the kidney is in continuous contact with the external environment between the glomerulus and the orifice of the urinary excretory organ. Also, the alveoli of the lungs, the skin, the vagina and the nasal passages are lined with epithelial cells.

Permeability across epithelial cells can be affected by the presence of influx or efflux transporters (saturable integral membrane proteins that catalyze the transfer of molecules through a biological membrane).[23] For example, in the gastrointestinal tract, effective permeability (P_{eff}) is measured by the rate of disappearance of a molecule from the lumen of the intestine (apical membrane penetration), which can be slowed if the molecule is a substrate for active efflux or accelerated if the molecule is a substrate for active influx. By contrast, apparent permeability (P_{app}) is measured as the rate of transfer of a drug molecule across a polarized epithelial cell monolayer into the systemic side (basolateral).[24] In both cases, the measurement can be affected by saturable transporters.

The partition coefficient is inversely proportional to paracellular permeability.[25] Molecules that are absorbed by a paracellular route move through the water-filled channels between cells. As we will discuss later in this chapter, permeability rate is a balance between lipophilicity (affinity for the lipid bilayer) and hydrogen bonding (noncovalent bond formation between molecules and the polar head groups of a biological membrane). Contrary to the view expressed in many publications, the partition coefficient should not always be considered as a surrogate for membrane permeability. The partition coefficient is proportional to membrane permeability by virtue of the fact that increasing lipophilicity will increase membrane penetration. However, as a surrogate, the octanol/water-partition coefficient lacks a measure of the hydrogen-bonding component of passive transcellular permeability and is inversely proportional to paracellular permeability.

2.2 pK_a

The ionization of monoprotic acids and bases and hence their solubility and absorption is dependent on their pK_a, the pH at which the drug is 50% ionized. First developed by Hammett and published in 1940,[26] the relation between the dissociation constants of benzoic acid derivatives and the longer range electronic (inductive, mesomeric and field) effects is linear and additive.

$$pK_a = pK_a^0 + \rho\Sigma\sigma \tag{1}$$

In this equation, pK_a^0 is the pK_a of the unsubstituted molecule, σ refers to the constant assigned to a respective substituent and ρ is a constant assigned to the particular ionizable group. The approach was later refined by Taft *et al.*[27] extending the theory and equation to aliphatic and other systems. The principles and methodologies of a classic experimental pK_a determination are well known.[28] Ionization constants for an *n*-protic weak acid may be represented by the following generic definitions:

$$HA^{-(n-1)} \Leftrightarrow H^+ + A^{-n}, \quad pK_a^1 = pH - \log\frac{[A^{-n}]}{[HA^{-(n-1)}]}$$

$$H_2A^{-(n-2)} \Leftrightarrow H^+ + HA^{-(n-1)}, \quad pK_a^2 = pH - \log\frac{[HA^{-(n-1)}]}{[H_2A^{-(n-2)}]} \tag{2}$$

$$H_nA \Leftrightarrow H^+ + H_{n-1}A^-, \quad pK_a^n = pH - \log\frac{[H_{n-1}A^-]}{[H_nA]}$$

To accurately measure these *n macroscopic* pK_a values by standard methods, it is necessary to obtain information about the ratio of equilibrium concentrations of the *n* ionized and one neutral form of the acid by the most noninvasive method possible, while monitoring pH, temperature and ionic strength of the solution. Ionization constants of bases and mixed systems are determined using the same principle. The problem of macroscopic pK_a determination thus becomes a problem of accurate measurement of relative concentrations in water solution. Widely applied methods for this purpose are:

- Acid–base titrations (classical and potentiometric)
- UV spectroscopy
- Capillary electrophoresis
- Chromatographic technique
- Solubility and lipophilicity profiles

An excellent review of all the above methods has been published recently.[29,30] *In vitro* models are readily adaptable as surrogates for determining *in vivo* ionization with one important caveat: experimental conditions must be in tune with the physiological environment, *i.e.*, ionic strength of the 0.15 M medium and average human body temperature of 37 °C.

It is important to note that $pK_a^{(k)}$ determined in one of the above-mentioned experiments only provides information on a species with *k* protons bound, but says nothing about *where* exactly these protons are located. In fact, with *n* ionization centers available there are $\binom{n}{k}$ possible *microspecies*, each holding *k* protons.[1] In some biochemical reactions, however, microspecies act individually[31] and it is desirable to resolve them. Terms like microspecies and *microconstants* apply to an accurate description of acid/base equilibria of multiprotic molecules, where individual protonation states are resolved.[32–34] pK_a macroconstants observed in classic experiments are thus an effect of statistical averaging of individual microdissociation events in Figures 1 and 2.

The concepts presented above are best illustrated by an example. The drug cetirizine (Figure 1) seems to be an ideal candidate. All 12 microconstants and eight

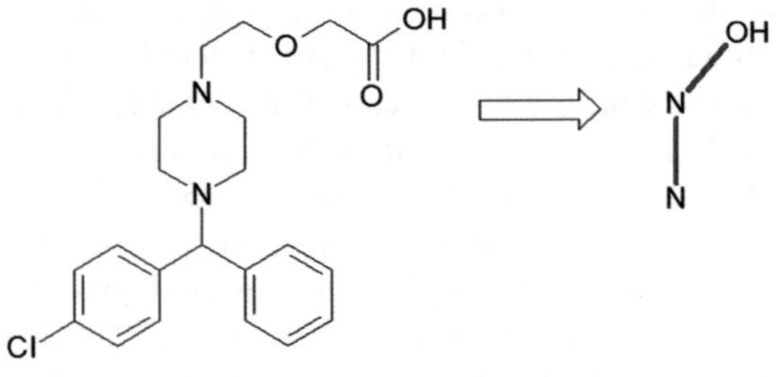

Figure 1 *Molecule of cetirizine and schematic representation of its protonation sites*

---

[1] $\binom{n}{k}$ symbolizes the number of distinct ways *k* objects may be placed among *n* sites, $k \leq n$

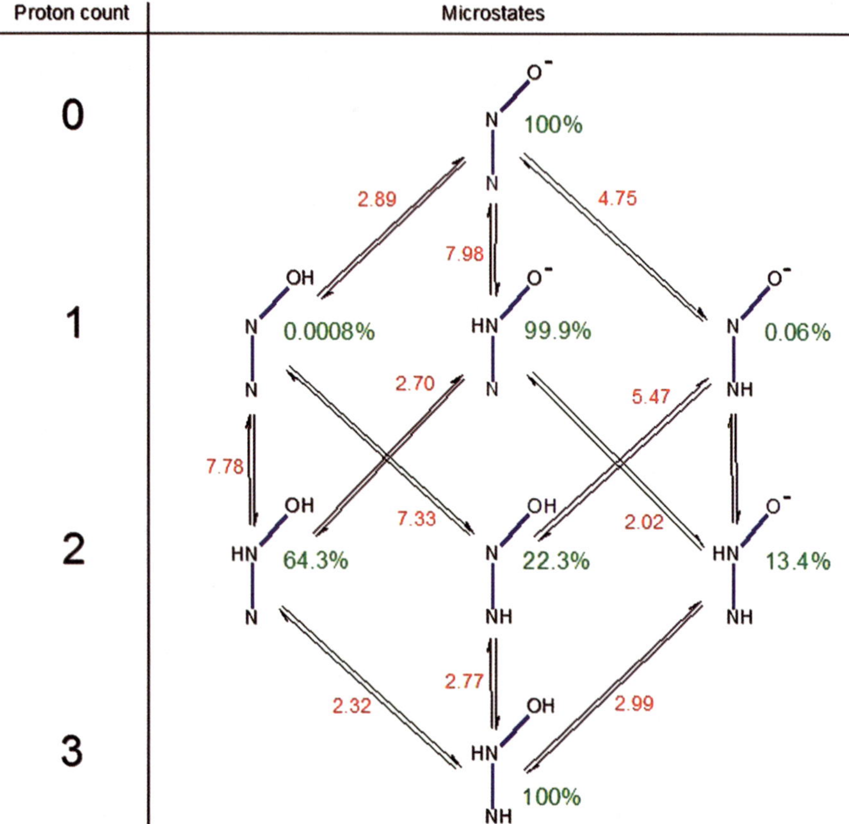

Figure 2 *Protonation microequilibria of cetirizine. Arrows are labeled by microconstants measured experimentally.[35] Protonation microstates are labeled by their relative contributions (probabilities) to the respective parent macrostate*

microspecies (protonation states) of this triprotic molecule have been determined experimentally,[35] along with three pK_a macroconstants: 7.98, 2.90 and 2.12. We already know that only the overall proton count in each of the macrostates can be deduced from titration experiments. In the neutral macrostate, there are three available sites leading to three neutral microstates. There are three different ways of distributing two protons among three sites providing another three microstates in the +1 macrostate. Together with the fully deprotonated and the fully protonated microstate, there are eight possible protonation microstates of cetirizine bound together by a network of 12 microequilibria as shown in Figure 2.

The existence of all the depicted microstates has been confirmed experimentally.[35] The neutral macrostate is strongly dominated by a zwitterionic microstate with one proton bound to the middle nitrogen. Therefore, the idea of the 7.98 macroconstant pK_a value being "assigned" to this amine group is a good approximation. However, "assigning" the remaining two constants, 2.90 and 2.12, to any functional group cannot be made, since the components of +1 macrostate (two protons bound) contribute with probabilities of the same order of magnitude. In general, the closer the micro-constants are to each other, the less accurate the sequential ionization picture becomes.

Recently, microconstant measurements have been validated for 25 drug molecules, based primarily on multiwavelength spectroscopy.[36] The applied methodologies, however, are difficult and limited up to triprotic molecules when no symmetry constraints are present.[33] In more complex cases, the microspecies resolution is still possible by applying some *in silico* methods for predicting ionization constants.

2.3 Solubility

Solubility is one of the most important parameters considered in screening drug candidates for oral absorption. Due to limited intestinal transit time, even highly permeable compounds can suffer from poor absorption if their solubility is low. On the other hand, high solubility allows a pharmaceutical researcher to concentrate on other aspects of absorption. These "low" and "high" solubility categories may not be sufficient in cases where a more precise quantitative solubility measure is needed.[37] The most useful information is the solubility of a drug candidate in gastrointestinal fluids under *in vivo* conditions.[38] For practical purposes, however, pharmaceutical industry researchers estimate solubility values employing a wide array of *in vitro* measurement procedures. Before discussing these techniques, it is important to understand the principles of solubility under thermodynamic equilibrium.

2.3.1 *Thermodynamic Solubility*

The theoretical solubility profile of a pure monoprotic base, B, provides a simple yet powerful illustration of equilibrium solubility. A saturated solution of B at relatively high pH values will contain dissolved compound, $B_{(aq)}$, in thermodynamic equilibrium with pure solid base, $B_{(s)}$:

$$B_{(s)} \Leftrightarrow B_{(aq)} \tag{3}$$

The corresponding equilibrium constant consists of the ratio of thermodynamic activities:

$$K' = \frac{a_{B_{(aq)}}}{a_{B_{(s)}}} \tag{4}$$

Activity of the pure solid base is constant and may be included in a new equilibrium constant, S_B, and the activity of aqueous B can be approximated by its concentration, $[B_{(aq)}]$ leading to:

$$S_B = [B_{(aq)}] \tag{5}$$

Hence, at high pH values, the aqueous concentration of a neutral monoprotic base is equal to a constant called the *intrinsic solubility*. But, this is not the only equilibrium the base is subjected to. One must also consider ionization of B:

$$BH^+_{(aq)} \Leftrightarrow B_{(aq)} + H^+ \qquad K_a = \frac{[B_{(aq)}][H^+]}{[BH^+_{(aq)}]} \tag{6}$$

The solubility profile equation of aqueous B at high-pH values is derived as follows:

$$C^{B}_{(aq)} = S_B \left(1 + \frac{[H^+]}{K_a} \right) \equiv S_B \, (1 + 10^{pK_a - pH}) \tag{7}$$

Where $C^{B}_{(aq)}$ denotes the *total* concentration of B at saturation, $[B_{(aq)}] + [BH^+_{(aq)}]$. Note the limited applicability of this equation: a weak base with $pK_a = 9$ would be about 10^8 times more soluble at pH $= 1$ than it is at pH $= 9$! Clearly, this is not consistent with observation. To understand what really happens at lower pH, one must consider exactly *how* the pH decrease in an aqueous solution of B can be achieved. In an idealized *quasi-static* experiment, a solution of strong acid, HX, is gradually added to the mixture, allowing the equilibrium to reestablish at each time point. New supplies of protons are bound by neutral B molecules yielding more and more BH^+ ions, while increasing the concentration of negative counterions, X^-. It is the presence and character of these counterions that determines the solubility of B at low-pH values. At a certain pH, the amount of $BH^{\pm}$ and X^- ions reaches saturation of the solid BHX salt. Further, quasi-static addition of HX will result in conversion of the remaining solid B into solid-salt BHX. At this point, the equilibrium expressed by Equation (3) is no longer valid and it must be replaced by the *solubility product*, K_{sp}, of the BHX salt:

$$BHX_{(s)} \Leftrightarrow BH^+_{(aq)} + X^-_{(aq)}, \quad K_{sp} = [BH^+_{(aq)}][X^-_{(aq)}] \tag{8}$$

The solubility profile curve at low pH is thus described by:

$$C^{B}_{(aq)} = \frac{K_{sp}}{[X^-]} (1 + 10^{pH - pK_a}) \tag{9}$$

There is a key difference between Equations (7) and (9). While solubility at high pH depends solely on B, at low pH the solubility also depends on the *concentration* and *character* of the negative counterions. The conditions of our experiment will be addressed as a *titration regime*.

Figure 3 shows both profile equations in action in a titration regime. The observed profile is a minimal combination of the neutral base and salt curves. Please note the dramatic decrease of solubility at pH$<$ 2. It is a direct consequence of the *common ion effect*, the reciprocal dependence on the counterion concentration in Equation (9).

The actual *in vitro* measurements of thermodynamic solubility correspond to the idealized titration regime conditions only to some extent. The closest method seems to be a labor-intensive shake-flask experiment requiring relatively large amounts of a dry crystalline drug and long equilibration times.[39] Moreover, each pH point requires a separate measurement in different buffer. Different buffers usually represent different ionic conditions and may lead to internal inconsistency of the solubility profile thus obtained. The buffer issue deserves an entire subsection of this review and will be discussed later. The real question is whether this investment of resources is worthwhile. The answer depends on who is asking the question. A drug-development

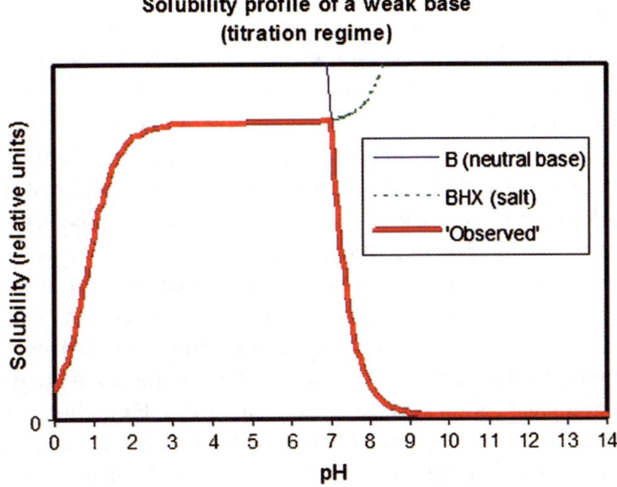

Figure 3 *Theoretical pH-solubility profile in titration regime for a weak base with pK_a=9.6 (thick line). It represents changes in solubility when a saturated solution of the base at pH=14 is slowly titrated with a strong acid HX to pH=0 (titration regime), saturation maintained at all times. The profile is a minimal combination of two curves represented by Equation (6) (free-base region) and Equation (8) (salt region), respectively. A sharp decrease in solubility at pH<1.5 occurs due to common ion effect*

scientist must know whether the drug in his tablet will dissolve in human gastrointestinal (GI) tract, or not, and how fast this process will take place. In this case, the answer is a resounding "yes"; thermodynamic solubility of drugs and excipients is obviously a vital parameter in drug development. The good news is that thermodynamic solubility can be approximately determined with methods faster than a shake-flask experiment (see Ref. 29 for an excellent review). A different answer – "preferably yes" – would be given by a drug-discovery scientist, who deals with DMSO stock solutions, small amounts of substance, and is interested in the fastest and cheapest solubility screen. Unfortunately, a method that is popular in discovery circles involves measuring so-called "kinetic solubility". This method does not have much in common with equilibrium solubility. Rather, this method utilizes turbidimetric detection of compound's precipitation from water–DMSO mixtures. However, a freshly formed, metastable fine precipitate has a higher solubility than crystalline solid.[29] In addition, the presence of even small amounts of DMSO significantly affects solubility of poorly soluble compounds.

2.3.2 To Buffer or not to Buffer

We devote this entire section to a subject that still does not receive due attention in spite of strong thermodynamic arguments, as well as experimental evidence. The fact is most of the existing *in vitro* equilibrium solubility protocols neglect the maintenance of a proper ionic composition of the dissolving medium, yielding results incompatible with *in vivo* conditions. It has to be noted that the situation in man is more complex, such that solubility provides an indicator rather than a concrete parameter. First, the medium in the GI tract is not a simple buffer, since it may have constituents that significantly affect the compound's solubility. In human and dog, for example, the gallbladder plays an important role because the release of its contents into the GI tract can significantly improve compound solubility and consequently absorption. This effect can account for observed species differences; for example, rats do not possess a gallbladder and instead secrete bile acid, lecithin and other constituents at low-level over time.[40]

Second, the pH in the gastrointestine varies considerably, from pH 1–3 in the stomach to pH 5–8 in the intestine, and thus affects where acidic and basic drugs are absorbed. This effect is especially evident in assessing the solubility of weak-base salts. Gastrointestinal fluids contain many counterions for bases, regionally maintained by physiological action at constant concentrations. Examination of pancreatic fluid reveals that it is composed of 160 mM Na^+, 50 mM Cl^-, 75 mM HCO_3^- and lesser quantities of SO_4^{-2} and HPO_4^{-2} at pH $= 8$.[41] In the duodenum, it is mixed with chloride-rich gastric contents. The net result is relatively constant ~140 mM intestinal concentration of Na^+ with Cl^- and HCO_3^- as its main counterions.[42] However, pH does change along the small intestine moving from the proximal region (duodenum) to the distal region (ileum). These facts have two consequences. First, the titration regime of *in vitro* solubility profile determination no longer applies to *in vivo* situations, where a relatively constant concentration of Cl^- is maintained. The difference is apparent in Figure 4, where solubility profile of a weak base is drawn for a *constant common ion regime* $[X^-] = 0.14$ M.

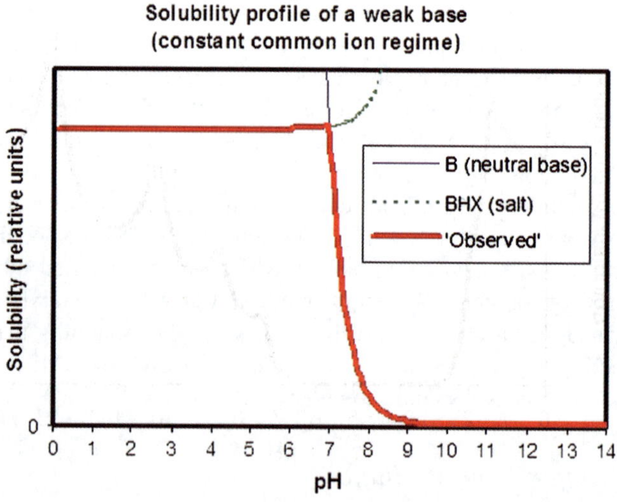

Figure 4 *Theoretical pH-solubility profile in constant common-ion regime of a weak base with $pK_a = 9.6$ (thick line). It represents changes in solubility when the base is dissolved to saturation in water solutions containing constant 0.14 M concentration of the dominant counterion X^-. Solid-salt BHX is formed at lower pH*

Generally, the solubility of a salt is considerably higher and remains flat going down the pH scale. As a second consequence, buffered dissolution media used *in vitro* contain considerable concentrations of foreign counterions, which may completely change salt solubility from that expected *in vivo*. For example, pharmaceutical scientists often employ simulated intestinal fluids, SIF and FaSSIF, containing as much as 8.7 g and 5.0 g of KH_2PO_4 per liter, respectively.[43] The importance of this observation is apparent from Equation (9) and has been experimentally proven by Streng *et al.*[44] who studied the effects of different counterions on the solubility profile of terfenadine. A sharp ~ 10 fold decrease in the peak solubility was observed in the following order of counterions: lactate> methanesulfonate > chloride > phosphate. The peak solubility of terfenadine phosphate was reported to be about 2.7 times lower than that of terfenadine hydrochloride. Differing cations also exert significant influence on the solubility of acidic salts.[45,46] Of course, the degree of complexity increases when compounds with multiple ionizable groups are measured in solutions with multiple-competing counterions.

Figure 4 also suggests that, throughout most of the pH scale, small pH changes are well tolerated and the dissolution medium does not have to be buffered. Monitoring pH with a pH-meter should be enough. In fact, just two solubility points, one in the salt region and one in the free base region plus known pK_a values are sufficient to reproduce the entire solubility profile of a monoprotic base.[47] In general, compounds with N ionizable groups are characterized by up to $N+1$ separate pH regions (one curve for neutral compound and N salt curves). Solubility profiles of these multiprotic compounds may look quite complicated (see Figure 5). However,

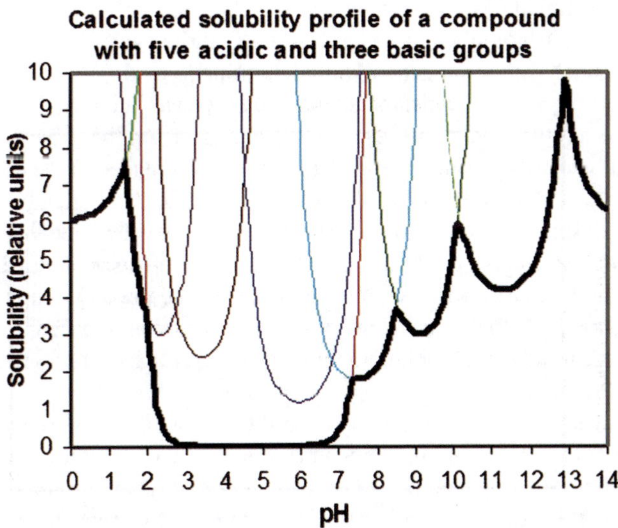

Figure 5 *Theoretical solubility profile (thick line) and its component curves (thin lines) of a compound containing five acidic and three basic groups. The pK_a values are: 2, 2.7, 4.1, 5.3, 6.6 (acids) and 12.7, 9.9, 8.3 (bases). Thick dot at pH=4 represents reference solubility of 0.006 units. Broad U-shaped curve in the $2.5 < pH < 7$ region represents solubility of the neutral compound, other components are salt curves. Note that not all of the salt curves would be experimentally observed*

no more than $N+1$ strategically selected pH-solubility points would be required to fully characterize such compounds; the actual number may be smaller due to regional overlap.

In conclusion, instead of fixing the pH, attention should be shifted toward defining and controlling the concentration of dominant counterions to resemble the average composition of human GI fluids. To a first approximation, the simplest dissolving medium would be 140 mM NaCl. After saturation equilibrium is reached, the pH should be measured along with solubility. pH may be then adjusted to a new value by addition of *known* amounts of concentrated HCl or NaOH, followed by equilibration and a new pH reading. The results then may be recalculated back to 140 mM NaCl. Of course, in the salt regions, appropriate solid hydrochloride or sodium salts of the compound should be used instead of the neutral material.

3 Permeability

Surrogates for *in vivo* permeability can be determined *in situ* or *in vitro*. Generally, these experiments are used to determine the permeability of drugs across the epithelial cell layer lining the GI tract.

In situ methods involve the external perfusion of an isolated segment of intestinal tissue.[48] The rat has been shown to be an excellent model for human intestinal permeability.[49,50] One of the principle advantages of the *in situ* method is that the intact rat intestinal tissue maintains a normal complement of influx and efflux transporters. Cephalosporins and beta-lactam antibiotics are known to be absorbed by oligopeptide transporters in the GI tract. The rat *in situ* permeability method was used to determine the saturable Michaelis–Menten-type kinetics for this class of drug molecules.[51,52] In like fashion, the permeabilities of thymidine anti-viral drugs, like AZT, involving the nucleoside transporter were determined *in situ*.[53]

In vitro methods to measure intestinal permeability fall into three classes. First are the methods that utilize small pieces of tissue excised from the experimental animal and mounted in a side-by-side Ussing chamber. Drug is placed in the donor well and the rate of appearance on the receiver well is determined analytically. Next are the methods that utilize a cellular monolayer grown on a semi-permeable filter. These methods are extremely popular in early ADME (Absorption, Distribution, Metabolism and Excrection) screening and can be adapted for high throughput. Third are the methods that utilize some form of an artificial membrane composed of isolated phospholipids or nonpolar organic molecules. Each of these methods will be discussed below.

In vitro measurement of intestinal permeability using excised segments from three regions of rat intestine: jejunum, ileum and colon for 19 drugs with different physicochemical properties was determined by Ungell *et al.*[54] They observed a significant decrease in permeability to hydrophilic drugs and a significant increase in permeability for hydrophobic drugs aborally to the small intestine ($P < 0.0001$). They found that for hydrophilic drugs (low permeability and low log D), the ratio of colon:jejunal permeability was less than one, while for hydrophobic drugs (higher permeability and higher log D), the ratio of colon:jejunal permeability is observed to be greater than one. At certain pH values, the permeability of small hydrophilic drugs may have a large paracellular component,[55] and it is well known that the transepithelial electrical resistance (TEER) of colon is much higher than small intestine. TEER increases as the width of tight junctions decrease. The tight junction width has been determined to be 8–12 0 A in jejunum, 4–5 A in ileum and 2–2.5 A in colon.[56] The narrower tight junctions in colon suggest that paracellular transport will be much less significant, which helps to explain the lower ratio of colon:jejunal permeability for hydrophilic drugs.

In vitro measurement of permeability using a cellular monolayer derived from human or canine epithelial cells is the most popular method for early ADME high-throughput-screening (HTS). These methods were pioneered by Hidalgo *et al.*[57] and Audus *et al.*[58] and have been adopted by all of the major pharmaceutical companies.[59] Most commonly, a human colonic cancer cell line called "Caco-2" is used as the monolayer on semi-permeable membranes that ride "piggy-back" on top of a normal Petri dish.[24,60] This configuration allows the drug to be added to the apical or the basolateral side of the monolayer to determine apparent permeability (P_{app}). Hildalgo and Li[61] have characterized and well described Caco-2 cells as being able to form intestine-like cellular monolayers that also provide membrane transporters and enzymes, which can significantly influence the absorption of many drugs. One of the most prominent efflux proteins, P-glycoprotein (PgP), is expressed in Caco-2 cells and has been well characterized. Caco-2 cell-based assays have been developed to identify compounds that are PgP substrates.[62]

Other cell lines that have been used for this purpose are the Mandin Darby Canine Kidney (MDCK) cell line from the distal-tubule of the kidney and the LLC-PK1 cell line from the proximal tubule of the kidney.[63] The Caco-2 cell line requires approximately 21 days of culture to achieve maturity, as measured by the size of the tight junctions between the cells, although shorter cell culture periods have been reported. Because they are derived from the colon, Caco-2 cells have very small tight junctions (4.3–4.5 A).[64] In contrast, MDCK cells do not form such tight junctions and the pore size between cells is 6–8 A.[65] Thus, we see that pore size increases in the order of colon < Caco-2 < MDCK < human jejunum. One implication of this fact is that permeability measured in Caco-2 cell lines for hydrophobic drugs should be closely correlated with human jejunum because the main mechanism for those compounds is passive transcellular permeability. However, the Caco-2 cell line is not a good model for the absorption of small polar molecules because of an exaggerated decrease in paracellular permeability compared to human jejunum. However, recent studies indicate that the permeability measured in Caco-2 cell lines can be corrected by calculating the paracellular contribution that would have been observed in an epithelial cell with larger pore size.[66]

The size and shape of a drug molecule, its acid and base dissociation constants, and the pH of the GI tract all influence the absorption rate constant for specific regions of the GI. Pade and Stavchansky[67] measured the Caco-2 cellular permeability for a diverse set of acidic and basic drug molecules at two pH values. They concluded that the permeability coefficients of the acidic drugs was greater at pH 5.4, whereas that of the basic drugs was greater at pH 7.2; they also found that the transcellular pathway was the favored pathway for most drugs, probably due to its larger accessible surface area. The paracellular permeability of the drugs was size and charge dependent. The permeability of the drugs through the tight junctions decreased with increasing molecular size. Furthermore, the pathway also appeared to be cation-selective, with the positively charged cations of weak bases permeating the aqueous pores of the paracellular pathway at a faster rate than the negatively charged anions of weak acids. Thus, the extent to which the paracellular and transcellular routes are utilized in drug transport is influenced by the fraction of ionized and unionized species (which in turn depends upon the pK_a of the drug and the pH of the solution), the intrinsic partition coefficient of the drug, the size of the molecule and its charge.

The Parallel Artificial Membrane Permeation Assay (PAMPA) method, introduced by Kansy *et al.*[68] is a simple, robust, HTS that has been shown to be predictive for passive diffusion through membranes. In addition, PAMPA can deliver information in parallel regarding the lipophilicity, the ionization state and the solubility of a compound. Sugano *et al.*[69] have published a systematic study of the optimal conditions for measuring drug permeability using the PAMPA assay. This group varied the chain length and charge of the lipid solution used to coat the semi-permeable membranes. Moreover, they found that the best correlation with the fraction of dose absorbed for 33 compounds was obtained when the donor medium was held at pH 5.5, in accordance with the pH-partition theory.[70] In a fashion similar to the "corrected" Caco-2 permeability method described above, PAMPA permeabilities can be corrected to include the permeability expected for the paracellular route.[71,72]

4   Metabolism

Numerous *in vitro* and/or *in situ* models have been developed to investigate drug metabolism. In order of decreasing complexity, they are: the isolated perfused liver, isolated liver slices, hepatocytes in coculture with epithelial cells or bacteria, hepatocytes in suspension and in primary culture, subcellular hepatic microsomal S9 fractions and high-speed pellet microsomes. The field of metabolism is immense and will not be covered in depth here. However, one should have a basic understanding of the benefits and drawbacks of each one of these methods.[73]

During oral drug absorption, metabolism can occur in the enterocyte cells that line the lumen of the intestine and in the liver before a drug enters the systemic circulation. Loss of drug in either of those locations constitutes "first pass elimination". This first-pass effect can be observed for any type of extravascular delivery (nasal, pulmonary, rectal, vaginal, etc.). The magnitude of the first-pass elimination can be quite different in different tissues. However, the most common site for first-pass metabolism is the gut and liver. Once a drug enters the systemic circulation, most of its metabolic elimination occurs in the liver. Consequently, much of the literature on metabolism centers on hepatic metabolism.

In situ liver perfusion is the most reliable method to determine the relative rate of metabolism and the expected metabolic products for the species being tested because it exposes the drug to the entire liver organ in a fashion that most closely resembles the *in vivo* condition.[74,75] However, the cost and time-consuming nature of the *in situ* experiment precludes rapid testing for metabolism during the discovery phase. Also, this method is not applicable to human experimentation for obvious reasons.

In vitro drug metabolism has been explored by using many different experimental methods. An excellent review covers much of the important information in this area.[76] Isolated liver slices, microsomal membrane preparations and expression of recombinant drug-metabolizing enzymes have all been used to measure metabolism *in vitro*. As the name suggests, liver slices are small pieces of fresh liver tissue, thin enough to allow drugs in the incubation medium to diffuse into and out of the hepatocytes. This type of preparation provides an estimate of metabolism by all of the enzymes in the liver. However, the liver slices have a short, viable lifetime due to impaired diffusion of nutrients and oxygen.

Classes of metabolic enzymes include: Phase I (esterases, hydrolases, oxidases), and Phase II conjugation enzymes (glucuronide, sulfate and glutathione transferases). The most common enzyme for Phase I drug metabolism is a mixed-function oxidase called Cytochrome P450 (CYP). These are potent oxidative enzymes, which can be classified in three major classes CYP1, CYP2 and CYP3, designated with a number (for class), letter (subclass) and number (for each gene) depending on sequence homology.[77] Only a small number of the enzymes are known to be directly involved in drug metabolism. Three of those, CYP 2C9, CYP 2D6 and CYP 3A4, account for over 90% of the drug oxidations in humans.[78] Interaction with cytochromes may result in an increased metabolic rate of the compound if it is a substrate or it may result in altered plasma levels of other drugs if it inhibits or induces cytochrome oxidase activity.

If liver cells (hepatocytes) are isolated and grown in culture, drugs are exposed to a similar array of enzymes. Because CYP enzymes are bound to the internal membrane fraction of hepatocytes, the liver can be homogenized and a preparation of vesicles of the hepatocyte endoplasmic reticulum called microsomes can be incubated with drug molecules. This preparation suffers because many of the soluble Phase II conjugation enzymes that are found in the cellular cytoplasm are lost. An alternative method for measuring microsomal metabolism involves isolation of the so-called "S9 fraction", which includes the cytosolic soluble conjugation enzymes.

Alternatively, during the early phase of drug design, cytochrome inhibition is regularly monitored using fluorescent *in vitro* assay techniques and recombinant isoenzymes from species of interest.[79,80] The study of drug metabolism using purified cytochrome isoenzymes, however, can only give part of the answer, as their metabolic potential is limited. In contrast, microsomal preparations, which provide additional metabolic enzymes, have been well established to provide an indication of the susceptibility of a drug to phase I hepatic metabolism[81] and as a consequence they are routinely used in pharmaceutical labs as the first-choice test for metabolism studies. The availability of cryopreserved hepatocytes[82] has facilitated determinations of *in vitro* stability determination, metabolite formation and identification of phase 2 reactions as well. The role of early metabolic stability assessment and the different *in vitro* approaches have been widely reviewed and are considered to be the state of the art[83–85] in preclinical drug profiling and candidate selection.

A number of chemicals and drugs, such as phenobarbital, strongly induce CYP activity and increase their own metabolism.[86] *In vivo* assessment of CYP induction is reserved for the later phase of drug development if indicated. It requires multiday treatment, an effort which is not justified for a compound in the early phase. A number of *in vitro* methods using cultures of primary hepatocytes in Matrigel matrixes are available and correlate well with data observed from *in vivo* experiments.[87]

In vitro drug metabolism is useless unless the results can be converted to a clearance (CL) parameter to be used in simulation of human pharmacokinetics. Numerous methods have been developed for calculation of human CL from *in vitro* metabolic turnover or half-life. In their definitive evaluation of 12 methods, Obach *et al.*[88] found that the best method produced estimates of human CL that are within 80% of the experimental values. These methods fall into three classes, *in vivo* allometric scaling, *in vitro* metabolism and physiological simulation. Allometric scaling typically focuses on interspecies relationships between CL and body weight. West *et al.*[89,90] have developed an interesting theory about why *in vivo* metabolic rates scale as the three-quarter power of body mass. They derived a general model that describes how essential materials are transported through space-filling fractal networks of branching tubes and conclude that allometric scaling relates to the common properties of circulatory systems in higher organisms.[91] Physiologically-based pharmacokinetics (PB/PK) simulations rely on a mathematical model of the organism parameterized with *in vitro* or *in silico* properties to predict human PK. The most successful work in this field involves merging these methods by using *in vitro* metabolism to correct the scaling derived from allometry.[92–94]

A discussion of *in vitro* metabolism would not be complete without a discussion of drug–drug interactions (DDIs) and polymorphic expression of drug-metabolizing enzymes. DDIs and drug–food interactions have been the subject of numerous publications and scientific meetings in recent years.[95–101] This is a critical issue, as the failure to understand these interactions has led to failed clinical trials and adverse events in patients, including death. DDIs occur as a result of one drug modifying the absorption and/or metabolism of another drug in the GI tract, or modifying the PK and/or pharmacodynamic (PD) effects of the other drug after it enters the blood. These interactions can be of a competitive nature or noncompetitive nature. A noncompetitive mechanism for DDI makes it difficult to predict from *in vitro* data alone, because the interaction is not governed by Michaelis–Menten kinetics. However, for competitive DDIs, a recent paper[98] provides a collation of 193 DDI studies involving inhibition of CYP3A4, CYP2D6 or CYP2C9, along with *in vitro* K_i values and PK parameters.

Cytochrome interaction is also one of the most common reasons for interindividual differences in drug metabolism. One example of an enzyme with polymorphic expression is CYP2C19, resulting in a significant population (2–6% of Asians) of "poor metabolizers" (PMs).[102] In "extensive metabolizers" (EMs), approximately 80% of doses of the proton pump inhibitors (PPIs), omeprazole, lansoprazole and pantoprazole seem to be cleared by CYP2C19, whereas CYP3A is more important in PMs. Five-fold higher exposure to these drugs is observed in PMs than in EMs with high levels of CYP2C19.[103]

Finally, we briefly mention metabonomics. It is well known that genomics and proteomics can provide vast amounts of genotype data regarding the transcription and expression of cellular proteins. However, they do not provide a measure of the resulting phenotype. High-resolution ^{1}H NMR spectroscopy used in conjunction with pattern recognition provides one such tool for defining the dynamic phenotype of a cell, organ or organism in terms of a metabolic phenotype.[104] In addition, there are benefits of this metabonomics/metabolomics approach to problems in toxicology. One of the major benefits of this approach is its high-throughput nature and cost-effectiveness on a per sample basis. Using such a method, the consortium for metabonomic toxicology (COMET) are currently investigating approximately 150 models of liver and kidney toxins.[105]

5 Protein Binding

Once absorbed into the organism, the drug has to reach its site of action where the free concentration of the active material will determine its effect. The free concentration of the drug is modified by the interaction of the compound with proteins, such as human serum albumin (HSA), which decreases the free concentration of the drug available for interaction with the target. Protein binding of drug molecules to plasma or tissue proteins means different things to various people in the drug discovery and development pipeline. To the discovery scientist, protein binding in tissues is usually high affinity, low-capacity binding to pharmacological receptor or enzyme targets. This binding is very important in optimizing the pharmacological response and helping to minimize the dose needed. Typically, this type of protein binding is characterized by an equilibrium dissociation constant (K_d ~ low-nM range) and a maximal binding capacity (B_{max}). Frequently, the concentration of a drug in tissues is higher than the concentration of the target protein, making equilibrium binding very important. Under these conditions, mass-action equations lead to 50% occupancy of the target protein when the drug is at a concentration equal to its K_d. Because the concentration of receptor protein is lower than the drug concentration, it then follows that the bound concentration is a very small fraction of total drug.

To the scientist working in drug metabolism and pharmacokinetics (DMPK), protein binding in tissues is much less important than protein binding in plasma because only the unbound drug can cross the cellular membranes and become accessible to the metabolic enzymes in tissues like liver, lung or intestines. The concentration of plasma proteins generally far exceeds the concentration of drug in the plasma and the affinity is low (K_d ~ µM to mM range) compared to binding at target sites. Thus, the interaction of a drug with a low affinity, high-capacity plasma protein results in "stoichiometric" binding and is usually characterized by the percent of the drug that is "unbound" with respect to plasma proteins. This free drug is able to partition into membranes of tissues and be metabolized or produce a therapeutic or toxicological response.

Example of binding to a target protein:

[Target] = 10 nM with 100 nM K_d:

[Protein] = 1×10^{-8} M, [Drug] = 1×10^{-7} M, $K_d = 1 \times 10^{-7}$ M

Percent of total protein that is bound would be 49%

Percent of total drug that is bound would be 5%

Whereas an example of plasma protein binding:

[Plasma Prot] = 0.6 mM with 10 µM K_d

[Protein] = 6×10^{-4} M, [Drug] = 1×10^{-7} M, $K_d = 1 \times 10^{-5}$ M

Percent of total protein that is bound would be 0.02%

Percent of total drug that is bound would be 98%

Thus, we see that for plasma–protein binding, even when the drug concentration is far below the K_d for binding, the percent of drug binding can be very large. Naturally, this means that only 2% of the drug is freely available to interact with cellular target tissue and doses need to be large to achieve the same target-site concentration. Therefore, unless drugs have a very low affinity for plasma protein ($\sim$ 1 mM), there will be an appreciable percent bound.

Protein binding to target tissues is measured *in vitro* using a radioligand or fluorescent ligand-binding assay. Protein binding to plasma protein is measured by using equilibrium dialysis.[106]

Protein binding can nowadays be easily assessed via HPLC chromatography using HSA chemically bonded to the stationary phase.[107–109] These HPLC-based methods to determine the binding of drug candidates to human plasma proteins have the advantage of being faster and more economical than traditional ultracentrifugation or ultrafiltration methods while providing higher precision.

6   Toxicity

Toxicity screening *in vivo* is required for preclinical studies of NCEs. This analysis usually consists of (a) acute rat and mouse toxicity, (b) repeating dose rat toxicity, (c) dog-rising dose tolerance, (d) genotoxicity and (e) male fertility. End points usually include clinical signs, bodyweight, food consumption, clinical pathology (including liver tests such as coagulation tests, bilirubin, transaminases and alkaline phosphatases) and necropsy with histopathologic examination of gross lesions. Major target organs are usually weighed, but histopathological evaluation is not performed routinely.[110] Hepatotoxicity is the most common reason for drug development to be discontinued. In 1985, toxicity as a whole generally accounted for only 11% of terminations. However, that increased to approximately 20% by 2000.[111] Also, it has been estimated that the correlation between liver toxicity in man and animals is only about 50%. Thus, the track record for toxicity testing in animals is not very good.

This backdrop sets the stage for *in vitro* testing because human tissues can be used and this type of surrogate may yield better predictive results than *in vivo* animal testing. Cytotoxicity or the measure of toxicity in cell culture is the most common *in vitro* method. Several end points for cytotoxicity tests have been developed.

6.1 Cell Viability: MTS Assay for In Vitro Cytotoxicity

A number of studies have reported that the MTS (3-(4,5-dimethylthiazol-2-yl)-5-(3-carboxymethoxyphenyl)-2-(4-sulfophenyl)-2H-tetrazolium) *in vitro* cytotoxicity assay is a convenient method for assessing cell viability. The main features found with this assay are its ease of use, accuracy and rapid indication of toxicity. This method is of interest particularly when exposure to unknown chemical substances requires the rapid detection and evaluation of toxic effects. Direct comparisons of the IC_{50} value determined in the MTS assay and *in vivo* cytotoxicity showed a significant, direct correlation. The highest correlation was found when the MTS assay was compared with test systems using human cell lines.[112] The MTS assay is based on the conversion of a tetrazolium salt into a colored, water-soluble formazan product by mitochondrial activity of viable cells at 37 °C. The amount of formazan produced by dehydrogenase enzymes is directly proportional to the number of living cells in culture and can be measured at 492 nm.

6.2 Membrane Damage: Release of LDH (Lactate Dehydrogenase)

Frequently, the first sign of hepatotoxicity is breakdown of the outer cell plasma membrane. If the cell membrane is full of "holes", it allows the internal contents of the hepatocyte to leak out into the culture medium. By taking an aliquot of the cell culture supernatant following exposure to a toxic chemical and incubating the sample with a reagent that recognizes LDH, the amount of LDH that leaks out of the damaged cell can be determined. This assay is based on a coupled enzymatic assay that results in the conversion of resazurin into the fluorescent compound, resorufin.

6.3 Induction of Apoptosis: Caspase Activity

Apoptosis or "programmed cell death" is a complicated process that is one of the mechanisms that can be responsible for hepatotoxicity. When a hepatotoxic drug that acts by initiating apoptosis is exposed to primary hepatocytes in cell culture, enzymes of the cysteine aspartic acid-specific protease (caspase) family are released. Induction of apoptosis and activation of caspases can result from a variety of stimuli, including growth factor withdrawal, exposure to radiation or chemotherapeutic agents or initiation of the Fas/Apo-1 receptor-mediated cell death process. Active caspases participate in a cascade of cleavage events that disable key homeostatic and repair enzymes and bring about systematic structural disassembly of dying cells. Commercial kits are available to measure the activities of caspase-3 and -7. These members of the caspase family play key effector roles in apoptosis in mammalian cells.

6.4 HERG Potassium Channel Interaction

Toxicological events and cardiovascular toxicity are of particular concern to the regulatory authorities. Torsade de points (TdP), an example of such adverse cardiovascular responses, is specifically monitored by the FDA. TdP results from a prolongation of the QT interval of the cardiovascular action potential and may be life threatening as it results in polymorphic ventricular tachycardia.[113] Specific cardiac-ion channels have shown to be implicated in QT interval prolongation, most prominently the human ether-a-go-go related potassium channel (hERG) and similar ion channels (KVLQT1, KVLQT1-MinK and KCNQ1-KCNE1).[114,115] In an attempt to improve the preclinical and clinical evaluation of proarrhythmogenic substances, European regulatory authorities have recently defined the need to assess more precisely and efficiently the potential for QT interval prolongation, especially by non-cardiovascular drugs. At the end of 1997, the Committee for Proprietary Medicinal Products (CPMP) published for the attention of the pharmaceutical industry a document entitled "Points to Consider: the Assessment of the Potential for QT interval Prolongation by Non-cardiovascular Medicinal Compounds" (CPMP 986/96). In June 2004, the FDA released guidelines (ICH S7B) on the nonclinical evaluation of the potential for delayed ventricular repolarization by human pharmaceuticals.

Efforts to predict the risk of potential QT prolongation over the last decade have resulted in the number of preclinical *in vitro* and *in vivo* methodologies. *In vitro* methods to predict QT liabilities range from simple recombinant cellular assays to disaggregated cells, isolated tissues and perfused heart preparations.[116]

Among the *in vitro* studies, radioligand-binding experiments,[117] ion flux measurements[118] and membrane-potential measurements using fluorescent dyes[114] have been well established and are widely used to predict potential liabilities. Certainly, radio-oligand-binding assays allow one to monitor the ligand-target interaction; however, they fail to discriminate between potassium channel blockers and openers as they do not provide a functional readout.

Recently predictive *in silico* modeling for hERG channel blockers has been described.[119] Different approaches have aimed primarily at filtering out potential hERG channel blockers in the context of combinatorial and virtual libraries and to elucidate structure-activity relationships. These new computational methods may predict trends, but are not as yet sufficiently precise to make valid predictions.

6.5 Microarrays

The deciphering of the humane genome and the development of microarray techniques have generated hope and opened new possibilities for assessing toxicological changes induced by drugs on the genomic level. However, it is important to note that just monitoring the up- and down-regulation of genes that may be associated with adverse effects is of little value by itself, without insight into mechanistic pathways or correlation to histological findings Only when the end points can be well attributed to specific mechanisms of action and pathways, do they become useful predictors of drug induced toxicity.

The number of useful biomarkers to predict neurotoxicity, hepatotoxicity or cardiovascular toxicity is still rather limited and they are not yet well established as tools in pharmaceutical laboratories. Even though some decent correlations between biomarkers and toxicological events have been demonstrated,[120] a significant amount of validation work still has to be performed. For example, while natriuretic peptides and troponin can be clinical markers for cardiovascular toxicity, preclinical use in nonhuman species has not yet proven to be reliable.[121] Therefore, despite the claimed value of biomarkers for drug profiling, caution has been advised.[122] Their use awaits further validation before they can be broadly applied to the preclinical evaluation of drug profiles and substituted for animal experiments, thus leaving scientists at present with standard *in vitro* and *in vivo* methods.

6.6   Recent Approaches

One drawback to the study of cytotoxicity in primary cell culture of a single human cell type is that the processes of metabolic activation of a toxic metabolite may not be observed and a toxic chemical might not be detected. Metabolic activation is required for many drug molecules. For example, acetaminophen is oxidized to a reactive benzoquinone imine intermediate that can react with proteins in the liver to cause cellular damage. An innovative *in vitro* technique has been developed that purports to facilitate identification of toxic chemicals that require activation or have sites of toxicity in a variety of organs. The integrated discrete multiorgan cell culture system (IdMOC), is based on the "wells within a well" concept, consisting of a cell-culture plate with larger wells, within each of which are multiple smaller wells. Cells from multiple organs can be cultured initially in the small wells (one organ per well, each in its specialized medium). On the day of toxicity testing, a volume of drug-containing medium is added to the larger well to flood all inner wells, thereby interconnecting all the small wells. After testing, the overlying medium is removed and each cell type is evaluated for toxicity using appropriate endpoints.[123]

7   Investigation of Compound Selectivity

Broad pharmacological drug profiling has been a selection filter in drug discovery for many years. The aim is to determine potential liabilities associated with off-target drug activities. A number of contract research organizations (CROs) perform these tests either *in vitro* or *in vivo*, and determine the interaction of drugs with a whole panel of defined molecular targets that have been reported to cause adverse effects in man. With the development of high-throughput technologies, it has become reasonable to profile a library or larger set of compounds against an important number of *in vitro* targets and use the resulting information to filter potential leads in order to select those candidates that may be best suited for further development.[5] The advantage is that key decisions are made with a comprehensive knowledge of the positive and negative attributes in each of the compound classes. This is to say that not all the off-target activities raise a red flag on the compounds. In some cases, for example, discovery of anti-tumor kinase inhibitors, these off-target interactions are considered useful. As multiple kinases regulate cell proliferation, pan-kinase inhibition may contribute to the drug's therapeutic profile.

8.   Conclusion and Outlook

The competitive nature of the pharmaceutical industry and the high costs associated with drug development has created great demands for improvements in the productivity and speed of drug discovery. As the demands on pharmaceutical companies to increase the output of their R&D pipeline grow stronger, so does the requirement for more stringent downstream filters to reduce attrition rates. The response has been to increase the stringency in selecting compounds as potential candidates, to identify and remove those that have intrinsic liabilities and are prone to fail in the development process. This earlier emphasis on "druggability" has become of fundamental importance not only for biotech companies, for which failure to bring a product to the market is a matter of survival, but also for the top pharma companies, which have to sustain profitability for reinvestment in further research and development.

Investment in leads for which these properties have not been measured runs the risk of overlooking compounds with better overall pharmaceutical profiles. Proactive early measurement of property information provides the opportunity to select advantageous HTS hits and leads. Accumulating relevant information on active compounds as early as possible during the discovery process using broad profiling for potential liabilities is a critical element of such a strategy. Early profiling in subsequent stages helps to understand the specific liabilities of pharmacophores, and to build favorable pharmaceutical properties into the structures of the lead series in parallel with activity optimization.

The application of these ideas on quality and drug-likeness of compounds, rather than trying to achieve the goal of reaching a clinical candidate by playing a numbers game and investing heavily in high-throughput technologies, has changed the drug-discovery process. The activities of screening, medicinal chemistry and drug profiling, which were previously performed in a sequential manner, have merged to an integrated activity package (Figure 6).

The hit-to-lead and lead-to-candidate process is no longer driven by a search for drug candidates with ever-higher activity. In contrast, early emphasis is put on aspects such as of drug-likeness, pharmacokinetics and metabolism as well as indicators of potential toxicological liabilities. In the drive to increase the speed of bringing new drugs to market while reducing the development costs and rate of candidate attrition, the pharmaceutical industry is seeking fast, accurate approaches for toxicity prediction to aid in the discovery and development processes and to reduce failure rates during the development process.

Current technologies, implemented in the laboratories, have already improved the knowledge basis in early discovery and development; however, more is to come. Genomic and *in silico* techniques, once validated, will be able to provide information economically and with a broader spectrum of detailed information on the mechanism or pathway affected.

Many players in the pharmaceutical industry expect that this parallel drug-discovery paradigm will shorten drug preclinical development time and reduce the attrition of late-stage drug candidates. As a result, early profiling approaches are an area of intense competition among pharmaceutical and biotechnology companies, all of whom want to decrease the escalating development cost for each new chemical entity. We all hope that early profiling will accelerate the conversion of screening hits into drugs on the market, which would ultimately be to everyone's benefit, including the patients.

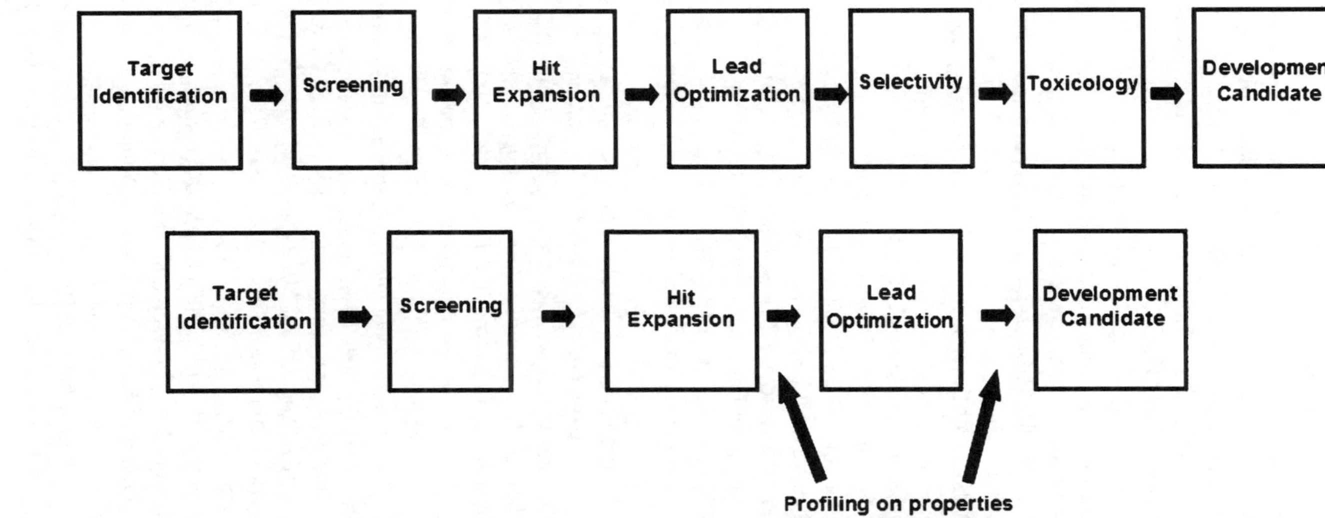

Figure 6 *Alternative approaches to the drug-discovery process. Upper panel: Sequential process with profiling of the final-development candidates only. Lower Panel: Parallel process including early profiling of lead candidates before starting the optimization and final-development candidates as well*

References

1. A.H. Maehle, C.R. Prull and R.F. Halliwel, *Nat. Rev. Drug. Discov.*, 2002, **1**, 637.
2. I. Kola and J. Landis, *Nat. Rev. Drug. Discov.*, 2004, **3**, 711.
3. T. Kennedy, *Drug Discov. Today*, 1997, **2**, 436.
4. R.A. Prentis, Y. Lis and S.R. Walker, *Br. J. Clin. Pharmacol.*, 1988, **25**, 387.
5. M. Entzeroth, B. Chapelain, J. Guilbert and V. Homon, *JALA*, 2000, **5**, 69.
6. C.A. Lipinski, *J. Pharmacol. Toxicol. Methods*, 2000, **44**, 235.
7. D.F. Veber, S.R. Johnson, H.Y. Cheng, B.R. Smith, K.W. Ward and K.D. Kopple, *J. Med. Chem.*, 2002, **45**, 2615.
8. M.B. Bolger, B. Agoram, R. Fraczkiewicz and B. Steere in *"Drug Bioavailability. Estimation of Solubility, Permeability and Bioavailability"*, H.v.d. Waterbeemd, H. Lennernäs and P. Artursson (eds), Wiley, Weinheim, 2003.
9. M.C. Wenlock, R. P. Austin, P. Barton, A.M. Davis and P.D. Leeson, *J. Med. Chem.*, 2003, **46**, 1250.
10. P. Beroza, H.O. Villar, M.M. Wick and G.R. Martin, *Drug. Discov. Today*, 2002, **7**, 807.
11. F. Sams-Dodd, *Drug. Discov. Today*, 2005, **10**, 139.
12. M.J. Valler and D. Green, *Drug Discov. Today*, 2000, **5**, 286.
13. P. Gribbon and A. Sewing, *Drug Discov. Today*, 2005, **10**, 17.
14. A. Leo, C. Hansch and D. Elkins, *Chem. Rev.*, 1971, **71**, 525.
15. L. Di and E. H. Kerns, *Curr. Opin. Chem. Biol.*, 2003, **7**, 402.
16. E.H. Kerns and L. Di, *Drug Discov. Today*, 2003, **8**, 316.
17. J. Hodgson, *Nat. Biotechnol.*, 2001, **19**, 722.
18. H. Van De Waterbeemd, D.A. Smith and B.C. Jones, *J. Comput. Aided. Mol. Des.*, 2001, **15**, 273.
19. A. Leo, P.Y. Jow, C. Silipo and C. Hansch, *J. Med. Chem.*, 1975, **18**, 865.
20. S.A. Wildman and G.M. Crippen, *J. Chem. Inf. Comput. Sci.*, 1999, **39**, 868.
21. R. Dvorsky, S. Balaz and R.J. Sawchuk, *J. Theor. Biol.*, 1997, **185**, 213.
22. E. Rodriguez-Boulan and C. Zurzolo, *J. Cell. Sci. Suppl.*, 1993, **17**, 9.
23. W. Sadee, V. Drubbisch and G.L. Amidon, *Pharm. Res.*, 1995, **12**, 1823.
24. P. Artursson and J. Karlsson, *Biochem. Biophys. Res. Commun.*, 1991, **175**, 880.
25. P. Artursson and C. Magnusson, *J. Pharm. Sci.*, 1990, **79**, 595.
26. L.P. Hammett, *"Physical Organic Chemistry: Reaction Rates, Equilibria, and Mechanisms"*, McGraw-Hill, New York, 1940.
27. R.W. Taft and I.C. Lewis, *J. Am. Chem. Soc.*, 1959, **81**, 5343; R.W. Taft and I.C. Lewis, *J. Am. Chem. Soc.*, 1958, **80**, 2436; R.W. Taft, S. Ehrenson, I.C. Lewis and R.E. Glick, *J. Am. Chem. Soc.*, 1959, **81**, 5352; R.W. Taft, *J. Am. Chem. Soc.*, 1953, **75**, 4231.
28. D.D. Perrin, B. Dempsey and E.P. Serjeant, *"pKa Prediction for Organic Acids and Bases"*, Chapman & Hall, London and New York, 1981.
29. A. Avdeef, *"Absorption and Drug Development. Solubility, Permeability, and Charge State"*, Wiley, Hoboken, NJ, 2003.
30. A. Avdeef, *Curr. Top. Med. Chem.*, 2001, **1**, 277.
31. B. Noszal, V. Scheller-Krattiger and R.B. Martin, *J. Am. Chem. Soc.*, 1982, **104**, 1078.

32. B. Noszal, *J. Phys. Chem.*, 1986, **90**, 4104.

33. Z. Szakacs and B. Noszal, *J. Math. Chem.*, 1999, **26**, 139.

34. M. Borkovec, M. Brynda, G.J.M. Koper and B. Spiess, *Chimia*, 2002, **56**, 695.

35. K.Y. Tam and L. Quere, *Anal. Sci.*, 2001, **17**, 1203.

36. K.Y. Tam and K. Takacs-Novak, *Anal. Chim. Acta*, 2001, **434**, 157.

37. B. Agoram, W.S. Woltosz and M.B. Bolger, *Adv. Drug Deliv. Rev.*, 2001, **50** (Suppl 1), 41.

38. E.S. Kostewicz, U. Brauns, R. Becker and J.B. Dressman, *Pharm. Res.*, 2002, **19**, 345.

39. W.H. Streng, *"Characterization of Compounds in Solution. Theory and Practice"*, Kluwer, New York, 2001.

40. C.A. Lipinski in *"Pharmaceutical Profiling in Drug Discovery for Lead Selection"*, R.T. Borchardt, E.H. Kerns, C.A. Lipinski, D.R. Thakker and B. Wang (eds), American Association of Pharmaceutical Scientists, Arlington, VA, 2004, 93.

41. W.F. Ganong, *"Review of Medical Physiology"*, Appleton & Lange, Stamford, CT, 1999.

42. L.R. Johnson, *"Gastrointestinal Physiology"*, Mosby, St. Louis, MO, 1989.

43. J. B. Dressman, G.L. Amidon, C. Reppas and V.P. Shah, *Pharm. Res.*, 1998, **15**, 11.

44. W. H. Streng, S.K. Hsi, P.E. Helms and H.G. Tan, *J. Pharm. Sci.*, 1984, **73**, 1679.

45. A.T. Serajuddin, P.C. Sheen and M.A. Augustine, *J. Pharm. Pharmacol.*, 1987, **39**, 587.

46. Z.T. Chowhan, *J. Pharm. Sci.*, 1978, **67**, 1257.

47. B.A. Hendriksen, M.V. Felix and M.B. Bolger, *AAPS Pharm. Sci.*, 2003, **5**, E4.

48. G.L. Amidon, J. Kou, R.L. Elliott and E.N. Lightfoot, *J. Pharm. Sci.*, 1980, **69**, 1369.

49. L. Salphati, K. Childers, L. Pan, K. Tsutsui and L. Takahashi, *J. Pharm. Pharmacol.*, 2001, **53**, 1007.

50. U. Fagerholm, M. Johansson and H. Lennernas, *Pharm. Res.*, 1996, **13**, 1336.

51. P.J. Sinko and G. L. Amidon, *Pharm. Res.*, 1988, **5**, 645.

52. P.J. Sinko and G.L. Amidon, *J. Pharm. Sci.*, 1989, **78**, 723.

53. G.B. Park and A.K. Mitr, *Pharm. Res.*, 1992, **9**, 326.

54. A.L. Ungell, S. Nylander, S. Bergstrand, A. Sjoberg and H. Lennernas, *J. Pharm. Sci.*, 1998, **87**, 360.

55. A. Adson, P.S. Burton, T.J. Raub, C.L. Barsuhn, K.L. Audus and N.F. Ho, *J. Pharm. Sci.*, 1995, **84**, 1197.

56. C.O. Billich and R. Levitan, *J. Clin. Invest.*, 1969, **48**, 1336.

57. I.J. Hidalgo, T.J. Raub and R.T. Borchardt, *Gastroenterology*, 1989, **96**, 736.

58. K.L. Audus, R.L. Bartel, I.J. Hidalgo and R.T. Borchardt, *Pharm. Res.*, 1990, **7**, 435.

59. R.T. Borchardt, P.L. Smith and G. Wilson, *Pharm. Biotechnol.*, 1996, **8**, 1.

60. P. Artursson and R.T. Borchardt, *Pharm. Res.*, 1997, **14**, 1655.

61. I.J. Hidalgo and J. Li, *Adv. Drug Deliv. Rev.*, 1996, **22**, 56.

62. S. Doppenschmitt, H. Spahn-Langguth, C.G. Regardh and P. Langguth, *J. Pharm. Sci.*, 1999, **88**, 1067.

63. T.K. Ip, P. Aebischer and P.M. Galletti, *ASAIO Trans.* 1988, **34**, 351.

64. C.J. Watson, M. Rowland and G. Warhurst, *Am. J. Physiol. Cell. Physiol.*, 2001, **281**, 388.

65. A. Adson, T.J. Raub, P.S. Burton, C.L. Barsuhn, A.R. Hilgers, K.L. Audus and N.F. Ho, *J. Pharm. Sci.*, 1994, **83**, 1529.

66. R. Saitoh, K. Sugano, N. Takata, T. Tachibana, A. Higashida, Y. Nabuchi and Y. Aso, *Pharm. Res.*, 2004, **21**, 749.

67. V. Pade and S. Stavchansky, *Pharm. Res.*, 1997, **14**, 1210.

68. M. Kansy, F. Senner and K. Gubernator, *J. Med. Chem.*, 1998, **41**, 1007.

69. K. Sugano, H. Hamada, M. Machida, and H. Ushio, *J. Biomol. Screen*, 2001, **6**, 189.

70. K. Sugano, H. Hamada, M. Machida, H. Ushio, K. Saitoh and K. Terada, *Int. J. Pharm.*, 2001, **228**, 181.

71. K. Sugano, N. Takata, M. Machida, K. Saitoh and K. Terada, *Int. J. Pharm.*, 2002, **241**, 241.

72. K. Sugano, Y. Nabuchi, M. Machida and Y. Aso, *Int. J. Pharm.*, 2003, **257**, 245.

73. E.F. Brandon, C.D. Raap, I. Meijerman, J.H. Beijnen and J.H. Schellens, *Toxicol. Appl. Pharmacol.*, 2003, **189**, 233.

74. M. Rowland, *Drug Metab. Rev.*, 1984, **15**, 55.

75. S. Sahin and M. Rowland, *J. Pharmacol. Toxicol. Methods*, 1998, **39**, 35.

76. S. Ekins, B.J. Ring, J. Grace, D.J. McRobie-Belle and S.A. Wrighton, *J. Pharmacol. Toxicol. Methods*, 2000, **44**, 313.

77. D. A. Flockhart and J.R. Oesterheld, *Child. Adolesc. Psychiatr. Clin. N. Am.*, 2000, **9**, 43.

78. P.B. Danielson, *Curr. Drug Metab.*, 2002, **3**, 561.

79. S. Ono, T. Hatanaka, S. Miyazawa, M. Tsutsui, T. Aoyama, F.J. Gonzalez and T. Satoh, *Xenobiotica*, 1996, **26**, 1155.

80. C.L. Crespi and B.W. Penman, *Adv. Pharmacol.*, 1997, **43**, 171.

81. A.P. Li, C. Lu, J.A. Brent, C. Pham, A. Fackett, C.E. Ruegg and P.M. Silber, *Chem. Biol. Interact.*, 1999, **121**, 17.

82. N.J. Hewitt, K.U. Buhring, J. Dasenbrock, J. Haunschild, B. Ladstetter and D. Utesch, *Drug. Metab. Dispos.*, 2001, **29**, 1042.

83. S. Ekins, B.J. Ring, J. Grace, D.J. McRobie-Belle and S.A. Wrighton, *J. Pharmacol. Toxicol. Methods*, 2000, **44**, 313.

84. T.N. Thomson, *Med. Res. Rev.*, 2001, **21**, 412.

85. T.N. Thomson, *Curr. Drug Metab.*, 2000, **1**, 215.

86. F.J. Gonzalez and C. B. Kasper, *Biochemistry*, 1980, **19**, 1790.

87. J.M. Silva, P.E. Morin, S.H. Day, B.P. Kennedy, P. Payette, T. Rushmore, J.A. Yergey and D.A. Nicoll-Griffith, *Drug Metab. Dispos.*, 1998, **26**, 490.

88. R.S. Obach, J.G. Baxter, T.E. Liston, B.M. Silber, B.C. Jones, F. MacIntyre, D.J. Rance and P. Wastall, *J. Pharmacol. Exp. Ther.*, 1997, **283**, 46.

89. G.B. West, J.H. Brown and B.J. Enquist, *Science*, 1997, **276**, 122.

90. G.B. West, J.H. Brown and B.J. Enquist, *Science*, 1999, **284**, 1677.

91. G.B. West, W.H. Woodruff and J.H. Brown, *Proc. Natl. Acad. Sci. USA*, 2002, **99** (Suppl 1), 2473.

92. T. Lave, A.H. Schmitt-Hoffmann, P. Coassolo, B. Valles, G. Ubeaud, B. Ba, R. Brandt and R.C. Chou, *Life. Sci.*, 1995, **56**, 473.

93. T. Lave, S. Dupin, C. Schmitt, R.C. Chou, D. Jaeck and P. Coassolo, *J. Pharm. Sci.*, 1997, **86**, 584.

94. T. Lave, P. Coassolo and B. Reigner, *Clin. Pharmacokinet.*, 1999, **36**, 211.

95. S. Ekins, C.L. Waller, P.W. Swaan, G. Cruciani, S.A. Wrighton and J.H. Wikel, *J. Pharmacol. Toxicol. Methods*, 2000, **44**, 251.

96. S.C. Armstrong, K.L. Cozza and N.B. Sandson, *Psychosomatics*, 2003, **44**, 255.

97. M. Attar and V.H. Lee, *Pharmacogenomics*, 2003, **4**, 443.

98. K. Ito, H.S. Brown and J.B. Houston, *Br. J. Clin. Pharmacol*, 2004, **57**, 473.

99. Y. Nozaki, H. Kusuhara, H. Endou and Y. Sugiyama, *J. Pharmacol. Exp. Ther.*, 2004, **309**, 226.

100. R.L. Robinson and M.S. Burk, *Am. J. Med.*, 2004, **116**, 357.

101. K. Watanabe, T. Sawano, T. Jinriki and J. Sato, *Biol. Pharm. Bull.*, 2004, **27**, 77.

102. D.A. Flockhart, *Clin. Pharmacokinet.*, 1995, **29** (Suppl 1), 45.

103. Z. Desta, X. Zhao, J.G. Shin and D.A. Flockhart, *Clin. Pharmacokinet.*, 2002, **41**, 913.

104. J.K. Nicholson, J.C. Lindon and E. Holmes, *Xenobiotica*, 1999, **29**, 1181.

105. J.L. Griffin and M.E. Bollard, *Curr. Drug Metab.*, 2004, **5**, 389.

106. T.H. Tsai, *J. Chromatogr. B*, 2003, **797**, 161.

107. T.A. Noctor, M.J. Diaz-Perez and I.W. Wainer, *J. Pharm. Sci.*, 1993, **82**, 675.

108. K. Valko, *J. Chromatogr. A*, 2004, **1037**, 299.

109. K. Valko, S. Nunhuck, C. Bevan, M.H. Abraham and D.P. Reynolds, *J. Pharm. Sci.*, 2003, **92**, 2236.

110. F. Ballet, *J. Hepatol.*, 1997, **26** (Suppl 2), 26.

111. I. Kola and J. Landis, *Nat. Rev. Drug Discov.*, 2004, **3**, 711.

112. G. Malich, B. Markovic and C. Winder, *Toxicology*, 1997, **124**, 179.

113. J.J. Salata, N.K. Jurkiewicz, A.A. Wallace, R.F. Stupienski III, P.J. Guinosso Jr. and J.J. Lynch Jr., *Circ. Res.*, 1995, **76**, 110.

114. R. Netzer, A. Ebneth, U. Bischoff and O. Pongs, *Drug Discov. Today*, 2001, **6**, 78.

115. W. Crumb and I. Cavero I, *Pharm. Sci. Technol. Today*, 1999, **2**, 270.

116. R. Webster, D. Leishman and D. Walker, *Curr. Opin. Drug Discov. Devel.*, 2002, **5**, 116.

117. Z. Zhou, Q. Gong, B. Ye, Z. Fan, J.C. Makielski, G.A. Robertson and C.T. January, *Biophys. J.*, 1998, **74**, 230.

118. W. Tang, J. Kang, X. Wu, D. Rampe, L. Wang, H. Shen, Z. Li, D. Dunnington and T. Garyantes, *J. Biomol. Screen.*, 2001, **6**, 325.

119. A.M. Aronov, *Drug Discov. Today*, 2005, **10**, 149.

120. F.M. Goodsaid, *J. Pharmacol. Toxicol. Methods*, 2004, **49**, 183.

121. J.H. Kim, S.J. Skates, T. Uede, K.K. Wong, J.O. Schorge, C.M. Feltmate, R.S. Berkowitz, D.W. Cramer and S.C. Mok, *J. Am. Med. Assoc.*, 2002, **287**, 1671.

122. L. Suter, L.E. Babiss and E.B. Wheeldon, *Chem. Biol.*, 2004, **11**, 161.

123. A.P. Li, C. Bode and Y. Sakai, *Chem. Biol. Interact.*, 2004, **150**, 129.

In Silico Surrogates for In Vivo Properties: Profiling for ADME and Toxicological Behavior

MICHAEL B. BOLGER, ROBERT FRACZKIEWICZ AND BOYD STEERE

Simulations Plus, Inc., 1220 W Avenue J, Lancaster, CA 93534, USA

1 *In Silico* Surrogates for *In Vivo* Properties

Drug discovery has become a competitive race for the first to reach the finish line bringing a new drug candidate to the market. Prior to 1997, the process was mostly impaired by failures of compounds to meet safety requirements and by inadequate pharmacokinetic properties. In addition, success in drug development has been hampered by toxicity and insufficient efficacy when development candidates are investigated in man.[1] With pharmacokinetics being the major challenge in this context, the industry has responded with early evaluation of hits or leads from screening in assay systems, which enable drug-like properties to be evaluated. *In silico* prediction of such parameters has become of increasing interest, as these properties are sometimes difficult to measure directly for larger numbers of compounds. On the other hand, *in silico* methods as surrogates for *in vivo* properties are more and more readily obtained as hardware performance has increased and as an understanding of structural modeling has improved.

In silico methods can be divided into three main groups. These are (a) statistical methods for property estimation, (b) direct calculation of properties using quantum mechanics, and (c) physiological simulation of integrated properties using numerical methods. The first method requires lots of high quality biological property data collected by using *in vivo* or *in vitro* methods. Often, the curation of a high-quality database is more time consuming that the development of the statistical models. The second method is based on calculating properties from first principles. There are examples in which these methods provide very accurate estimates of physical properties such as pK_a;[2] however, when it comes to complex biological properties like permeability and fraction absorbed, quantum mechanical (QM; See Table 2 for commonly used abbreviations and acronyms) methods must be combined with some empirical data to provide reasonable results for diverse molecules. Physiological simulation is a great tool for integration of the *in vivo*, *in vitro*, and *in silico* properties. Simulation allows one to do hypothesis testing, sensitivity analysis, and virtual clinical trials. The first step in building a statistical model of a biological property (the dependent variable) is to break down the structures of the molecules that have been tested into small calculated properties called descriptors that can be used as independent variables in model building.

1.1 Molecular Descriptor Generation

The human mind is truly an amazing instrument, especially in its ability to synthesize. For an experienced chemist, it is sufficient to just look at a set of drawings of molecular structures to say which ones are similar and to estimate their expected properties intuitively. On the other hand, computers are incredibly fast and tireless in performing tedious data processing tasks, like predicting biological and chemical properties from molecular structure. Since computers cannot "look" at molecules in a synthetic way, we must translate structures into the language of numbers and symbols. In general, a number derived from molecular structure through some logical or mathematical function is called a "molecular descriptor". To serve a useful purpose, a single structure must be encoded by many, preferably hundreds, of molecular descriptors. These descriptors are then taken as numerical inputs into mathematical predictive models. Two general classes of descriptors exist: molecular and atomic. The former are functions of the whole molecule and are good in modeling whole molecule properties like solubility or lipophilicity. The latter class should be used in predicting localized properties like ionization constants.

Many thousands of descriptor types have been developed over the years of chemical research. This short review is not intended to summarize these efforts in detail. Todeschini and Consonni have already provided a reference book for this purpose.[3] Commercial and academic software for generating molecular descriptors has proliferated over the recent years. Todeschini's book lists software packages known in the year 2000. Speed of calculation, descriptor quality and diversity should be considered in selecting software. For example, absorption, distribution, metabolism, excretion, toxicity (ADMET) Predictor™ (Simulations Plus, Inc.) works at a rate of ~250,000 molecules per hour calculating 272 molecular and 44 atomic descriptors in the following categories:

- simple constitutional descriptors (MWt, number of nitrogens, *etc.*)
- heteroatomic functional groups (counts of primary and secondary amines)
- topological indices (specialized calculation of molecular branching)
- atom-type electrotopological state indices (charge associated with atom types)
- ionization in water (fraction cationic, fraction anionic, *etc.*)
- molecular pattern flags (recognition of primary alpha amino acids)
- electronic properties (sums of partial charges on specific atoms)

- hydrogen bonding (counts of donors, acceptors, and charges associated)
- molecular size and shape (moments of inertia, surface areas, *etc.*)
- solvation effects (solvation energy, polarizability, and dipole moments)
- protein recognition (specific molecular patterns common to proteins).

1.2 Modeling Methods

Descriptor generation deals with the first half of the structure–property relationship and modeling methods address the second. A model is a mathematical function that takes the descriptor values as inputs and produces a prediction of a property. A wide variety of model types exists, and this section summarizes a particular class of them called empirical models. In empirical modeling, data from a set of training compounds are used to fit the parameters in the model function, and no prior knowledge of the underlying relationship between structure and property is assumed. In all of the following cases, the training data set is composed of the molecular descriptors and the experimental results associated with a group of molecules.

In addition to a training data set, a model requires other groups of compounds that are not used to fit its parameters. A second, usually smaller group is called the "test set" and is essential for validating the quality of the model's predictions. Only predictions for compounds that had no role to play in model design can indicate its "generalizability" – its ability to accurately predict the property of interest for any given molecule that falls into the chemical space of its training set.

The principal reason that a test set is necessary for validation is that empirical model-building methods cannot readily distinguish between noise and information in data sets, so the methods are prone to adjusting the model parameters to reduce error beyond the point warranted by the information contained in the data. This problem is called "overtraining" and can be countered by a variety of techniques such as descriptor reduction and early stopping, and readers interested in those topics are referred to the more detailed reviews of numerical methods cited in each of the following sections.

After a model has been built, its creator should compare its performance against that of the same model built on randomly scrambled training data. This validation step is necessary to avoid a problem described by Topliss and Edwards.[4] in which models based on a sufficiently high number of training descriptors sometimes happen upon a useless solution by a chance combination of variables that are actually unrelated to the property of interest.

1.3 Multiple Linear Regression

One of the earliest methods used for building models is linear regression, first employed as "the method of least squares" by Legendre and Gauss in the 17th Century to study the orbits of celestial bodies. An extension of linear regression using more than one variable is called multiple linear regression (MLR).

In MLR, the model is a straightforward linear combination of descriptors or functions of descriptors. Because MLR is based on a linear equation, it is called a "linear" model (as opposed to the "non-linear" models to be discussed later in this chapter). The most salient feature of linear models is that the influences of the descriptors are independent from each other, ruling out cooperative relationships.

This inherent simplicity is valuable in two ways. First, the number of parameters to be fit is minimal, avoiding the common problem of trying to solve too many variables with too little data. This problem is often referred to as the "Curse of Dimensionality" and results in a degradation of a model's generalizability. Second, MLR models are easy to interpret. Each descriptor used is accompanied by a coefficient and a sign, and this information provides the relative weight and direction of each descriptor's contribution to the property of interest.

MLR is most suited to the modeling of simple physical properties for which the contributions of the descriptors are more likely to be independent than cooperative. It is also the preferred technique when the amount of available data is extremely limited. MLR is less suited for modeling complex physical or biological processes, as these tend to be non-linear in nature.

1.4　Partial Least Squares

Another well-established method for building predictive models is partial least squares (PLS) regression. PLS is a modern relative of MLR, having been established in the 1960's by Wold.[5] The method reaches beyond linear regression by replacing the descriptors with a matrix of "latent variables" distilled from both the structural features of the training compounds and their experimental results. In PLS, the use of the term latent variables differs from its formal definition in other regression methods.[6,7]

The use of these latent variables provides PLS with a greater capacity than MLR to distill most of the information contained in a large number of descriptors and properties into a smaller number of factors and thereby avoid some of the ravages of the Curse of Dimensionality. It is especially suited to dealing with systems containing many highly correlated descriptors. PLS is related to the methods of principal components analysis (PCA) and maximum redundancy analysis (MRA), in that all three methods augment the raw descriptors with matrices derived from variance found in the descriptors themselves (PCA), the properties to be modeled (MRA), or a combination of both (PLS).[7]

The performance of PLS is comparable to that of methods such as ridge regression[8] and neural networks,[9] particularly when it is used to model properties that have a principally linear relationship to combinations of the descriptors. An excellent example of the use of PLS in the interpretation of the spectra of chemical mixtures is provided by Tobias.[7]

1.5　Artificial Neural Network

In an effort to simulate the theoretical behavior of neuron cells, researchers in the 1940's and the 1950's such as McCullogh, Pitts, and Rosenblatt developed functions that mimicked the "threshold" response of a synapse. Their theory of synapses held that the output of a neuron cell was at or near zero until the sum of all of the input potentials connected to it passed a certain threshold, at which point the cell's output would be at or near one. Mathematically, their "neuron functions" yielded a binary or a sigmoidal response to a linear combination of its inputs, and could be connected to each other input-to-output to build layers of neurons. Models built on these highly interconnected groups of neurons were called "multi-layer perceptrons", or MLP's, and neural nets became the subject of intense interest for artificial intelligence applications during the 1960s.[10]

Artificial neural networks (ANN's) were effectively set aside for 15 years after a 1969 study by Minksy and Papert demonstrated their failure to correctly model a simple exclusive OR (XOR) function.[11] The XOR function describes the result of an operation involving two bits (1 or 0). A simple OR function produces a value of 1 if either bit or both bits have a value of 1. The XOR differs from an OR function in the output of an operation on two bits of value 1. The XOR function will yield a 0 while the OR function will yield a 1. Interest in ANN's resumed in the 1980's after modifications were made to the layering of their neurons that allowed them to overcome the XOR test as well as a wide variety of other non-linear modeling challenges.

Predictive models are built with ANN's in much the same way as they are with MLR and PLS methods: descriptors and experimental data are used to fit (or "train" in machine-learning nomenclature) the parameters of the functions until the performance error is minimized. Neural networks differ from the previous two methods in that (1) the sigmoidal shapes of the neurons' output equations better allow them to model non-linear systems and (2) they are "subsymbolic", which is to say that the information in the descriptors is effectively scrambled once the internal weights and thresholds of the neurons are trained, making it difficult to examine the final equations to interpret the influences of the descriptors on the property of interest.

Unlike PLS and principal components methods, ANN's do not intrinsically convert data into latent variables during training. This leaves them vulnerable to overfitting and the Curse of Dimensionality if they are used to model large numbers of inputs to limited amounts of data. Early stopping or the application of descriptor reduction methods prior to training is essential for building generalizable ANN models under such conditions.

The cooperative relationship between the structural features of molecules and many physiological processes makes artificial neural network models a frequent choice for predicting the ADMET properties of drug candidates.[12]

1.6 Support Vector Machines

The most recent advance in machine-learning modeling to garner widespread application by fields outside of artificial intelligence itself is the support vector machine (SVM). SVM's were first developed by Vapnik in 1992.[13]

SVM's are an outgrowth of kernel methods. In such methods, the data is transformed with a kernel equation (such as a radial basis function) and it is in this mathematical space that the model is built. Care is taken in the construction of the kernel that it has a sufficiently high dimensionality that the data become linearly separable within it. A critical subset of transformed data points, the "support vectors", are then used to specify a hyperplane called a large-margin discriminator that effectively serves as a linear model within this non-linear space. An introductory exploration of SVM's is provided by Cristianini and Shawe-Taylor[14] and a thorough examination of their mathematical basis is presented by Schölkopf and Smola.[15]

The principal advantages that SVM's have over ANN's are that (1) the solution for the large-margin discriminator has a single minimum, so the system cannot fall into a sub-optimal solution provided by a "local minimum" like a neural net can, (2) the SVM will not overtrain. Nevertheless, an SVM can still be overfit by providing it with a data set that is too small or too noisy to relate the training compounds to the property of interest, so a test set is still necessary for validation. Typically, if the number of support vectors required by an SVM model is equal or nearly equal to the number of training compounds, a researcher should either seek more data or more informative descriptors, and (3) SVM models do not suffer from the susceptibility of some ANN models with too few neurons to cluster predictions at certain values or truncate their numerical range of coverage.

SVM's have been shown to have superior performance to ANN's in non-linear modeling studies such as optical character recognition.[16] In our hands, initial applications of SVM's to the prediction of ADMET properties from molecular structure indicate that their overall error performance is similar to that of ANN's, but that the predicted values have a smoother distribution and a slightly less-compressed range.

2   Estimation of Biopharmaceutical Properties

2.1   Partition Coefficient

Partition coefficient was the focus of *in silico* modeling efforts for many years before the term *in silico* was introduced. Al Leo and Corwin Hansch published some of the earliest attempts to estimate the octanol/water partition coefficient (P_{ow}) for complete molecules from individual substituent parameters measured *in vitro*.[17] These investigators introduced the "fragment method" for estimating log P and the estimated values have become know as Clog$P$.[18,19] Various fragment methods all suffer from a common deficiency, in that they do not rely on fundamental properties of a molecule to estimate log P. They all require the inclusion of an experimental value of a partition coefficient for the fragmental portion of the molecule in order to derive a composite log P. These methods have been the subject of rigorous reliability evaluation and have found to be unreliable for certain situations.[20,21]

Estimation of log P by using quantitative structure property relationships (QSPR) modeling and molecular descriptors (described above) has resulted in a number of highly accurate methods.[22,23] Methods involving MLR, PLS, and artificial neural network ensembles (ANNE) modeling have been reviewed.[24,25] In summary, estimation of partition coefficient has now reached a stage where the error associated with estimation is approximately equal to experimental error and reliable estimates can be obtained *in silico*.

2.2   pK_a

Measuring pK_a has become easier and more convenient over the recent years. However, it requires either purchase, or synthesis and purification of chemical compounds. This may not always be cost-effective, particularly for early drug-discovery scientists who wish to screen up to millions of compounds in chemical libraries. In situations like this, *in silico* estimation of pK_a is the only way to obtain results. This review will be focused only on pK_a prediction by empirically based computational methods.

Microspeciation explained in the preceding chapter of this book has profound consequences for *in silico* prediction of ionization constants. Since macroscopic pK_a values are predicted from molecular structure, having a uniquely defined structure with unambiguously positioned protons is a necessary condition of a successful prediction. Unfortunately, unique structures are characteristic of microstates, not macrostates. Consequently, macroconstants cannot be derived directly from the structure. On the other hand, experimental microconstants are available only for a handful of compounds, while the scientific literature is rich in macroconstants for tens of thousands of chemicals.

Predictive models may be built indirectly against observed macroconstants via rigorous application of the microequilibria mathematics, but the difficulty is in the exponential growth in the number of microconstants and microstates as a function of the number of ionizable groups per molecule. Many prediction methods avoid this challenge by neglecting microspeciation altogether or by using various approximation schemes. Fortunately, in drug discovery and development applications, the investigated molecules rarely possess high numbers of ionizable groups. Therefore, exact resolution of complete sets of ionization microequilibria is computationally feasible and there is no need to introduce additional sources of error by using approximate algorithms.

Empirical methods of pK_a estimation can be tentatively divided into three groups based on the approach used:

(1) linear free energy relationships (LFER) – methods utilizing empirical relations of Hammett and Taft
(2) QSPR – methods correlating calculated structural descriptors with pK_a
(3) database lookup – methods searching of similar structures in a predetermined database of molecules with known experimental pK_a.

Historically, LFER-based methods are the oldest. As early as 1935, Hammett first discovered the simple linear relationship between pK_as of benzoic acid (^{0}) and its *para*-substituted derivative (^{S}):[26]

$$pK_a^S = pK_a^0 - \rho\sigma_S \tag{1}$$

where σ_s is a Hammett constant characteristic to a given substituent S and ρ is a constant for the reaction class considered ($\rho = 1$ for para-substituted benzoic acids). Ionization constants are proportional to Gibbs free energies of ionization reactions, hence the abbreviation LFER. Taft observed similar relations for aromatic and aliphatic systems.[27–30] By the early 1980s, LFER methods had been studied in depth and developed into an elaborate prediction system summarized in the well-known book by Perrin *et al.*[31] Thousands of Hammett and Taft reaction constants were determined experimentally for many systems.

Cases in which LFER breaks down have been identified. In general, when presented with a new molecule, LFER-based methods must first find the most suitable parent molecule from a large database of experimental pK_a. Of course, the word "suitable" is the key to this step. The entire parent molecule must be a structural fragment of the molecule in question, since it provides the value of the reaction constant. Prediction success depends on whether the experimental pK_a° of the parent molecule contains the bulk of the information associated with the pK_a in question. If yes, then the distinct part of the new molecule, which is divided into a set of "substituents", exerts a small perturbation on pK_a°. In this case, application of corrections similar to those shown in Equation 9 may result in highly accurate estimation. In LFER-based methods, microspeciation may be applied to multiprotic systems, but it requires that parameters for charged fragments be present in the database. In this arena, two commercially available programs lead the way: ACD/pK_a DB from Advanced Chemistry Development and Pallas[32,33] from CompuDrug (Table 1). Academic software SPARC,[34–36] partially funded by US EPA, also belongs in this category, although pK_a perturbations are split into types of interactions (field, mesomeric, induction, solvation, etc.) vastly increasing the number of constants per substituent.

Newer methods of the second group (Marvin, ADMET Predictor, and CspK_a) apply methods based on correlations between calculated descriptors and observed pK_a values. The QSPR approach has proven successful in predicting whole molecule properties. The main problem here is choice of proper descriptors. Catalan et al[37] expressed gas-phase proton affinities of oxygen, nitrogen, and carbon bases as multivariate linear functions of 1s orbital binding energies and first ionization potentials. Quantum-mechanical descriptors were also used by other research groups in developing QSPR correlations for either gas-phase proton affinities, or aqueous pK_a values of narrow classes of compounds.[38–48]

Unfortunately, quantum descriptors are computationally too expensive to be used in high-throughput *in silico* prediction. Gasteiger and Hutchings used empirically calculated residual electronegativities and effective polarizabilities of a protonated group to predict gas-phase proton affinities of alcohols, ethers, their thio analogues, and carbonyl compounds.[49,50] This approach was later extended to ionization in aqueous phase.[51] Thus, they invented QSPR prediction of pK_a based on purely empirical atomic descriptors. Recently, Gasteiger reported the use of artificial neural networks, based on an extended set of atomic descriptors, to successfully predict pK_a values for monoprotic -OH and -SH acids and monoprotic nitrogen bases.[52] Empirically calculated partial atomic charges have also been utilized as atomic descriptors in QSPR models of pK_a of -OH acids.[53] Partial charges alone, however, may not be sufficient to build a universal pK_a model applicable to a wide class of compounds. For example, in a series of alkylamines NR_xH_{3-x} (R = Me, Et) the partial charge on the N atom increases with x implying smaller electron density and diminishing proton affinity.[54] The observed proton affinity trend, however, is quite opposite. In an interesting approach Xing and co-workers[55,56] invented molecular tree structured fingerprints, a novel class of atomic descriptors based on atom types, to build models predicting pK_a of ~1000 acids and bases.

Similar to LFER approaches, QSPR models may or may not take microspeciation into account. Microspecies are distinguished naturally if atomic descriptors take into account the protonation state of other groups. Although quite challenging to model developers, exact and complete inclusion of microspeciation rewards the programer with better pK_a prediction for multiprotic systems. Two commercial software

Table 1 *Software for empirical pK_a prediction*

Software product	Vendor	Method	Treatment of microstates	URL
ACD/pK_a DB	Advanced Chemistry Development	LFER	Partial[a]	www.acdlabs.com
Pallas	Compudrug	LFER	None	www.compudrug.com
SPARC	University of Georgia	LFER	Complete	ibmlc2.chem.uga.edu/sparc/index.cfm
Marvin	ChemAxon	QSPR	Complete	www.chemaxon.com
ADMET Predictor	Simulations Plus	QSPR	Complete	www.simulations-plus.com
CSpK_a	ChemSilico	QSPR[b]	None	www.chemsilico.com
ADME Boxes	Pharma Algorithms	Database[c]	None	www.ap-algorithms.com
DISCON	Russian Academy of Medical Sciences	Database	None	www.ibmh.msk.su/molpro/discon.html

[a] Limited to diprotic and perfectly symmetric systems only ("apparent exact" constants in ACD terminology)
[b] Limited to 5 ionization centers per molecule only
[c] Estimates only principal acid and base pK_a

packages: Marvin from ChemAxon and ADMET Predictor from Simulations Plus (Table 1) feature universal QSPR-based predictive models of pK_a with exact treatment of microspeciation. Both programs calculate multiprotic pK_a for a wide class of organic compounds. The difference between the two is in the scope of atomic descriptors and model complexity. Marvin uses just three atomic descriptors (partial charge, atomic polarizability, and intramolecular hydrogen-bond indicator) in simple linear or nonlinear relations for microconstants.[57] In ADMET Predictor, up to 44 proprietary atomic descriptors are fed into artificial neural network ensembles to calculate microconstants. $CSpK_a$ from ChemSilico also applies artificial neural networks as the modeling method using topological and electrotopological descriptors. It is not clear whether these descriptors are of atomic level. $CSpK_a$ limits prediction up to 5 macroscopic pK_a per molecule and ignores microspeciation.

Sayle has noticed that direct assignment of "typical" pK_a values according to detected ionizable atom types predicts pK_a values for simple molecules with standard deviation of 0.95 log units.[58] The observation that similar molecules, or similar molecular fragments, have similar pK_a is the basis of the last class of methods. A database containing either complete structures or structural fragments and experimental pK_a values is searched for the nearest neighbors of the compound in question. Correction factors may be eventually applied to the observed pK_a values. Of course, treatment of microspeciation is not possible in this approach. Software in this group

Table 2 *Definition of acronyms and abbreviations*

Acryonym or abbreviation	Definition
ACAT	Advanced compartmental absorption and transfer
ADMET	Absorption, distribution, metabolism, excretion, toxicity
ANNE	Artificial neural network ensemble
CL	Clearance
CRADA	Cooperative research and development agreement
CYP_{450}	Cytochrome P450
EPA	Environmental protection agency
FDA	Food and drug administration
H-Peff	Human effective permeability
LFER	Linear free energy relationships
MAE	Mean absolute error
MLR	Multiple linear regression
MRA	Maximum redundancy analysis
P_{ow}	Octanol/water partition coefficient
PCA	Principle component analysis
PLS	Partial least squares
QM	Quantum mechanics
QSPR	Quantitative structure property relationships
RMSE	Root mean square error
R-Papp	Rat apparent permeability
SA	Structure alerts
SD	Standard deviation
SVM	Support vector machine
Vd/VDss	Volume of distribution/Vd steady-state
XOR	Exclusive OR

includes ADME Boxes from Pharma Algorithms and DISCON from the Russian Academy of Medical Sciences.

2.3 Permeability

Computational methods for estimation of permeability have been reviewed by Lipinski.[59,60] In order to develop *in silico* models of permeability, one needs a source of high quality *in vitro* or *in vivo* permeability data. The US Food and Drug administration (US FDA) funded research on *in vivo* human jejunum effective permeability has provided a small database of high quality data in normal volunteers.[61–63] *In situ* and *in vitro* rat intestinal permeability, and *in vitro* cell culture permeability were discussed earlier in this chapter.

Lennernas *et al.*[61] have developed a method for measuring human effective permeability (H-Peff) using a regional intestinal perfusion technique. In this method, a perfusion apparatus consisting of a multichannel tube with two inflatable balloons (10 cm apart) is swallowed by the patient and eventually located in the proximal jejunum. Dilute solutions of the test drug are introduced at the inlet located at the center of the 10 cm section, and the loss of drug is determined from the concentration in the outlet intestinal perfusate. In such a fashion, the H-Peff for 22 carefully selected drug molecules has been determined and a theoretical model of H-Peff has been developed.[64] The small size of the published H-Peff database is most likely due to the expense of the human measurement.

Our group has developed a model of human jejunum effective permeability (H-Peff) and *in situ* rat wall permeability (R-Papp) measurements. The *in situ* R-Papp values were converted to H-Peff values using a linear relationship.[65] Data from 44 examples (38 drugs and 6 primary alpha amino acids) were used for model building. We found that molecular descriptors relating to lipophilicity, hydrogen-bond donors and acceptors, and hydrogen-bond topology were selected using sensitivity analysis. The performance of the model is shown in Figure 1. When testing an external test set, the mean absolute error (MAE) was 0.19 log units, the root mean square error (RMSE) was 0.23 log units, and the explained variance (R_2) was 74%.

Several authors have reviewed the strengths and weaknesses of various modeling approaches for predictive ADMET in early discovery.[66–68] These approaches involve computational models for permeability *in vivo*, *in vitro*, and *in situ*. In general such models have broad applicability for estimation of both passive transcellular and paracellular permeability. However, we still have not developed a general model that accurately predicts the role of transport proteins in active and facilitated uptake of drugs following oral administration.

2.4 Solubility

Next to permeability, aqueous solubility is the most important biopharmaceutical property associated with oral drug absorption. These two properties have established the basis for a biopharmaceutical classification system (BCS) and have become the subject of an FDA guidance entitled "Waiver of *In Vivo* Bioavailability and Bioequivalence Studies for Immediate-Release Solid Oral Dosage Forms Based on a

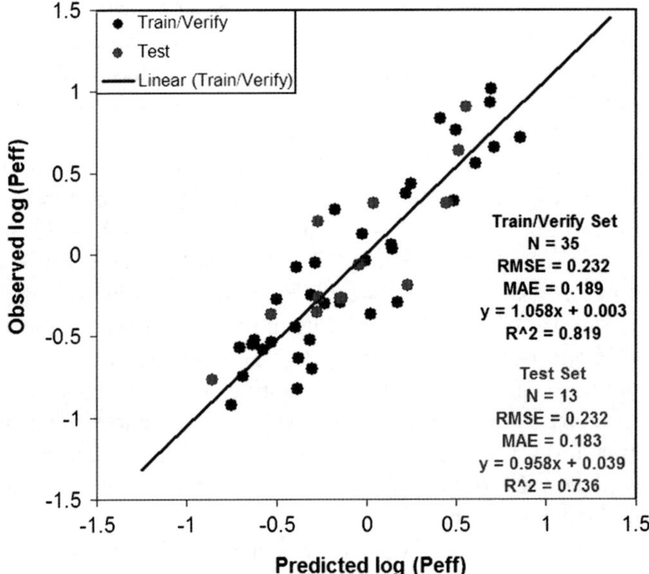

Figure 1 *ANNE model of human effective permeability. Data for the model was obtained from published sources or provided through an agreement with Therapeutic Systems Research Laboratories (http://www.tsrlinc.com). Black dots represent the training and verification data set and the red dots represent the external verification data set*

Biopharmaceutics Classification System".[62] The fundamental parameters that define oral drug absorption in humans are used as a basis for this classification scheme. These Biopharmaceutic Drug Classes are defined as: Case 1: high solubility–high permeability drugs, Case 2: low solubility–high permeability drugs, Case 3: high solubility–low permeability drugs, and Case 4: low solubility–low permeability drugs. If one is able to accurately estimate permeability and solubility from molecular structure then it becomes feasible to estimate the BCS classification and to begin to develop *in silico* methods for simulation of gastrointestinal absorption.[69–71]

Theoretical models of aqueous solubility are very difficult to develop and even more difficult for the user of such a model to evaluate. Quite often the statistician developing the model does not know the source and exact experimental conditions for solubility measurements. Consequently, the resulting models give highly specific results that relate primarily to the data sets used in the development. This does not mean that solubility models cannot be generalized to diverse structural types, but it does mean that statistical estimates of solubility may be quite different, depending on the nature of the experimental conditions used to collect the solubility data for a given training set.

Yalkowsky and Banerjee[72] have published an extensive review of methods for estimating aqueous water solubility of organic compounds. Many methods have been developed based on measured properties such as partition coefficients, chromatographic parameters, and activity coefficients.[73–75] Purely *in silico* methods are based on LFER, and a variety of geometric, electronic, and topological molecular descriptors.[76–83]

The earliest models of aqueous solubility were based on the activity coefficient of the hydrophobic portion of a molecule, its surface area, and its interfacial tension.[73] The solvatochromic method of estimating water solubility has been compared with the relationship between solubility and the octanol–water partition coefficient, and it has been demonstrated that the latter provides a better estimation.[74,84] When applied to a diverse series of over 100 small non-electrolyte organic molecules, the aqueous solubility and octanol–water partition coefficient ($\log P_{ow}$) were found to be related by the simple equation $\log Sw = -1.016 \log PC + 0.515$, where Sw is the molar solubility of liquid solutes in water and PC is the experimental partition coefficient of the solutes in the octanol–water system.[85] Other models using this approach have been developed for various small sets of drug molecules.[86] Meylan compiled a large database of 1450 aqueous solubilities, including over 100 drug molecules, and developed a very predictive model using $\log P_{ow}$, melting point, molecular weight, and 12 correction factors based on structure.[75] This model was able to estimate the solubility of an external test set of 85 molecules with a standard deviation (SD) of 0.96 log units. If a computed value of $\log P_{ow}$ is available, this method of calculating solubility can be used without measured values.

We have been very interested in models of aqueous solubility solely based on *in silico* parameters. Klopman developed a method for estimating water solubility based on the group contribution approach.[76] Two models were developed, based on 21 organic compounds, with SD of 0.58 log units for the test first test set of 13 compounds and SD of 1.25 log units for the second test set of all 21 compounds. Using topological descriptors (connectivity indices, kappa indices, and electrotopological state indices) and two small databases of steroids and barbituric acids, an ANN model of aqueous solubility was developed.[77] Separate models were built for each class of drugs using back-propagation neural networks with one hidden layer and five topological indices as input parameters. The results indicate that neural networks can produce useful models of the aqueous solubility of a congeneric set of compounds, even with simple structural parameters. When applied to a much larger data set ($N = 210$) of primarily drug molecules, the same method produced a model that was able to estimate an external test set ($N = 51$) with a SD of 0.58 log units.[79] Genetic algorithm and simulated annealing routines, in conjunction with MLR and ANNs, have been used to develop a model of aqueous solubility from 332 small organic molecules.[78] This model was able to predict an external test set of 32 molecules with RMSE of 0.34 log units. More recently Lobell and Sivarajah[83] have compared their model of aqueous solubility to ten other published and commercial models. They conclude that prediction of the aqueous solubility of neutral molecules is fairly accurate but that the solubility estimation of zwitterions and molecules that are charged at the pH used for the measurement are more problematic. Thus, it would appear that purely *in silico* methods might be successfully used to estimate aqueous solubility.

2.5 Protein Binding

When building *in silico* models for biopharmaceutical properties, one must understand the molecular nature of the property in order to select molecular descriptors that have relevance to the problem. Plasma protein binding involves the reversible interaction of a drug with three main types of protein. Serum albumin, which is the most abundant protein in blood plasma (35–50 mg ml^{-1}), α1-acid glycoprotein (0.5–1 mg ml^{-1} blood plasma), and lipoproteins are the important plasma proteins. Acidic and neutral drugs are generally bound more extensively to albumin, and basic drugs to α1-acid glycoprotein and lipoproteins. Only unbound drug is available for diffusion to the disease target site where it can induce a pharmacological effect.[83] Lobell and Sivarajah have reported a simple fast method for estimating volume of distribution (Vd), and human plasma protein binding. They transformed the fraction unbound (fu) in human plasma into log((1−fu)/fu). This log construct translates fu into a scale, which ensures that small differences at the high end of plasma protein binding affinities (fu close to 1) are treated with equal importance, compared to larger differences in the medium range. They also found that AlogP98 was highly correlated with log(1−fu)/fu) for uncharged compounds. Another general rule that emerged was that for compounds of equal hydrophobicity (provided AlogP98 is in the range of −1–6) the relative magnitude of human plasma protein binding in different charge groups descends in the following order: plasma protein binding of negatively charged compounds > uncharged compounds > positively charged compounds > permanently positively charged compounds. Plasma protein binding of zwitterionic compounds was observed to be generally low.

3 Estimation of Pharmacokinetic Properties

3.1 Clearance

One of the earliest examples of *in silico* modeling for human clearance (CL) and Vd involved the use of back-propagation artificial neural networks.[87] The models with highest prediction for CL and Vd were obtained when rat and dog CL and Vd, combined with *in vitro* protein binding and partition coefficients of the drugs, were used as input parameters. In general, human *in vivo* clearance is a difficult property to estimate from structure alone. A discussion of progress in estimating metabolism, which would be the first step in predicting clearance, is described below.

3.2 Volume of Distribution

The estimation of Vd prior to *in vivo* studies involves two distinct approaches. Statistical models based on a database of measured values have been shown to provide good estimates. Lobell derived some general rules from *in silico* modeling.[83] The relative magnitude of Vd steady-state (VDss) depends on the predominant charge state at pH 7.4: VDss of compounds with positive AlogP98: positively charged > uncharged > permanently positively charged > negatively charged. VDss of compounds with negative AlogP98: uncharged $\geq$ positively charged > negatively charged = zwitterionic. VDss depends on a drug's affinity to bind to plasma proteins and tissue components and its ability to cross tissue membranes. Predominantly negatively charged compounds (acidic compounds with a pK_a < 7.4) tend to be highly bound to serum albumin due to both ion pair and hydrophobic interactions. In addition, they have unfavorable charge–charge interactions with negatively charged phospholipids in tissue membranes, which largely limit their ability to cross cell membranes. As a consequence, the tendency of negatively charged compounds to penetrate into cells is very low; they are therefore mostly confined to plasma and extracellular fluid, resulting in a low VDss.

Poulin and Theil[88] have developed a mechanistic model for estimating the Vd based on physiologically based pharmacokinetics (PBPK). For this method, the tissue: plasma partition coefficient for each organ of the body is calculated by consideration of the volume fraction of neutral and phospholipids and water found in the tissues of a particular organ. For example, the volume fraction of neutral lipids in human adipose tissue is 0.79 whereas the volume fraction of neutral lipids in cardiac tissue is 0.0115.[89] By contrast the volume fraction of water in adipose and heart are 0.18 and 0.76 respectively. Combined with the P_{ow}, these volume fractions are used to estimate the distribution of a drug molecule into each tissue. Summation of the product of tissue volume and tissue/plasma partition coefficient produces the estimate of Vd.[88]

3.3 Metabolism

In silico methods to estimate cellular metabolism and *in vivo* clearance based on the chemical structure have not been widely successful. However, initial efforts to develop computational methods have been applied to this field in the following areas: (a) prediction of metabolite formation, (b) estimation of regiospecific metabolic lability, (c) prediction of CYP enzyme subfamilies responsible for particular drug metabolism, (d) structure-based docking of drug molecules to CYP enzymes, (e) estimation of binding affinity (Km) to CYP3A4, and (f) genome-wide prediction of metabolic pathways. Langowski[90] has reviewed the field of computational methods in prediction of xenobiotics metabolism.

One of the earliest attempts to model chemical transformations in a living system was carried out in 1987. This system consists of a biotransformation database and one or more logic-based prediction tools.[91] This system and other knowledge-based systems provide a branching tree of possible metabolites but provide no information on likelihood or quantitative rates of production.

Korzekwa *et al.*[92] applied semi-empirical AM1 calculations to the prediction of the regiospecificity of Cytochrome P450 (CYP_{450}) metabolism. This method produces an approximation of the transition state energy for hydrogen-atom abstraction from potential metabolic sites on a drug molecule.[93] Because the rate-limiting step in P450 metabolism is hydrogen atom abstraction, one can determine the relative metabolic lability of a drug molecule. This type of calculation has not been successfully applied to estimation of *in vivo* clearance, but when it is combined with structure-based models of selected CYP_{450}s, it can provide useful rank ordering of metabolism for early drug discovery.[94–97] This method works best with substrates of CYP3A4 because of the promiscuous nature of the substrate-binding site. Since the substrates can achieve a variety of orientations in the binding site, the region of lowest hydrogen abstraction energy will be the most frequently oxidized. Semi-empirical AM1 calculations can be time consuming and are not amenable to high throughput calculations. Singh *et al.* carried out AM1 and trend vector calculations on 50 CYP3A4 substrates whose major sites of metabolism are known in the literature. They found that for these 50 substrates, only those hydrogen atoms with solvent accessible surface area exposure ≥ 8.0 A(2) are susceptible to CYP3A4-mediated metabolism. This approach formed the basis for a general model that predicts sites on drugs that are susceptible to CYP_{450} 3A4-mediated hydrogen radical abstraction followed by a hydroxylation reaction.[98]

Since the semi-empirical method works best for substrates of CYP3A4, it would be convenient to be able to pre-screen substrates to determine the CYP family specificity to help in determining the reliability of the predictions. Another group compiled a database of known human CYP_{450} substrates, products, and non-substrates for 38 enzyme-specific groups (total of 2200 compounds). Second, they determined the cytochrome-mediated metabolic reactions most typical for each group and examined the substrates and products of these reactions. To assess the probability of P450 transformations of novel compounds, they built a non-linear quantitative structure metabolism relationships (QSMR) model based on Kohonen self-organizing maps (SOM).[99] The Kohonen SOM was able to cluster substrates into regions of the self organizing map that corresponded to specific CYP families. The same group also applied ANN technology to predict the magnitude of binding affinity for CYP enzymes.[100]

Structure-based molecular modeling is another promising approach to improvement of the prediction of region-specificity of P450 metabolism. De Groot developed pharmacophore models and three-dimensional quantitative structure–activity relationships either alone or in combination with protein homology models to provide substrate-binding specificity information for CYP_{450}s.[101]

Most recently, *in silico* methods have been applied to the systems biology of metabolic pathways.[102] An analysis of the human genome assigns 2709 human enzymes to 896 bioreactions; 622 of the enzymes are assigned roles in 135 predicted metabolic pathways. The predicted pathways closely match the known nutritional requirements of humans. This analysis identifies probable omissions in the human genome annotation in the form of 203 pathway holes (missing enzymes within the predicted pathways). The predicted human metabolic map is described by a Pathway/Genome Database called HumanCyc, which is available at http://HumanCyc.org/.

4 Estimation of Toxicological Properties

Computational modeling of toxicology in drug development has become a hot topic in the drug industry. Large amounts of data are available from free online government sources in the EPA and FDA. The problem is that most of the chemical space covered by these databases is not drug-like. This makes the problem of toxicity prediction much more difficult for drug development.[103] Specifically in the area of mutagenicity and carcinogenicity, the databases rely heavily, although not exclusively, on bacterial mutagenicity data of non-pharmaceutical-type molecules as the primary knowledge base. In order to address the question of reliability of predictions, a panel of 394 marketed pharmaceuticals with Ames Salmonella reversion assay and other genetic toxicology findings was extracted from the 2000 to 2002 Physicians' Desk Reference and evaluated using MCASE, TOPKAT, and DEREK, the three most commonly used computational databases.[104] The evaluation of these programs indicated a generally poor sensitivity of all systems for predicting Ames positivity and even poorer sensitivity in prediction of other genotoxicities. As might be expected, all three programs were more highly predictive for molecules containing carcinogenicity structural alerts than for those without such alerts. Taking all genotoxicity assay findings into consideration, there were 84 instances in which positive genotoxicity results could not be explained in terms of structural alerts, suggesting the possibility of alternative mechanisms of genotoxicity not relating to covalent drug-DNA interaction.

On the positive side, with funding from a cooperative research and development agreement (CRADA) between the US FDA and the owners of MultiCase™, an optimized version of MCASE was developed with good predictive capability for carcinogenic potential. These groups demonstrated that the standard MCASE system derived from a small data set ($N = 319$) detected few structure alerts (SA) for carcinogenicity ($n = 17$), and had poor coverage for beta-test compounds. Conversely, the new, optimized FDA-OTR/MCASE system based on a large dataset ($n = 934$) detected many SA ($N = 58$) and had good coverage (94%). In addition, the study showed that the standard MCASE(A07-9) software had poor predictive value for carcinogens and specificity for non-carcinogens (50 and 42%), detected many false positives (58%), and exhibited poor concordance (46%). Conversely, the new, FDA-OTR/MCASE(AF5-8) system demonstrated excellent predictive value for carcinogens and specificity for non-carcinogens (97 and 98%), detected only one false positive (2%), and exhibited good concordance (75%). The dramatic improvements in the performance of the MCASE were due to numerous modifications. Finally, recent developments in prediction of toxicity from chemical structure have been reviewed.[105]

5   Integration of Surrogate Data and Estimations with Physiological Simulation

A recent FDA white paper entitled "Challenge and Opportunity on the Critical Path to New Medical Products" (FDA, March 2004) states that "*We need to make the effort required to create better tools for developing medical technologies*", and "*...there is hope that greater predictive power may be obtained from in silico (computer modeling) analyses … . Some believe that extensive use of in silico technologies could reduce the overall cost of drug development by as much as 50 percent*".

In the preceding sections of this chapter, we have discussed numerous biopharmaceutical properties and methods, which have the potential to deliver a mind-numbing avalanche of data. Simple databases and tables of properties combined with simple rules like the "Rule-Of-Five" do not provide the platform or integration-of-information needed to make decisions.

In our experience with sophisticated physiologically based simulations using GastroPlus™, we have seen that detailed mechanistic simulations can explain quantitatively complex interactions that govern the behavior of many single drugs that had not been well understood previously. We do this, when data are sufficient, by attempting to find a single set of model parameters that adequately explain all observations across different dosing levels. When such a model can be found, the insight and understanding into the interacting mechanisms that govern the behavior of the drug is unmatched. Providing such insight has enabled researchers to make better project decisions in drug development, reducing risks, cost, and time in further drug development. A detailed discussion of the Advanced Compartmental Absorption and Transit (ACAT) model and its application in drug discovery and development is beyond the scope of this chapter. The ACAT model has been well described in the literature.[106–109]

Reference

1. T. Kennedy, *Drug Discov. Today*, 1997, **2**, 436.
2. K.N. Rogstad, Y.H. Jang, L.C. Sowers and W.A. Goddard, *Chem. Res. Toxicol.*, 2003, **16**, 1455.
3. R. Todeschini and V. Consonni, *Handbook of Molecular Descriptors*, Wiley-VCH, Verlag GmbH, D-69469 Weinheim, Germany, 2000.
4. J.G. Topliss and R.P. Edwards, *J. Med. Chem.*, 1979, **22**, 1238.
5. H. Wold, *The Fix-Point Approach to Interdependent Systems*, North-Holland, Amsterdam, North Holland, 1981.
6. T. Dijkstra, *Latent Variables in Linear Stochastic Models*: *Reflections on Maximum Likelihood and Partial Least Squaresmethods*, Sociometric Research Foundation, Amsterdam, The Netherlands, 1985.
7. R. Tobias, *The Twentieth Annual SAD Users Group International Conference*, Cary, North Carolina, 1995.
8. A. Hoerl and R. Kennard, *Technometrics*, 1970, **12**, 55.
9. S. Haykin, *Neural Networks : A Comprehensive Foundation*, Prentice Hall, Upper Saddle River, NJ, 1999.
10. F. Rosenblatt, *Psychol. Rev.*, 1958, **65**, 386.
11. S.A. Papert, *Perceptrons*, MIT Press, Boston, MA, 1969.
12. J. Zupan and J. Gasteiger, *Neural Networks in Chemistry and Drug Design*, Wiley-VCH, Weinheim, Germany, 1999.
13. B.E. Boser, I.M. Guyon and V.N. Vapnik in *5th Annual ACM Workshop on COLT*, D. Haussler (ed), ACM Press, Pittsburgh, PA, 1992, 144.
14. N. Cristianini and J. Shawe-Taylor, *An Introduction to Support Vector Machines and Other Kernel-Based Learning Methods*, Cambridge University Press, Cambridge, MA, 2000.
15. B. Schölkopf and A.J. Smola, *Learning with Kernels*, MIT Press, Cambridge, MA, 2002.
16. S.J. Russell and P. Norvig, *Artificial Intelligence: A Modern Approach*, 2nd edn, Pearson Education, NJ, 2003, 749.
17. A. Leo, P.Y. Jow, C. Silipo and C. Hansch, *J. Med. Chem.*, 1975, **18**, 865.
18. D.J. Abraham and A.J. Leo, *Proteins*, 1987, **2**, 130.
19. A.J. Leo, *Prog. Clin. Biol. Res.*, 1989, **291**, 53.
20. R.F. Rekker, A.M. ter Laak and R. Mannhold, *Quant. Struct. Act. Rel.*, 1993, **12**, 152.
21. V.K. Gombar and K. Enslein, *J. Chem. Inf. Comp. Sci.*, 1996, **36**, 1127.
22. W.M. Meylan and P.H. Howard, *J. Pharm. Sci.*, 1995, **84**, 83.

23. J.J. Huuskonen, A.E. Villa and I.V. Tetko, *J. Pharm. Sci.*, 1999, **88**, 229.

24. D. Eros, I. Kovesdi, L. Orfi, K. Takacs-Novak, G. Acsady and G. Keri, *Curr. Med. Chem.*, 2002, **9**, 1819.

25. R. Mannhold and H. van de Waterbeemd, *J. Comput. Aided Mol. Des.*, 2001, **15**, 337.

26. L.P. Hammett, *Physical Organic Chemistry*, McGraw-Hill, New York, 1940.

27. R.W. Taft and I.C. Lewis, *J. Am. Chem. Soc.*, 1959, **81**, 5343.

28. R.W. Taft and I.C. Lewis *J. Am. Chem. Soc.*, 1958, **80**, 2436.

29. R.W. Taft, S. Ehrenson, I.C. Lewis and R.E. Glick, *J. Am. Chem. Soc.*, 1959, **81**, 5352.

30. R.W. Taft, *J. Am. Chem. Soc.*, 1953, **75**, 4231.

31. D.D. Perrin, B. Dempsey and E.P. Serjeant, *pKa Prediction for Organic Acids and Bases*, Chapman & Hall, London and New York, 1981.

32. F. Csizmadia, A. Tsantili-Kakoulidou, I. Panderi and F. Darvas, *J. Pharm. Sci.*, 1997, **86**, 865.

33. A. Tsantili-Kakoulidou, I. Panderi, F. Csizmadia and F. Darvas, *J. Pharm. Sci.*, 1997, **86**, 1173.

34. S.H. Hilal, S.W. Karickhoff and L.A. Carreira., *Quant. Struct. Act. Relat.*, 1995, **14**, 348.

35. S.H. Hilal, Y. El-Shabrawy, L.A. Carreira, S.W. Karickhoff, S.S. Toubar and M. Rizk, *Talanta*, 1996, **43**, 607.

36. S.H. Hilal, L.A. Carreira and S.W. Karickhoff, *Talanta*, 1999, **50**, 827.

37. J. Catalan, O. Mo, P. Perez and M. Yanez, *J. Chem. Soc. Perkin Trans. 2*, 1982, 1409.

38. C. Grüber and V. Buß, *Chemosphere*, 1989, **19**, 1595.

39. R. Karaman, J.-T.L. Huang and J.L. Fry, *J. Comp. Chem.*, 1990, **11**, 1009.

40. M.J. Citra, *Chemosphere*, 1999, **38**, 191.

41. K.C. Gross, P.G. Seybold, Z. Peralta-Inga, J.S. Murray and P. Politzer, *J. Org. Chem.*, 2001, **66**, 6919.

42. K.C. Gross and P.G. Seybold, *Int. J. Quantum Chem.*, 2001, **85**, 569.

43. K.H. Kim and Y.C. Martin, *J. Org. Chem.*, 1991, **56**, 2723.

44. K.H. Kim and Y.C. Martin, *J. Med. Chem.*, 1991, **34**, 2056.

45. J. Spanget-Larsen, *J. Phys. Org. Chem.*, 1995, **8**, 496.

46. R. Gargallo, C.A. Sotriffer, K.R. Liedl and B.M. Rode, *J. Comp. Aided Mol. Design*, 1999, **13**, 611.

47. B.R. Hollingsworth, P.G. Seybold and C.M. Hadad, *Int. J. Quantum Chem.*, 2002, **90**, 1396.

48. U.A. Chaudry and P.L.A. Popelier, *J. Org. Chem.*, 2004, **69**, 233.

49. J. Gasteiger and M.G. Hutchings, *J. Am. Chem. Soc.*, 1984, **106**, 6489.

50. M.G. Hutchings and J. Gasteiger, *J. Chem. Soc. Perkin Trans. 2*, 1986, 447.

51. M.G. Hutchings and J. Gasteiger, *J. Chem. Soc. Perkin Trans.*, 2, 1986, 455.

52. J. Gasteiger, A.-X. Yan and T. Kleinöder in Chemical information and computation 2002. *223rd ACS National Meeting and Exposition*, Orlando, FL, April

7–11, 2002, W.A. Warr (ed), Wendy Warr & Associates, Cheshire, UK, 2002, 98.

53. S.L. Dixon and P.C. Jurs, *J. Comp. Chem.*, 1992, **13**, 492.

54. Z.B. Maksic and R. Vianello, *J. Phys. Chem. A*, 2002, **106**, 419.

55. L. Xing and R.C. Glen, *J. Chem. Inf. Comp. Sci.*, 2002, **42**, 796.

56. L. Xing, R.C. Glen and R.D. Clark, *J. Chem. Inf. Comp. Sci.*, 2003, **43**, 870.

57. J. Szegezdi and F. Csizmadia, *227th National Meeting of the American Chemical Society*, Anaheim, CA, 2004.

58. R. Sayle, Physiological ionization and pKa prediction, http://www.daylight.com/meetings/emug00/Sayle/pkapredict.html, Metaphorics LLC, 2000.

59. C.A. Lipinski, F. Lombardo, B.W. Dominy and P.J. Feeney, *Adv. Drug Deliv. Rev.*, 1997, **23**, 3.

60. C.A. Lipinski, F. Lombardo, B.W. Dominy and P.J. Feeney, *Adv. Drug Deliv. Rev.*, 2001, **46**, 3.

61. H. Lennernas, O. Ahrenstedt, R. Hallgren, L. Knutson, M. Ryde and L.K. Paalzow, *Pharm. Res.*, 1992, **9**, 1243.

62. G.L. Amidon, H. Lennernas, V.P. Shah and J.R. Crison, *Pharm. Res.*, 1995, **12**, 413.

63. H. Lennernas, *J. Pharm. Sci.*, 1998, **87**, 403.

64. S. Winiwarter, N.M. Bonham, F. Ax, A. Hallberg, H. Lennernas and A. Karlen, *J. Med. Chem.*, 1998, **41**, 4939.

65. U. Fagerholm, M. Johansson and H. Lennernas, *Pharm. Res.*, 1996, **13**, 1336.

66. S. Ekins, C.L. Waller, P.W. Swaan, G. Cruciani, S.A. Wrighton and J.H. Wikel, *J. Pharmacol. Toxicol. Methods*, 2000, **44**, 251.

67. J.E. Penzotti, G.A. Landrum and S. Putta, *Curr. Opin. Drug Discov. Dev.*, 2004, **7**, 49.

68. W.J. Egan and G. Lauri, *Adv. Drug. Deliv. Rev.*, 2002, **54**, 273.

69. A. Avdeef, *Curr. Top. Med. Chem.*, 2001, **1**, 277.

70. L.X. Yu, G.L. Amidon, J.E. Polli, H. Zhao, M.U. Mehta, D.P. Conner, V.P. Shah, L.J. Lesko, M.L. Chen, V.L. Lee and A.S. Hussain, *Pharm. Res.*, 2002, **19**, 921.

71. N.A. Kasim, M. Whitehouse, C. Ramachandran, M. Bermejo, H. Lennernas, A.S. Hussain, H.E. Junginger, S.A. Stavchansky, K.K. Midha, V.P. Shah and G.L. Amidon, *Mol. Pharmaceut.*, 2004, **1**, 85.

72. S.H. Yalkowsky and S. Banerjee, *Aqueous Solubility. Methods of Estimation for Organic Compounds*, Marcel Dekker, New York, NY, USA., 1992.

73. G.L. Amidon, S.H. Yalkowsky and S. Leung, *J. Pharm. Sci.*, 1974, **63**, 1858.

74. S.H. Yalkowsky, R. Pinal and S. Banerjee, *J. Pharm. Sci.*, 1988, **77**, 74.

75. W.M. Meylan, P.H. Howard and R.S. Boethling, *Environ. Toxicol. Chem.*, 1996, **15**, 100.

76. G. Klopman, S. Wang and D.M. Balthasar, *J. Chem. Inf. Comput. Sci.*, 1992, **32**, 474.

77. J. Huuskonen, M. Salo and J. Taskinen, *J. Pharm. Sci.*, 1997, **86**, 450.

78. M.E. Brooke and P.C. Jurs, *J. Chem. Inf. Comput. Sci.*, 1998, **38**, 489.

79. J. Huuskonen, M. Salo and J. Taskinen, *J. Chem. Inf. Comput. Sci.*, 1998, **38**, 450.

80. N.R. McElroy and P.C. Jurs, *J. Chem. Inf. Comput. Sci.*, 2001, **41**, 1237.

81. X. Xia, E. Maliski, J. Cheetham and L. Poppe, *Pharm. Res.*, 2003, **20**, 1634.

82. J.R. Votano, M. Parham, L.H. Hall and L.B. Kier, *Mol. Divers.*, 2004, **8**, 379.

83. M. Lobell and V. Sivarajah, *Mol. Divers.*, 2003, **7**, 69.

84. Y. Ran and S.H. Yalkowsky, *J. Chem. Inf. Comput. Sci.*, 2001, **41**, 354.

85. S.C. Valvani, S.H. Yalkowsky and T.J. Roseman, *J. Pharm. Sci.*, 1981, **70**, 502.

86. R. Pinal and S.H. Yalkowsky, *J. Pharm. Sci.*, 1987, **76**, 75.

87. W.A. Ritschel, R. Akileswaran and A.S. Hussain, *Methods Find. Exp. Clin. Pharmacol.*, 1995, **17**, 629.

88. P. Poulin and F.P. Theil, *J. Pharm. Sci.*, 2002, **91**, 1358.

89. P. Poulin, K. Schoenlein and F.P. Theil, *J. Pharm. Sci.*, 2001, **90**, 436.

90. J. Langowski and A. Long, *Adv. Drug Deliv. Rev.*, 2002, **54**, 407.

91. F. Darvas, *J. Mol. Graph.*, 1988, **6**, 80.

92. K.R. Korzekwa, J.P. Jones and J.R. Gillette, *J. Am. Chem. Soc.*, 1990, **112**, 7042.

93. K.R. Korzekwa, W.F. Trager, S.J. Smith, Y. Osawa and J.R. Gillette, *Biochemistry*, 1991, **30**, 6155.

94. J. Grogan, S.C. DeVito, R.S. Pearlman and K.R. Korzekwa, *Chem. Res. Toxicol.*, 1992, **5**, 548.

95. K.R. Korzekwa and J.P. Jones, *Pharmacogenetics*, 1993, **3**, 1.

96. J.P. Jones and K.R. Korzekwa, *Methods Enzymol.*, 1996, **272**, 326.

97. L. Higgins, K.R. Korzekwa, S. Rao, M. Shou and J.P. Jones, *Arch. Biochem. Biophys.*, 2001, **385**, 220.

98. S.B. Singh, L.Q. Shen, M.J. Walker and R.P Sheridan, *J. Med. Chem.*, 2003, **46**, 1330.

99. D. Korolev, K.V. Balakin, Y. Nikolsky, E. Kirillov, Y.A. Ivanenkov, N.P. Savchuk, A.A. Ivashchenko and T. Nikolskaya, *J. Med. Chem.*, 2003, **46**, 3631.

100. K.V. Balakin, S. Ekins, A. Bugrim, Y.A. Ivanenkov, D. Korolev, Y.V. Nikolsky, A.V. Skorenko, A.A. Ivashchenko, N.P. Savchuk and T. Nikolskaya, *Drug Metab. Dispos.*, 2004, **32**, 1183.

101. M.J. de Groot and S. Ekins, *Adv. Drug Deliv. Rev.*, 2002, **54**, 367.

102. P. Romero, J. Wagg, M.L. Green, D. Kaiser, M. Krummenacker and P.D. Karp, *Genome Biol.*, 2005, **6**, R2.

103. R.D. Clark, P.R. Wolohan, E.E. Hodgkin, J.H. Kelly and N.L. Sussman, *J. Mol. Graph. Model.*, 2004, **22**, 487.

104. R.D. Snyder, G.S. Pearl, G. Mandakas, W.N. Choy, F. Goodsaid and I.Y. Rosenblum, *Environ. Mol. Mutagen.*, 2004, **43**, 143.

105. M.D. Barratt and R.A. Rodford, *Curr. Opin. Chem. Biol.*, 2001, **5**, 383.

106. M.B. Bolger, B. Agoram, R. Fraczkiewicz and B. Steere in *Drug Bioavailability. Estimation of Solubility, Permeability and Bioavailability*, H.v.d. Waterbeemd, H. Lennernäs and P. Artursson (eds), Wiley, 2003.

107. M.B. Bolger, T.M. Gilman, R. Fraczkiewicz, B. Steere and W.S. Woltosz in *Cell Culture Models of Biological Barriers: In Vitro Test Systems for Drug Absorption and Delivery*, C.M. Lehr (ed), Taylor & Francis, Saarbrücken, 2002.

108. B. Agoram, W.S. Woltosz and M.B. Bolger, *Adv. Drug Deliv. Rev.*, 2001, **50** (Suppl 1), 41.

109. R.M. Dannenfelser, H. He, Y. Joshi, S. Bateman and A.T. Serajuddin, *J. Pharm. Sci.*, 2004, **93**, 1165.

Uses of High Content Screening in Chemical Optimization

FRANCESCA CASANO, ZHUYIN LI AND
TINA GARYANTES

sanofi-aventis, Mail Stop: JR1–303D, P.O. Box 6800, Route 202/206, Bridgewater, NJ 08807, USA

1 Introduction

High-content screening, HCS, is screening where spatially resolved information, be it microscopy-based images or interpolated scans, are collected from each well of a microtiter plate.[1] HCS contrasts with high throughput screening, HTS, where measurements averaged over the well are made. The most common form of HCS is microscopy-based screening, in which pictures of cells in culture are evaluated for physiological changes in order to deduce the biological effect of chemicals in the cellular environment. A typical HCS system consists of a microscope with autofocus and stage movement, a camera, and an automated analysis system for interpreting the image.

Typically, HCS is used either for the assessment of small or targeted chemical libraries or after an initial traditional screen has focused attention on the most promising compounds. However, as faster systems are developed, larger screens may become routine.

The advantage of HCS is the ability to immediately separate compounds that are giving the desired biological effect from those which are producing a non-specific or uninteresting effect, such as toxicity. In addition, HCS allows us to study sub-populations of cells in culture, for example: only cells in a given phase of cell division, only those that are expressing a chosen marker, or a single cell type in a mixed co-culture.[2] We can also distinguish mature from immature cells, cells from cell fragments, *etc.* Questions that require the analysis of sub-cellular compartments or the translocation of proteins within the cell can also be answered. Essentially, any movement or morphological change can be assessed on a specific cell population as long as it can be seen in a microscope and selected with a computer algorithm.

On the other hand, HCS is slow and data intensive. Traditional HTS can be extremely rapid, with whole plates of 384 or 1,536 compounds often measured in less than a minute with bright fluorescent dyes and creating less than a kilobyte of data. In contrast, HCS often takes many minutes to read an entire plate and can rapidly produce giga- to terabytes of information. In addition, enough single cells need to be measured for statistical significance. Therefore, experiments are generally limited to 96- or 384-well plates. Large surface area 1536-wells, such as on virtual well plates, may also be used, but may require more than one image per well.

2 When is HCS Used

Generally, there are three scenarios for which HCS should be used: instead of HTS,[2–5] after a traditional HTS,[6] and during chemical optimization and beyond.[7,8] HCS is also used with siRNA to identify new targets or to better define the role of genes.[9–11]

HCS is rarely recommended instead of an HTS unless there is a strong, clear advantage of an HCS approach compared to the increased speed and general decrease in workload of HTS. Generally, this means HCS must be enabling to be used for primary screening. Some examples of enabling would be: when a co-culture is required, when only cells in a given growth phase are relevant but you cannot synchronize the cells, or when there is a morphological change that is not accompanied by a total change in fluorescence or luminescence. If the potential advantage of running an HCS is to rule out something that can be ruled out with a counterscreen or selectivity assay, it is rarely advantageous enough. However, if you are looking for a very rare event, such as selectivity between two receptors with very high homology, the advantage of being able to pick out a specific agonist or antagonist from a noisy primary screen more than makes up for the lack of speed or ease.

In the second situation, using HCS after a traditional HTS campaign, our ability to sift through a list of hundreds to thousands of primary hits rapidly is critical. In this case, the goal should be to determine the compounds that truly interact with the target and filter out those that appear to interact but are really false positives because of cytotoxicity, micelle formation, lack of specificity, *etc.* If we can also determine which compounds have attractive drug-like structures at the same time, so much the better. This is where HCS often comes in, since we can learn about toxicity, efficacy, and mode of action simultaneously. This added information often more than makes up for the speed and difficulty of HCS.

In the third scenario, during chemical optimization, high-quality data that probes biological activity directly is critical for assessing the retention or improvement of potency as a chemical series is being optimized. If only a limited number of compounds are being evaluated per month, a high-quality assay with limited throughput is ideal. Generally, it is used with a primary biochemical assay to direct structure-activity relationships (SAR) although it can be used alone effectively, particularly within a chemical series. This situation is perfect for HCS.

Because HCS enables the screener to quickly sort through compounds to find the truly interesting one rapidly, HCS can be a critical tool for improving the chemical diversity of the output of a screen. For instance, if at the end of a screen there are 15 potentially attractive chemical series with a size range from 1 to 300 members and a potency range of 10 to 5 μM, most chemists would start by evaluating the larger chemical families (due to better-developed SAR straight out of the screen) and most potent series (since there is less work needed to improve potency). Based on these criteria, we are likely to work in a chemical space that is very close to what has been optimized in the past. However, HCS can focus our attention to novel mechanisms and/or bioavailable compounds early in the project, reordering the attractiveness of the chemical starting points.

3   HCS Systems

HCS systems typically contain an autofocus, fluorescent microscope with an automated microplate stage, a camera, and dedicated signal processing hardware and software. Examples of such systems are Cellomics' ArrayScan, GE's INCell, Evotec's Opera, *etc.* There are also systems that are not true imaging systems but rather perform laser scanning of the microplates, typically with laser excitation and PMT detection of one or more wavelengths through optical fibers. The position of fluorescent signals is then interpolated from the intensity of the signal during the scan. The closer the scans and the more measurements per scan, the higher the resolution. The Automation Partnership's Acuman and Applied Biosystem's FMAT are examples of such systems. Most HCS is done with fluorescent signals, but it can also be done with other modalities such as phase contrast or Nmarski, if available.

3.1   Limitations of HCS

The use of HCS systems for HTS is typically limited by either speed or resolution. For scanning systems, the reason for this relationship is relatively straightforward, since the resolution is directly correlated with the scan density: the more scans, the longer it takes to read a plate. For microscopy systems, the relationship still holds, because the number of cells that can be studied per field decreases as the magnification increases. In addition, greater acuity of the focus is required at higher resolution.

The time required for automatic focusing of the microscope is often a major contributor to the time required to read a plate. Most microtiter plates are not truly flat but rather vary by hundreds of microns between and within wells across a plate. Sometimes the variation is a smooth transition from thick to thin from one side to another of the plate, or from the outside to the inside. For other plates, the variation is within the wells with the center of the well higher or lower than the edges. In general, the HCS measurements will go faster if the plate is flatter because autofocus time will decrease. The best of microtiter plates can hold a tolerance of about 100 μm flatness over the entire plate (108 mm) or ~0.1%.

Another factor that ultimately limits the speed of HCS systems is how rapidly the image can be interpreted and the data stored and retrieved in the system. The amount of information generated by an HCS assay is tremendous. For a 384-well plate with one field of cells per well and a 4-megapixel camera, the raw image data for a single wavelength is about 1.6 GB. Many assays require multiple wavelengths; one for identifying the cell, one for identifying the nucleus and maybe one for measuring translocation of a protein within a cell. Often more than one field of cells per well is required to get a statistically valid sampling. A 100,000-compound screen, with a 4-megapixel camera, 2 wavelengths, and an average of 2 fields per reading, would produce 1.9 terabytes of raw data. To get around this bottleneck, some systems store only the analyzed data, some immediately identify the cells from background and only store the pixels that represent cells, and others just store the raw data. These different approaches are trading information for speed.

The use of HCS systems for lead optimization is generally limited more by resolution than by speed, since the number of compounds studied is smaller. However, even in lead optimization, speed can be a significant limitation and should not be discounted entirely. For the greatest resolution, microscopy-based systems are generally preferred over scanning systems.

3.2 How to Pick an HCS System

The decision of which HCS system to purchase should really be dominated by comparing the trade-offs among the ability of the software to interpret image features as relevant biological events, the time required to develop an assay, the optical resolution, the throughput, and the balance of cost/reliability/service. It is impossible to optimize for all of these axes simultaneously, and therefore, different systems have prioritized different features.

The minimum requirement for considering a system is can it do your assays? Specifically, does the system have sufficient resolution? Can the software interpret the images? Does the system have the excitation and emission wavelengths and the environmental controls required? For some whole-cell assays, such as the coexpression of two surface markers, a resolution of 1–2 µm is adequate. For most assays, standard fluorescent microscopes with 10–40x objectives will suffice. For true fine structure analysis of subcellular localization, growth cone structure, or chromosome structure, for instance, confocal systems will likely be required. True confocal systems restrict the excitation and emission light through a pinhole so that only light that is in focus is detected. Sometimes, however, confocality can actually be detrimental. For example, if the cellular phenomenon of interest has a greater depth than the confocal Z plane, then you may not see it in a confocal mode. In addition, if light is limiting then the image may become too dim to interpret.

Almost all HCS assays will require at least two fluorescent dyes. One is often to identify which objects are cells or nuclei and a second is to look for specific subcellular expression or localization of a protein or morphological change. A third dye can be used for additional information, such as cell type, co-localization of a second protein, *etc.* Systems that simultaneously measure two or more wavelengths are advantageous not only for speed but also to enable live cell assays and the measurement of kinetic response to the addition of a reagent. Environmental control is sometimes required for these live-cell assays. Most higher throughput applications are done with fixed cells for convenience.

There are many other factors that should be considered when picking an HCS system. In particular, software ease of use is paramount. When assessing ease of use, consider the ease of assay development, the flexibility of the picking cell populations and responses, whether the software architecture is open or closed, and the availability of canned algorithms. Some systems require basic programming skills. Other systems provide prepackaged algorithms to cover most scenarios. Much of the value in an HCS system is in the image processing, so carefully consider your options.

Given factors discussed above, the choice of HCS system should probably be driven by software quality, then resolution and speed. Once the list is narrowed, quality of service, data storage facility, cost, ability to be automated, and reliability need to be factored into the final decision.

4 Examples Show the Power of HCS

Examples best illustrate the power of HCS. There is a tremendous variety of assays possible. This includes chemotaxis, morphological changes, nuclear translocation, subcellular localization, cell-to-cell communication, cell viability, toxicity, micronuclei formation, cell cycle arrest, and receptor internalization. An NF-κB nuclear translocation assay, an apoptosis assay, and a gap junction screen are described below. Only the gap junction assay was performed at sanofi-aventis; the other two examples were taken from an author's experience at Prelux with the development of the precursor to the INCell 3000.

4.1 Example 1: NF-kB Nuclear Translocation Assay (from Prelux)

The transcription factor NF-κB controls many diverse cellular processes, including growth, development, inflammation, immune response, apoptosis, and oncogenesis.[12,13] NF-κB induces the expression of inflammatory cytokines and chemokines and, in turn, is induced by them. When this positive feedback mechanism is deregulated, NF-κB becomes aberrantly active producing the chronic or excessive inflammation associated with rheumatoid arthritis, asthma, and inflammatory bowel disease. In addition, it is well established that NF-κB plays an important role in some virally induced diseases and has been implicated in pathological conditions such as Alzheimer's disease. For all these reasons, there has been much interest, especially in the pharmaceutical and biotechnology sectors, in developing inhibitors of NF-κB that will function without significant side effects, especially over a long-term course of treatment.

NF-κB is a collective name for the complexes formed by the multigene NF-κB-Rel family. In mammalian cells, there are five NF-κB subunits, RelA (p65), RelB, c-Rel, p105-p50, and p100-p52, all containing a ~300aa region of extensive homology in their N-termini, termed the Rel homology domain (RHD), which mediates their DNA-binding and dimerization. Most combinations of homo- and hetero-dimers are possible, but the most abundant forms of NF-κB are p65/p50 heterodimers and p65/p65 homodimers.

Typically, in unstimulated normal cells, NF-κB subunits are held in an inactive cytoplasmic form bound to a member of the IκB protein family, IκBα being the most important one. In response to stimulation, including exposure to the inflammatory cytokines TNF-α or IL-1, physical stress such as UV- or γ-irradiation, reactive oxygen intermediates, and virus-derived dsDNAs or RNAs, signaling pathways are initiated that result in the activation of the IκB kinase (IKK) complex. IKK phosphorylates a member of the IκB family, promoting their ubiquitination and degradation by the proteosome. The degradation process releases NF-κB, allowing it to translocate to the nucleus where it can regulate gene expression.

To show activation of NF-κB, electrophoretic mobility shift assays are typically performed to look at the specific binding of activated NF-κB to DNA. This technique requires relatively large numbers of cells, is laborious, is not performed in intact cells, and is subject to artifacts. Another typical cellular assay measures translational regulation of gene reporter constructs in transfected cells occurring hours after cellular activation.

Because protein translocation from the cytoplasm to the nucleus can be readily visualized by immunocytochemistry, the NF-κB nuclear translocation assay was the first application that was developed in 1998 as a validation for High Content Analysis on the ArrayScan.[14] This assay has now been validated on many automated imaging platforms, on a wide variety of untransfected cells as a fixed-end-point cellular assay using fixation and staining of the endogenous protein and detection by fluorescent-labeled secondary antibody. Furthermore, the intracellular dynamic distribution of NF-κB in living cells has been performed by recombinant expression of an EGFP-p65RelA fusion protein in a variety of cell lines.[15] The wide range of expression levels in transiently transfected cells has allowed the study of NF-κB regulation as a function of its concentration in cells. In fibroblasts, when >10-fold over-expressed, the protein is either concentrated in the cytoplasm but not available for nuclear translocation, or it is aberrantly concentrated in the nucleus and the impaired nuclear translocation correlates with the loss of anti-apoptotic function. In addition, cells overexpressing EGFP-RelA can be fixed and stained with an antibody against endogenous p65 and a secondary fluorescent antibody (for example, a Texas Red-conjugate) to correlate the relative expression levels of EGFP-RelA and endogenous RelA.

Immunofluorescence staining of p65 in unstimulated HeLa cells shows largely cytoplasmic localization and significantly less nuclear fluorescence, due to the continuous shuttling between cytosol and nucleus even in nonactivated cells.[16] Following a 20-min stimulation with TNF-α, a large amount of p65 is translocated into the nuclei, although significant amounts of p65 still remain in the cytoplasm (Figure 1). Software has been developed to measure fluorescence in both the cytoplasm and the nucleus of each individual cell (Figure 2). The software utilizes the nucleus as a tool to identify cells in each well; for this reason, cells are stained with Hoechst 33342. In all cases, the automated imaging platforms focuses on the stained nuclei and the nuclear boundaries are drawn by detecting a threshold fluorescence intensity above the background. A mask, sampling the nuclear area, is formed by eroding[*] these objects. A region sampling the cytoplasmic area is formed by dilation[†] of these objects. The pixel intensities in each region are binned[‡] and the nuclear:cytoplasm

---

[*] Eroding – when analyzing an image, the image is broken up into equal bits or pixels. Objects are then defined as being a group of pixels that share some property or pattern, such as intensity over a threshold. The objects can then be labeled and analyzed. Sometimes it is interesting to shrink the object by disregarding the outer most pixels; this is called "eroding" the object.

[†] Dilation – In an object as described above, sometimes it is interesting to include some extra pixels surrounding the object in the analysis. The process of systematically adding a thin layer of pixels to the circumference of an object is called dilation. It is common to subtract an "eroded" object from a "dilated" object to get an annulus that defines the membrane of a cell, for instance.

[‡] Binned – Binning is the process of grouping adjacent pixels together so that the groups are measured rather than the individual pixels. This process lowers the resolution of the image, averages the intensity of the pixels, decreases the noise associated with reading the pixels from a CCD camera, decreases the size of the data file, and speeds up data acquisition.

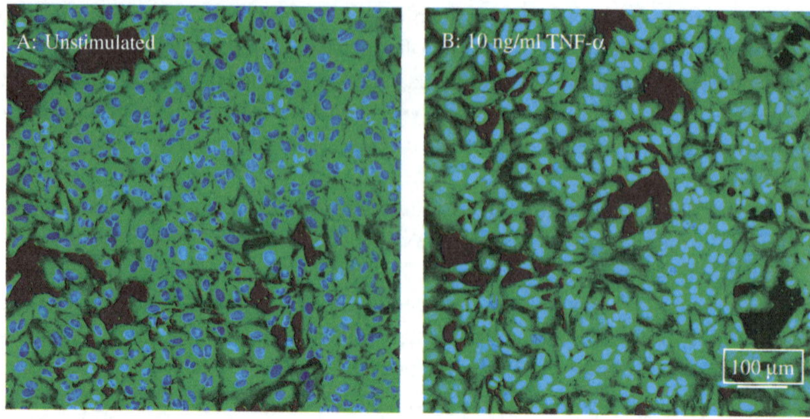

Figure 1 *Trafficking of NF-κB – In HeLa cells, NF-κB is translocated from the cytoplasm to the nucleus in response to stimulation with 10 ng/ml TNF-α. The nuclei are stained with Hoechst 33342 (blue) and NF-κB is detected by indirect immunofluorescence staining with an Oregon Green-labeled anti-mouse IgG. (A) In the unstimulated image, the nuclei are devoid of NF-κB and thus appear blue. (B) Upon stimulation, the nuclei appear green-blue due to the co-localization of the two fluorophores. The full image size is 0.75 × 0.75 mm, 1280x1280 pixels @ 0.6 μm pixilation, ~300-500 cells/image (HeLa cells plated at 15,000/well in 96-well plates for 24 hrs prior to imaging). Imaging time ~4.5 s/image (2 s/pass, one pass at 488 nm and the second at 364 nm excitation) on the INCell 3000 (4)*

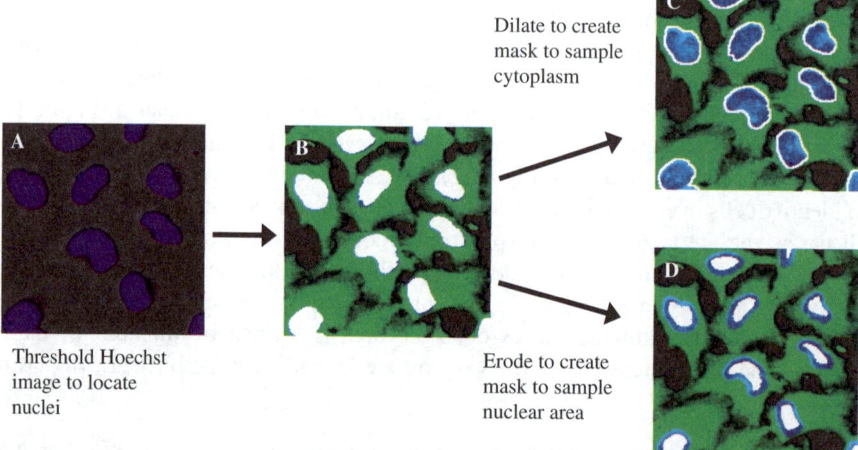

Figure 2 *Nuclear trafficking imaging analysis – (A) Using the blue nuclear marker Hoechst 33342, objects/nuclei are defined as contiguous groups of pixels brighter than a user defined threshold.(B) A nuclear bitmap is created (white). (C) The ring from the edge of each nuclear region is dilated* a user-defined number of pixels to define cytoplasm. Average signal intensity in cytoplasm region is calculated per cell. (D) The nuclear region is eroded* a user-defined number of pixels to sample the nuclear area. Average signal intensity in eroded nuclear region is calculated per cell. The ratio of the average intensity in the nuclear region to the average intensity in the cytoplasm region is calculated per cell as a measure of the degree of protein translocation*

ratio is used as a measure of nuclear translocation. If the nuclear:cytoplasm ratio is low, the protein is mostly in the cytoplasm. If the nuclear:cytoplasm ratio is high, the protein is mostly in the nucleus.

The protocol used to determine the nuclear:cytoplasm ratio is described here briefly. Unstimulated and stimulated (10 ng/ml rhTNF-α, 20 min.) HeLa cells are fixed in 4% formalin, washed in D-PBS and permeabilized with 0.1% Triton X-100/0.5% BSA in D-PBS for 10 min. at RT. After blocking non-specific antigenic sites with 2% BSA/2% Normal Goat Serum in D-PBS for 1 h at RT, cells are incubated for 1 h with an anti-p65 NF-κB antibody diluted in 2% BSA/2% Normal Goat Serum in D-PBS. After washing with 2% BSA/2% Normal Goat Serum in D-PBS, cells are stained with Oregon Green-conjugated secondary antibody for 45 min. at RT After three washings in D-PBS, nuclei are stained with 1 μM Hoechst 33342 for 30 min. Samples were imaged on the INCell 3000 (GE Healthcare), a high throughput, automated line scanning confocal fluorescence microscope. Excitation from an Ar+ laser – 364 nm for Hoechst 33342 and 488 nm for Oregon Green – is focused to a line on the sample, and fluorescence emission from the entire line is imaged through a confocal slit mask and separated onto cameras operating in the blue (420–480 nm bandpass filter) and green (515–555 nm bandpass filter) while scanning the sample over the objective. The objective is a Nikon ELWD 40X/0.6NA; the

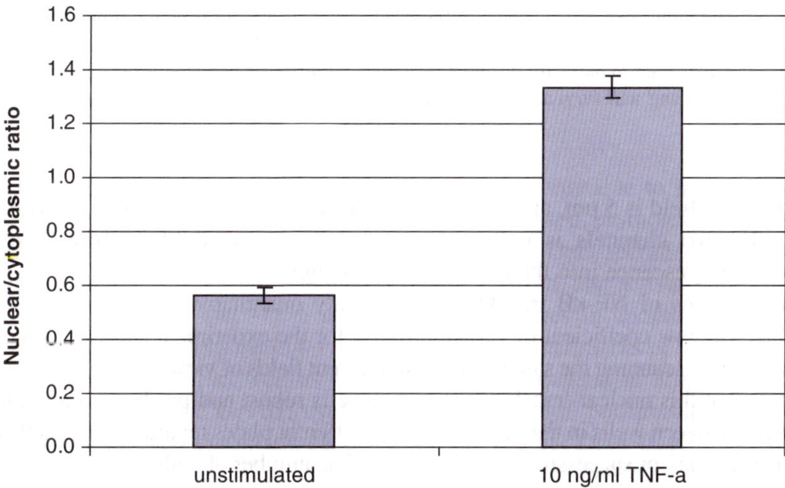

Figure 3 *Quantifying nuclear trafficking – Experimental ratios of the amount of NF-κB antibody staining in the nucleus as compared to the cytoplasm, for unstimulated and stimulated (10 ng/ml TNF-α) HeLa cells. The Z-factor[§] for the assay was 0.73. The CV[¶] was 6.3%*

$$^{\S}Z-\text{factor} = 1 - 3\left(\frac{\sigma_{\text{stimulated}} + \sigma_{\text{control}}}{\text{stimulated} - \text{control}}\right)$$

$$^{\P}\text{CV} = \text{coefficient of variation} = \frac{\text{standard deviation difference}}{\text{difference}} = \frac{\sqrt{\sigma^2_{\text{stimulated}} + \sigma^2_{\text{control}}}}{\text{stimulated} - \text{control}}$$

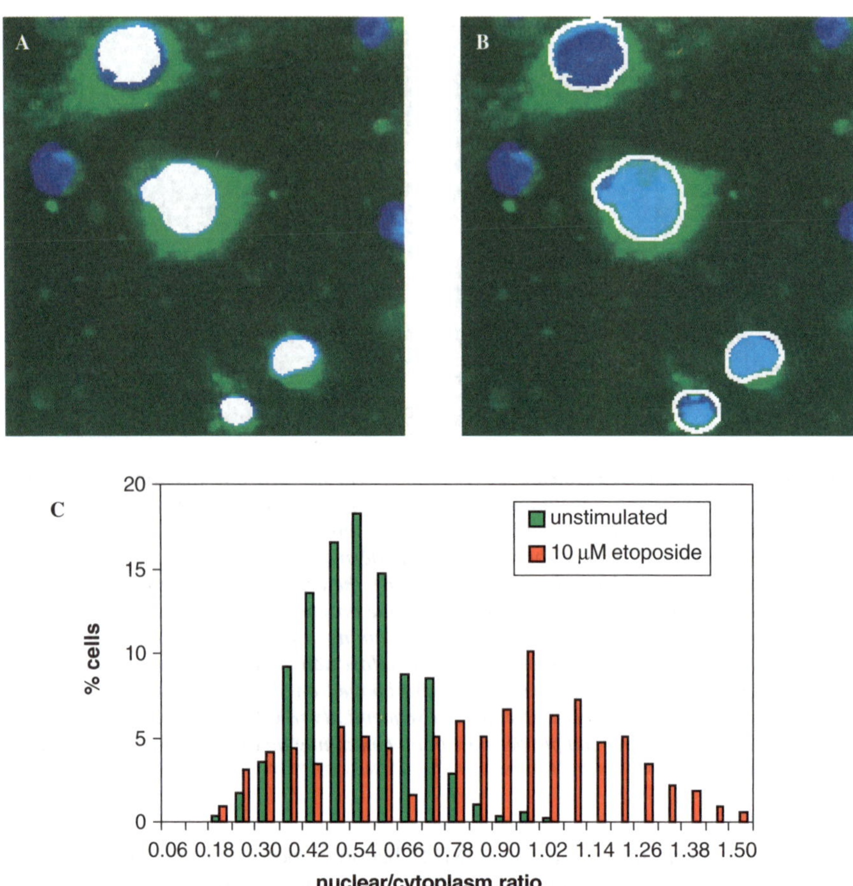

Figure 6 *Image analysis – Cytochrome - c-GFP localization and nuclear fragmentation – The nuclei are identified using Hoechst 33342 signal (A). A mask sampling the nuclear area is formed by eroding these objects (A, white), and a region sampling the cytoplasm area is formed by dilation of the nuclear mask and then the subtraction of the original nuclear mask (B, white): this is the same type of analysis commonly used to detect nuclear trafficking of transcription factors. The pixel intensities in each region are binned, and the ratio of the brightest region of cytoplasm to the dimmest region in the nuclear area is used as a measure of nuclear exclusion of the GFP (C)*

that have released cytochrome *c* but not yet had significant caspase activity show smoothing of the GFP pattern, extension of the GFP out to the cell edge, and spread of the GFP into the nucleus. Cells that have experienced caspase activity are rounded up and loosing adherence. In cells with such compact morphology, it is difficult to make reliable measurements of the staining pattern in the cytoplasm. Rather, the most reliable signal for detection of cytochrome *c*-GFP release is invasion of the nucleus by the GFP.

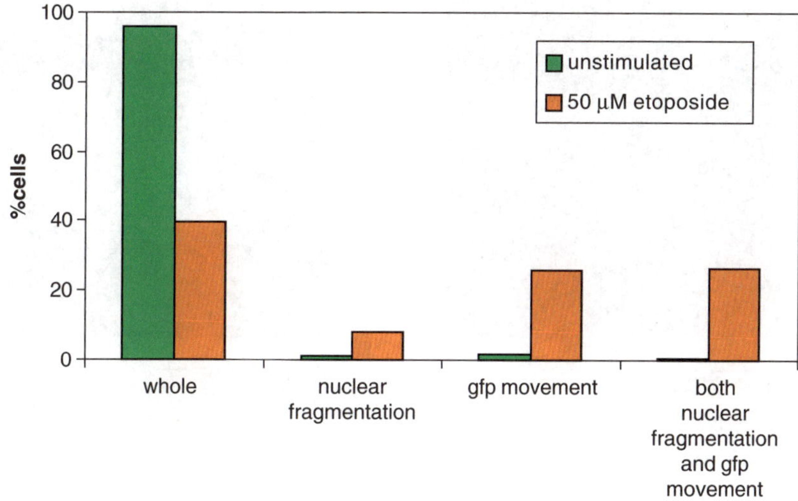

Figure 7 *Image analysis – cell classification – the result of measuring both % of nuclear stain in 5–6 µm grains (nuclear fragmentation) and an increase in the ratio of nuclear to cytoplasmic staining (GFP movement) for cytochrome-c-GFP is plotted as a percent of the total cells imaged. Unstimulated cells do not show either GFP translocation into the nucleus or nuclear fragmentation, which both become prominent upon stimulation with 50 µM etoposide*

Nuclear condensation is one of many morphological changes that can be used to detect caspase activity in the cell. However, cell shrinkage, loss of adherence, and blebbing, which are standard hallmarks of apoptosis, are somewhat ambiguous and can be caused by toxic effects other than caspase activation. Nuclear condensation and fragmentation, however, are generally only associated with caspase activity (caspase-3) and can be measured by analyzing the morphology of the nuclear staining. The fact that no other reagent is necessary is a substantial advantage for the long-time course in this study, although a third channel (such as a loadable fluorogenic caspase substrate) might provide a more specific measure of caspase activity; however, at the same time, it would create additional possibilities of toxicity and other off-target effects.

The details of the image analysis are shown schematically in Figure 6. The nuclei are identified using the Hoechst 33342 signal. A mask sampling the nuclear area is formed by eroding these objects, and a region sampling the cytoplasm area is formed by dilation. This analysis is of the same type commonly used to detect nuclear trafficking of transcription factors. The pixel intensities in each region are binned, and the ratio of the brightest region of cytoplasm to the dimmest region in the nuclear area is used as a measure of nuclear exclusion of the GFP.

Apoptotic nuclei may be either fragmented or simply condensed down to a small grain. Grains of 5–6 µm within the nuclei are identified, and the fraction of the Hoechst 33342 intensity contained in these grains is used as the measure of nuclear condensation. For both measurements, a threshold is set based on the distribution

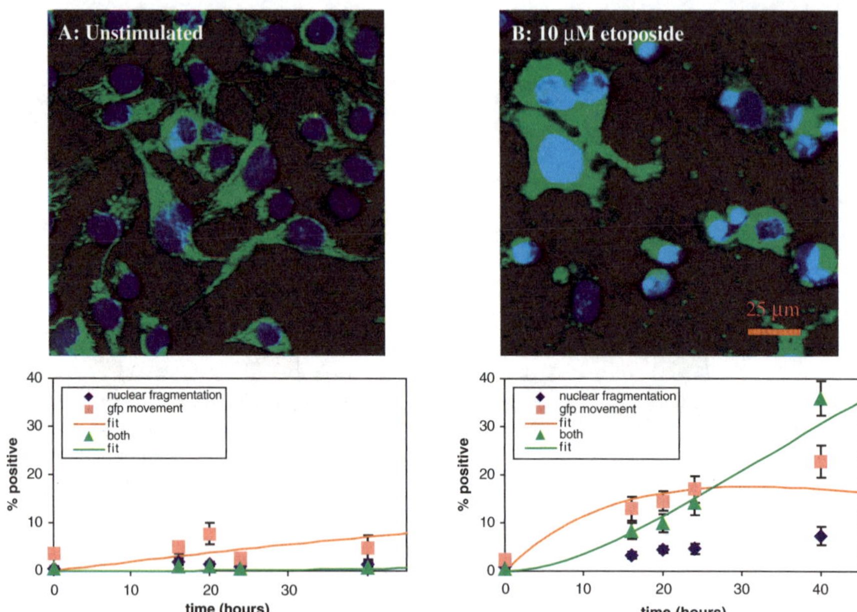

Figure 8 *Apoptosis in HT1080-cyt-c-GFP induced with etoposide – the progress of the population through apoptosis is measured by multiple scans of the same, live sample. Panels (A) and (B) show the time course for cells uninduced and induced with 10 µM etoposide,respectively*

observed in un-stimulated *vs.* stimulated samples, so that each cell can be assigned to one of four classes:

1. GFP localization unchanged and nucleus whole (the expected starting state).
2. Nucleus condensed without a change in GFP localization.
3. GFP localization changed and nucleus whole.
4. GFP localization changed and nucleus condensed (the expected end-point).

The result of each measurement (two fields in a well in a 96-well microtiter plate) is the fraction of cells in each state, as shown in Figure 7.[17] The progress of the population through apoptosis is measured by multiple scans of the same, live sample, generating the time course plotted in Figure 8. Panels (A) and (B) of Figure 8 show the time course for cells uninduced and induced with 10 µM etoposide. It is apparent from this data that the qualitative time course is that expected for apoptosis occurring along the intrinsic pathway. In other words, the population of cells which show cytochrome *c* release increases first, followed by cells which show both cytochrome *c* release and nuclear fragmentation, indicating caspase activation. Relatively few cells show nuclear fragmentation with no cytochrome *c* release. Ideally, this population might be expected to be completely absent. However, for both measurements, the distribution of apoptotic and non-apoptotic cells is not fully resolved, so there will be some uncertainty in the classification. Also, in the samples, there will be a low rate of

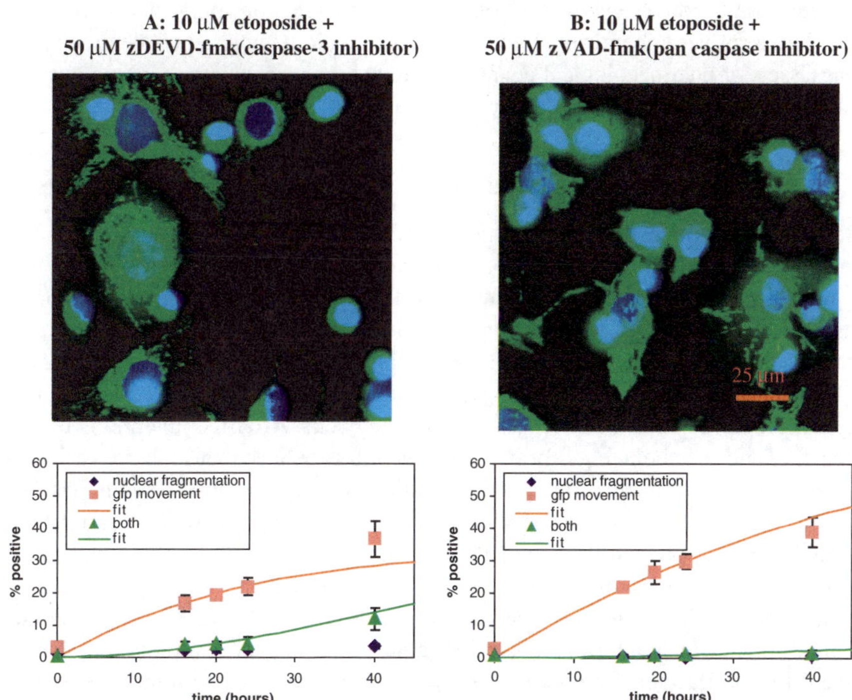

Figure 9 *Apoptosis in HT1080-cyt-c-GFP induced with etoposide + caspase inhibitors – the time course shows the effects of the caspase-3 specific inhibitor zDEVD-fmk (A) and the pan caspase inhibitor zVAD-fmk (B) on etoposide induced apoptosis in the HT1080-cyt-c-GFP cell line. The qualitative results fit the classic model of the intrinsic pathway. Cytochrome c release is essentially caspase independent and proceeds even in the presence of the pan caspase inhibitor zVAD-fmk. Nuclear condensation is slowed down by the caspase-3 specific inhibitor zDEVD-fmk and completely prevented by the pan caspase inhibitor zVAD-fmk*

spontaneous cell death, not necessarily along the etoposide pathway. The time course showing the effects of the caspase-3-specific inhibitor zDEVD-fmk and the pan caspase inhibitor zVAD-fmk on etoposide-induced apoptosis in the HT1080-cyt c-GFP cell line are shown in panels (A) and (B) of Figure 9. Again, the qualitative results fit the classic model of the intrinsic pathway. Cytochrome *c* release is essentially caspase-independent and proceeds even in the presence of the pan caspase inhibitor zVAD-fmk. Nuclear condensation is slowed down by the caspase-3-specific inhibitor zDEVD-fmk and completely prevented by the pan caspase inhibitor zVAD-fmk. However, images of zVAD treated cells show that tracking these two traits does not account for all the events associated with etoposide-induced apoptosis. Even with caspase activity fully inhibited, there are still morphological changes (rounding, partial loss of adherence), which are characteristic of apoptosis but not necessarily caspase dependent.

In conclusion, the relative ease of sample preparation, once the engineered cell lines are developed, makes such image-based studies suitable for extension to characterization of panels of chemotherapeutic reagents or apoptosis inhibitors. In addition, a forward chemical genetics approach could be applied to identify molecular targets of pharmacological agents with unknown mechanism of action to search for inducers of apoptosis selectively in cancer cells.

4.3 Example 3: Gap Junction Inhibitors (from sanofi-aventis)

Gap junction proteins have 4 transmembrane domains. Six of such proteins form a channel, and two of the channels from neighboring cells form a junction. These junctions differ from orthodox ion channels in that they span two plasma membranes instead of one. Gap junctions are permeable to important secondary messengers, such as cAMP, Ca^{++} and IP_3, thus coordinating electrical and metabolic signal synchronization between cells. Before gap junction formation, electrical and metabolic signals are not equilibrated between neighboring cells. After the junction is formed, electrical and metabolic equilibrium is reached. It is believe that such cell–cell communication may lead to propagation of injury during ischemia, and it has been hypothesized that compounds that specifically inhibit gap junction communication could be used to prevent the propagation of injury.[21]

A high-content-screening-based gap junction communication assay measures the transfer of calcein dye between neighboring cells (Figure 10).[2] Because calcein dye is transported to neighboring cells through gap junctions more rapidly than by passive diffusion through membranes, an assay which looks at dye transfer is specific to the formation of active gap junctions. Therefore, at 4 hr, activated calcein dye is transported to neighboring cells primarily through gap junctions. However, this type of dye translocation cannot be detected by conventional assay technology, since the fluorescence intensity is exactly the same before and after dye translocation.

The assay method was as follows: rat glioma C6 cells stably expressing gap junction proteins were stained with Hoechst 33342 dye for nuclear staining, and calcein AM dye for the cytosol. These calcein-labeled cells were named donor cells. They were mixed with the same cells stained only with nuclear dyes, termed acceptor cells, at various ratios, then plated in 384-well tissue culture-treated Falcon plates at various cell densities. Cells were allowed to attach to the plate and to activate the calcein AM dye for 60 min. At the end of the incubation, test compounds or control compounds were added. A stop solution was added after a fixed time. Images were then acquired and analyzed. Live cells were used for imaging since cell fixation caused leakage of calcein dye through membranes and weakened signals.

ArrayScan II was used for image acquisition and data analysis. For image acquisition, one field per well with dual wavelengths, one for the nuclear Hoechst dye (λ_{ex} = 365 nm, λ_{em} = 460 nm), and one for the cytoplasmic calcein dye (λ_{ex} = 480 nm, λ_{em} = 520 nm), was taken during primary screening. Multiple fields per well were acquired during IC_{50} and follow up studies to improve the data confidence. Exposure time was fixed at 0.1 s. Subsequently, images were analyzed off-line using a proprietary algorithm developed based on the ArrayScan's toolbox. Cell perimeter and area, fluorescence intensity of the cell, nuclear size, and co-localization within the plasma are examples of parameters calculated and used to restrict the counting of only calcein-loaded cells.

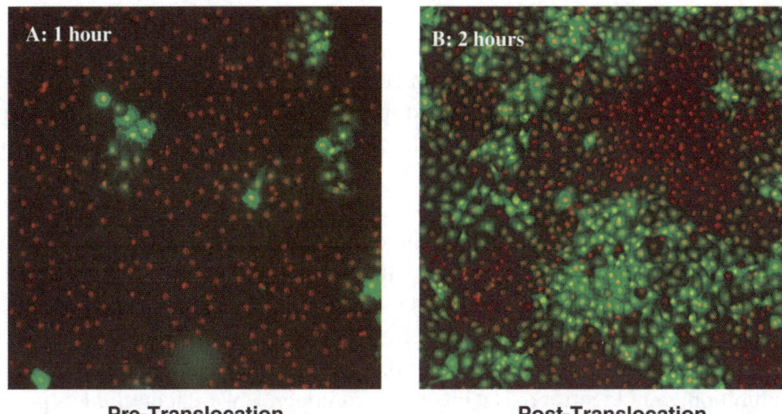

Figure 10 *HCS-enabled screening of gap junctions – Rat glioma C6 cells stably expressing gap junction proteins are stained with Hoechst 33342 dye (red nuclear stain, λex = 365 nm) and mixed with a small population of the same cells that had also been stained with a cytoplasmic dye, calcein AM (green, λex = 480 nm). After 1 h, the cells have settled and the two populations are generally easily distinguished (A). Only a few of the non-calcein dyed cells have picked up any calcein. After 2 hr of incubation (B), the calcein has spread extensively throughout the culture and this change in the distribution of dye is easily imaged on the ArrayScan II*

Because the ArrayScan is based on digital fluorescence microscopy imaging technology, its digital image resolution and imaging field size are dependent on the system magnification. At a given cell density per well, low resolution gives a larger field size, permitting more cells to be sampled; however, the resolution may not be sufficient to analyze individual cells. On the other hand, high magnification results in high resolution, a smaller field size, and fewer cells sampled. Consequently, more images per sample are required to minimize variation in the sampling. These requirements of additional field numbers and tightened focusing standards reduce the throughput of HCS when high magnification is applied. Because the assay only requires whole-cell resolution, a 10X objective created a good balance between resolution and throughput for the screen.

Because the calcein-transfer assay measured the cell-to-cell communication between neighboring cells connected by gap junctions, confluent monolayers of cells within each assay well are essential. The influences of cell density and donor to acceptor ratio on assay quality were assessed. Donor–acceptor ratio was varied between 1:10 and 1:80. It was found that a ratio of 1:20 gave the best signal to background ratio at an appropriate cell density. At a given donor–acceptor ratio, between 12,000 and 20,000 cells per well of a 384-well plate – pseudo-monolayer – formed, which facilitated the optimal junctional communication. When cell density was greater than 20,000 per well, multiple layers of cells formed, interfering with the imaging analysis. When cell density was lower than 12,000, cell patches formed, and junctional communication was restricted. Therefore, for primary screening, a cell density of 18,000 was chosen. It was found that calcein dye transfer between cells followed a pseudo-first-order reaction kinetic between 0 and 210 min, reaching a limit after 210 min due to saturation of the cells in each well by the calcein dye. A reaction time of 120 min after compound addition was chosen for the screen.

The assay was validated and adapted for primary screening using 384-well plates. There were two factors that limited the throughput of the primary screening. First, the imaging acquisition speed was about 43 min per plate without online data analysis. Second, because live cells were used for imaging, it was important to have consistent incubation times for all plates. Therefore, only 12 384-well plates could be processed during a normal working day without full automation. To compensate for the time limitation and to accelerate the screening, primary screening was conducted using a mixture of 4 compounds per well. 486,000 compounds were screened. Z- and Z'-factors were in the range of 0.4 to 0.6. Compounds that showed greater than 50% inhibition were identified as a hit. 1,515 compounds were identified as primary hits, 103 of them were confirmed and considered interesting, and 53 compounds were active at less < 30 μM. Liquid-handling errors, edge effects, and poor cell adhesion were the primary reasons for unconfirmed hits.

Unlike conventional assays for which additional assays are necessary to eliminate false positives or to understand the mechanism of action, high-content screening technology enables us to eliminate unwanted compounds without the need to perform additional assays. After primary screening, we reexamined some of the compounds that reduced the number of calcein-labeled cells but failed to meet other criteria to be identified as gap junction blockers. Figure 11 gives some examples of such compounds. Panel (A) is an image of an uninhibited translocation of dye. Panel (B) shows a strong gap junction blocker that prevented dye from translocating between cells. Fewer cells are stained, and the fluorescence intensity of the donor cells remains. Panel (C) shows that a strong cytotoxic compound killed the cells. Panel (D) illustrates a compound that caused cell membrane leakage, which reduced the fluorescence intensity of individual cells and increased the fluorescence background of the well. (E) shows a strongly fluorescent compound that cannot be washed away and which caused everything in the well to become fluorescent. (F) is an example of a compound that changed the morphology of the cells.

5 Summary

HCS should be one technique in the toolbox of potential assays for confirming functional activity and drug suitability post-HTS. The strength of HCS is its ability to enable biologists to evaluate multiple, diverse biological phenotypes in a population of cells during screening, immediately post-screening or during chemical optimization. Any phenomenon that can be seen in a microscope can now be assessed at medium throughput to support compound optimization. This includes subcellular localization and transportation of proteins, as with NF-κB; complex cellular phenotypes, as with apoptosis; and intercellular communication, as with gap junctions. With any model system, we are generally limited by the availability and relevance of the cells; this also holds true for HCS. However, as drug discovery scientists move toward an increased biological relevance of screens, HCS will become more and more important.

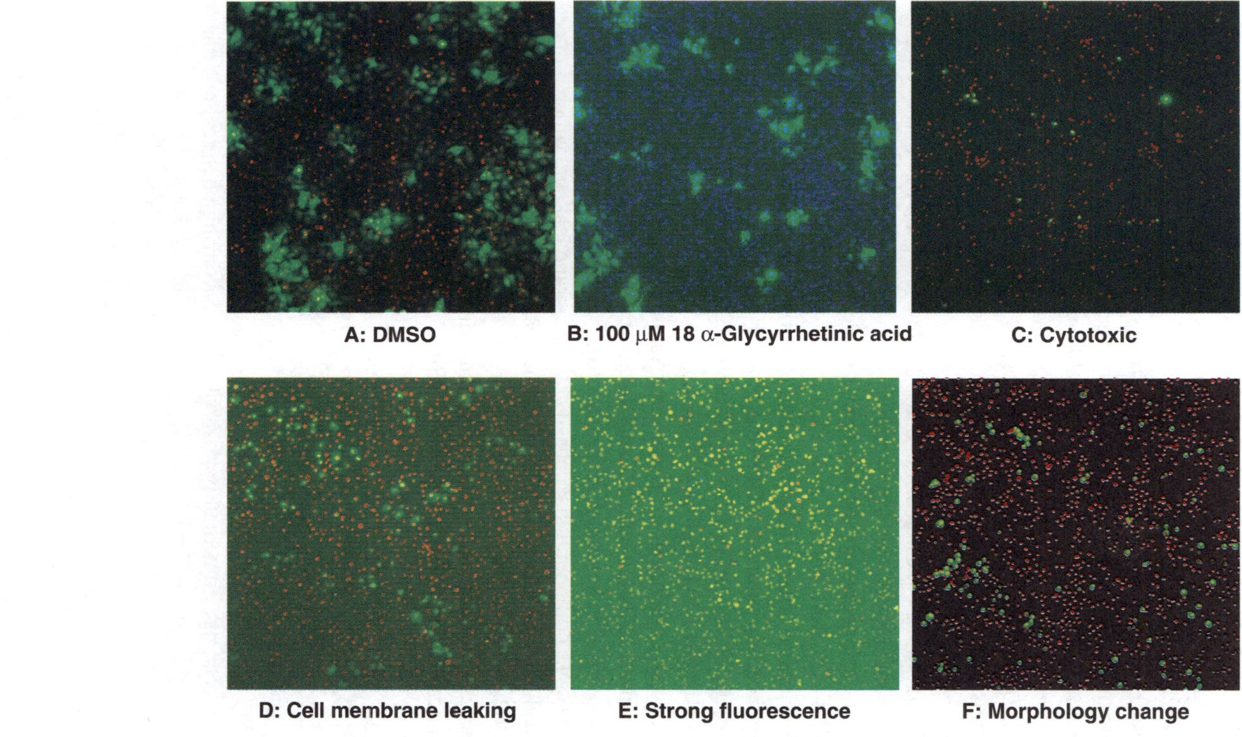

Figure 11 *HCS technology permitted the elimination of false positives – Uninteresting compounds are automatically identified and eliminated by the ArrayScan software as opposed to normal uninhibited (A) or inhibited (B) samples. (C) shows a strong cytotoxic compound that killed the cells. (D) shows a compound that caused cell membrane leakage, which reduced the fluorescence intensity of individual cells and increased the fluorescence background of the well. (E) shows a strong fluorescence compound that cannot be washed away and which caused everything in the well to become fluorescent. (F) is an example of a compound that changed the morphology of cells*

References

1. K.A. Giuliano and D.L. Taylor, *Trends Biotechnol.*, 1998, **16**(3), 135–140.
2. Z. Li, Y. Yan, E.A. Powers, X. Ying, K. Janjua, T. Garyantes and B. Baron, *J. Biomol. Screen.*, 2003, **8**(5), 489–499.
3. K.M. Borchert, R.J. Galvin, C.A. Frolik, L.V. Hale, D.L. Halladay, R.J. Gonyier, O.J. Trask, D.R. Nickischer and K.A. Houck, *Assay Drug Dev. Technol.*, 2005, **2**, 133–141.
4. D. Gurwitz and R. Haring, *Drug Discov. Today*, 2003, **8**(24), 1108–1109.
5. G. Milligan, *Drug Discov. Today*, 2003, **8**(13), 579–585.
6. Z.E. Perlman, T.J. Mitchison and T.U. Mayer, *Chembiochem*, 2005, **6**(1), 145–151.
7. R. Koop, *Drug Discov. Today*, 2005, **10**(11),781–788.
8. M. Bertelsen and A. Sanfridson, *Assay Drug Dev. Technol.*, 2005, **3**(3), 261–271.
9. K.A. Giuliano, Y.T. Chen and D.L. Taylor, *J. Biomol. Screen.*, 2004, **9**(7),557–568.
10. H. Erfle, J.C. Simpson, P.I. Bastiaens and R. Pepperkok, *Biotechniques*, 2004, **37**(3), 454–458, 460, 462.
11. V. Starkuviene, U. Liebel, J.C. Simpson, H. Erfle, A. Poustka, S. Wiemann and R. Pepperkok, *Genome Res.*, 2004, **14**(10A), 1948–1956.
12. V. Pande and M.J. Ramos, *Curr. Med. Chem.*, 2005,**12**(3), 357–374.
13. H. Clevers, *Cell*, 2004, **118**(6), 671–674.
14. G.J. Ding, P.A. Fischer, R.C. Boltz, J.A. Schmidt, J.J. Colaianne, A. Gough, R.A. Rubin and D.K. Miller, *J. Biol. Chem.*, 1998, **273**(44), 28897–28905.
15. F. Carlotti, S.K. Dower and E.E. Qwarnstrom, *J. Biol. Chem.*, 1999, **274**(53), 37941–37949.
16. A. Birbach, P. Gold, B.R. Binder, E. Hofer, R. de Martin and J.A. Schmidt, *JBC*, 2002, **277**(13), 10842–10851.
17. F. Casano, J. Colonell and J. Painter, unpublished results.
18. T. Defrance, *Transplantation*, 2005, **79**, (Suppl. 3), S4–S7.
19. J.C. Reed, *Oncology(Williston Park)*, 2004, **13**(Suppl. 10), 11–20.
20. J.C. Reed, K.S. Doctor and A. Godzik, *Sci. STKE*, 2004, **239**, re9.
21. C.J. Wei, X. Xu and C.W. Lo, *Ann. Rev. Cell Dev. Biol.*, 2004, **20**, 811–838.

Subject Index